Lecture Notes in Civil Engineering 423

Lecture Notes in Civil Engineering (LNCE) publishes the latest developments in Civil Engineering—quickly, informally and in top quality. Though original research reported in proceedings and post-proceedings represents the core of LNCE, edited volumes of exceptionally high quality and interest may also be considered for publication. Volumes published in LNCE embrace all aspects and subfields of, as well as new challenges in, Civil Engineering. Topics in the series include:

- Construction and Structural Mechanics
- Building Materials
- Concrete, Steel and Timber Structures
- Geotechnical Engineering
- Earthquake Engineering
- Coastal Engineering
- Ocean and Offshore Engineering; Ships and Floating Structures
- Hydraulics, Hydrology and Water Resources Engineering
- Environmental Engineering and Sustainability
- Structural Health and Monitoring
- Surveying and Geographical Information Systems
- Indoor Environments
- Transportation and Traffic
- Risk Analysis
- Safety and Security

To submit a proposal or request further information, please contact the appropriate Springer Editor:

- Pierpaolo Riva at pierpaolo.riva@springer.com (Europe and Americas);
- Swati Meherishi at swati.meherishi@springer.com (Asia—except China, Australia, and New Zealand);
- Wayne Hu at wayne.hu@springer.com (China).

All books in the series now indexed by Scopus and EI Compendex database!

Dong-Sheng Jeng · Baoping Cai
Editors

Proceedings of 11th International Conference on Coastal and Ocean Engineering

Proceedings of ICCOE 2024

Springer

Editors
Dong-Sheng Jeng
School of Civil Engineering
Griffith University
Southport, QLD, Australia

Baoping Cai
College of Mechanical and Electronic Engineering
China University of Petroleum
Qingdao, Shandong, China

ISSN 2366-2557 ISSN 2366-2565 (electronic)
Lecture Notes in Civil Engineering
ISBN 978-981-96-0180-6 ISBN 978-981-96-0181-3 (eBook)
https://doi.org/10.1007/978-981-96-0181-3

This Springer imprint is published by the registered company Springer Nature Singapore Pte Ltd.
The registered company address is: 152 Beach Road, #21-01/04 Gateway East, Singapore 189721, Singapore

Preface

The Proceedings of the 2024 11th International Conference on Coastal and Ocean Engineering (ICCOE 2024) reflect the latest advancements and research contributions in the field of coastal and ocean engineering. This prestigious conference, sponsored by China University of Petroleum (East China), Shandong, China, and co-sponsored by Qingdao University of Technology, China, and Yantai Research Institute of Harbin Engineering University, China, was held in Yantai, China, from August 9 to 11, 2024. The conference also benefited from the technical support of Griffith University, Australia.

ICCOE 2024 brings together innovative academics and industrial experts in the field of Coastal and Ocean Engineering to a common forum. Nowadays, demand for marine energy is growing. Conventional fossil resources (oil, coal and natural gas) face supply constraints and volatile markets, and concerns about economics and climate change have become widespread. Furthermore, an abundant and promising source of marine energy exists in waves, tides and offshore currents and swells.

The call for papers generated nearly 100 submissions, from which the Program Committee selected 49 papers for inclusion in the conference proceedings. The diverse contributions from leading experts and emerging scholars alike reflect the dynamic nature of structural mechanics, hydrodynamics, fluid mechanics, offshore wind power generation, deep sea equipment and ship engineering and its ongoing evolution. We are confident that the collection of papers will have a positive impact on the field of Coastal and Ocean Engineering and provide invaluable insights and experiences for experts and scholars in the field.

Putting together ICCOE 2024 was a team effort. We are grateful to the program committee, who worked very hard in reviewing papers and providing technical support for authors. We should also express our gratitude to the keynote speakers for their insightful presentations, to the conference moderators for keeping the sessions on track, and to the session chairs for their efforts in evaluating the oral presentations.

We hope ICCOE 2024 leaves a lasting impression on you, and let's look forward to an even more exciting conference next year.

Qingdao, China
Southport, Australia

Prof. Baoping Cai
Prof. Dong-Sheng Jeng
Proceedings Editors

Committee

Conference General Chair

Baoping Cai	China University of Petroleum (East China), China

Conference Co-chairs

Dong-Sheng Jeng	Griffith University, Australia; Qingdao University of Technology, China
Shuzheng Sun	Harbin Engineering University (Yantai) Graduate School, China

Conference Advisory Chair

Qihua Zuo	Zhejiang University, China

Conference Program Chairs

Decheng Wan	Shanghai Jiao Tong University, China
Bin Teng	Dalian University of Technology, China

Conference Program Co-chairs

Min Lou	China University of Petroleum (East China), China
Yong Ma	Wuhan University of Technology, China
Binzhen Zhou	South China University of Technology, China
Linjian Ma	Army Engineering University of PLA, China
Zhenchang Zhu	Guangdong University of Technology, China

Conference Regional Chairs

Technical Committee

Contents

Structural Mechanics and Structural Health Monitoring of Marine Structures

An Improved Design Wave Analysis Method for Offshore Floating Truss Platform

Boqian Yu(✉) and Aijun Ling

Offshore Engineering Technology Center of CCS, Tianjin TEDA 300457, China
bqyu@ccs.org.cn

Abstract. In recent years, a large number of floating platforms similar to semi-submersible platforms but using a large number of medium-sized components for frame support have appeared in offshore engineering. Such platforms are used in floating wind power, Marine aquaculture platforms, sightseeing and tourism platforms. Design wave analysis is required for conventional offshore floating platform, but the traditional design wave analysis method is not fully applicable to trussed platform due to the large differences in structural forms. In order to accurately evaluate structural safety, traditional design wave analysis will lead to too many calculation conditions and too optimistic calculation results. In this paper, an improved design wave analysis method is proposed to solve the structural design problem of floating truss platform in the ocean, and the technical route of this method is summarized.

Keywords: Trussed platform · floating platform · design wave analysis method · improved design wave analysis method

1 Introduction

The design wave [1] analysis method is an important step in the design of offshore floating platform. It mainly determines a most dangerous condition by designing wave analysis method [3], and then simplifies the structural analysis under complex sea state into several fixed sea state structural analysis. This analysis method has great advantages and high accuracy in dealing with the platform structures with large scale, low structure complexity and less overall deformation, such as traditional semi-submersible platform and ship structure. For many years, it has been used as the main analytical means in Marine engineering [2]. Although the application of design wave analysis methods varies among different designers and classification societies, this way is basically used to evaluate Marine structures.

However, in recent years, a large number of floating truss platforms have appeared, and have been widely used in wind power, Marine aquaculture and other fields. This kind of platform is fundamentally different from the traditional Marine semi-submersible platform and ship structure in the force transmission path (see Fig. 1). The floating truss platform is more complicated in the force transmission path, and the wave load and the buoyancy load will be transmitted through different force transmission paths when they

D. Jeng and B. Cai (Eds.): ICCOE 2024, LNCE 423, pp. 3–10, 2025.
https://doi.org/10.1007/978-981-96-0181-3_1

act on the same position. This leads to a limitation of the profile load screening of the most dangerous conditions which was used in the traditional design wave analysis method. When the structure has diagonal bracing and multiple transverse bracing arranged in one section, the traditional section load will not be able to evaluate the maximum structural response in this section (see Fig. 2).

Therefore, it is necessary to propose a method that can find out the most dangerous conditions of support structures that are not on rigid surfaces, such as diagonal braces. The improved design wave analysis method is to add a simplified structural stiffness equation on the traditional design wave analysis method. The most dangerous condition of the corresponding member is solved by the structural stiffness matrix, and the design wave structure analysis is carried out on this basis load case (see Fig. 3).

Fig. 1. The traditional platform assumes cross-section.

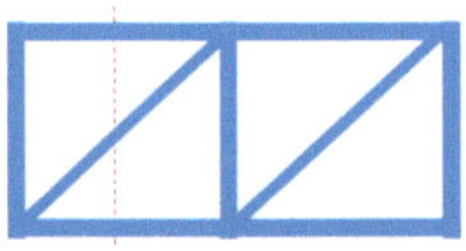

Fig. 2. Assumed section of the platform when trussed.

Fig. 3. Improved design wave analysis method.

2 Traditional Design Wave Analysis Method

The rigid body assumption is used in global motion analysis, and the rigid body motion equation is used.

$$F = m\ a$$

In the case of generalized external force F and generalized acceleration a obtained by motion analysis, the rigid body is artificially divided into two parts with their own generalized mass matrix.

$$m = m1 + m2$$

The acceleration is also transformed by motion to correspond to a1 and a2, the external global load is integrated into F_1 and F_2, respectively, and the interface load Fsec is integrated by the motion equation.

$$F_1 + F_{sec} = m1\ a1$$

The interface load F_{sec} can be inversely calculated by the rigid body motion assumption, and the most dangerous operating condition and the maximum interface load F_{sec} can be obtained according to the analysis of the interface load in the frequency domain. But the trussed platform cannot simply apply the above theory, because the support structure of the trussed platform is usually indeterminate structure. In this way, a problem is faced in the process of solving, the most dangerous conditions of section load F_{sec} and some major force transfer structures do not occur under the environmental conditions of the maximum section load.

3 Improved Design Wave Analysis Method

The traditional design wave analysis method mainly includes the following steps: calculating the hydrodynamic load, calculating the maximum section force, carrying out load prediction and screening design wave, carrying out load transfer and structural calculation.

The improved design wave analysis method is used to solve the quasi-static simplified structure model after calculating the hydrodynamic load. The element stress σ^e is obtained by finite element method.

$$F + F_a = K\Delta$$

$$K^e\delta = \sigma^e$$

Through the summary of the structural bending moment shear stress obtained by the simplified model, the prediction and design wave screening of the structural bending moment shear stress are carried out, and the load transfer structure is calculated.

The two design methods are basically the same in the hydrodynamic calculation and load transfer solution. The difference is that the traditional design wave analysis uses the profile load assumed by the rigid body for load prediction and design wave screening, while the improved design wave analysis method uses the bending moment shear on each structure obtained by the simplified structural analysis model for load prediction and design wave screening. Then the maximum response was predicted by wave spectrum and response spectrum. Then the design wave is filtered by the maximum response.

$$S_{resp} = S_{wave}|R_{rao}|^2$$

$$X_{max} = (2M_0 \ln (10800/(2\pi(M_0/M_2))))^{0.5}$$

$$H_{design} = X_{max}/R_{x,T0}$$

The following takes a semi-submersible truss platform in a certain sea state as an example to compare the difference between the improved design wave analysis method and the traditional design wave analysis method.

4 An Improved Design Wave Analysis Method Taking a Truss Platform as an Example

The basic situation of a platform is as follows (see Fig. 4).: the column area of the platform and the bottom float area are suitable for the plate and shell model, and a large number of rod and cross braces are arranged between the columns.

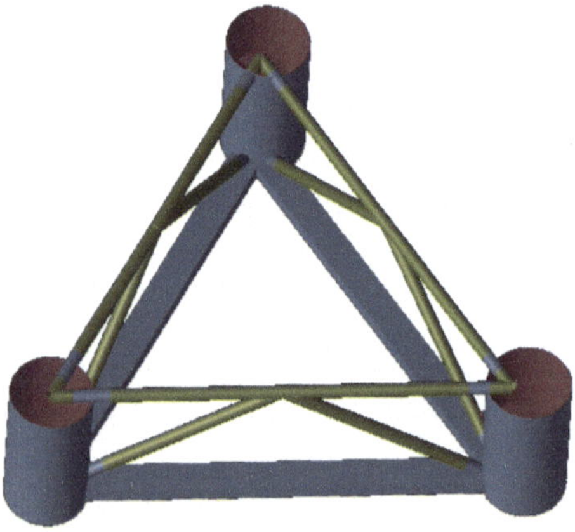

Fig. 4. Basic model of structure.

Due to the existence of morison structure and plate and shell structure at the same time, in the overall analysis, respectively modeling and building a beam model to simplify the evaluation of the overall structural strength.

The analysis steps are as follows:

1. Select shell model and morison model according to the overall situation of the platform
2. A simplified beam model representing structural characteristics is established in the overall model to form a brief stiffness matrix
3. the simplified model and shell model through the block connection binding to form a coupling design model
4. Hydrodynamic analysis
5. The bending moment shear response rao on each beam was obtained by solving the full-wave period simplified stiffness matrix
6. Make long-term response prediction according to the bending moment shear response rao on each beam
7. According to the bending moment shear response rao on each beam and the long-term forecast results, the design wave screening is carried out to determine the dangerous conditions

After the dangerous conditions were screened out, the design wave work was basically completed, and then the conventional strength buckling analysis was carried out by the structure designers. I will not go into details here.

The calculation and reading results of some structural response load curves are as follows (see Fig. 5).

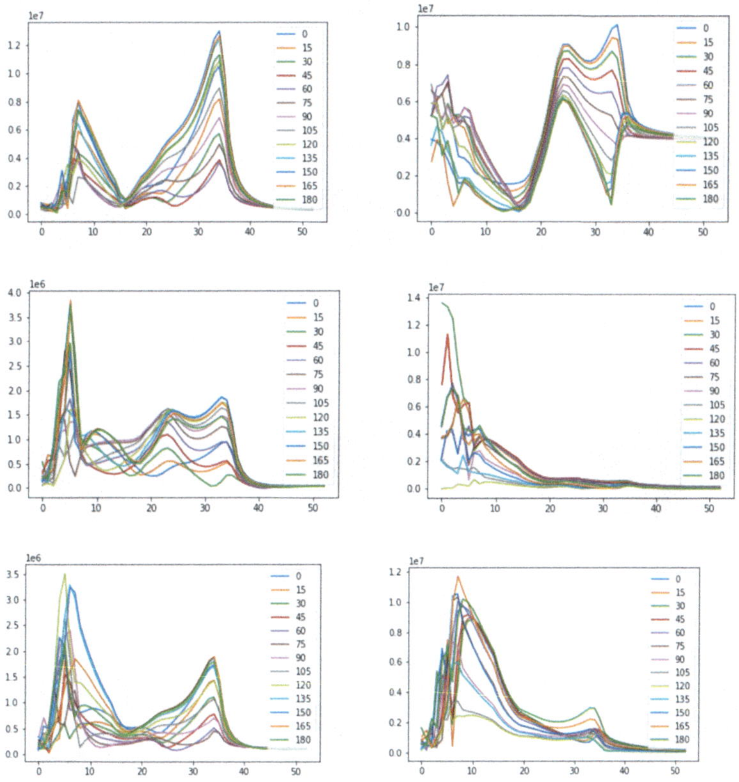

Fig. 5. Structural response load curves.

The structural stress response of this type of platform is very complex, and the design wave and response prediction by several sets of curves alone are not enough to meet the needs of the whole ship structural analysis. In order to highlight the load characteristics, the structure position, the response period, the unit number and the maximum are mapped with the three-dimensional curve of the response envelope value of the full direction load. The results are shown in the figure below (see Fig. 6).

According to the stress RAO obtained from the calculated response as the input parameter of the design wave, the long-term response is predicted by the wave spectrum and motion response, and the design wave parameters are determined by the maximum response and maximum RAO as follows (see Table 1).

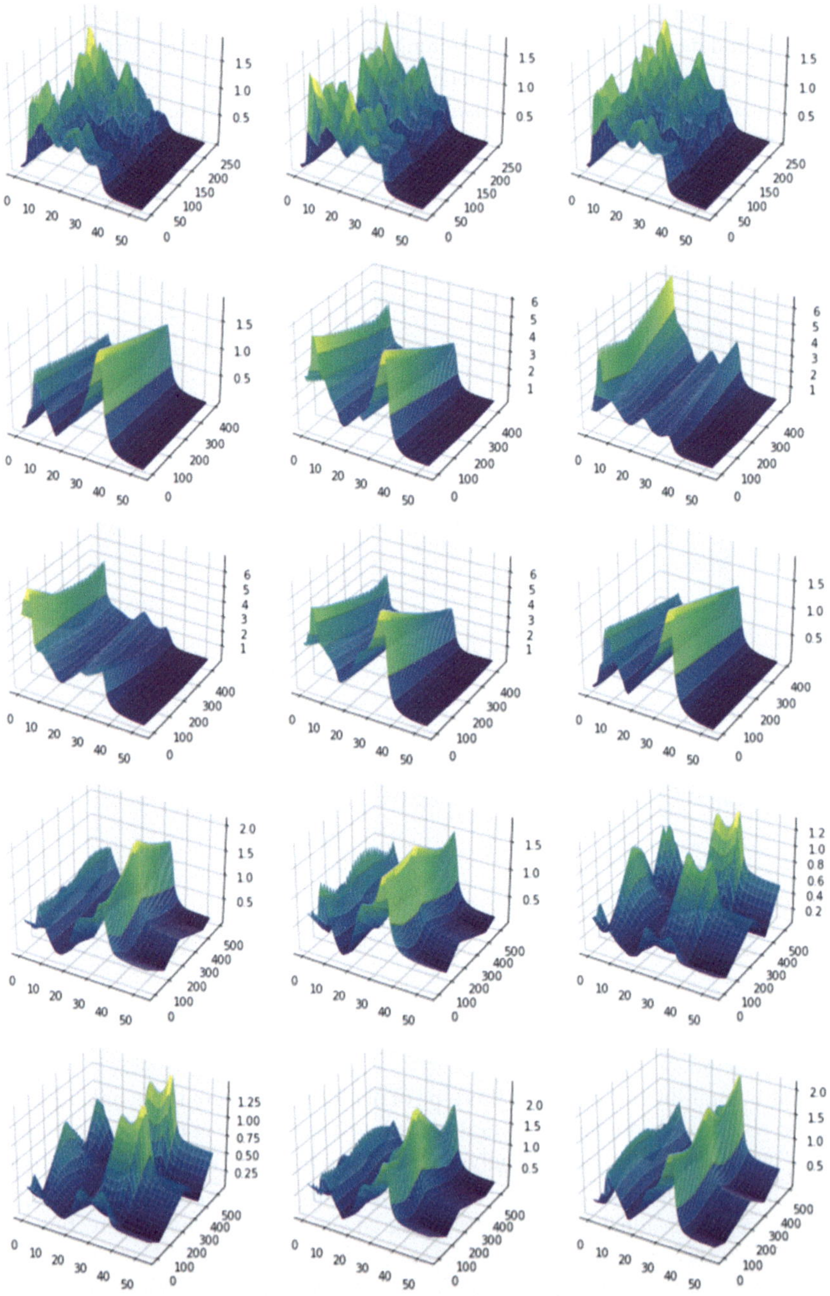

Fig. 6. The three-dimensional curve of the response envelope value in all directions.

Table 1. Result of improved design wave.

Heading direction	T/s	WaveH/m	Heading direction	T/s	WaveH/m
180	8	17.71	90	8	18.72
180	9	19.22	90	7	8.08
165	9	18.14	75	9	18.74
165	6	9.96	75	8	18.44
165	7	10.55	75	6	9.83
150	8	13.93	45	7	15.50
150	7	14.99	45	8	17.51
135	8	14.89	60	6	8.46
120	5	7.23	60	8	18.19
105	6	8.79	30	7	16.16
0	8	17.13	30	8	18.83

It can be seen from the selected design wave results that the design wave load is significantly different from that of the conventional profile (see Table 2), and the wave difference in different directions and the mutual coupling of the structure are more obvious, which can better reflect the impact of environmental conditions on each component. If the conventional design wave profile is used to screen the load, only the design waves in the direction of 30°, 75° and 120° can be screened, and the dangerous working conditions on the inclined member cannot be sufficiently understood.

Table 2. Result of design wave.

Heading direction	Perison	Hs/m
30	8	17.43
75	8	15.82
120	7.5	15.38

5 Conclusions

The traditional design wave analysis method is assumed section, which has good applicability for large-scale full-plate shell structures. However, when the structural form changes and a large number of cross and diagonal braces appear, the platform form is fundamentally different from that of conventional semi-submersible platforms. At this time, if the section assumption is still applied, there will be insufficient evaluation of the inclined beam and intermediate cross bracing structure. There are also big problems in the design wave screening, and the selected design wave conditions cannot effectively evaluate all dangerous conditions. If the strength analysis unilaterally relies on these limited design wave conditions to evaluate, there will be a greater safety risk in strength. When entering the fatigue analysis stage, the dangerous areas are not screened thoroughly, or a large number of fatigue conditions are not satisfied.

Based on the existing design wave analysis logic, a simplified model representing the overall stiffness distribution of the structure is added, and the stress response of the simplified model is filtered to form an improved design wave analysis method, which solves the problem that the traditional design wave profile load is too rough to evaluate the structural internal force. For trussed platform, the improved design wave analysis method can be used to evaluate the dangerous condition of structure more effectively. Using the improved design wave analysis method, the design optimization of the whole project is more accurate and effective.

References

1. Jin-Feng, W., Ren-Chang, L., Yong-Gang, W., et al.: Global strength analysis of an ultra-deep water drill ship based on the design wave method. Ship and Ocean Engineering. (2019)
2. Takashima, H., Tsubokawa, Y., Endo, T., et al.: Evaluation method of offshore design wave considering long period swell. In: Japanese Conference on Coastal Engineering. Japan Society of Civil Engineers (2015)
3. Dao, N.M., Tran, T.H., Nguyen, V.Q.: Calculation of design wave loads using a constrained new wave methodology. i hc quc gia hà ni. (2006)

Application of the Fast-Marching Algorithm for Multi-sensor Network Data Analysis

Xuan-Kien Dang, Hoang-Minh Luu(✉), and Viet-Dung Do

AIT Research Group, Ho Chi Minh City University of Transport, Ho Chi Minh, Vietnam
minh.luu@ut.edu.vn

Abstract. Today, Multi-Sensor Networks (MSN) have numerous applications in areas such as facility condition monitoring, disaster management, equipment inspection, and resource management. Structural displacement data are parameters collected from multiple sensors, necessitating a method for accurate and efficient analysis to assess the structural state effectively. This paper presents an analysis of multi-sensor network (MSN) data gathered from marine structures, such as the Offshore Jacket Platform (OJP). The paper aims to determine the structural displacement resulting from environmental impacts by constructing a data-link ridge progression graph of sensor nodes. The authors propose a two-step approach to identify the optimal trajectory progression for ridges. Firstly, the Dijkstra algorithm is employed within a multi-mesh sensor network to establish connections between points and determine the progression path of the building health diagram. Subsequently, a fast-marching algorithm is applied to optimize the path progression into a ridge trajectory, thus creating a health diagram for marine structures. The effectiveness of the proposed solution is demonstrated through its application in two test scenarios, yielding promising initial results.

Keywords: Data analysis · Fast-marching algorithm · Health diagram · Multi-sensor networks · Offshore jacket platform

1 Introduction

Many studies on monitoring the health condition of marine structures have been conducted since the 1970s. However, these studies remain largely theoretical and lack significant practical applications. Monitoring the condition of marine structures involves various applications with distinct requirements. These are broadly categorized into three types: monitoring the operational status of machinery, overseeing the entire structural integrity, and localized monitoring of specific structural components. It is evident that monitoring the condition of individual machinery parts and localized structural elements is less effective than comprehensive monitoring of the entire structure. Because the root causes may not be directly linked to the overall structural condition. However, comprehensive monitoring poses significant challenges due to substantial investment costs, the requirement for advanced equipment, and the need for accurate processing of extensive data sources. As a result, finding solutions for comprehensive monitoring is an urgent

D. Jeng and B. Cai (Eds.): ICCOE 2024, LNCE 423, pp. 11–21, 2025.
https://doi.org/10.1007/978-981-96-0181-3_2

matter that needs to be addressed [1, 2]. Marine structures exposed to severe impacts from the operating environment will experience structural changes that degrade their quality [2, 4]. The fixed truss is frequently battered by waves and currents, leading to fluctuations or subsidence, resulting in the tilting of the working platform [5–6]. Additionally, the wharves will be impacted by varying loads and weather conditions, which can lead to the development of cracks and structural failure [7]. Even offshore structures can experience shaking from external forces like wind, waves, or collisions, leading to the potential collapse of facilities such as drilling rigs [8]. The vibrations of offshore structures are frequently due to environmental loads, which can reduce equipment performance, pose a danger to structures, and impact the functionality of construction over the last few decades [9, 10]. Reducing the vibrations in offshore structures, large deformations, and nonlinear feedback pose major challenges today. Haeri et al. (2017) introduced an inverse vibration engineering algorithm to develop a simplified two-dimensional and three-dimensional reference model to monitor the condition of fixed truss structures [11]. The benefit of this approach is that it enables the early detection of damage. However, it is essential to monitor the condition of the mechanical structure to prevent potential failures. Mousavi et al. (2021) applied deep learning neural networks (DNNs) to detect damage-sensitive points based on the frequency data of vibration signals [12]. Thereby, the DNN solution detects damage in mechanical systems under uncertain operating conditions [13], accounting for the measurement errors.

Deep learning (DL) is also widely applied in damage identification of offshore structures through measured vibration reflection from randomly impacted structures [14, 15]. On the other hand, the combination of Internet of Things (IoT) technology and Artificial Intelligence (AI) is also applied in the design of anomaly detection systems for bridge construction [16]. This solution showed promising results during the experimental scenario at Binh Loi Bridge, providing insight for future studies. During the operation of marine works, imbalance is also the main cause of failure; motion control solutions have been applied to stabilize the proposed lifting process [17]. In general, the studies have offered solutions to address the issue of stabilizing the operational process and warning about damage states in the structure. However, these solutions are largely focused on the development of control theory, which empirically is initially alerted at local monitoring points. Thus, we should consider developing solutions to monitor the overall health of the structure. This paper represents a new method for measuring the displacement of the OJP model using a sensor network. The sensors are arranged at key locations to measure vibration amplitude and the inclination of the overall structure. In addition, the supervisory station accesses the data array over time and applies a fast-marching algorithm to draw the progression of the structure's condition. The feasible results regarding the health status of the OJP model prove the effectiveness of the proposed solution.

The paper is structured as follows: Sect. 2 develops the MSN architecture and discusses the crucial challenges in sensor networks. Section 3 presents the development of fast-marching solutions in data analysis to assess the structure condition. Section 4 showcases the outcomes obtained through two test scenarios on the OJP model. Some conclusions and potential future directions are outlined in Sect. 5.

2 Multi-sensor Network System

Real-time data collection and transmission are essential for effective monitoring of marine structures. In this study, the authors collected data on vibration amplitude and inclination causing displacement to evaluate the stability of the OJP structure (Fig. 1). The data is then transmitted to a central controller for analysis, allowing for timely warning in case of abnormalities.

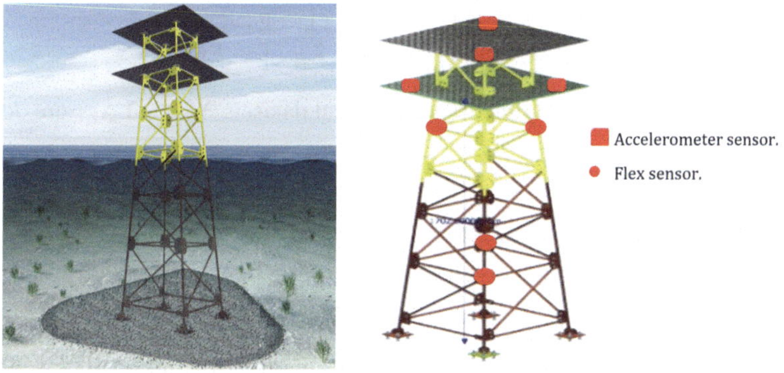

Fig. 1. Overview of MSN installation for OJP model.

Suppose the sensor network has N nodes, the MSN is expressed as [18].

$$X_{k+1} = f(X_k) + \omega_k, Z_k^n = h^n(X_k) + v_k^n \tag{1}$$

in which, $X_k \in \Re^m$ denotes the state vector at time k and $Z_k^n \in \Re^l$ is the measurement vector. We have, $n = 1, \ldots, N$ denotes different sensor nodes and $N \geq 2$. $f(.)$ and $h^n(.)$ correspond to the nonlinear state transition and measurement function vector. Besides, ω_k and v_k^n represent the force of interaction of Gaussian interference. $\omega_k \sim N(0, Q_k)$ represents process noise with an average value in the limit domain 0 to sub variance Q_k. $v_k^n \sim N\left(0, \Re_k^n\right)$ is a measurement noise with an average value in the value domain 0 to sub variance $\Re_k^n$. $E\left[\omega_k \left(v_k^n\right)^T\right] = S_k^n$ represents the domain of interaction of par variances. The bonding mechanism between adjacent sensor nodes is selected for MSN system. Therefore, the scalar graph of multipoint sensor networked systems is defined by $\Psi(\mathbb{N}, \varepsilon)$, where $\mathbb{N} = \{1, \ldots, N\}$ is the set of sensor nodes. If node j can receive data from node n, then $(n, j) \in \varepsilon$. The adjacent set of nodes is defined as $\mathbb{N}_n\{j | (j, n) \in \varepsilon\}$ for node n. If there is no adjacent node for n nodes, then $\mathbb{N}_n = \emptyset$. Point symbol $\mathbb{N}'_n = \mathbb{N}_n \cup \{n\}$. The measurement is determined from the sensor nodes, and is modeled by Eq. (2) as follows:

$$y_k^n = \lambda_{0,k} Z_k^n + \Lambda_{1,k} \lambda_{d_k,k} Z_{k-d_k}^n + \Lambda_{2,k} v_k^n \tag{2}$$

We have, $\Lambda_{1,k} = (1 - \lambda_{0,k})$, $\Lambda_{2,k} = \left(1 - \lambda_{0,k}\right)\left(1 - \lambda_{d_k,k}\right)$, $d_k \geq 1$ are the delayed steps of the sensor at time k. The variables $\lambda_{0,k}$ and $\lambda_{d_k,k}$ are independent of each other and satisfy Bernoulli's law of distribution. If $\|y_k^n\| \leq \Theta_0$ then the measurement is not achieved at time k, then $y_k^n = v_k^n$ where $\lambda_{0,\,k} = 0$, $\lambda_{d_k,k} = 0$ and $d_k = 1$. If $\|y_k^n\| > \Theta_0$ and $\hat{d}_k \neq 0$ normal measurement, then $y_k^n = v_k^n$where $\lambda_{0,\,k} = 1$ and $d_k = 1$. If $\|y_k^n\| > \Theta_0$ and $\hat{d}_k \neq 0$ the resulting measurement is a delay, then $y_k^n = z_{k-d_k}^n$ where $\lambda_{d_k,k} = 1, \lambda_{0,\,k} = 0$ and $d_k = \hat{d}_k$. The properties of $\lambda_{0,\,k}$ and $\lambda_{d_k,k}$are defined [19].

$$\begin{cases} Prob\{\lambda_{0,k} = 1\} = E[\lambda_{0,k}] = \alpha, \\ Prob\{\lambda_{0,k} = 0\} = E[1 - \lambda_{0,k}] = 1 - \alpha, \\ Prob\{\lambda_{d_k,k} = 1\} = E[\lambda_{d_k,k}] = \mu, \\ Prob\{\lambda_{d_k,k} = 0\} = E[1 - \lambda_{d_k,k}] = 1 - \mu \end{cases} \tag{3}$$

for $\alpha, \mu \in [0, 1]$ are scalar quantities. The statistical datasets $\Lambda_{1,k}$ and $\Lambda_{2,k}$ are defined.

$$\mathrm{E}[\Lambda_{1,k}] = \overline{\Lambda}_1, \mathrm{E}[\Lambda_2, k] = \overline{\Lambda}_2 \tag{4}$$

where $\overline{\Lambda}_1 = 1 - \alpha$, $\overline{\Lambda}_\mathbf{2} = (1 - \alpha)(1 - \mu)$. The data array from the MSN is the input value for the data analysis system to determine the structure health condition.

The contents to be addressed in this study are as follows: (i) Developing an overview of the MSN system for the OJP model. (ii) Applying the FM algorithm to analyze the MSN data, thereby determining the health state of marine structures. (iii) Carrying out experiments on the fast-marching (FM) algorithm for MNS data chains installed on the OJP model to demonstrate the effectiveness of the proposed solution.

3 Data Analysis Based on FM Algorithm

The measurement signal of MSN is decoded through the integrated analog-digital converter on Kit STM32F411RE. Data is transmitted to the Cloud for extraction and analysis using the FM algorithm at the supervisor station. The analysis process of measurement data represented in Fig. 2 includes two stages as follows:

- *Stage 1:* Establishing sensor nodes on the grid of points (mesh), then applying the Dijkstra algorithm to determine the sub-path between points.
- *Stage 2:* Marching the sub-paths into a ridge through the Eikonal equation.

The Dijkstra algorithm works in a grid of points by accessing sub-paths starting from the first node to the waypoint. The process works according to two basic principles [20]: firstly, the secondary route of the shortest path is determined by the shortest distance; and secondly, with the given shortest distance x, between points A and C, a path from point A to C through point three B will always have a distance greater than or equal to x. For the problem of optimizing the actual route distance, let G be a grid of reference points as a graph system from the starting point P_0 via waypoints $P_{1,j(1)}, P_{2,j(2)}, \ldots, P_{N-1,j(N-1)}$ to the set point P_N, where $j(n)$ is the index of the reference point that passes optimally in the nth stage of the planned route.

We have, $CP_{i,j}$ indicates the total value of consumption (e.g., energy consumption, cumulative danger, or delay time) that the route transfers to the waypoint $P_{i,j}$ from all reference points of previous periods. Let M be the number of waypoints at each stage,

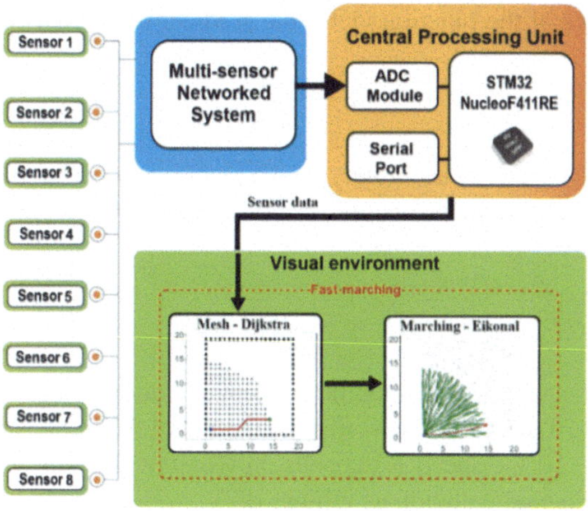

Fig. 2. Processing of the MSN data by using the FM algorithm.

Δ_{C_k} represents the instantaneous computational function that moves from the reference point j in the previous period $P_{i-1,j}$ to the current reference point $P_{i,j}$, and the required travel time is denoted as Δ_{t_k}. The function Δ_{C_k} can be computed as.

$$\Delta_{C_k} = f\left(U\left(P_{i-1,k}\right), W\left(P_{i-1,k}\right)\right)\Delta_{t_k} \tag{5}$$

The function that evaluates movement along different sub-paths to the current reference point $P_{i,j}$ can be rewritten as follows:

$$CP_{i,j} = \begin{bmatrix} cp_{i-1,1} + \Delta_{C_1} \\ cp_{i-1,2} + \Delta_{C_2} \\ \vdots \\ cp_{i-1,M} + \Delta_{C_M} \end{bmatrix} \tag{6}$$

where $cp_{i-1,k}$ is the total rating for moving to the mark $P_{i-1,k}$. Minimizing n evaluated in Eq. (5) is selected as the secondary path to the coordinate point $P_{i,j}$. In the same approach, it is possible to find the optimal secondary path with minimal calculations for all other reference points in the same i^{th} period as $P_{i,j}$. After that, the sub-path with a minimum rating from the starting point to the ith is saved. Repeat this process until it reaches the setpoint, that is, the minimum rated route in the grid system is initiated.

The FM is an arithmetic algorithm that tracks the progression of a partial motion curve ridge with a boundary value (according to the Eikonal equation). This ridge curve can assume sharp angles and split into many separate curves as well as combine into a larger curve. Describing the position of the previous point in the expansion process involves calculating the time to T, whereas the previous point reaches each value in the basic mathematical interface space. For one-dimensional price array, we can easily obtain the time T and the speed F function, and practically consider the distance θ as [21].

$$\theta = F.T \tag{7}$$

The spatial derivative of the function yields the gradient function as follows:

$$1 = F\frac{dT}{d\theta} \tag{8}$$

The ridge curve is drawn more accurately with a greater number of grid points selected, as the magnitude of the gradient function in $T(\theta)$ is inversely proportional to the speed F.

$$\frac{1}{F} = |\nabla T| \tag{9}$$

In multidimensional space, combining the concept of orthogonal gradients with sets of levels of functions to $T(\theta)$, we can describe the motion of the front as the solution under boundary conditions. If the speed F depends only on the position, then Eq. (9) can be formatted into the Eikonal equation as follows [22]:

$$|\nabla T|F = 1 \tag{10}$$

The main equipment in the proposed solution process (Fig. 3) are as follows:

1. The STM32F411 kit acquires measurement signals from the sensor network and converts them from analog to digital.
2. The sensor network includes 4 acceleration sensors and 4 flex sensors that measure the displacement of the structure.
3. In the supervisory station, the computer uses the ST-LINK/V2 communication standard to connect with STM32 and retrieve data for processing. It also draws the ridges of OJP model structure status.
4. The ESP32 DEVKIT V1 aim to gather data using the STM32F411 kit, and then transmit the data to the cloud for processing and monitoring.

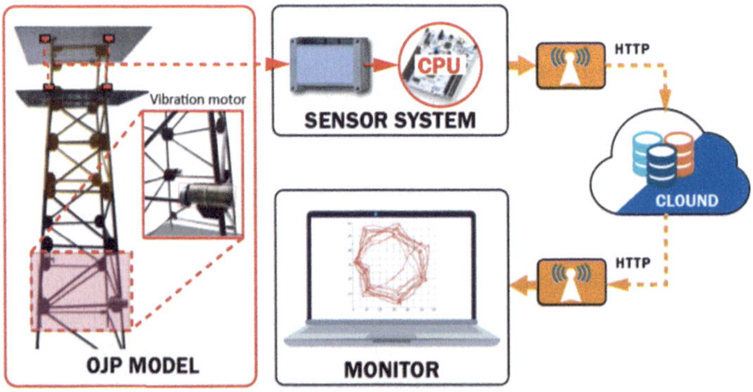

Fig. 3. Overall configuration of the multi-sensor network data analysis system.

4 Results and Evaluations

In this paper, the authors replicated the impact of the actual environment on the OJP model structure by adjusting the frequency of the vibration motor installed on the truss frame. Displacement data measured from MSNs were synthesized and analyzed using the FM algorithm in two experimental scenarios presented in the following subsections.

4.1 Experimental Scenario 1

The first scenario involves converting displacements into sub-ridges that link each pair of sensor nodes. This is accomplished by simultaneously considering each set of 8 sensor points, resulting in 8 sub-ridges connecting θ_i (i= 1,2,...,8), and processing them together using the FM algorithm (10).

In Fig. 4, there are up to seven sub-ridges with straight lines, but only one twisted sub-ridge is visible from node 2 to node 3. So this indicates that the structure of the OJP model is in a stiffness state under environmental influence. However, some abnormalities have been detected in the structural area from nodes 2 to 3, requiring investigation to prevent possible serious damage. Using the ridges combination to monitor the structural displacement over time can help to more accurately assess the overall structural health, which is discussed in the next section.

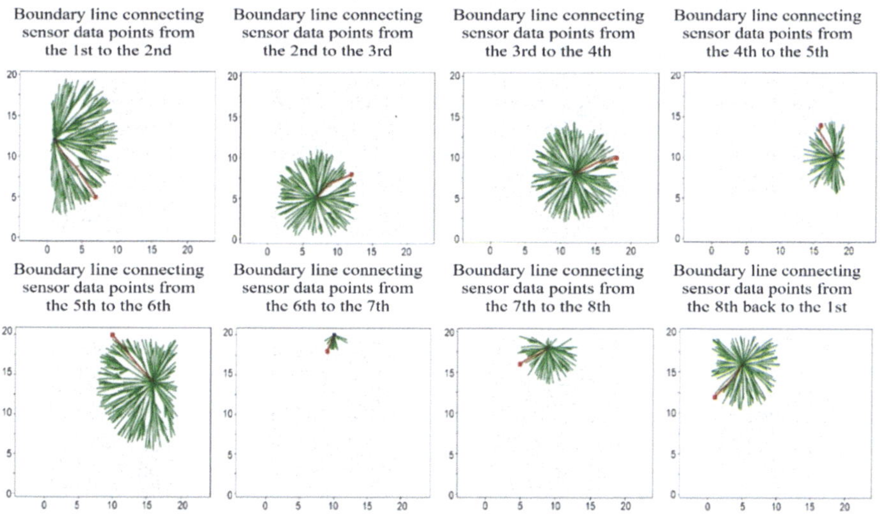

Fig. 4. Results for identifying sub-ridges in experimental scenario 1.

4.2 Experimental Scenario 2

The second scenario regards studying the movement of marine structures over time. Each data series displays the highest displacement value recorded by the sensor nodes per measurement, resulting in 24 data series observed daily. The authors collected data for a month, identifying 720 series that reflect the state of the OJP model over time. This paper describes the analysis of displacement data for 8 measurement times using the FM algorithm, with results indicated in Fig. 5. Additionally, Table 1 represents the displacement data series gathered and stored in the cloud over time. Furthermore, the compilation of ridges over time is illustrated in Fig. 6, verifying that the variation of ridges accurately evaluates the structure condition.

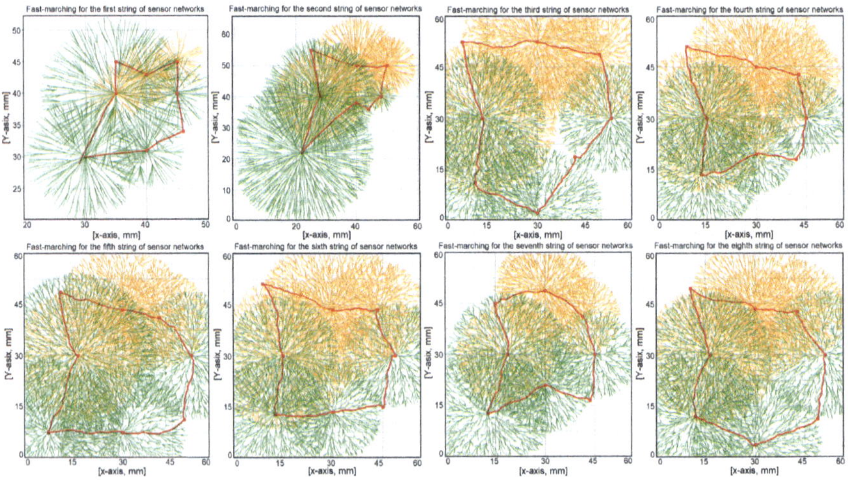

Fig. 5. The ridges depict the condition of the OJP model structure over time.

Table 1. Series data of sensor (S) nodes extracted from 8 measurements

Series	S1[*mm*]	S2[*mm*]	S3[*mm*]	S4[*mm*]	S5[*mm*]	S6[*mm*]	S7[*mm*]	S8[*mm*]
1	2.29	2.94	1.11	1.25	2.20	2.95	1.61	2.93
2	1.23	2.69	2.88	0.86	3.01	1.72	1.06	1.33
3	1.01	0.86	2.50	3.01	1.18	1.69	3.50	2.47
4	1.99	0.67	2.72	2.33	1.94	0.03	3.21	2.45
5	0.74	0.01	3.65	2.64	2.08	0.45	3.06	3.52
6	1.96	0.23	3.01	2.52	1.89	0.59	3.28	2.82
7	1.96	0.84	2.84	2.38	2.06	1.01	2.69	2.55
8	1.89	0.47	3.04	2.67	1.96	0.32	2.91	2.50

The experimental results that monitor the structure state based on the evolution of the ridge are displayed in Fig. 5. The displacement amplitude of the radial ridge peak with a narrower radius distance indicates low structural fluctuations. This means the structure does not pose any damage and meets the safety limits. Furthermore, the

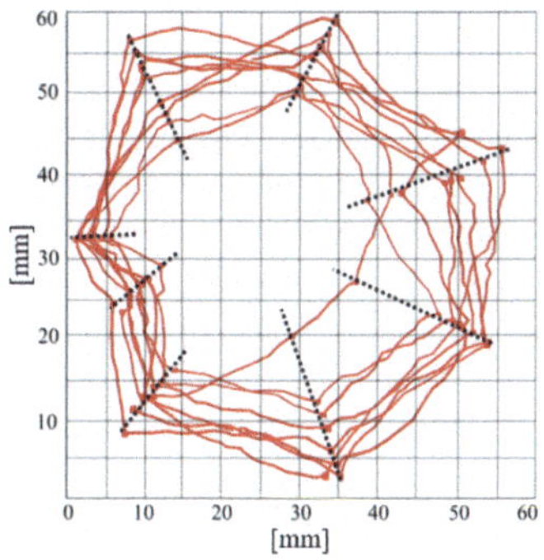

Fig. 6. Monitoring diagram of OJP model structure state in experimental scenario 2.

structural integrity of marine structures depends on loading conditions, the durability of the structure itself, and the severity of environmental conditions. The overall condition diagram of the OJP model structure is depicted in Fig. 6, which illustrates ridge maps without fracture lines representing the structural body and a safe working floor system during operation. Accurate assessment of structural conditions in harsh environmental conditions plays a critical role in limiting maritime accidents.

5 Conclusions

In this paper, we introduce a method for measuring the displacement of the OJP structure using MSN data analysis. Our approach involves applying the FM algorithm to analyze displacement datasets, allowing us to track the evolution of ridges that indicate the health status of marine structures. The effectiveness of our proposed method is demonstrated through practical results obtained from two scenarios regarding sensor networks installed on the OJP experimental model. It is essential to select an industrial sensor type and establish a wireless data communication standard between the sensor and the central processor in order to enhance the system's accuracy for practical applications. Future studies will focus on directly processing dataset of the sensor network in a time-series data format to improve response speed and accuracy.

References

1. Montero, G., Barra, C.D., Novoa, N.: Seismic response modification factors for wharf structures in Chile. In: Proceedings of Coastal Structures 2019, pp. 663–673, Hannover, Germany (2019). https://doi.org/10.18451/978-3-939230-64-9_066
2. Pezeshki, H., Adeli, H., Pavlou, D., Siriwardane, S.C.: State of the art in structural health monitoring of offshore and marine structures. Proceedings of the Institution of Civil Engineers: Maritime Engineering. **176**(2), 89–108 (2023). https://doi.org/10.1680/jmaen.2022.027
3. Chen, X. D., Li, J. J., Huo, Z. Y.: Study on safety monitoring model of crack opening displacement for high-pile wharf. IOP Conference Series: Earth and Environmental Science, pp. 12065–12075. Macao, China, 16–19 July 2019. https://doi.org/10.1088/1755-1315/344/1/012065
4. Li, M., Kefal, A., Oterkus, E., Oterkus, S.: Structural health monitoring of an offshore wind turbine tower using iFEM methodology. Ocean Eng. **204**, 107291 (2020). https://doi.org/10.1016/j.oceaneng.2020.107291
5. Malekzehtab, H., Golafshani, A. A.: Damage detection in an offshore jacket platform using genetic algorithm based finite element model updating with Noisy modal data. The 2nd International Conference on Rehabilitation and Maintenance in Civil Engineering, pp. 480–490. Java, Indonesia, 8–10 Mar 2013. https://doi.org/10.1016/j.proeng.2013.03.044
6. Aghaeidoost, V., Afshar, S., Tajaddod, N.Z., Asgarian, B., Rahman, S.H.: Damage detection in jacket-type offshore platforms via generalized flexibility matrix and optimal genetic algorithm (GFM-OGA). Ocean Eng. **281**, 114841 (2023). https://doi.org/10.1016/j.oceaneng.2023.114841
7. Roeder, C.W., Graff, R., Soderstrom, J., Yoo, J.H.: Seismic performance of pile-wharf connections. J. Struct. Eng. **131**(3), 428–437 (2005). https://doi.org/10.1061/(asce)0733-9445(2005)131:3(428)
8. Tran, T. D., Do, V. D., Dang, X. K., Mai, B. L.: Improving the control performance of jacking system of Jack-up rig using self-adaptive fuzzy controller based on particle swarm optimization. 8th EAI International Conference, INISCOM 2022, pp. 184–200. Virtual Conference, 21–22 April 2022. https://doi.org/10.1007/978-3-031-08878-0_13
9. Kandasamy, R., Cui, F., Townsend, N., Foo, C.C., Guo, J., Shenoi, A., Xiong, Y.: A review of vibration control methods for marine offshore structures. Ocean Eng. **127**, 279–297 (2016). https://doi.org/10.1016/j.oceaneng.2016.10.001
10. Hang, Z., Binrong, W., Xinliang, T., Xin, L., Yehong, D., Manchang, W., Zhike, P.: Experimental study on mitigating vibration of floating offshore wind turbine using tuned mass damper. Ocean Eng. **288**(1), 115974 (2023). https://doi.org/10.1016/j.oceaneng.2023.115974
11. Haeri, M.H., Lotfi, A., Dolatshahi, K.M., Golafshani, A.A.: Inverse vibration technique for structural health monitoring of offshore jacket platforms. Appl. Ocean Res. **62**, 181–198 (2017). https://doi.org/10.1016/j.apor.2016.11.010
12. Mousavi, Z., Varahram, S., Ettefagh, M.M., Sadeghi, M.H., Razavi, S.N.: Deep neural networks–based damage detection using vibration signals of finite element model and real intact state: an evaluation via a lab-scale offshore jacket structure. Struct. Health Monit. **20**(1), 379–405 (2021). https://doi.org/10.1177/1475921720932614
13. Qui, X.L.: A novel multistage damage detection method for trusses using time-history data based on model order reduction and deep neural network. Mech. Syst. Signal Process. **200**, 110635 (2023). https://doi.org/10.1016/j.ymssp.2023.110635
14. Bao, X., Fan, T., Shi, C., Yang, G.: Deep learning methods for damage detection of jacket-type offshore platforms. Process Saf. Environ. Prot. **154**, 249–261 (2021). https://doi.org/10.1016/j.psep.2021.08.031

15. Xiang, Z.Q., Wang, J.T., Wang, W., Pan, J.W., Liu, J.F., Le, Z.J., Cai, X.Y.: Vibration-based health monitoring of the offshore wind turbine tower using machine learning with Bayesian optimisation. Ocean Eng. **292**, 116513–116536 (2024). https://doi.org/10.1016/j.oceaneng.2023.116513
16. Dang, X.K., Ho, L.A.H., Nguyen, X.P., Mai, B.L.: Applying artificial intelligence for the application of bridges deterioration detection system. Telecommunication Computing Electronics and Control. **20**(1), 149–157 (2022). https://doi.org/10.12928/TELKOMNIKA.v20i1.20783
17. Do, V. D., Dang, X. K., Le, A. T.: Fuzzy adaptive interactive algorithm for rig balancing optimization. 2017 International Conference on Recent Advances in Signal Processing, Telecommunications & Computing (SigTelCom), pp. 143–148. Da Nang, Vietnam 09–11 Jan 2017. https://doi.org/10.1109/SIGTELCOM.2017.7849812
18. Xiong, J., Shu, L., Wang, Q., Xu, W., Zhu, C.: A scheme on indoor tracking of ship dynamic positioning based on distributed multi-sensor data fusion. IEEE Access, 1–13 (2016). https://doi.org/10.1109/ACCESS.2016.2607232
19. Jiahao, Z., Shesheng, G., Guo, L., Juan, X., Xiaomin, Q., Bingbing, G.: Distributed recursive filtering for multi-sensor networked systems with multi-step sensor delays, missing measurements and correlated noise. Signal Process. **181**, 107868–107892 (2021). https://doi.org/10.1016/j.sigpro.2020.107868
20. Wang, H., Mao, W., Eriksson, L.: A three-dimensional Dijkstra's algorithm for multi-objective ship voyage optimization. Ocean Eng. **186**, 106131–106146 (2019). https://doi.org/10.1016/j.oceaneng.2019.106131
21. Garrido, S., Malfaz, M., Blanco, D.: Application of the fast-marching method for outdoor motion planning in robotics. Robot. Auton. Syst. **61**(2), 106–114 (2013). https://doi.org/10.1016/j.robot.2012.10.012
22. Treister, E., Haber, E.: A fast marching algorithm for the factored eikonal equation. J. Comput. Phys. **324**, 210–225 (2016). https://doi.org/10.1016/j.jcp.2016.08.012

Parameter Identification of Upper Platform Structure of Offshore Substation Based on Data-Driven Stochastic Subspace Identification

Xinlei Ban[1], Yicheng Peng[1], and Kangqian Xu[2](✉)

[1] Shanghai Electric Power Engineering Co., Ltd., Shanghai 200025, China
[2] Qingdao University of Technology, Qingdao 266520, Shandong, China
banxl@sepd.com.cn

Abstract. Offshore substation serves as a pivotal structure in offshore wind farms. However, its harsh operational environment exposes it to unpredictable safety issues under seismic motions. In order to clarify the damage status of the upper structure of the offshore substation in the earthquake, it is necessary to systematically identify the structure. This study establishes a finite element (FE) model of an offshore substation, uses typical seismic waves as input to obtain the acceleration time history responses at the beam-column joints, and adopts the data-driven stochastic subspace identification (SSI) method to identify the structural modal parameters. The identified natural frequencies and modes are compared with the modal parameters obtained by finite element analysis (FEA). The results show that the data-driven SSI method can accurately estimate the modal parameters of the structure under unknown input conditions. Under seismic loads, the SSI method can identify global modes without requiring a high model order, providing an effective method for the vibration monitoring of offshore wind farms.

Keywords: Offshore Substation · Finite Element Model · Stochastic Subspace Identification

1 Introduction

Wind energy, as a clean and renewable resource, has garnered significant attention worldwide [1, 2]. Among the various forms of wind energy, offshore wind power stands out as one of the most promising renewable energy sources and has experienced rapid and robust development in China [3]. Offshore substations plays a crucial role in efficiently delivering the electricity generated by wind turbines to the mainland. As indispensable core structures in offshore wind farms, they face increasingly harsh service environments and complex loads as wind farms expand into deeper waters. Continuous exposure to random loads such as wind, waves, currents, and earthquakes [4–7] can cause these structures to vibrate, leading to unpredictable safety problems and the risk of significant economic losses [8]. Therefore, it is essential to conduct targeted health monitoring

D. Jeng and B. Cai (Eds.): ICCOE 2024, LNCE 423, pp. 22–31, 2025.
https://doi.org/10.1007/978-981-96-0181-3_3

and system identification of offshore substations. Currently, research and engineering practices related to the health monitoring of such structures are in the stages of rapid development [9–11]. Given the distinct "top-heavy" characteristic of offshore substations and the complexity of the marine environment, it is challenging to obtain the actual external excitation acting on these structures. As a result, traditional methods rely on structural input signals and output responses are not feasible [12]. Instead, output-only structural system identification methods must be employed, but their effectiveness for offshore substations has not yet been confirmed. The model order of the SSI method is inherently an assumption. By using different orders for identification and eliminating the modal parameters identified at higher orders through the formula, more accurate parameter identification values can be obtained by taking the average of the results.

This study conducts a finite element numerical simulation of an offshore substation structure. Using SSI method based on output-only responses under seismic loads, the modal parameters of the entire structure are identified. The identified results are then compared with FEA results to validate the feasibility of using the SSI method for modal parameter identification of offshore substation structures.

2 Numerical Simulation of an Offshore Substation

2.1 Project Overview Introduction

This paper focuses on the offshore substation of a specific offshore wind power project, with a voltage level of 220 kV. The substation consists of an upper module and a lower jacket support structure. The upper module is composed of a four-story steel frame, with floor heights of +12.0 m, +18.0 m, +23.0 m, and +28.0 m, and a main plan dimension of 35 m × 23 m. It uses a spatial steel frame structure, as shown in Fig. 1. The x-axis horizontal direction contains 7 frames (axis labels 1–7), while the z-axis horizontal direction contains 5 frames (axis labels A-E). The lower structure employs a four-legged straight-pile jacket foundation, connected to the main columns of the upper platform structure (labeled A2, E2, A6, E6). The design service life of the offshore substation structure is 50 years.

2.2 Finite Element Model

This study utilizes the general finite element analysis software ABAQUS for numerical simulations. The finite element model of the offshore substation is depicted in Fig. 1. Structural components in the model are represented using B31 beam elements. A seismic time history analysis is conducted on the offshore substation using El Centro earthquake record as the input excitation. The power spectrum of the earthquake wave is shown in Fig. 2. From the seismic response analysis, acceleration time histories at partial nodes in the structural model can be obtained. These responses serve as output data, facilitating the subsequent structural parameter identification using the SSI method.

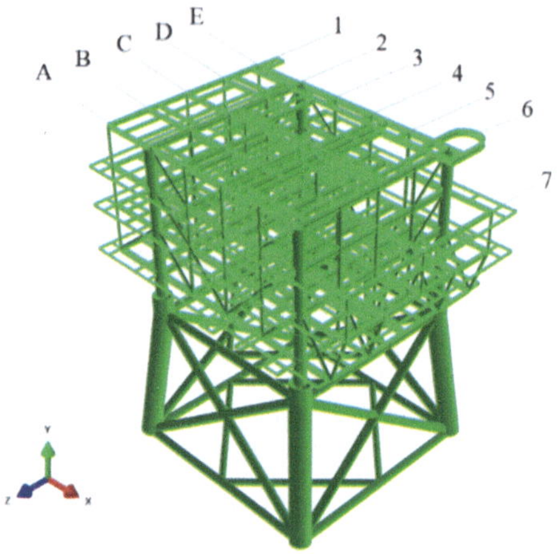

Fig. 1. FE model of the offshore substation

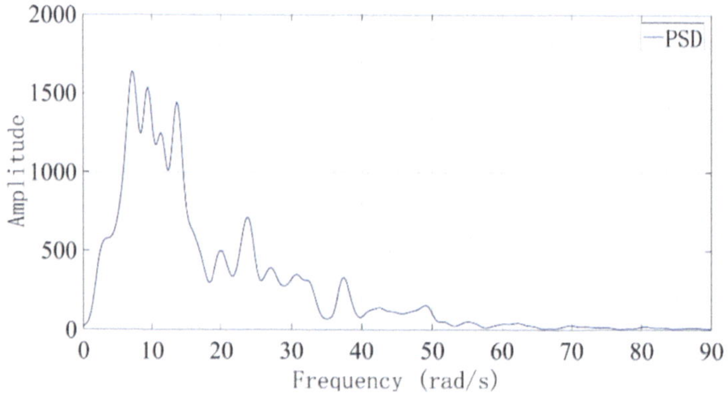

Fig. 2. Power spectrum of El Centro wave

3 SSI Based on Output-Only Structural Responses

The SSI method is a mainstream and advanced modal identification technique. Its principle is to identify structural modal parameters based only on the vibration responses of the structure. This method includes the covariance-driven stochastic subspace method (Cov-SSI) and the data-driven stochastic subspace method (Data-SSI). According to the literature [13], the Data-SSI method has higher identification accuracy compared to the Cov-SSI method, making it widely used in engineering. Therefore, this study employs the Data-SSI method to identify the modal parameters of the offshore substation.

The El Centro wave is used to excite the structure, and the dynamic analysis of the structure is conducted to provide the vibration responses at all nodes of the upper structure. For the collection of structural responses, A-axis plane accelerometers arrangement shown in Fig. 3 are used as an example. Accelerometers are placed on the A-axis plane, C-axis plane and E-axis plane (a total of 36 points) along the x-axis direction of the upper platform structure of the offshore substation to obtain the acceleration time histories in the x-axis direction at the corresponding nodes, which are used as output response data for modal identification. Additionally, the modal analysis in ABAQUS is used to calculate the natural frequencies and mode shapes of the structure using the Lanczos algorithm. These results are used to compare with the SSI identification results to verify the applicability of the SSI method. In the structural identification and analysis, the global modes of the structure are particularly important (e.g., changes in the natural frequencies of the global modes can reflect the damage state of the structure). Since the amplitudes in the x-direction at beam-column nodes on the A-axis are never simultaneously zero, the sensors arrangement in Fig. 3 can identify the natural frequencies and amplitudes of all global modes at the corresponding nodes. However, the FEA can obtain local modes caused by the deformation of single or partial components, leading to discrepancies in the order of modes identified by the FEA method and the SSI method. Considering that such local modes contribute little to the overall structural analysis and have limited practical application value, they are ignored in this paper.

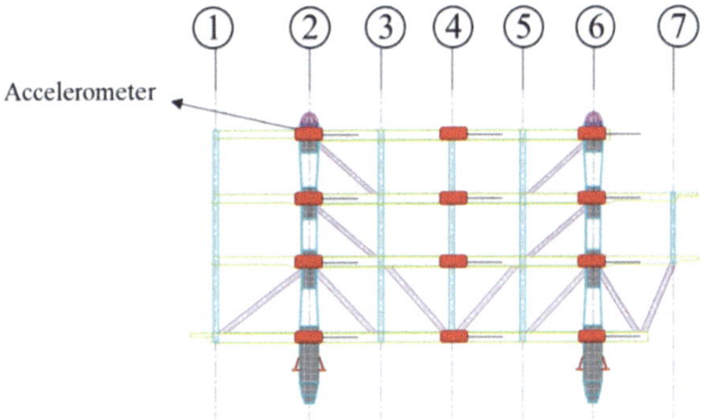

Fig. 3. Measuring points of modal parameter identification

3.1 Comparison of Natural Frequency

First, the natural frequencies identified using the SSI method are compared and validated. Table 1 summarizes the first nine frequencies identified by SSI method and their corresponding frequencies obtained from FEA (where the 3rd, 6th and 8th modes in the FEA are local modes caused by individual or partial components). All frequencies are expressed in terms of angular frequency (rad/s) in this paper. The natural frequencies identified by SSI are denoted as ω_{SSI}, and those obtained from FEA are denoted as ω_{FEA}. To quantify the differences between the two, the relative error ε_{ω} is used. The relative error of the natural frequency is calculated using Eq. (1). A result close to zero indicates a good match between the two sets of frequencies.

$$\varepsilon_{\omega} = \frac{|\omega_{\mathrm{FEM}} - \omega_{\mathrm{SSI}}|}{\omega_{\mathrm{FEM}}} \times 100\% \quad (1)$$

As shown in Table 1, the relative error between the frequencies identified by the SSI method and those obtained from FEA for the corresponding mode shapes is within 3% for the first nine modes, with the maximum error being only 2.31%. This demonstrates that SSI method can accurately identify the natural frequencies of the offshore substation structure. Further examination indicates that the relative error increases with higher mode orders, but the overall match remains satisfactory. Moreover, in structural analysis, higher modes contribute less to the structural response compared to lower modes, which means they have a minor impact on practical applications. Thus, the SSI method is adequate for practical engineering use.

Table 1. Comparison of structural frequencies between SSI and FEA.

SSI		ε_ω (%)	FEA	
Mode	Natural Frequency (rad/s)		Mode	Natural frequency (rad/s)
1	18.34	0.96	1	18.52
2	19.00	0.25	2	18.96
3	22.69	2.31	4	23.23
4	28.09	1.02	5	28.38
5	31.48	0.80	7	31.74
7	33.70	1.12	9	34.08
8	37.49	1.38	11	38.01
9	39.14	2.30	14	40.06

3.2 Comparison of Mode Shapes

Next, the identified mode shapes by SSI method are compared and validated. Figure 4 displays the finite element mode shapes corresponding to the first four modes identified by the SSI method. From Fig. 4, it can be observed that the first mode shape involves overall translation of the structure along the x-direction. The natural frequencies of the first three modes are relatively close. It can be found that the first mode is the global mode of the superstructure, the second mode is the local mode, and the third mode is the local mode with only a single-member mode. The fourth mode is the global mode.

Figure 5 compares the deformations along the x-axis direction of the offshore substation structural A-axis, C-axis and E-axis plane, under the mode shapes identified by the SSI method, with the results obtained from finite element simulations. Through comparison, it is evident that there is a high degree of conformity between the deformations of the two method. Figure 5 illustrates that the SSI method can identify global modes based on structural vibration responses. Nevertheless, when it concerns local modes that only involve individual members and pile foundations (e.g. 4th mode of FEA), the lack of accelerometer positioned at those specific locations hinders the identification of

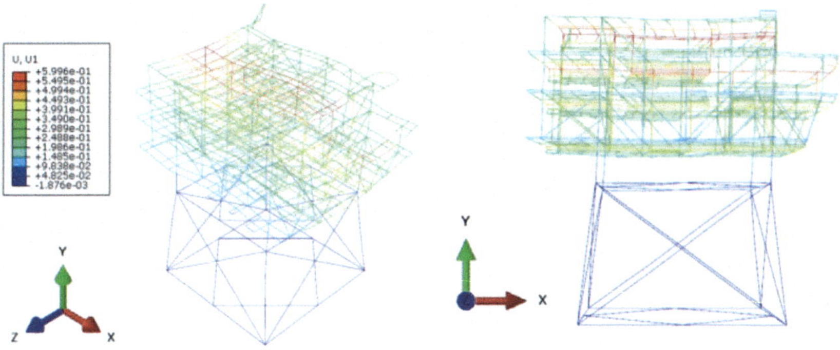

(a)1st mode (ω_{FEA} =18.52 rad/s, global mode)

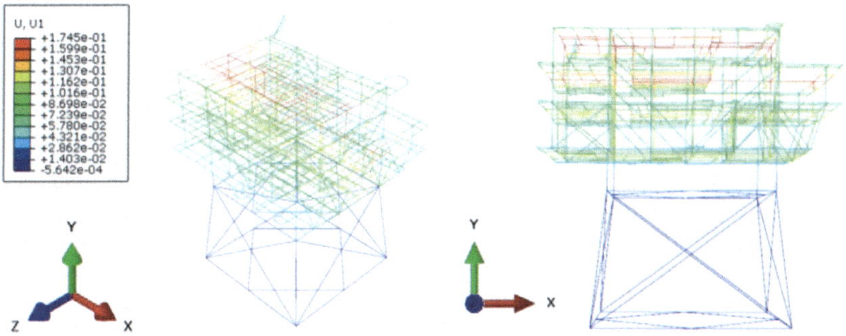

(b)2nd mode (ω_{FEA} =18.96 rad/s, local mode)

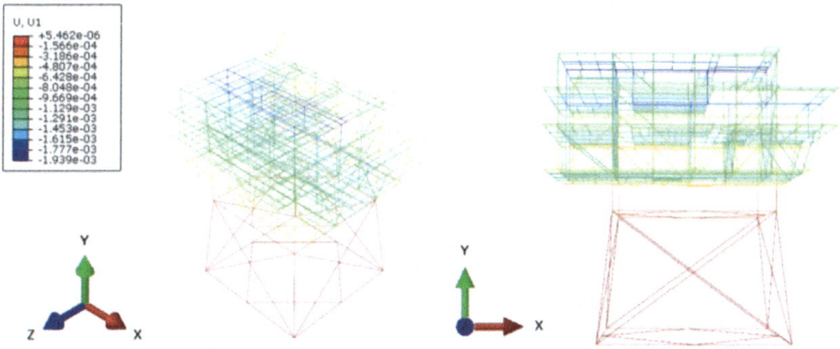

(c)3rd mode (ω_{FEA} =19.29 rad/s, single-member mode)

Fig. 4. Mode shape of finite element model

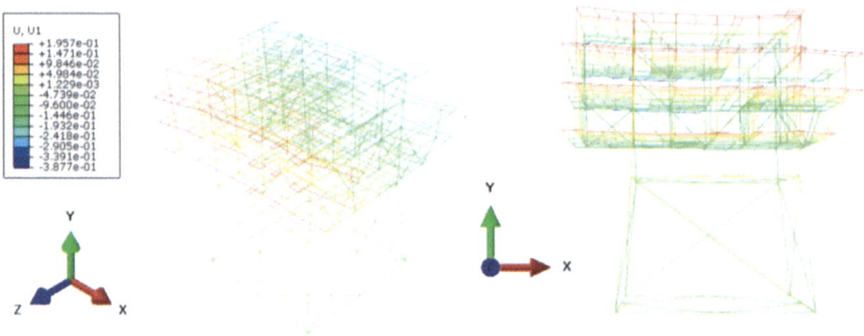

(d)4th mode (ω_{FEA} =23.23 rad/s, global mode)

Fig. 4. *(continued)*

corresponding modes. To quantitatively compare the differences between FEA and SSI, the Modal Assurance Criterion (MAC) is employed to quantify the comparison of mode shapes. The mode shapes identified by the SSI method are denoted as ϕ_{SSI}, while those obtained from FEA are denoted as ϕ_{FEA}. The determination of MAC is calculated using Eq. (2). A value approaching 1 indicates a high degree of similarity between the two sets of mode shapes.

Table 2 summarizes the MAC results for the first five modes of the structure obtained from FEA method, in comparison with the mode shapes identified by the SSI method. From Table 2, with the exception of the third mode, it is evident that the calculated MAC values are all very close to 1, with a maximum error of only 0.18% for the second mode. This indicates a high level of agreement. However, SSI method fails to identify the third mode due to its mode as a single-member mode, where no accelerometers have been set. Combining the results from Fig. 5 and Table 2, it can be concluded that the SSI method effectively identifies the amplitudes of various modes at measurement locations on the offshore substation. Hence, by placing accelerometers along critical structural locations in the desired direction, and collecting acceleration time histories under seismic loads, the SSI method can accurately identify the amplitudes at these locations under different modes.

$$\mathrm{MAC} = \frac{\phi_{\mathrm{FEM}}^{\mathrm{T}}\phi_{\mathrm{SSI}}}{\sqrt{\phi_{\mathrm{FEM}}^{\mathrm{T}}\phi_{\mathrm{FEM}}}\sqrt{\phi_{\mathrm{SSI}}^{\mathrm{T}}\phi_{\mathrm{SSI}}}} \tag{2}$$

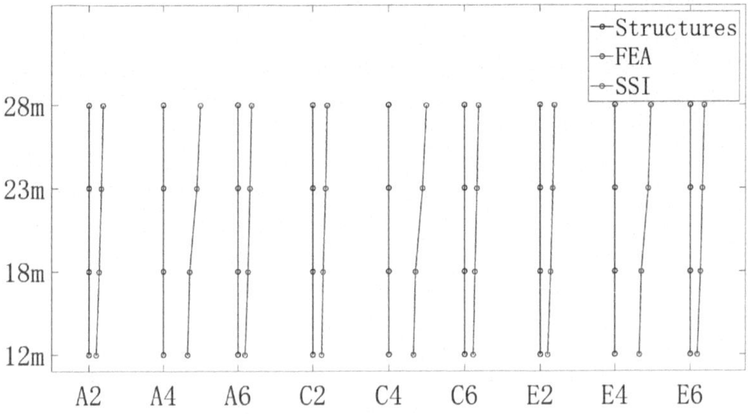

(a) 1st mode (SSI)-1st mode (FEA)

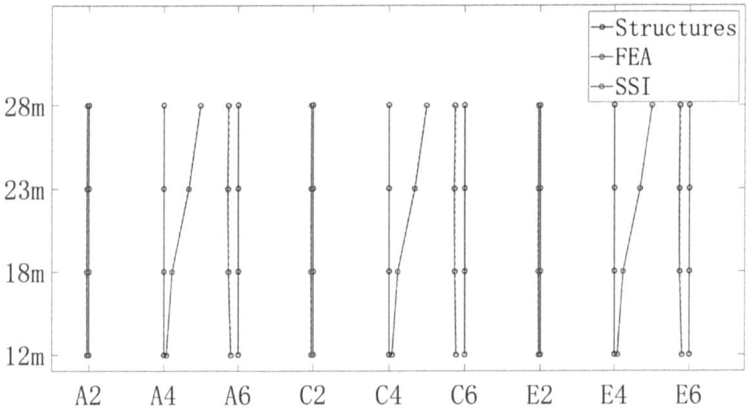

(b) 4th mode (SSI)-5th mode (FEA)

Fig. 5. Comparison of the mode shapes by SSI and FEA in x-axis direction

Table 2. MAC results of SSI and FEA.

Mode (SSI)	Mode (FEA)	MAC
1	1	1.0000
2	2	0.9982
3	4	0.2104
4	5	0.9997
5	7	0.9994

4 Conclusions

This paper establishes a finite element model of offshore substation structures and addresses the structural response of offshore substations under seismic loads. Due to the asymmetric nature of the structural responses in offshore substations, seismic loads effectively stimulate various structural modes. With the application of the stochastic subspace model, the identification of global modes can be achieved without necessitating a high mode during seismic loads. By employing the data-driven SSI method based on only output responses, this paper conducts modal parameters identification for the overall structure. The identified natural frequencies and mode shapes closely match those obtained from FE numerical simulations, validating the efficacy of the data-driven SSI method in accurately predicting the natural frequencies and mode shapes of offshore substations under unknown and complex environmental loads. This method can be applied to structural system identification of offshore substations, thus providing technical support for the development of real-time structural health monitoring systems for offshore wind farms in the future.

The offshore substation is a relatively new concept that differs from the commonly seen onshore structures. At this stage, research primarily involves numerical simulation, which uses relatively simple assumptions, such as stiffness reduction to simulate damage. The actual conditions of the structure and complex damage patterns will be further explored in subsequent research.

Acknowledgements. This study was sponsored by Shanghai Pujiang Program [Grant No. 23PJ1421100], Research Project of Shanghai Electric Power Engineering Co., Ltd. [Grant No. KY-2023-01].

The authors sincerely thank the Pacific Earthquake Engineering Research Center (PEER) from University of California, Berkeley for providing service of searching ground motion records for this study.

The authors also sincerely thank Qingdao University of Technology for providing software and compute server for the FEA part of this study.

All of the support is gratefully acknowledged.

References

1. Peng, Q.: Practice and analysis of structural health monitoring engineering of offshore wind power. Ship Eng. **43**(S1), 81–86 (2021)
2. Li, R.N., Chen, L., Zhang, H., et al.: Bearing capacity of finned pile foundation for offshore wind turbine. Henan Sci. **39**(6), 948–953 (2021)
3. Zhao, L.: Statistics on the scale target of offshore wind power development in China during the '14th five-year plan.' Wind Energy **6**, 44–46 (2022)
4. Liao, Y.D., Wang, L., Zhang, P.C., Chen, D.: Dynamic response analysis of offshore wind turbine structures under combined environmental loads. Water Resour. Hydropower Eng. **52**(3), 206–216 (2021)
5. Lu, G.S., et al.: Damage detection of offshore wind power structures considering pile-soil interaction and scour. J. Vib. Shock **42**(01), 105–114 (2023)
6. Zuo, J.J., Li, X., Wang, W.H., Lou, Y.K., Zhong, Y.: Safety assessment of offshore substation platform under extreme ice conditions. Acta Energiae Solaris Sinica **39**(9), 2508–2514 (2018)

7. Du, P.C., Zhang, F.Y., Ge, K., Wei, Y.C.: Dynamic response of offshore substation structure under long-period ground motion. Henan Sci. **41**(2), 248–255 (2023)
8. Chen, H.Y., Xi, S.T.: Cost structure and price mechanism of offshore wind power. Wind Energy **2022**(1), 12–15 (2022)
9. Weijtjens, W., Verbelen, T., Sitter, G.D., Devriendt, C.: Foundation structural health monitoring of an offshore wind turbine-a full-scale case study. Struct. Health Monit. **15**(4), 389–402 (2016)
10. Song, B., Zhao, W.N., Shuang, M.: Analysis of the influence of scour depth on the dynamic response of offshore wind turbine towers under earthquake action. Chin. J. Eng. **41**(10), 1351–1359 (2019)
11. Zhu, S.Y., Zhu, Z.M., Ke, S.T.: A review of failure analyses and structural health monitoring techniques for offshore wind turbines. South. Energy Constr. **5**(2), 47–59 (2018)
12. Ljung, L.: Perspectives on system identification. Ann. Rev. Control. **34**(1), 1–12 (2010)
13. Xin, J.F., Shen, J.L., Zhang, Y.B.: Relation between noise and matrix dimension of data-driven stochastic subspace identification method. J. Vibr. Shock **32**(16), 152–157 (2013)

Static Pushover Analysis of Offshore Substation Under Rare Earthquake

Wang Liao[1,2(✉)]

[1] Shanghai Investigation, Design and Research Institute Co., Ltd., Shanghai 200335, China
liaowang@sidri.com

[2] Hydraulic Engineering Postdoctoral Research Station, Dalian University of Technology, Dalian 116024, Liaoning, China

Abstract. As the key structure of offshore wind farm, offshore substation must ensure that no structural collapse occurs even under extremely large earthquakes. In this study, static pushover analysis method is used to evaluate the elastoplastic performance and seismic resilience of a real offshore substation located in high seismic intensity area in China. A finite element calculation model considering pile-soil interaction is established for the pushover analysis. The result demonstrates that the specific offshore substation structure maintains its elastic integrity during rare earthquakes which may occur in the site, indicating no threat of structural failure or collapse. By amplifying the pushover force, it can be observed that the jacket foundation yields first and forms plastic hinges near the seabed, ultimately leading to the collapse of the structure.

Keywords: Offshore Substation · Static Pushover Analysis · Seismic Performance

1 Introduction

Offshore wind power is a promising renewable energy source in the pursuit of carbon neutralization. Its rapid evolution in China underscores its potential. Offshore substation is the core structure of an offshore wind farm. It can connect dozens of offshore wind turbines and integrate the functions of gathering, boosting and transmission to complete the important task of outputting offshore wind power to the land.

Currently, offshore substation is mainly constructed as fixed jacket steel platform structure. Due to its high functional demands and structural importance, especially affected by the geological conditions of China's coastal sites, most of the offshore wind farm sites are located in high seismic intensity areas, the seismic resilience of offshore substation must be ensured. Therefore, investigating the seismic performance of offshore substation is a pressing matter of the current era.

To ensure the stability of structures under rare earthquakes, the API Recommended Practices [1, 2] stipulate that the structure must meet ductility requirements, allowing for damage to the structure and foundation but prohibiting collapse. However, the

D. Jeng and B. Cai (Eds.): ICCOE 2024, LNCE 423, pp. 32–41, 2025.
https://doi.org/10.1007/978-981-96-0181-3_4

Recommendations do not include the evaluation of the elastoplastic seismic performance and potential failure mechanisms of structure under rare earthquakes. This study focuses on offshore substation structures and static pushover analysis is used to assess the elastoplastic seismic performance of structure under rare earthquakes.

2 Engineering Overview and Modeling of the Offshore Substation

2.1 Engineering Overview

The offshore substation consists of an upper platform structure and a lower jacket foundation. The platform structure comprises three stories, with dimensions of 24.6 m by 32.7 m and a height of approximately 16 m. The vertical distance between the upper and lower horizontal braces of the jacket foundation is 17.4 m. The pile foundation is made up of four steel pipe piles, each with a diameter of 2.2 m. The piles have a length of about 99 m, with the pile tips reaching a depth of approximately 79.5 m. The elevation of the seabed is -8.9m, and the elevation at the bottom of the pile is −88.0 m.

The structural design lifespan of the offshore substation is 50 years, and it is designed to withstand a seismic fortification intensity of 8 degrees in Chinese seismic design code [3]. According to the seismic safety assessment report for the project site, the horizontal peak ground acceleration for a rare earthquake is 0.536g, and the characteristic period of the ground acceleration response spectrum is 0.70s. The design response spectrum of rare earthquakes in the site is shown in Fig. 1.

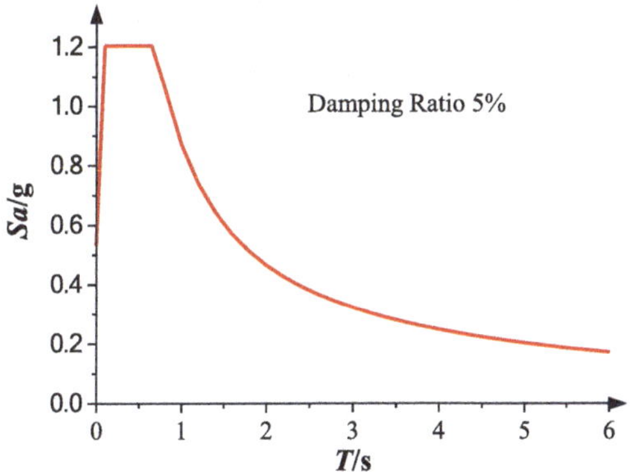

Fig. 1. Design spectrum of rare earthquakes in the site

2.2 Finite Element Model of the Offshore Substation

Figure 2 depicts the finite element model of the offshore substation. The structural members of the substation are simulated by beam elements. The material composition of the entire model includes DH36, Q355C, and Q235B steel. Hydrodynamic pressure acting on the jacket foundation submerged in water is simulated using the Morison formula [4, 5]. The structural damping of the model adopts Rayleigh damping, and the damping ratio of the whole structure is calculated as 5%.

The simulation of pile-soil interaction is based on the calculation model from the API recommendations [1, 6]. The *p-y* curve, *t-z* curve and *Q-z* curve of the soil are transformed into nonlinear springs on the pile nodes of the model.

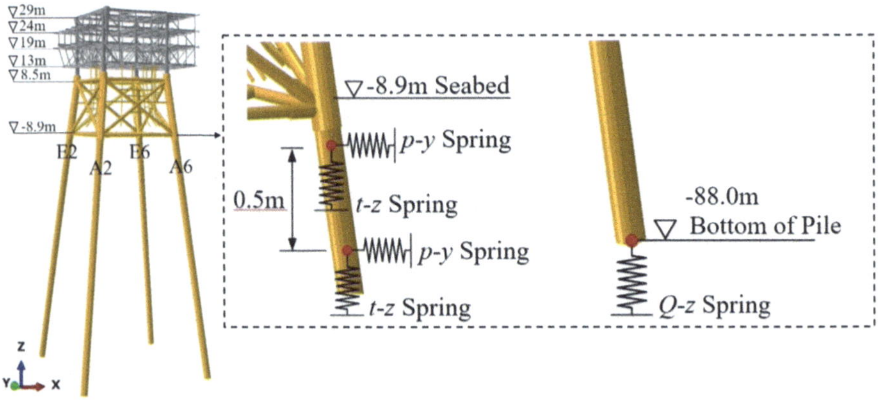

Fig. 2. Finite element model of the offshore substation

3 Static Pushover Analysis of the Offshore Substation

3.1 Introduction of Static Pushover Analysis

Static pushover analysis by capacity spectrum method was first developed by Freeman [7]. Fundamentally, this method consists of overlaying the demand curve with the structural capacity curve within a unified coordinate system to obtain the performance point, thereby enabling the evaluation of the structural response under specified seismic conditions. Fajfar [8] improved the capacity spectrum method based on elastoplastic response spectrum. Due to its clarity in concept and intuitive visualization, the capacity spectrum method has applicated widely. In this study, capacity spectrum method is adopted and the main procedure of the analysis approach is shown in Fig. 3.

The step-by-step procedure of the seismic safety evaluation conducted in this study is shown as follows:

(1) Structural Model Creation: Establishing the finite element model of the offshore substation, including geometry, material properties, pile-soil interactions and boundary conditions (Fig. 2).
(2) Lateral Force Distribution Definition: Determining the pattern of the lateral force distribution based on the characteristics of the offshore substation structure and the requirements of the design code.
(3) Static Pushover Analysis: Applying a monotonically increasing lateral force to the structure and analyzing the response until the target displacement or failure point is reached.

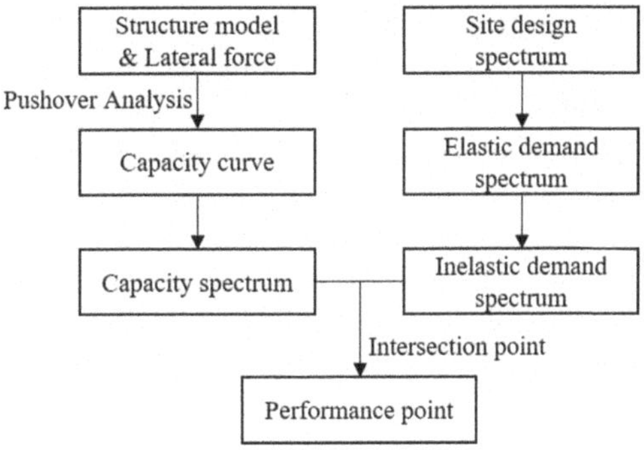

Fig. 3. Procedure of static pushover analysis by capacity spectrum method

(4) Capacity Spectrum Development: Developing the capacity curve of the structure, which represents the relationship between the lateral force and the displacement at the roof of the offshore substation and converting the curve into capacity spectrum.
(5) Demand Spectrum Development: Calculating the design seismic response spectrum of the engineering site and converting it to the demand spectrum.
(6) Seismic Safety Evaluation: Overlaying the demand spectrum with the structure's capacity curve to find the performance point and evaluating the result to determine if the offshore substation structure meets the anti-seismic demand.

In the overall analysis process, the main sources of potential error are likely to come from the definition of the lateral force distribution. Due to the uneven mass distribution characteristic of the offshore substation structure, traditional triangular distribution and uniform distribution are not suitable for the static pushover analysis.

3.2 Capacity Curve of the Offshore Substation

Determination of the Lateral Force

Considering the distinct structural characteristics of offshore substations, which feature a heavy upper platform and a light foundation, they differ significantly from typical buildings. Therefore, it is crucial to ensure that the distribution of lateral forces is closer to real distribution of the structure during an earthquake. In this study, the modal loading method is used to determine the distribution form of the lateral forces on the four main columns of the offshore substation. This method assumes that the seismic force of each story of the structure is related to the mode participation value, so the lateral force increment of the i story of the structure can be expressed as follow.

$$\Delta F_{ji} = \frac{W_i\sqrt{\sum_{j=1}^{m}(\phi_{ij}\gamma_j)^2}}{\sum_{k=1}^{N} W_k\sqrt{\sum_{j=1}^{m}(\phi_{kj}\gamma_j)^2}}\Delta V_b \tag{1}$$

$$\gamma_j = \sum_{i=1}^{n}(m_i\varphi_{ij}) \Big/ \sum_{i=1}^{n}(m_i\varphi_{ij}^2) \tag{2}$$

In Eqs. 1 and 2, ΔF_{ji} is the lateral force increment; ΔV_b is the base shear force caused by seismic force; $Ø_{ij}$ is the i story j order modal of the structure; Wi is the mass of i story; γ_j is the j modal participation coefficient.

Based on the modal response spectrum analysis, the shear force distribution of the offshore substation structure is determined, and the lateral force distribution is obtained by calculating the difference between the shear forces of adjacent layers. In this study, the equivalent pile length linear model proposed in the previous study [9] is used for the response spectrum analysis to obtain the lateral force distribution of four main columns of the offshore substation along the elevation direction. The calculation results are shown in Table 1.

Table 1. Lateral force distribution of the static pushover analysis

Elevation /m	Lateral force distribution			
	A2 column /kN	E2 column /kN	A6 column /kN	E6 column /kN
−8.9	6931.02	6719.37	4655.39	4560.41
8.5	5537.08	5388.55	5419.89	5406.69
13	1942.44	2046.68	589.83	535.61
19	1672.38	1450.39	2314.10	2073.07
24	1745.02	1605.19	1645.05	1591.25
29	886.29	835.44	980.63	835.44

Obtaining Capacity Curve through Pushover Analysis

The lateral forces distributed as Table 1 are applied to the finite element model of the offshore substation shown in Fig. 2. By gradually increasing the lateral forces, the components of the structure yield sequentially until the structural displacement exceeds the limit, revealing the elastoplastic behavior of the structure under seismic motions. Figure 4 shows the relationship between the base shear force and the top displacement ($S_{\text{base}} - \Delta_{\text{roof}}$ curve) for the substation structure, which is defined as the capacity curve.

3.3 Calculation of the Demand Spectrum

The demand spectrum refers to the curve that converts the seismic response spectrum to the acceleration-displacement relationship with the acceleration-period relationship. When calculating the demand spectrum curve, the acceleration response spectrum and the displacement response spectrum are calculated respectively, and the spectral acceleration is used as the ordinate and the spectral displacement is used as the abscissa.

At first, the design response spectrum shown in Fig. 1 is converted into elastic demand spectrum by Eq. 3.

$$\begin{cases} S_{aE} = \beta \cdot \text{PGA} \\ S_{dE} = \frac{T^2}{4\pi^2} S_{aE} \end{cases} \tag{3}$$

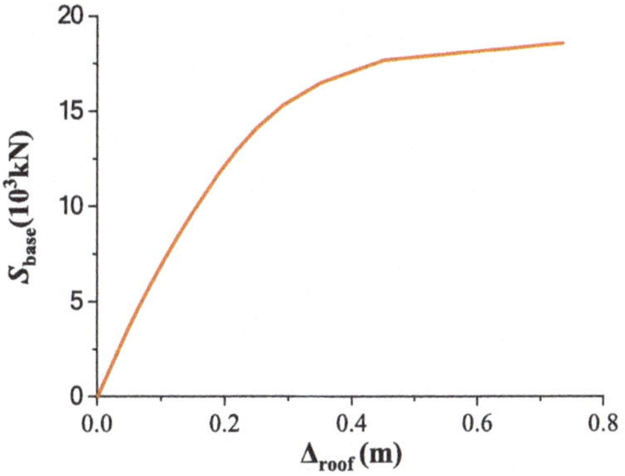

Fig. 4. Pushover capacity curve of the offshore substation

In Eq. 3, S_{aE} is the acceleration of the elastic demand spectrum; β is the seismic amplification factor; PGA is the peak ground acceleration; T is the characteristic period of the structure.

Through the conversion, the elastic demand spectrum curve of the offshore substation site under the rare earthquake action is obtained as shown in Fig. 5.

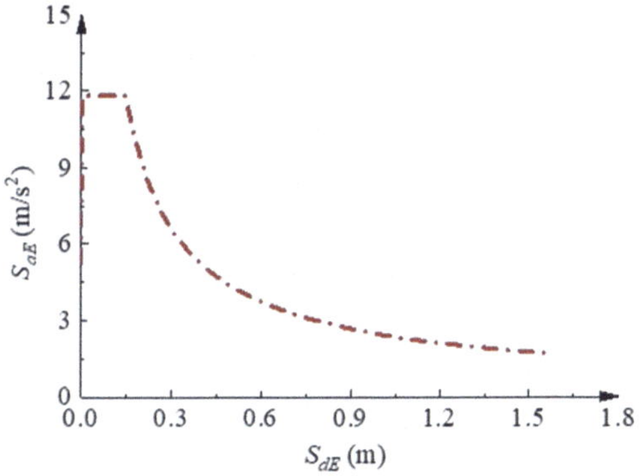

Fig. 5. Elastic demand spectrum of the offshore substation

After obtaining the elastic demand spectrum, it is reduced by the reduction coefficient to obtain the inelastic demand spectrum of the structure. The calculation method is detailed in Eq. 4.

$$\begin{cases} S_{aI} = S_{aE}/R \\ S_{dI} = \mu \frac{T^2}{4\pi^2} S_{aI} \end{cases} \quad (4)$$

In Eq. 4, S_{aI} is the acceleration of the inelastic demand spectrum; S_{dI} is the displacement of the inelastic demand spectrum; μ is the ductility factor of the platform structure; R is the reduction factor applied to the inelastic demand spectrum. The reduction factor

R can be calculated using the R-μ-T reduction model introduced by Vidic [10]. The detailed expression for this model is provided as follows:

$$R = \begin{cases} c_1(\mu - 1)^{CR}\frac{T}{T_0} + 1 & T \le T_0 \\ c_1(\mu - 1)^{CR} + 1 & T > T_0 \end{cases} \tag{5}$$

$$T_0 = c_2\mu^{CT}T \tag{6}$$

In Eqs. 5 and 6, the reduction factors are set as $c_1 = 1$, $c_R = 1$, $c_2 = 0.65$, $c_T = 0.3$ according to Ref. [10].

Utilizing the Eqs. 4–6, the inelastic demand spectrum curve for the site of the offshore substation subjected to rare earthquakes is derived and depicted in Fig. 6.

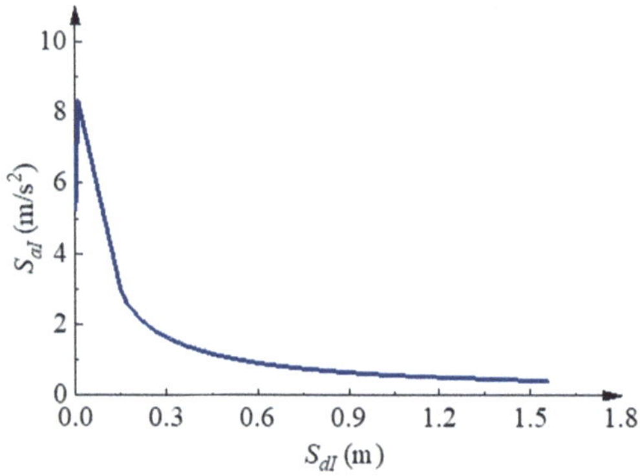

Fig. 6. Inelastic demand spectrum of the offshore substation

3.4 Seismic Safety Evaluation by the Performance Point

To determine the performance point of the offshore substation structure under the action of rare earthquakes for the assessment of seismic performance, the first step involves converting the structural capacity curve into a capacity spectrum.

$$S_{aC} = S_{\text{base}} / (a_1 M) \tag{7}$$

$$S_{dC} = \Delta_{\text{roof}} / (\gamma_1 X_{1,\text{roof}}) \tag{8}$$

In Eqs. 7 and 8, S_{aC} is the acceleration of the capacity spectrum; S_{dC} is the displacement of the capacity spectrum; M is the weight of the platform structure; a_1 is the mass participation coefficient of 1st order modal; γ_1 is the participation coefficient of 1st order modal; $X_{1,\text{roof}}$ is the top displacement of 1st order modal.

The orange curve in Fig. 7 is the capacity spectrum of the offshore substation. By plotting the capacity spectrum curve and the demand spectrum curve together, their intersection point represents the performance point of the structure. The result shows that the capacity spectrum curve of the offshore substation intersects with the inelastic demand spectrum, indicating the existence of a performance point under rare earthquakes. This performance point is located in the elastic region of the capacity curve of the offshore substation, suggesting that the structure remains in the elastic deformation stage during rare earthquakes in this site. Therefore, the structural safety and seismic performance meet the seismic design requirements. Under rare earthquakes with horizontal peak acceleration of 0.536g, the horizontal displacement at the pile top is 0.0899 m and the base shear force is 5911.5 kN.

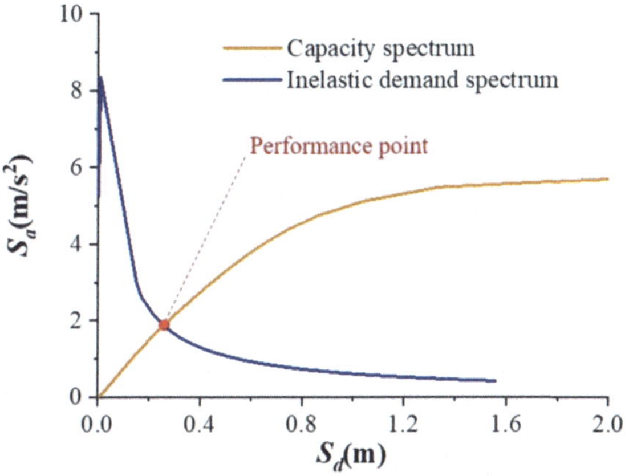

Fig. 7. Performance point of the offshore substation

4 Failure Mode of the Offshore Substation Under Extremely Rare Earthquake

Finally, this study investigates the collapse failure mode of offshore substation structure by continuously amplifying the lateral forces. During the gradual amplification of lateral seismic forces, plastic deformation occurs in the structure, and the location where plastic strain occurs first represents the weak part in the substation structure under extremely rare earthquakes. The failure mode of the structure then extends from the weak part. Therefore, based on the finite element analysis results of the static pushover analysis, the plastic deformation of the offshore substation structure during the initial plastic deformation and the final collapse deformation is depicted in Fig. 8. The results indicate that when the lateral forces are amplified by 2.681 times, the A6 and E6 pile of the jacket foundation yield first due to excessive lateral bending loads, forming plastic hinges. As the pushover load further increases, the plastic deformation regions of the four piles of the offshore substation expand gradually, ultimately leading to the collapse and failure of the structure.

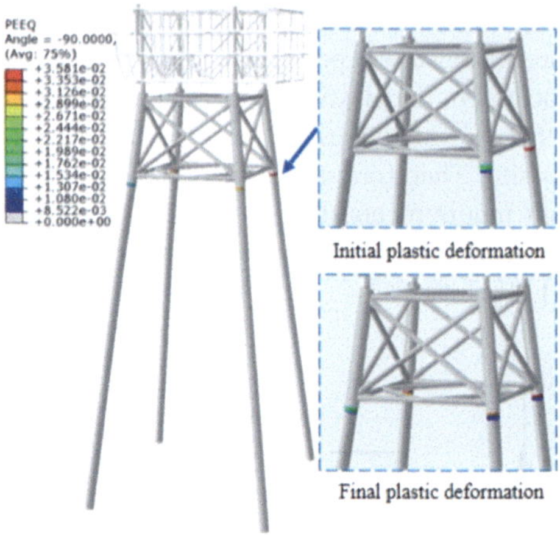

Fig. 8. Damage form of the offshore substation under extremely rare earthquakes

5 Conclusions

This study utilizes the static pushover analysis method to assess the elastoplastic performance and seismic resilience of a real offshore substation located in a high seismic intensity region in China.

The analysis results reveal that the specific offshore substation structure preserves its elastic integrity during rare earthquakes that may occur in this site, suggesting no immediate risk of structural failure or collapse.

Nevertheless, as the pushover force is progressively increased, it is observed that the jacket foundation yields first, forming plastic hinges near the mudline. The yielding of the piles eventually results in the collapse of the structure.

These results emphasize the importance of the seismic design of the jacket foundation and the piles to enhance the seismic stability and safety of offshore substations in regions with high seismic intensity.

Acknowledgements. This study was sponsored by Shanghai Pujiang Program [Grant No. 22PJD097] and Research Project of Shanghai Investigation, Design & Research Institute Co., Ltd. [Grant No. 2022QT(831)-007].

The author sincerely thanks Dalian University of Technology for providing software and compute server for the FEA part of this study.

All of the support is gratefully acknowledged.

References

1. API RP 2A-WSD: Planning, designing, and constructing fixed offshore platforms - Working stress design, 22nd edn. American Petroleum Institute, USA (2014)
2. ANSI/API RP 2EQ: Seismic design procedures and criteria for offshore structures. American Petroleum Institute, USA (2014)

3. JTS146-2012: Code for seismic design of water transport engineering. China Communications Press, Beijing (2012). (in Chinese)
4. Morison, J., Johnson, J., Schaaf, S.: The force exerted by surface waves on piles. J. Petrol. Technol. **2**(5), 149–154 (1950)
5. Penzien, J., Kaul, M.K.: Response of offshore towers to strong motion earthquakes. Earthq. Eng. Struct. Dyn. **1**(1), 55–68 (1972)
6. API RP 2GEO: Geotechnical and foundation design considerations. American Petroleum Institute, USA (2014)
7. Freeman, S.A., Nicoletti, J.P., Tyrell, J.V.: Evaluations of existing buildings for seismic risk-a case study of puget sound naval shipyard. In: Proceedings of 1st U.S. National Conference on Earthquake Engineering, pp. 113–122. ERRI, Berkeley (1975)
8. Fajfar, P., Gaspersic, P.: The N2 method for the seismic damage analysis of RC buildings. Earthq. Eng. Struct. Dyn. **25**(1), 31–46 (1996)
9. Liao, W., Jiao, Y.Q., Jiang, J., Chen, L., Wang, L.J., Qiao, D.S.: Seismic performance analyses on offshore substation with different pile-soil interaction. Ship Ocean Eng. **53**(2), 108–114 (2024). (in Chinese)
10. Vidic, T., Fajfar, P., Fischinger, M.: Consistent inelastic design spectra: strength and displacement. Earthq. Eng. Struct. Dyn.Dyn. **23**(5), 507–521 (1994)

Application of Deep Convolutional Neural Network for Assessing Fracture Risks of Coastal Construction

Viet-Dung Do[1], Xuan-Kien Dang[1(✉)], Tien-Dat Tran[1], Soi Ly[1], and Khai-Hoan Nhu[2]

[1] Artificial Intelligent in Transportation Research Group, Ho Chi Minh City University of Transport, Ho Chi Minh 700000, Vietnam
kien.dang@ut.edu.vn

[2] Faculty of Electrical and Electronic Engineering, Nha Trang University, Nha Trang 650000, Vietnam

Abstract. Coastal constructions (CCs) operate in harsh environmental conditions under many external forces. Therefore, the coastal construction damaged detectors based on the evaluation of cracks' progression play a crucial role in ensuring the safety of people and facilities. The proposed method analyzes and monitors the sequence of consecutive images collected by a camera system mounted at locations where cracks occur in the critical structure. A deep convolutional neural network (DCNN) is applied to identify and analyze crack evolution over time. By using parameters such as crack width, boundary, and progression, the proposed system assesses the crack progression of the constructions over time, thereby improving the level of safety and guaranteeing stable operation of the CCs in working conditions.

Keywords: Coastal Constructions · Convolutional Neural Network · Crack Width Progression · Structure Health Monitoring System

1 Introduction

Coastal construction damaged detectors (CCDD) are of great importance in exploiting the marine structures [1, 2]. But in reality, the implementation of experimental techniques faces such difficulties in maritime environments (shown in Fig. 1). The CCDD system applied to the monitoring process will increase safety conditions for both people and the CCs while also helping to reduce maintenance costs. The CCs such as wharves, submarines [3], and coating structures [4] are always affected by harsh conditions such as earthquakes, and ship collisions. Besides, the characteristics of seawater also cause corrosion and erosion to increase rapidly. Erosion phenomena that rapidly increase in scale can lead to structural displacement as well as crack [5–7]. Thus, maintaining the initial reliability of the structure throughout its useful life is significantly improved through the construction health monitoring.

D. Jeng and B. Cai (Eds.): ICCOE 2024, LNCE 423, pp. 42–53, 2025.
https://doi.org/10.1007/978-981-96-0181-3_5

In recent years, intelligent control techniques [8, 9] have been widely applied in identifying and providing early warnings for the health status of marine constructions. Simple two-dimensional and three-dimensional reference mathematical models combined with inverse vibration algorithms early identify problems that may occur in the CCs due to vibration factors [10–12]. In addition, artificial intelligence (AI) has also been initially integrated into the monitoring and anomaly detection system for civil constructions, based on wireless sensor connection technology bringing feasible results [13–17]. For the CCs, the stability and certainty of the structure directly affect the quality of the exploitation process. Zhang et al. (2024) used deep neural network techniques to identify possible fracture locations of mechanical structures caused by uncertain components [16]. In addition, artificial neural networks are used in estimating the state of marine environmental impacts based on a data set measuring the activity over time of structures [18, 19]. Overall, these proposed methods focused on developing intelligence theory have achieved feasible results, but monitoring is only in a few local areas.

Fig. 1. Damage of wharf construction in Vung Tau City, Vietnam.

Structural health monitoring methods involve observing and analyzing the structures over time to determine changes in properties [20, 21], leading to deterioration in performance [22]. Thus, integrating a CCDD helps optimize design, operation, and maintenance. Construction health monitoring is also used to predict when damage to structures occurs and to make decisions such as shutting down for maintenance. A CCDD system employs a multipoint sensor network to capture response values from the overall structure, the local structural response, and structural integrity management, which monitors the environmental factors, vibration, acceleration, and pressure. In particular, in the CCs, the factors cause cracks at key points because of reasons such as foundation settlement, climate change, erosion, and overload. Within the scope, this study focuses on analyzing the impact of cracks on construction health. From there, we propose a method to identify and process observed data from crack locations.

In this study, the authors developed a CCDD model that applies neural networks to evaluate the CCs damage state, specifically applying DCNN to analyze cracking conditions for the wharf. The paper contributions are presented as follows: *i)* establishing the problem formulation of a crack monitoring system in weak areas to collect images of the progression of crack changes over time, afterward, the remark is given; *ii)* developing a DCNN algorithm that analyzes a series of consecutive crack images to determine parameters related to the crack; and *iii)* synthesizing crack progression and width data to assess the overall fracture condition and provide damage warnings.

The paper is organized as follows: Sect. 2 discusses the challenges of monitoring wharf structure cracking. Next, building a solution to evaluate the work on developing the DCNN method in analyzing crack image sequences is carried out in Sect. 3. After that, Sect. 4 presents the results obtained in experimental cases with evaluations. Conclusions and future directions are given in Sect. 5.

2 Problem Formulation

Installation of sensor networks for monitoring CCs' fracture is an urgent process along with designing the construction structure. Quality monitoring data helps accurately assess the stage of the structure, which provides design methods based on the balance between economy and safety. The deterioration in the quality of the CCs is caused by several factors, which require specific types of sensors to collect data on numerous parameters. The quality of construction is evaluated by comparing the stability of the foundation with the designed values and the ability to work effectively. Thus, the construction design is guaranteed to operate optimally and safely, meeting stable and sustainable quality working standards over time. Figure 2 represents an overview of the monitoring system for the CCs structure, which guarantees the structural construction safety against sudden fractures.

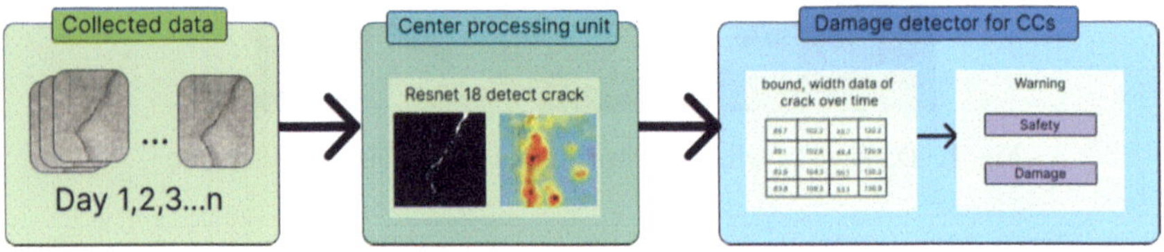

Fig. 2. Overview diagram of the proposed CCDD.

Remark 1: Crack locations that progress slowly with small widths do not significantly affect the stability of the building structure. However, suppose there is a sudden impact of a large amplitude external force due to a collision with the hull of a ship docked at the wharf or an earthquake. In that scenario, the width of the crack exceeds the safety threshold, resulting in structural damage. Therefore, in the case of using a manual assessment method, it will be difficult to detect the risk of fracture early.

Current methods for detecting coastal structural fractures do not meet the requirements for long-term and real-time tracking, due to the low level of automation. Within the scope of the study, we focused on analyzing cracks in the CCs based on a camera system located at the areas where cracks appear on the structure. The collected image sequences are processed with the DCNN technique to extract quality parameters (width, boundary, and progression) and send them to the central processor installed on the CCs for recording and providing status warnings. Figure 3 describes the experimental process of monitoring the progression of crack appearance areas.

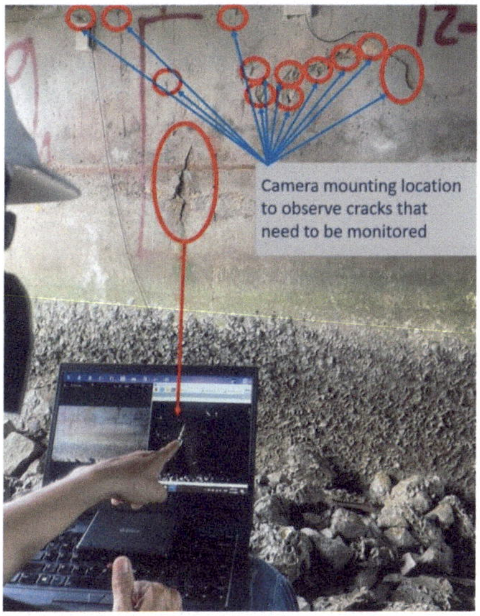

Fig. 3. Experimental detecting of crack progression at the wharf.

3 Methods and Materials

3.1 The Proposed Operation Mechanism

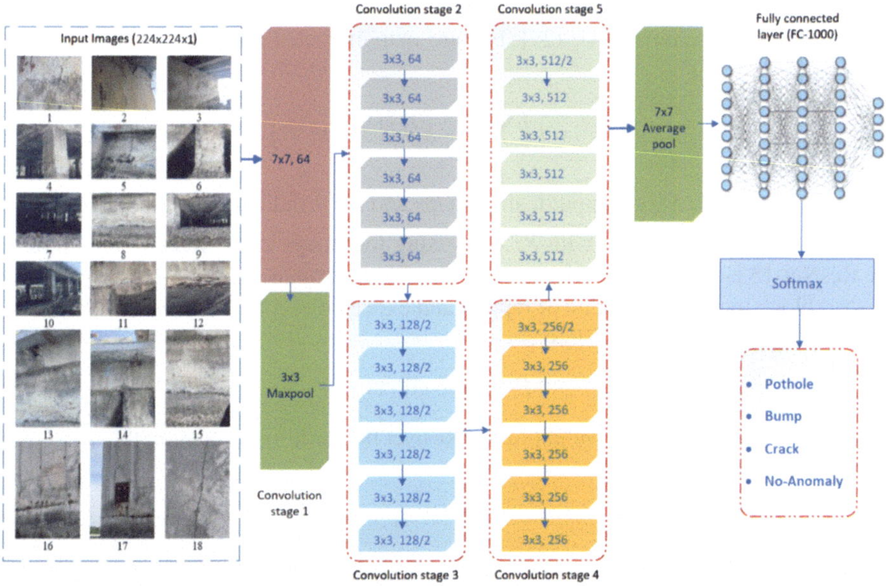

Fig. 4. Structure of ResNet-18 for crack identification system.

The wharf construction has installed cameras to monitor the progression of the crack boundary. In this study, ResNet-18 is employed to analyze images of cracks and determine crack progression parameters [23]. The ResNet-18 network architecture, shown in detail in Fig. 4, recognizes object images based on extracting deep features from training image data and applying convolutional filters to detect objects through 4 phases as follows:

- *Phase 1:* The input image is scaled to 224 × 224 × 1 size as input to the ResNet-18;
- *Phase 2:* Each image has the default boxes, then out images with resolution bounding boxes;
- *Phase 3:* Apply threshold to extract deep feature information through 5 convolutional layers (1);
- *Phase 4:* Extract crack progression value through feature evaluation with Loss function (4) and determine state with Softmax function (8).

Remark 2: Monitoring the health of structures primarily involves assessing the stability and displacement of the structure [24]. However, concrete structures exhibit slow crack progression, but the impact of cracking on the structure appears suddenly. Therefore, it is crucial to have suitable evaluating solutions for fractures in coastal structures.

3.2 Crack Analysis with DCNN Algorithm

The DCNN response provides the assessment results of the crack progression for the analysis frame. In network architecture, the map layers with multiple features scan crack images through a convolution kernel p. This extracts maximum feature information from neighboring areas. In this paper, we employ f as the activation function, which is a rectifier linear unit (ReLU) [25]. Feature map extracted after executing activation function [26] as

$$X_j^{l+1} = f\left(\sum\nolimits_{i \in M_j} X_i^l k_{ij}^{l+1} + b_i^{l+1}\right) \tag{1}$$

where M_j represents a set of input feature maps and X_j^{l+1} indicates the jth feature of the $(l+1)^{th}$ convolutional layer with X_i^l expressing the image characteristic. In addition, the ResNet-18 is in deeper CNNs, so it gets better performance [25, 26]. In general, the ResNet block includes the value of input x and expected output $H(x)$, which obtain the residual mapping function [27] as

$$F = W_2 \sigma(W_1 x) \tag{2}$$

therein, the W_i indicates the weight of the ResNet network. By executing the activation function σ, we get [28]

$$y = F(x, \{W_i\}) + x \tag{3}$$

Besides, with $(*)$ being a convolution operation, the loss function ls_k of the corresponding crack width state as [29]

$$ls_k = \sum_{k=0}^{n^{-1}} y_k * lnP_k \tag{4}$$

Algorithm 1: Assessing fracture risk of wharf construction using DCNN

Input Sequence images crack in wharf.
Output Bounding box of cracks, mean width of cracks.
1. **Initialize** Labeling dataset.
2. **Initialize** Splitting the dataset into training and testing sets.
3. **Initialize** Setting up ResNet18 model on Matlab Environment; EpochMax $\leftarrow$ **5**; LearningRate $\leftarrow$ **0.0002**.
4. **while** *epoch* $\leq$ EpochMax **do**
5. Training the ResNet18 model on Matlab.
6. Evaluating the assessment model with the Loss function by Eq.4, and compare the latest profits to the termination condition.
7. **if** the latest profits meet termination condition **then**
8. save assessing model.
9. **else**
10. $epoch \leftarrow epoch + 1$ and go to Step 5.
11. **end if**
12. **end while**
13. Extracting saved model to determine the bounding boxes base on heatmap of the crack [30]
14. $x_{min} \leftarrow min(x_{min1}, x_{min2})$
15. $y_{min} \leftarrow min(y_{min1}, y_{min2})$ (5)
16. $x_{max} \leftarrow max(x_{max1}, x_{max2})$
17. $y_{max} \leftarrow max(y_{max1}, y_{max2})$
18. Dividing the crack's bounding boxes into n parts and computing the width $width_i$ as
19. $$width_i = \frac{x_{max} \cdot y_{max} - x_{min} \cdot y_{min}}{x_{max} - x_{min}} \tag{6}$$
20. Computing mean width of crack
21. $$\text{MeanWidth} = \sum_{i=1}^{n} \frac{width_i}{n} \tag{7}$$
22. Exporting Bounding boxes and MeanWidth of crack progression
23. **end**

The label y represents the crack boundary type, which is 1, and the remaining positions are 0. In addition, n denotes the classification number of the crack image, and P_k expresses the probability that the crack sample belongs to the kth value. Label prediction with the softmax function is represented by

$$P_k = softmax(lg_k) = \frac{e^{lg_k}}{\sum_{k=0}^{n-1} e^{lg_k}} \tag{8}$$

where lg_k indicates the kth value of the output vector, showing the probability of this classification outcome. The output returns a binary label for image pixels. We expressed the pixel in a crack region as positive (P) and non-crack region as negative (N), To evaluate the accuracy of the ResNet-18 by Eq. 9, whereas True Positive (TP) indicates the case of correct identification of cracks and True Negative (TN) represent the case of correct identification of non-cracks.

$$Accuracy = \frac{TP + TN}{Total\ sample} \tag{9}$$

To increase the reliability and accuracy of the evaluation model, training data were collected at the quay of Vung Tau Port. Images were taken at monitoring points every

30 min in 180 days. The series of crack images are captured periodically by the camera via a Raspberry Pi-embedded computer. The DCNN input will be set to grey image samples created by adding noise to the original image. The process of training the assessing model is expressed in Algorithm 1.

4 Results and Evaluation

4.1 Configuration Parameters

We collected the images of crack point data for six months at the wharf of the Vung Tau port and applied the DCNN method to compute parameters related to the crack (measure width, evaluate crack boundary) to assess the operational state of the construction directly. The wharf's condition is assessed based on the development of cracks over time, which is evaluated at various points in the structure. The proposed system provides the most comprehensive view of the structural condition.

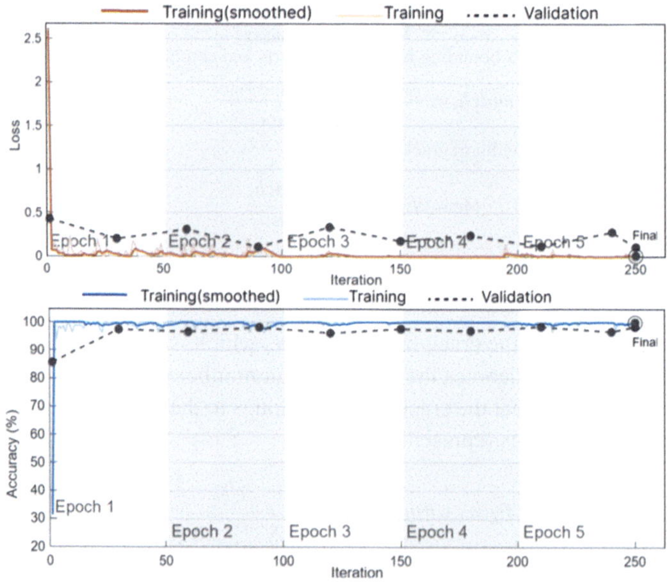

Fig. 5. Results of the ResNet-18 training process.

The authors apply the ResNet-18 to build the CCDD assessment model, implemented in the Matlab 2023b environment. Crack image samples include 8,640 images with a resolution of 224 × 224 × 1 pixels, which are divided into datasets of cracked and non-cracked samples. The training process was performed 250-iterations with a learning rate of 0.0002. The training results are shown in Fig. 5, whereas the accuracy after training is 99.85% and the loss function value reaches 0.098. Besides, the crack width is also determined from the width histogram presented in Figs. 6, 7, and 8 using pixel values, providing accurate fracture risk assessments.

Besides, the authors also compare the performance of DCNN with others, and the results are shown in Table 1. The accuracy of the proposed solution is higher than the VGG16 at 11.85%, the DenseNet169 at 9.95%, the InceptionV3 at 11.45%, and MobileNet at 2.16%. However, the training time of the proposed solutions is limited due to the need to analyze a large number of image samples, specifically, 3 h for ResNet18, 1.2 h for MobileNet, 1.7 h for InceptionV3, 2.5 h for DenseNet169, and 2.2 h for VGG16.

Table 1. Synthesis of parameters of CNN for image analysis.

Model	Training Time (hours)	Input Size	Layer	Accuracy (%)
VGG16 [31]	2.2	224 × 224 × 1	16	88.0
DenseNet169 [32]	2.5	224 × 224 × 1	169	89.9
InceptionV3 [33]	1.7	224 × 224 × 1	48	88.4
MobileNet [34]	1.2	224 × 224 × 1	128	97.69
Proposed ResNet18	**3**	**224 x 224 x 1**	**18**	**99.85**

4.2 Experimental Results

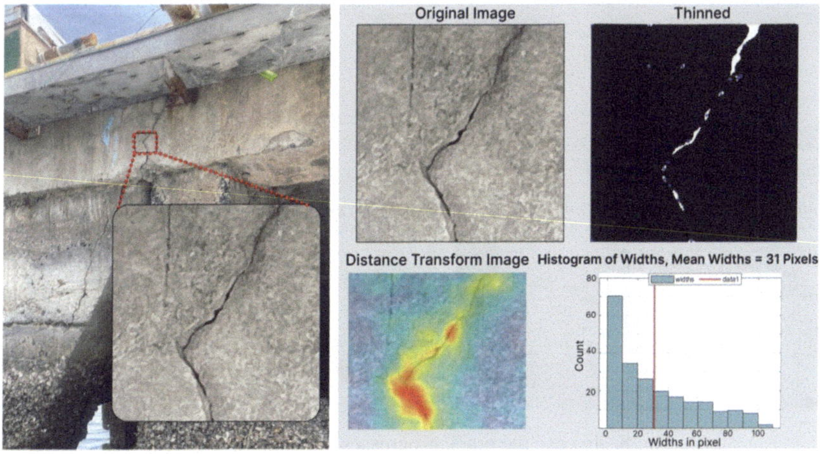

Fig. 6. In case of the crack begin appears on the wharf construction.

The results in Fig. 6 show a safe state with a crack width within the allowable limit of 31.0 pixels and low progression to increase the knot pattern. Under the environmental impact, the crack width will increase to 123.5 pixels, approaching the dangerous limit (shown in Fig. 7). A warning should be issued so the management unit can have a timely maintenance plan. Under harsh operating conditions or vibrations due to a vehicle collision with the CCs structure, it can lead to an abnormal increase in the crack width of 148 pixels, shown in Fig. 8, causing a damaged state for construction operations. The operator needs to stop working to prevent incidents from occurring. Early assessment of the fractures' possibility helps the construction safety under external forces conditions, solving the problems mentioned in *Remarks 1 and 2*.

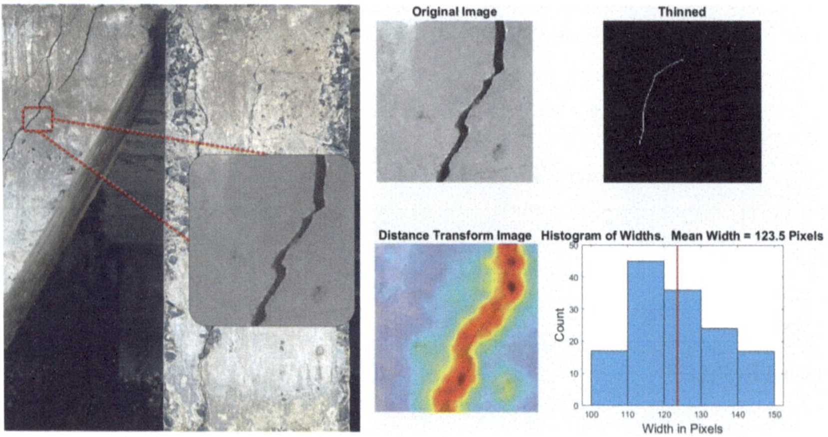

Fig. 7. The crack progresses to a warning level.

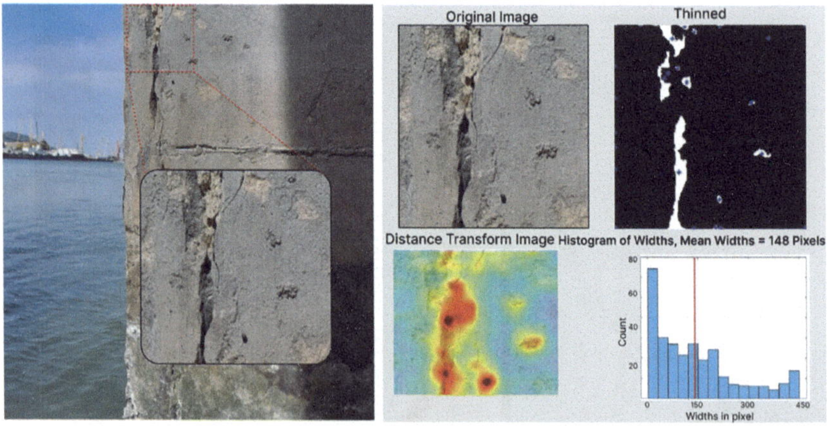

Fig. 8. In case the crack progresses to a damaged state.

To assess the overall state of the CCs, we establish cameras to observe the progression of cracks over time at key locations when cracks appear that can lead to fracture. Initially, this study limited the number of crack-monitoring camera locations to the eight weakest areas of the structure. Crack displacement states are analyzed with the DCNN to determine the average width, summarized in Table 2. Based on crack progression data over time as the basis for development solutions for overall assessment of the CCs' health state.

Table 2. Histogram data array of widths from 8 crack image series.

Monitoring times	Array data of cracks along with average width value (*pixel*)							
	1	2	3	4	5	6	7	8
1	88,1	45,2	122,4	71,4	45,2	38,1	62,2	71,4
2	89,2	48,5	123,3	67,1	47,5	39,2	69,2	79,3
3	89,4	48,5	124,3	69,1	49,3	39,4	69,4	79,9
4	91,7	54,6	124,6	71,24	61,2	51,7	69,7	81,8
5	93,5	66,3	135	89,6	69,63	63,5	73,3	83,5
6	98,34	70,3	137,8	93,8	103,81	88,34	78,4	88,6
7	137,6	78,4	148,1	97,6	107,62	97,6	87,8	97,6
8	168,8	98,1	26,5	124,3	154,5	98,8	88,8	128,2

5 Conclusion

In this study, the authors have successfully applied a DCNN for assessing the crack state (parameters of width, boundary, and progression). Based on the analysis of the crack image dataset, the operator receives queries about safety in order to manage the wharf construction effectively. Furthermore, the established multi-point crack monitoring system (8 crack points) helps the operator understand the overall health of the structure. In the future, the crack progression will be analyzed with an overall pattern using the fast-matching algorithm to define the safety state of the CCs.

References

1. Anastasios, S.K., Nicholas, E.S., Konstantinos, N.A.: Structural health monitoring for corrosion induced thickness loss in marine plates subjected to random loads. Ocean Eng. **273**, 14037 (2023)
2. Jin, Z.: Review on structural health monitoring of offshore platform. J. Phys.: Conf. Ser. (2014). In: The 10th International Conference on Engineering Mathematics and Physics, pp. 1–7 (2021)
3. Do, V.D., Dang, X.K., Ly, S., Nguyen, N.T.: Positioning the semi-submersible platform using model predictive control with thruster constraints. In: Conference Proceedings - 2023 IEEE Asia Meeting on Environment and Electrical Engineering, EEE-AM 2023, pp. 01–05. Institute of Electrical and Electronics Engineers Inc. (2023)
4. Do, V.D., Dang, X.K., Tran, T.D., Pham, T.D.A.: Jacking and energy consumption control over network for jack-up rig: simulation and experiment. Pol. Maritime Res. **29**(3), 89–98 (2022)
5. Yuhao, C., et al.: Analysis of factors affecting the severity of marine accidents using a data-driven Bayesian network. Ocean Eng. **269**, 113563 (2023)
6. Yuhao, C., Xinjian, W., Zaili, Y., Jin, W., Huanxin, W., Zhengjiang, L.: Research in marine accidents: a bibliometric analysis, systematic review and future directions. Ocean Eng. **284**, 115048–115067 (2023)

7. Kai, Y., Yanhui, W., Limin, J., Wenhao, W., Shengli, L., Yanbin, G.: A content-aware corpus-based model for analysis of marine accidents. Accid. Anal. Prev. **184**, 106991 (2023)
8. Mishra, M., Jain, V., Singh, S.K., Maity, D.: Two-stage method based on the you only look once framework and image segmentation for crack detection in concrete structures. Archit. Struct. Constr. **3**(4), 429–446 (2023)
9. Merabet, G.H., et al.: Intelligent building control systems for thermal comfort and energy-efficiency: a systematic review of artificial intelligence-assisted techniques. Renew. Sustain. Energy Rev. **144**, 110969 (2021)
10. Jiaxuan, L., et al.: Damage detection of offshore jacket structures using structural vibration measurements: application of a new hybrid machine learning method. Ocean Eng. **288**(2), 116078–116096 (2023)
11. Zhi-Qian, X., et al.: Vibration-based health monitoring of the offshore wind turbine tower using machine learning with Bayesian optimisation. Ocean Eng. **292**, 116513–116536 (2024)
12. Rosette, N., Wei, J., Elyse, M., Jean, M., Ferdinand, N., Ritha, N.: Intelligent damage diagnosis in bridges using vibration-based monitoring approaches and machine learning: a systematic review. Results Eng. **16**, 100761–100772 (2022)
13. Dang, X.K., Ho, L.A.H., Nguyen, X.P., Mai, B.L.: Applying artificial intelligence for the application of bridge deterioration detection system. Telecommun. Comput. Electron. Control (TELKOMNIKA) **20**(1), 149–157 (2022)
14. Zhang, Y., Yuen, K.V.: Review of artificial intelligence-based bridge damage detection. Adv. Mech. Eng. **14**(9), 1–21 (2022)
15. Jiale, H., et al.: Deep learning and data augmentation-based data imputation for structural health monitoring system in multi-sensor damaged state. Measurement **196**, 111206–111222 (2022)
16. Zhang, J., Huang, M., Wan, N., Deng, Z., He, Z., Luo, J.: Missing measurement data recovery methods in structural health monitoring: the state, challenges and case study. Measurement **231**, 114528 (2024)
17. Chambon, A., Bellaouchou, A., Atamuradov, V., Vitillo, F., Plana, R.: Deep learning based crack growth analysis for structural health monitoring. IFAC-PapersOnLine **55**(10), 3268–3273 (2022)
18. Yee, X.E., Mohamed, M.A.W., Montasir, O.A.: Application of artificial neural network on health monitoring of offshore mooring system. IOP Conf. Ser. Mater. Sci. Eng. **1144**(1), 012035–012043 (2021)
19. Pham, T.A., Dang, X.K., Koboević, Ž, Do, V.D., Pham, T.D.A.: Maritime data mining for marine safety based on deep learning: southern Vietnam case study. NAŠE MORE **71**(1), 21–29 (2024)
20. Ho, L.A.H., Do, V.D., Dang, X.K., Pham, T.D.A.: Early state prediction model for offshore jacket platform structural using EfficientNet-B0 neural network. EAI Endors. Trans. Ind. Netw. Intell. Syst. **11**(2), 1–10 (2024)
21. Zahoor, R., et al.: Lamb wave detection for structural health monitoring using a ϕ-OTDR system. Sensors **22**(16), 5962–5973 (2022)
22. Iraniparast, M., Ranjbar, S., Rahai, M., Nejad, F.M.: Surface concrete cracks detection and segmentation using transfer learning and multi-resolution image processing. Structures **54**, 386–398 (2023)
23. Singh, T., Sehgal, S.: Structural health monitoring of composite materials. Arch. Comput. Methods Eng. **29**, 1997–2017 (2022)
24. Bibi, R., et al.: Edge AI-based automated detection and classification of road anomalies in VANET using deep learning. Comput. Intell. Neurosci. **2021**, 1–16 (2021)
25. Jing, E., Zhang, H., Li, Z.G., Liu, Y., Ji, Z., Ganchev, I.: ECG heartbeat classification based on an improved ResNet-18 model. Comput. Math. Methods Med. 1–13 (2021)

26. Aydogdu, M.F., Celik, V., Demirci, M.F.: Comparison of three different CNN architectures for age classification. In: 2017 IEEE 11th International Conference on Semantic Computing (ICSC), pp. 372–377 (2017)
27. Dahl, G.E., Sainath, T.N., Hinton, G.E.: Improving deep neural networks for LVCSR using rectified linear units and dropout. In: IEEE International Conference on Acoustics, Speech and Signal Processing, pp. 8609–8613 (2013)
28. LeCun, Y., Bengio, Y.: Convolutional networks for images, speech, and time series. Handb. Brain Theory Neural Netw. **3361**(10), 255–258 (1995)
29. He, K., Zhang, X., Ren, S., Sun, J.: Deep residual learning for image recognition. In: The IEEE Computer Society Conference on Computer Vision and Pattern, vol. 34, pp. 770–778 (2016)
30. Alfasly, S., Liu, B., Hu, Y., Wang, Y., Li, C.T.: Auto-zooming CNN-based framework for real-time pedestrian detection in outdoor surveillance videos. IEEE Access **7**, 105816–105826 (2019)
31. Liu, S., Deng, W.: Very deep convolutional neural network based image classification using small training sample size. In: 2015 3rd IAPR Asian Conference on Pattern Recognition (ACPR), pp. 730–734 (2015)
32. Huang, G., Liu, Z., Van Der Maaten, L., Weinberger, K.Q.: Densely connected convolutional networks. In: 2017 IEEE Conference on Computer Vision and Pattern Recognition (CVPR), pp. 2261–2269 (2017)
33. Szegedy, C., Vanhoucke, V., Ioffe, S., Shlens, J., Wojna, Z.: Rethinking the inception architecture for computer vision. In: 2016 IEEE Conference on Computer Vision and Pattern Recognition (CVPR), pp. 2818–2826 (2016)
34. Pan, H., Pang, Z., Wang, Y., Wang, Y., Chen, L.: A new image recognition and classification method combining transfer learning algorithm and MobileNet model for welding defects. IEEE Access **8**, 119951–119960 (2020)

Research on Dynamic Responses of the Floating Bridge Under Motorcade Loads Based on the Kane Method

Song Ji, Junyi Liu, Xujun Chen(✉), Heng Huang, and Guizhang Hu

College of Field Engineering, Army Engineering University of PLA, Nanjing 210007, China
chenxujun213@sina.com

Abstract. To investigate the impact of motorcade loads on the dynamic responses of the floating bridge, this paper establishes the numerical model of the hinged floating bridge based on the Kane method. The accuracy and rationality of the proposed method are validated by comparing the numerical results with those in the related literature. Furthermore, the numerical model for the elastic hinged floating bridge is developed, and the influence of hinge stiffness on the dynamic responses of the floating bridge is analyzed. Subsequently, the case of motorcade loads passing through the floating bridge is considered, and the impact of load distance on the dynamic responses of the floating bridge is investigated. The study reveals that increasing the hinge stiffness can improve the smoothness of the load traveling through the floating bridge, but it also leads to an increase in the bending moments of the connector. The superposition effect of the loads can be utilized to prevent excessive forces on the floating bridge connectors by choosing an appropriate load distance.

Keywords: Floating Bridge · the Kane Method · Dynamic Response Analysis · Motorcade Loads

1 Introduction

Since ancient times, human civilization has been closely linked to water, and how to cross wide water areas is the crucial bottleneck encountered in human production and daily life. In good sheltered coastal areas, the construction of floating bridges to ensure the transportation of people and vehicles can significantly enhance transportation efficiency and effectively address the transportation issues between islands, in comparison to the existing method of using small barges. Many studies have been conducted in recent years to investigate the dynamic response characteristics of the floating bridge in different operating conditions.

Sedarat et al. [1] developed a three-dimensional finite element model of the Homer M. Hadley Memorial Bridge and analyzed the dynamic responses of the floating bridge under the light rail vehicle loads. Wang et al. [2] modeled a ribbon floating bridge using the finite element method, and verified the accuracy of the method in predicting the hydroelastic responses of the ribbon floating bridge in waves by comparing it with

D. Jeng and B. Cai (Eds.): ICCOE 2024, LNCE 423, pp. 54–64, 2025.
https://doi.org/10.1007/978-981-96-0181-3_6

the experimental results measured in the wave basin. Russell and Thrall [3] reviewed the development of military floating bridges in the United States from 1940 to 2000, emphasizing that the reliability of the floating bridges under high sea conditions remains a major challenge for the future. Huang et al. [4] proposed a theoretical method of multi-segment elastic foundation beams connected by elastic hinges, which can consider both hinged and continuous system floating bridges as well as irregular bridge spans. The correctness of the method was verified through comparisons with relevant numerical and experimental results. Viuff et al. [5] conducted a comparative analysis of the natural frequencies, wave-induced motions, and other characteristics of the end-anchored floating bridge in waves using OrcaFlex and SIMO-RIFLEX software. The results showed that both methods can simulate the dynamic responses of the floating bridge well, and the calculation deviation between the software is around 5%~15%. Yan et al. [6] took the end-anchored floating bridge as the research object and carried out an experimental study on the dynamic responses of the floating bridge under the combined action of earthquakes and waves with a 1:100 scaling ratio. The study found that the coupling effect of earthquakes and waves is related to the earthquake intensity and the sea state levels. Sekulic et al. [7] established a coupled analysis model of vehicle-bridge-wind to investigate the vehicle stability on the floating bridge under specific storm conditions. Wang et al. [8] used OrcaFlex software to establish a numerical model of the floating bridge and conducted a study on the dynamic responses of the floating bridge under the winds, waves, and ship collisions. And subsequently analyzed the effects of ship mass, collision position and other factors on the dynamic responses of the floating bridge. Wang and Jin [9] established a three-dimensional finite element model of a novel gasbag-type floating bridge based on the arbitrary Lagrangian-Eulerian (ALE) method and penalty-based method, realizing the coupling effect between the gasbag and external water as well as internal air. The dynamic responses of the floating bridge under different vehicle loads were further investigated.

Above all, the majority of existing research focuses on the dynamic responses of the floating bridge under the single vehicle load, while there are fewer studies on the interaction between motorcade loads and the floating bridge. In practical engineering, the floating bridge needs to ensure the safe traffic of a high volume of vehicles, making it necessary to investigate the dynamic responses of the floating bridge under motorcade loads. This paper establishes the numerical calculation model for the hinged floating bridge based on the Kane method and verifies it through comparisons with computational results in the literature [10]. Furthermore, the cases of the motorcade loads passing through the elastic hinged floating bridge are considered, and the effects of hinge stiffness and moving load distance on the dynamic responses of the floating bridge are analyzed.

2 Mathematical Formulation

2.1 Simplified Physical Model, Coordinate System and Numbering

For the convenience of calculation, the floating bridge needs to be appropriately simplified for modeling. As illustrated in Fig. 1, the floating bridge is assembled by N pontoons that can be assumed as rigid bodies by neglecting their own deformations. The top left edge of the floating bridge is set to be hinged to the bank edge, and the top right edge is connected to the bank edge by an equivalent spring.

As depicted in Fig. 1, two Cartesian coordinate systems are employed to represent the motions of the floating body. The global coordinate system *OXYZ* is fixed in space, with the *OXY* plane coinciding with the undisturbed water surface and the *OZ* axis pointed straight up. The local coordinate system $o_k x_k y_k z_k$ ($k = 1,2, \ldots, N$) moves with the floating body, with the $o_k x_k y_k$ plane set on the top deck surface of the floating body, and the $o_k z_k$ axis at the centerline of the left-end section of the floating body and vertically upward. At the initial moment, all floating bodies are in an upright floating state, numbered from left to right as P_1 to P_N.

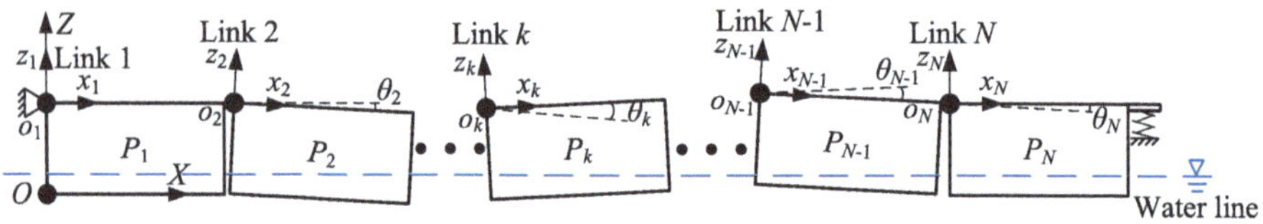

Fig. 1. Coordinates and numbers of the floating bodies.

2.2 Generalized Coordinates and Generalized Speeds

In the Kane method, the generalized coordinate is defined as the relative rotation angle θ_k ($k = 1,2, \ldots, N$) between the local coordinate system of the floating body and the neighboring low-order one, where the global coordinate system is considered as the floating body numbered P_0. The generalized speed is defined as the derivative of the generalized coordinate with respect to time, which can be described by the relative angular velocity ω_k ($k = 1, 2, \ldots, N$) of the floating body with respect to the neighboring low-order one.

Based on the above definitions, it is convenient to consider the hinged connections between the floating bodies. For the multiple floating body system shown in Fig. 1, the generalized coordinates and generalized speeds can be uniformly expressed as:

$$x_l = \begin{cases} \theta_k, & l = 3(k-1)+2 \\ 0, & \text{other} \end{cases} \tag{1}$$

$$\dot{x}_l = \begin{cases} \omega_k, & l = 3(k-1)+2 \\ 0, & \text{other} \end{cases} \tag{2}$$

where $l = 1,2, \ldots, 6N$ is the number of degrees of freedom of the floating body.

2.3 Vehicle Loads

In the process of calculation, the impact effect generated during vehicle traveling is neglected, and the influence of vehicle weight distribution is disregarded, simplifying it as the moving concentrated loads. As a result, the force F_{v}^k and moment T_{v}^k acting on the k-th floating body at each moment can be obtained:

$$F_{\mathrm{v}}^k = \sum_{i=1}^{i_0} p_i \tag{3}$$

$$T_{\mathrm{v}}^k = \sum_{i=1}^{i_0} p_i \delta_i(T) \tag{4}$$

where i is the moving load number, i_0 is the total number of the moving loads, p_i is the force of the i-th vehicle load, and $\delta_i(T)$ is the position function associated with time T and vehicle speed v, which can be expressed as:

$$\delta_i(T) = \begin{cases} 0, & T < T_i^k \\ 0.5l - v\left(T - T_i^k\right), & T_i^k \le T < T_i^{k+1} \\ 0, & T_i^{k+1} \le T \end{cases} \tag{5}$$

here, l represents the length of the floating body, while T_i^k and T_{i+1}^k indicate the time at which the i-th load moves to the k-th and $(k + 1)$-th hinge connector positions (with the $(N + 1)$-th hinge connector considered as the rightmost end of the N-th floating body). The distance between loads can be considered by adjusting the initial time of the moving loads.

2.4 The Kane Kinematic Equation

For the whole system, the sum of the generalized active force and the generalized inertial force is zero, and the expression for the components of the l-th degree of freedom is given as:

$$F_l^* + F_l = v_{klm} F_{km}^* + \omega_{klm} T_{km}^* + v_{klm} F_{km} + \omega_{klm} T_{km} = 0 \tag{6}$$

here the inertia force F_{km}^* and the inertia moment T_{km}^* on the k-th floating body are given as:

$$F_{km}^* = -M_k (v_{klm} \ddot{x}_l + \dot{v}_{klm} \dot{x}_l) \tag{7}$$

$$T_{km}^* = -I_{kmn} (\omega_{klm} \ddot{x}_l + \dot{\omega}_{klm} \dot{x}_l) - e_{rsm} \omega_{klr} \omega_{kpn} I_{ksn} \dot{x}_l \dot{x}_p \tag{8}$$

The union of Eqs. (6), (7), and (8) yields a unified formulation of the Kane equation:

$$A_{lp} \ddot{x}_p = f_l \tag{9}$$

where

$$A_{lp} = M_k v_{klm} v_{kpm} + I_{kmn} \omega_{klm} \omega_{kpm} \tag{10}$$

$$f_l = v_{klm} F_{km} + \omega_{klm} T_{km} - M_k v_{klm} \dot{v}_{kpm} \dot{x}_p - I_{kmn} \omega_{klm} \dot{\omega}_{kpm} \dot{x}_p$$

$$- I_{ksn}\omega_{klm}e_{rsm}\omega_{kpr}\omega_{kqn}\dot{x}_p\dot{x}_q \tag{11}$$

In Eqs. (6)–(11), m, n, r, and s are dummy indexes. l, p, and q are the degrees of freedom numbers. ω_{klm}, ω_{klr} and ω_{kpn} are the partial angular velocity matrix elements. $\dot{\omega}_{klm}$ and $\dot{\omega}_{kpm}$ are the derivative of the partial angular velocity matrix elements. V_{klm} is the partial velocity matrix element. $\dot{v}_{klm}$ and $\dot{v}_{kpm}$ are the derivative of the partial velocity matrix elements. F_{km} and T_{km} are the total external force and moment components on the k-th floating body, including gravity, buoyancy, hydrostatic restoring forces, vehicle loads and connection forces. M_k is the mass of the floating body, and e_{rsm} is the coordinate ordering matrix element. $\dot{x}_l$, $\dot{x}_p$, and $\dot{x}_q$ are the generalized speed matrix elements of the system, and I_{kmn} and I_{ksn} are the rotational moment of inertia matrix elements of the floating body.

3 Numerical Modeling and Validation

3.1 Model Parameters

The hinged floating bridge shown in Fig. 2 is assembled by five pontoons with the same size of length $l = 15$ m, width $b = 6$ m, and height $h = 5$ m. The pontoons are numbered from left to right as P_1 to P_5, and the hinge connectors are set up on the top deck at both ends of the pontoons, allowing relative pitching motions between pontoons, and the connectors are numbered from left to right as W_1 to W_4. A load of $p = 300$ kN starts from the left edge of the hinged floating bridge at time $T = 0$ and moves along the bridge axis at a constant speed of $v = 5$ m/s towards the right edge.

As seen in the cross-section diagram in Fig. 2, the pontoons are designed as hollow structures with a thickness of $t = 0.06$ m. Therefore, the weight and moment of inertia of each pontoon can be calculated using the following equations:

$$M = 2\rho t(lb + bh + lh) \tag{12}$$

$$I_{11} = \frac{\rho t}{6}\left[(3lb + bh + lh)h^2 + (lb + bh + 3lh)b^2\right] \tag{13}$$

$$I_{22} = \frac{\rho t}{6}\left[(3lb + bh + lh)h^2 + (lb + 3bh + lh)l^2\right] \tag{14}$$

$$I_{33} = \frac{\rho t}{6}\left[(lb + bh + 3lh)b^2 + (lb + 3bh + lh)l^2\right] \tag{15}$$

where ρ is the density of the pontoon, which is considered in the calculations to be made of uniformly poured concrete and has a value of 2658 kg/m^3.

Therefore, the main parameters of the pontoon are calculated as shown in Table 1.

3.2 Model Validation

Raftoyiannis et al. [10] considered the displacements at the location of the connector between pontoons as the variables to be solved and established the differential motion

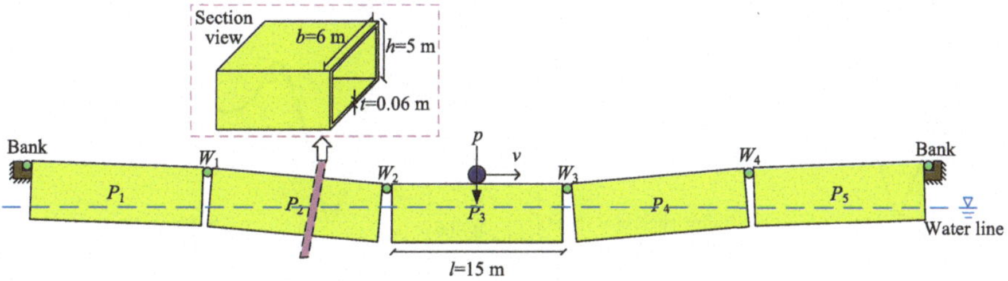

Fig. 2. Diagram of the hinged floating bridge

Table 1. Main parameters of the single pontoon

Parameter	Value	Parameter	Value
Length (m)	15	Mass (kg)	61207
Width (m)	6	I_{11} (kg•m^2)	579311
Height (m)	5	I_{22} (kg•m^2)	1774215
Draft (m)	0.68	I_{33} (kg•m^2)	1855151

equations of a hinged floating bridge (as shown in Fig. 2) based on the Lagrange method. The motion responses of the hinged floating bridge under the action of a single moving load with constant velocity were calculated.

In this paper, the parameters listed in Table 1 are used to calculate the vertical displacement curves of the hinged floating bridge at the locations of $W_1 \sim W_4$ connectors by the Kane method, as indicated by the solid black line in Fig. 3. Comparing the numerical results of this method with those of Raftoyiannis et al. [10], it can be seen that the deviation between the two methods is small, indicating that the Kane method adopted in this paper can effectively simulate the motion response characteristics of the hinged floating bridge.

Furthermore, as can be observed from Fig. 3, during the process of the moving load traveling from pontoon P_1 to P_5, the negative peaks of the vertical displacement reach at different times with $W_1 \sim W_4$ connectors, and the negative peaks at W_1 and W_4 are smaller than those at W_2 and W_3. This is due to the boundary conditions of the floating bridge, where the P_1 and P_5 pontoons at both ends are connected to the banks, resulting in smaller negative peaks of the vertical displacement at W_1 and W_4, which are close to them. Regarding the vertical displacements at W_2 and W_3, although certain positive displacements occur before and after the moving load passes through the connecting positions, their impact on the motion responses of the floating bridge is relatively minor compared to the negative displacements. This indicates that the negative peaks of vertical displacement should be emphasized during the process of the moving load passing through the floating bridge.

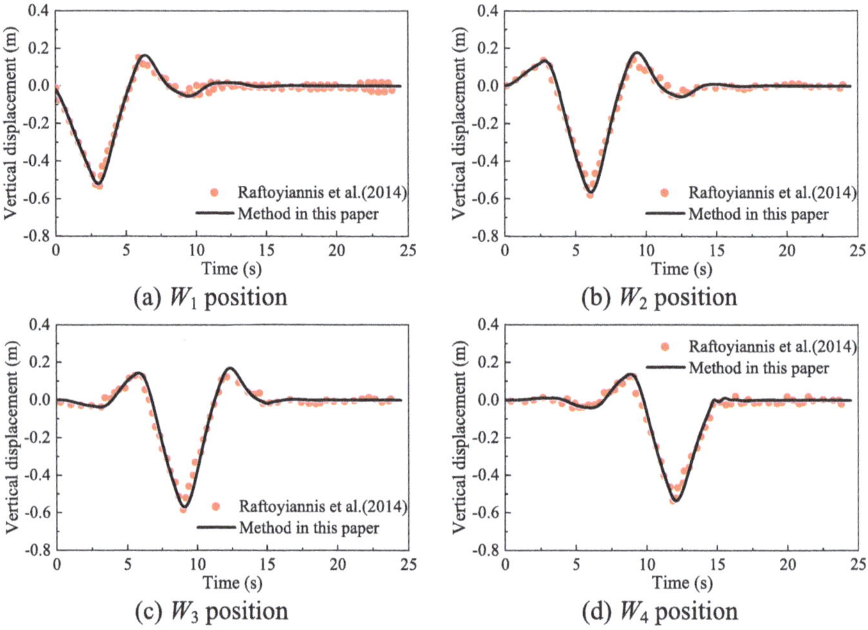

Fig. 3. Vertical displacement time histories of the floating bridge at different positions

4 Analysis and Discussion

In this section, the floating bridge is further extended to the structural type of 10 pontoons connected by elastic hinges as shown in Fig. 4, The geometric and mechanical parameters of the pontoons, as well as the boundary conditions of the floating bridge are the same as those in Sect. 3. The effect of hinge stiffness on the dynamic responses of the floating bridge is first investigated in the case of a single moving load passing through the floating bridge. Subsequently, the effect of moving load distance on the dynamic responses of the floating bridge is investigated by considering the case of motorcade loads passing through the floating bridge.

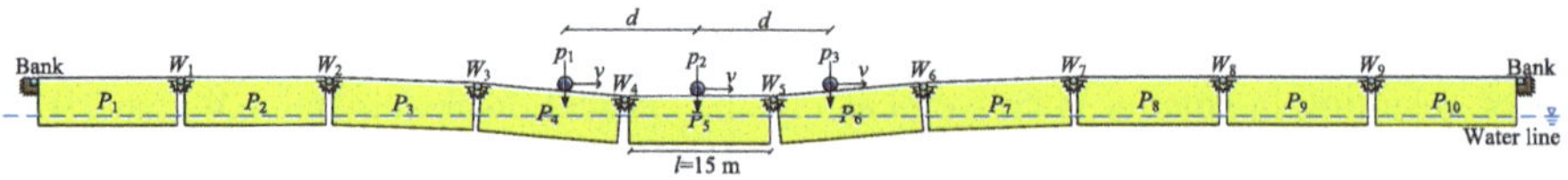

Fig. 4. Diagram of the elastic hinged floating bridge

4.1 Influence of Hinge Stiffness

To investigate the influence of hinge stiffness on the dynamic responses of the floating bridge, four kinds of hinge stiffnesses, $k = 0$, 5×10^6, 1×10^7, and 5×10^7 N•m/rad, which may be chosen in practical engineering, are selected. At $T = 0$ s, a single moving load of 300 kN traveled from pontoon P_1 to pontoon P_{10} at a constant speed of $v = 5$ m/s. The vertical displacements, vertical forces, and bending moments at position W_5 are calculated and plotted in Figs. 5, 6 and 7.

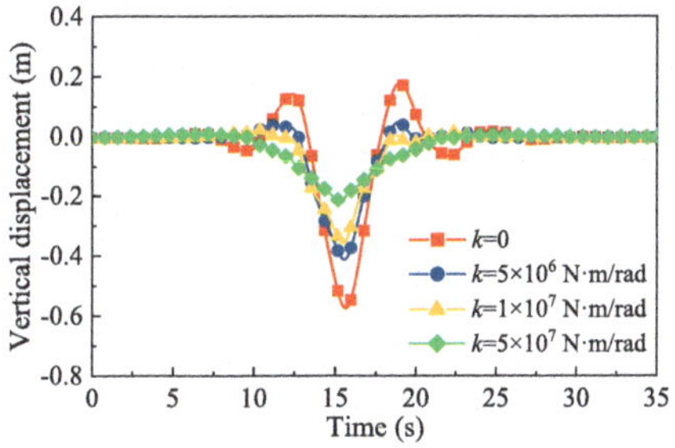

Fig. 5. Vertical displacement of the floating bridge at position W_5 for different hinge stiffnesses

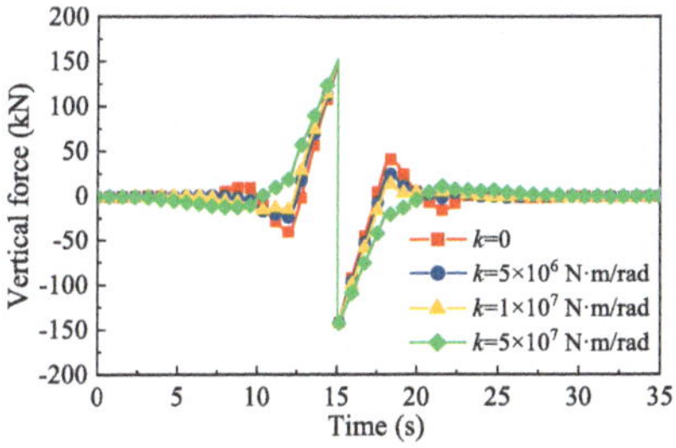

Fig. 6. Vertical force of the W_5 connector for different hinge stiffnesses

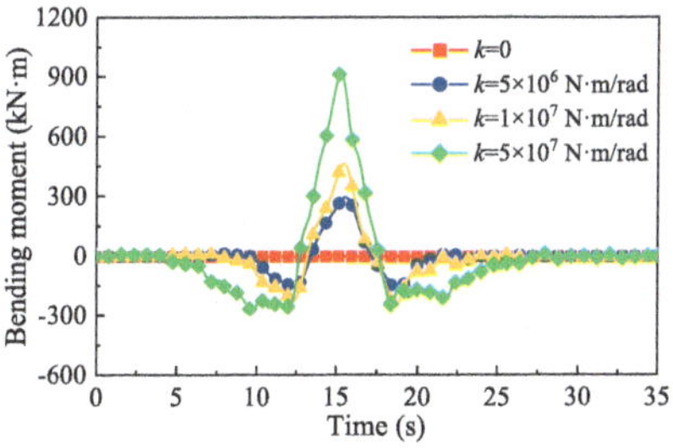

Fig. 7. Bending moment of the W_5 connector for different hinge stiffnesses

As evident from the vertical displacement curves at position W_5 in Fig. 5, the amplitudes of the motions (both positive and negative) of the floating bridge decrease with growing hinge stiffness during the passing progress of the moving load. Notably, when the hinge stiffness is 5×10^7 N•m/rad, there is almost no positive displacement but only negative displacement at the W_5 position. This suggests that increasing the hinge stiffness benefits the smoothness and stability of vehicles passing through the floating bridge. Regarding the vertical force curves at the W_5 connector shown in Fig. 6, a sudden change in vertical force occurs when the load moves on the position W_5. It can also be observed that the variation in hinge stiffness has little effect on the maximum vertical forces.

The curves of the bending moment at the W_5 connector in Fig. 7 exhibit a similar pattern to the displacement curves in Fig. 5. Before the load reaches position W_5, negative bending moments are first generated, followed by positive bending moments when the load arrives at W_5 position, and then generates the negative bending moments after driving away from the W_5 connector. The hinge stiffness has a minor impact on the peak negative bending moments, affecting only its appearance time, as the floating bridge with higher stiffness tends to generate negative bending moments earlier. However, the peak positive bending moments increase significantly with increasing hinge stiffness. By comparing the values of the positive and negative bending moments, it is evident that the positive bending moments generated during the load moving through the floating bridge should be paid attention.

4.2 Influence of Moving Load Distance

To consider the interaction between the motorcade loads and the floating bridge, three moving loads of weight 300 kN are selected to travel from pontoon P_1 to P_{10}, starting from time $T = 0$, at a constant speed of $v = 5$ m/s. The impact of the load distance on the dynamic responses of the floating bridge is investigated by varying the moving load distance $d = 20$, 30, and 40 m, with a single moving load used for comparison. The stiffness of the elastic hinge on the floating bridge is set to 1×10^7 N•m/rad, and the vertical displacements of the floating bridge at the position W_5, as well as the vertical forces and bending moments at the W_5 connector, are calculated and plotted in Figs. 8, 9 and 10.

By comparing the vertical displacement curves of the floating bridge in Fig. 8, it can be observed that when the load distance exceeds 30 m, the maximum vertical displacement of the floating bridge is almost the same as that of the single load. However, when the distance is 20 m, the maximum vertical displacement is slightly increased. Additionally, the displacement curves with the 20 m distance are found to have a superposition effect on the moving loads, indicating that when the distance is small, the sinking displacement caused by the previous load has not fully recovered to its balanced position before the next load arrives, thus increasing the displacement responses of the system.

The vertical force curves at the W_5 connector presented in Fig. 9 show that the maximum vertical force of the connector under different load distances has little difference from that of the single load, which indicates that the impact of the motorcade loads on the vertical forces is limited. On the other hand, the bending moment curves at the W_5 connector in Fig. 9 reveal that compared to the single load, the maximum bending moment of the connector is significantly reduced when the load distance is 20 m, while such a phenomenon is not found when the load distance is greater than 30 m. Combined with the bending moment curves with hinge stiffness of 1×10^7 N•m/rad in Fig. 7, it can be seen that when the single load moves on the W_5 connector around 15 s, negative bending moments occur during the periods of 7–12 s and 18–23 s. Calculations show that keeping the load distance at 15–24 m at this speed is beneficial for reducing the bending moments at the connector W_5, due to the superposition effect of the motorcade loads, wherein the positive bending moments caused by one load are offset by the negative bending moments of the neighboring load. This explains the reduction in connector bending moments observed at 20 m moving load distance, which is crucial for understanding the behavior of the floating bridge under vehicles, indicating that controlling the load distance can prevent excessive bending moments on the connectors.

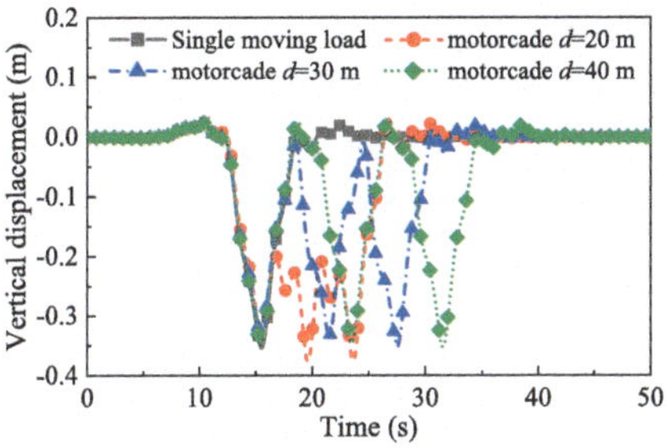

Fig. 8. Vertical displacement of the floating bridge at position W_5 for different moving load distances

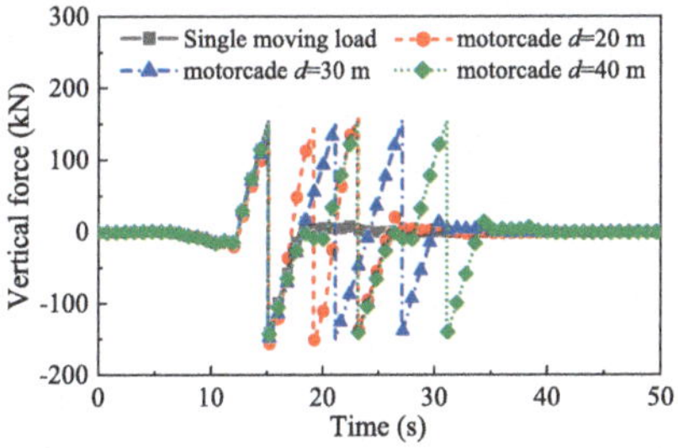

Fig. 9. Vertical force of the W_5 connector for different moving load distances

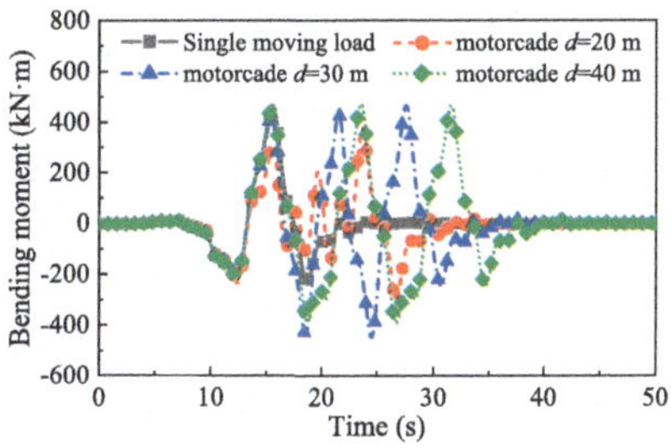

Fig. 10. Bending moment of the W_5 connector for different moving load distances

5 Conclusions

In this paper, the numerical calculation models of the hinged floating bridge and elastic hinged floating bridge are established based on the Kane method, and the relevant calculation results have been validated by the numerical results in existing literature. The study analyzes the influence of hinge stiffness and moving load distance on the dynamic responses of the floating bridge, leading to the following key conclusions:

(1) The method adopted in this paper can effectively predict the motion characteristics of the floating bridge under the moving loads, which can be further applied to the research of the floating bridge with elastic hinge connectors.
(2) Increasing the hinge connection stiffness can reduce the maximum displacement responses of the floating bridge, benefiting the smoothness and stability of the vehicles passing through the floating bridge.
(3) The displacement of the floating bridge caused by the motorcade loads and the single load is basically the same for the load distance greater than 30 m, while the maximum displacement of the floating bridge increases slightly due to the superposition effect of the loads for the 20 m distance.
(4) Under special conditions, the balance between positive and negative bending moments can be utilized to improve the load carrying capacity of the floating bridge by setting the load distance reasonably.

Acknowledgment. This work was supported by the National Natural Science Foundation of China (Grant No. 52371300 and Grant No. 52471309).

References

1. Sedarat, H., et al.: Dynamic response of a floating bridge to a moving light rail train. In: Structures Congress 2013, pp. 514–530. American Society of Civil Engineers, Pittsburgh, USA (2013)
2. Wang, C., Fu, S.X., Cui, W.C., Lin, Z.M.: Hydroelastic analysis of a ribbon bridge under wave actions. J. Ship Mech. **10**(6), 61–75 (2006)
3. Russell, B.R., Thrall, A.P.: Portable and rapidly deployable bridges: historical perspective and recent technology developments. J. Bridg. Eng. **18**(10), 1074–1085 (2013)
4. Huang, H., Chen, X.J., Liu, J.Y., Chen, Q.F.: A theoretical calculation method of irregular bridge span pontoon connected by elastic hinges. Mar. Struct. **76**, 102929 (2021)
5. Viuff, T., Xiang, X., Leira, B.J., Øiseth, O.: Software-to-software comparison of end-anchored floating bridge global analysis. J. Bridg. Eng. **25**(5), 04020022 (2020)
6. Yan, J.L., Liu, J.B., Liu, Z.C., Li, H., Guo, A.X.: Experimental study on the dynamic responses of the end-anchored floating bridge subjected to joint actions of earthquakes and water waves. Earthq. Eng. Struct. Dyn. **52**(10), 2945–2965 (2023)
7. Sekulic, D., Vdovin, A., Jacobson, B., Sebben, S., Johannesen, S.M.: Effects of wind loads and floating bridge motion on intercity bus lateral stability. J. Wind Eng. Ind. Aerodyn. **212**, 104589 (2021)
8. Wang, Z.H., Sha, Y.Y., Jakobsen, J.B.: Floating bridge response under combined ship collision, wind and wave loads. Ships Offshore Struct. 1–18 (2023). https://doi.org/10.1080/17445302.2023.2195242
9. Wang, H.H., Jin, X.L.: Dynamic analysis of maritime gasbag-type floating bridge subjected to moving loads. Int. J. Naval Archit. Ocean Eng. **8**(2), 137–152 (2016)
10. Raftoyiannis, I.G., Avraam, T.P., Michaltsos, G.T.: Analytical models of floating bridges under moving loads. Eng. Struct. **68**, 144–154 (2014)

Hydrodynamics, Fluid Mechanics, and Fluid-Structure Interaction

Experimental Study of the Load Exerted by Internal Solitary Waves on an Immersed Square Column

Guanjing Wang, Hui Du(✉), Jianfang Fei, Shaodong Wang, Pai Peng, Pu Xuan, and Zheyu Lu

College of Meteorology and Oceanography, National University of Defense Technology, Changsha 410005, China
duhui17@nudt.edu.cn

Abstract. The safety of marine structures and underwater vehicles is threatened by the internal solitary waves (ISWs). The experimental study of the load characteristics exerted by ISWs on an immersed square column has been carried out in a stratified fluid flume. A new type of submerged load measurement platform is proposed to measure the changes in load exerting on the square column, and an array of conductivity probes is used to obtain the characteristics of ISWs. The influence of wave-flow fields of ISWs and depth of square column on the load characteristics is comprehensively analyzed. The results show that the load on the square column is mainly determined by the ISW-induced flow field. The intensity and trend of the forces at different depths are different, and the intensity of forces increases with the increase of wave amplitude. Square column will experience roll and pitch when encountering ISWs during the diving process, with the pycnocline being the region with the strongest rolling and pitching.

Keywords: Internal solitary waves · Stratified fluid · Load on square column · Wave-flow fields · Submerged load measurement platform

1 Introduction

Internal solitary waves (ISW) in the ocean are a common type of wave that occurs within density-stratified seawater. ISWs usually carry enormous energy during propagation, which can cause convergence, divergence, and sudden currents of surrounding water bodies [1, 2]. The flow direction of the upper and lower layers of the ISW is opposite, forming a great shear effect in the pycnocline [3, 4]. The impact load generated by the shear effect can not only cause failure or even damage of marine structures [5, 6], but also lead to instability or even loss of control in the movement of submarine [7, 8]. Therefore, the study of the load characteristics exerted by ISW on the underwater structures has become an important topic in the fields of marine engineering and hydrodynamics in recent years.

In terms of theoretical research, the classic internal waves (IWs) theory combined with Morison's formula is used to estimate the forces exerted by IWs on simple structures

D. Jeng and B. Cai (Eds.): ICCOE 2024, LNCE 423, pp. 67–75, 2025.
https://doi.org/10.1007/978-981-96-0181-3_7

such as cylinders, spheres, and thin plates [9–13]. As for the numerical simulation, source functions, incident boundary conditions, and simulated physics wave making technology are used to simulate IWs. The load problem of the interaction between IWs and underwater structures is mainly numerically simulated and analyzed by researchers [14, 18].

Experimental study is an important way to objectively understand the load exerted by ISWs on underwater structures. Du et al. [7] measured the horizontal and vertical forces on slender body exerted by ISWs propagating along the slope, indicating that topographic effect directly determined the magnitude of the forces. Ji et al. [19] studied the load characteristics of slender body exerted by ISWs through experiments and theory, and pointed out that different incident angles of ISW, wave amplitudes, and the thickness ratio of the upper and lower fluids have a significant impact on the load characteristics. Xuan et al. [20] compared theoretical models and physical experiments to demonstrate that the horizontal force on slender body exerted by ISWs is composed of horizontal dynamic pressure and frictional force. Tian et al. [21] used experimental methods to study the induced pressure of ISWs on slope, and found that the polarity of induced pressure changes with the propagation of ISWs. Wang et al. [22] conducted an experimental study on the load characteristics exerted by ISW on a horizontal transverse cylinder and found that the peak of vertical force on the cylinder increases with the increase of wave amplitude. At present, the problem of the load exerted by ISW on structures mainly focuses on the study of horizontal and vertical forces, with unclear analysis of torque. Moreover, there is insufficient experimental research considering the effect of wave-flow fields of ISWs on square column.

This article used the gravity collapse method to generate ISWs in a stratified fluid flume. Conductivity probes and submerged load measurement platform were used to measure the flow field characteristics of ISWs and the load characteristics of square column, respectively. The specific arrangement of the paper is as follows: Section 1 is the introduction, Section 2 is the experimental methods, Section 3 is the results and analysis, and Sect. 4 is the conclusion.

2 Experimental Methods

The experiment of the load characteristics exerted by ISWs on square column was conducted in a stratified fluid flume with a main scale of 1200 cm × 40 cm × 60 cm (length × width × height). The experimental setup is shown in Fig. 1. A coordinate system xoz for the right-hand rule was established, with the origin o located at the junction of the leftmost end of the flume and the stationary free water surface. The horizontal positive x-axis runs from left to right along the length of the flume, and the vertical positive z-axis runs from bottom to top along the depth of the flume. The two ends of the flume were respectively equipped with ISW generation and absorber devices. The wave generation adopted the principle of gravity collapse [23], where L_0 and η_0 were the step length and step depth, respectively, and can generate concave ISWs at any position of the pycnocline. The wave absorber adopted a triangular wedge structure, which can effectively eliminate the influence of reflected waves by adjusting the wedge angle.

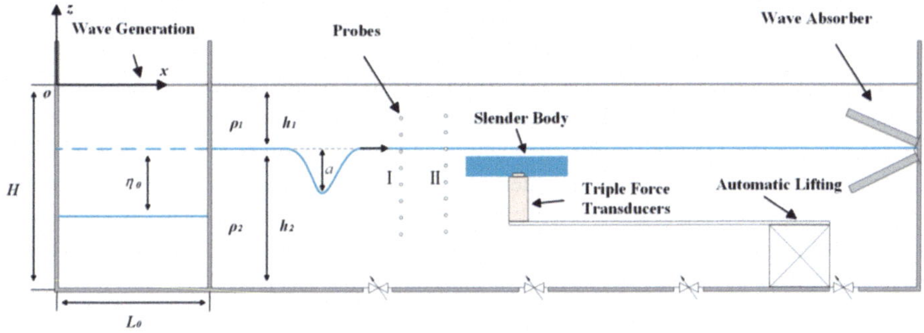

Fig. 1. A schematic of the experimental setup

The stratified environment was prepared using the "double tube" method. The saline water with a thickness of $h_2 = 40$ cm and a density of $\rho_2 = 1018.2$ kg/m^3 was injected into the lower layer of flume, and the fresh water with a thickness of $h_1 = 10$ cm and a density of $\rho_1 = 1003.6$ kg/m^3 was slowly injected into the upper layer. The two layers of fluid naturally mixed at the interface to form a pycnocline with a thickness of ~6 cm, as shown in Fig. 2. Define the buoyancy frequency $N(z) = \sqrt{-g\frac{d\rho(z)}{dz}}$, where $\rho(z)$ is the vertical distribution of density and g is the gravitational acceleration. The figure shows that the maximum buoyancy frequency is located in the middle of the pycnocline.

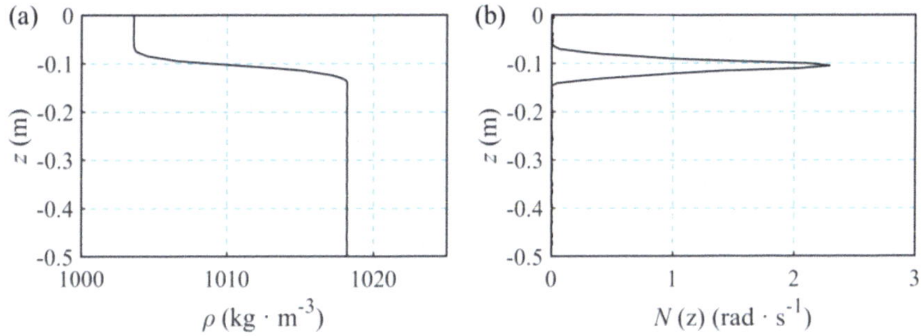

Fig. 2. Distribution of density and Brunt-Vaisala (B-V) frequency

In order to measure the wave-flow fields of ISWs, conductivity probe arrays were arranged at positions I and II, as shown in Fig. 1. Eight probes were arranged vertically at each position, with a spacing of 2 cm. The probes at positions I and II were used to measure the wave field of ISW, and the vertical displacement of the isodensity line of ISW at the highest floating frequency was defined as the amplitude of the ISW. The time interval Δt between the peak values of conductivity disturbance when the ISW passes through probe I and probe II was recorded. The experimental value of the propagation speed of the ISW can be obtained from equation $c = (x_2 - x_1)/\Delta t$.

The submerged load measurement platform was placed in the flume, which includes an automatic lifting device, connecting shaft, triple force transducers, and a square column. The connecting shaft is sufficiently long and rigid, which can effectively reduce the platform's interference with ISWs. The size of square column is 30 cm × 6 cm × 6 cm (length × width × height). The horizontal position of the platform was adjusted to place the model in the middle of the flume. In the experiment, the automatic lifting device was adjusted using a remote control to change the vertical position of the square column. Triple force transducers were used to measure the forces on the square column, Including x-axis horizontal force, z-axis vertical force, and xoz-plane torque. The experimental settings are shown in Table 1. Experiments were conducted on small, medium, and large wave amplitudes at each depth where the square column was located.

Table 1. Specific parameters of experimental settings. Depth refers to the distance between the center of gravity of square column and the water surface

Depth/m	ρ_1/kg*m^{-3}	ρ_2/kg*m^{-3}	h_1/m	h_2/m	a^*
0.05					0.105,0.143,0.167
0.105	1003.6	1018.2	0.10	0.40	0.113,0.138,0.165
0.169					0.097,0.136,0.162
0.253					0.108,0.147,0.171

The dimensionless form of physical parameters in the experiment is as follows: $a^* = a/H$, $c_x = \frac{F_x}{\rho(z)g\beta l}$, $c_z = \frac{F_z}{\rho(z)g\beta l}$, $M_{xz} = \frac{m_{xz}}{\rho(z)g\beta ls}$. where a is the amplitude of the ISW (m), H is the total depth of the fluid (m), F_x, F_z are the measured horizontal and vertical forces (N), m_{xz} is the measured xoz-plane torque, β is the horizontal fluid facing area of the square column (m^2), $\rho(z)$ is the fluid density at the depth of the square column (kg/m3), g is the gravity acceleration (m/s^2), l and s are the length and width of the square column, respectively.

3 Results and Analysis

The load characteristics exerted by ISW on square column are determined by the ISW-induced flow field. Figure 3 (a) shows the flow field measurement results of the ISW at the center profile position of view obtained for the typical working condition $a^* = 0.105$, The entire flow field of the ISW exhibits a wave packet shape, with an obvious vortex structure appearing near the trough. Its direction is counterclockwise, determined by the propagation direction of the ISW. Further decompose the flow field to obtain the structural characteristics of horizontal and vertical flow field s, as shown in Figs. 3 (b) (c). According to the distribution of horizontal flow field in Fig. 3 (b), the shear effect is formed at the pycnocline. The direction of horizontal velocity above the pycnocline is consistent with the direction of the ISW, while the horizontal velocity below the pycnocline is opposite and smaller. The horizontal velocity is close to zero near the pycnocline. According to the distribution of vertical flow field in Fig. 3 (c), the vertical field exhibits the characteristics of sinking flow at the windward of wave, upwelling flow at the backward of wave, and approaching zero at the trough, with the sinking and upwelling flow reaching their maximum values near the middle of the wave surface, respectively.

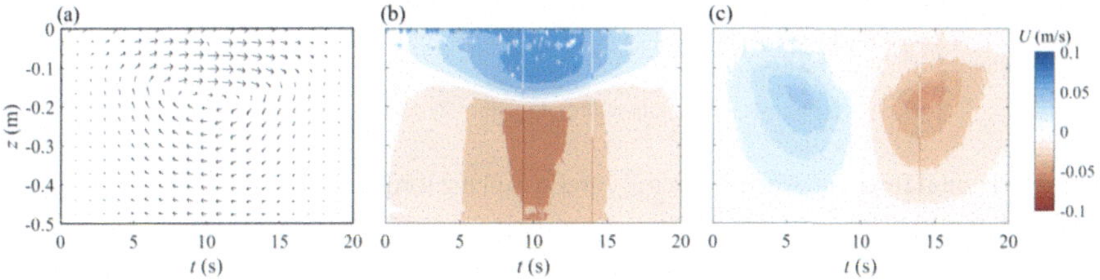

Fig. 3 (a) A schematic diagram of flow field of ISW, (b) Time series of horizontal velocity, (c) Time series of vertical velocity

According to the measurement results of the flow field, it can be seen that as the depth changes, the ISW-induced flow field change significantly, resulting in different forces exert on square column at different depths. Therefore, in experimenting the depth of the square column is changed, then ISWs of small, medium, and large amplitudes are generated for each depth. The load changes of the square column are measured to obtain the variation law of load with the depth of the square column and the wave amplitude.

Figure 4 shows the measurement results of the horizontal and vertical forces and *xoz*-plane torque when the square column above the pycnocline. Three conditions with wave amplitudes of 0.105, 0.143, and 0.167 are showed. According to Fig. 4 (a), it can be seen that during the process of encountering ISWs, the horizontal force on the square column is first positive and then negative. The dynamic characteristics show that the force on square column first is a pushing force along the propagation direction of wave, then transforms into a pulling force, and the pushing force is significantly smaller than the pulling force. As the wave amplitude increases, the trend of horizontal force remains unchanged. But the intensity increases, the increase in forward pushing force is significantly smaller than that in backward pulling force. Figure 4 (b) shows that the vertical force on square column is first negative and then positive. The dynamic characteristics show that the force on square column first is a downward pulling force, then gradually becomes an upward dragging force. The maximum value of the pulling force is smaller than the dragging force. As the wave amplitude increases, the trend of vertical force remains unchanged. But the magnitude of vertical force significantly increases, and the increase in the pull force is smaller than that of drag force. Figure 4 (c) shows that the *xoz*-plane torque on the square column is first negative and then positive. The dynamic characteristics show that there is a downward trend on the left side of the square column first, and then gradually becomes an upward trend, with the upward trend being much greater than the downward trend. The downward trend does not change much with the increase of wave amplitude, while the upward trend increases with the increase of wave amplitude.

Figure 5 shows the measurement results of the horizontal and vertical forces and *xoz*-plane torque when the depth of square column is consistent with the pycnocline. Three conditions with wave amplitudes of 0.113, 0.138, and 0.165 were showed. Figure 5 shows a significant difference in the variation trend of the force on the square column compared

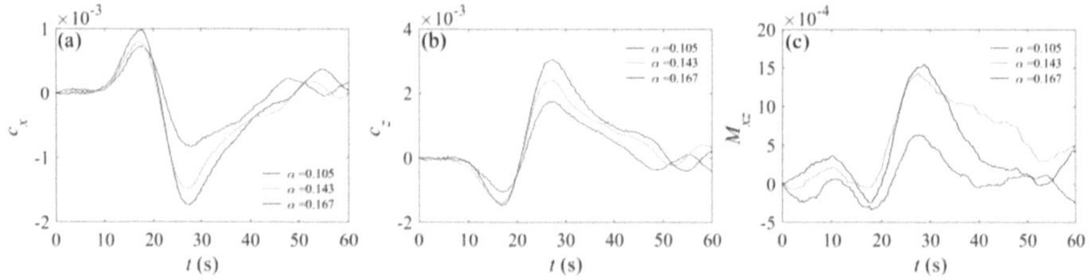

Fig. 4 Horizontal force c_x, vertical force c_z, and *xoz*-plane torque M_{xz} on the square column with a depth of 0.05 m

to Fig. 4. The horizontal force first increases positively and then decreases to 0, indicating that pushing force is exerted on the square column along the propagation direction of wave. As the wave amplitude increases, the trend of horizontal force remains unchanged, and the intensity slightly increases. The vertical force first increases negatively and then decreases to 0, indicating that downward pulling force is exerted on the square column. When the wave amplitude increases, the trend of vertical force remains unchanged and the maximum intensity of vertical force slightly increases. The *xoz*-plane torque first increases negatively and then decreases to 0, it indicates that there is a down-ward trend on the left side of the square column, and the downward trend does not change significantly with the change of wave amplitude. In addition, due to the stronger flow field, the force at this position is significantly greater than that in Fig. 4.

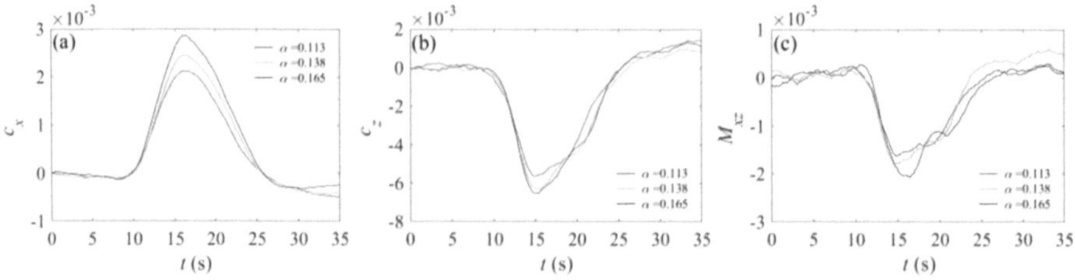

Fig. 5 Horizontal force c_x, vertical force c_z, and *xoz*-plane torque M_{xz} on the square column with a depth of 0.105 m

Figure 6 shows the measurement results of the horizontal and vertical forces and *xoz*-plane torque when the square column beneath the pycnocline. Three conditions with wave amplitudes of 0.097, 0.136, and 0.162 were showed. Figure 6 shows that the variation trend of the force on the square column is consistent with that in Fig. 5, except that the intensity and the influence of wave amplitude is greater. This is because the fluid density at the position of the square column is different. In addition, this position is close to the trough, which makes the force more affected by the wave amplitude.

Figure 7 shows the measurement results of the horizontal and vertical forces and *xoz*-plane torque when the square column beneath the wave packet. Three conditions with wave amplitudes of 0.108, 0.147, and 0.171 were showed. As shown in Fig. 7, the trend of changes in force and torque is opposite to Fig. 4, which is caused by the opposite direction of the upper and lower flow fields of the ISW. Due to the square column being

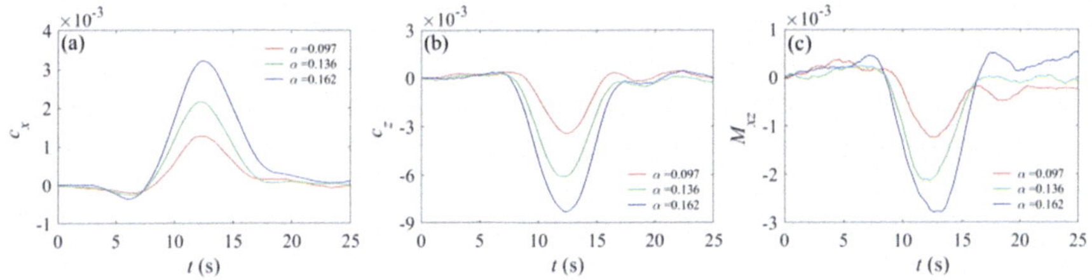

Fig. 6 Horizontal force c_x, vertical force c_z, and *xoz*-plane torque M_{xz} on the square column with a depth of 0.169 m

far away from the ISW and the small flow field intensity at its position, casing the force is relatively small compared to the above experimental conditions.

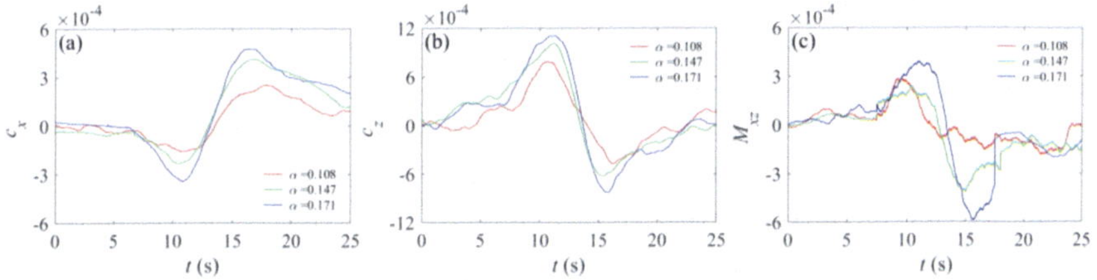

Fig. 7 Horizontal force c_x, vertical force c_z, and *xoz*-plane torque M_{xz} on the square column with a depth of 0.253 m

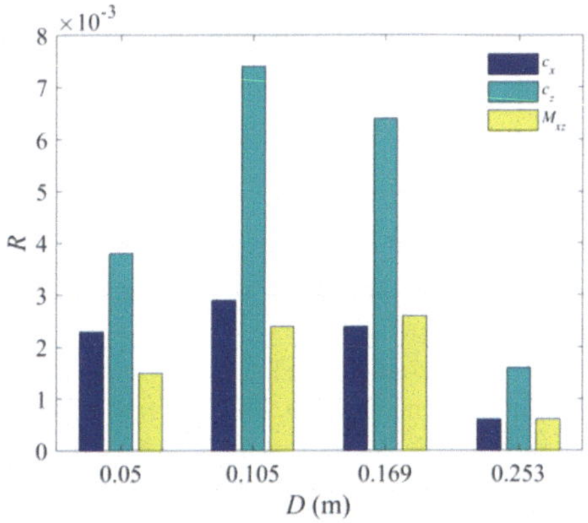

Fig. 8. The range of horizontal force c_x, vertical forces c_z, and *xoz*-plane torque M_{xz} on square column

In order to further analyze the variation characteristics of the load on square column, Fig. 8 shows the range of horizontal force c_x, vertical forces c_z, and *xoz*-plane torque on square column at different depths. As shown in the figure, at the same depth, the range of horizontal force and *xoz*-plane torque is comparable, while the range of vertical force is relatively large. When the depth of square column changes, the degree of variation in the range of horizontal force and *xoz*-plane torque is small, while the degree of variation in the range of vertical force is large. Based on the trend of force changes, it can be revealed that the motion trend during the square column's diving process is as follows. The roll and pitch increase first and then decrease, and the rate of decrease rapidly increases after reaching a certain depth. and the maximum values is near the pycnocline. The influence of pitch on the square column was stronger and should be taken seriously.

4 Conclusions

The experimental study of the load characteristics exerted by ISWs on a horizontal transverse square column is conducted in a stratified fluid flume, and the main conclusions obtained are as follows.

1. A new type of submerged load measurement platform has been proposed. The automatic lifting technology and triple force transducers are used to measure the load on the square column, and the depth of the square column can be controlled by a remote control, which reduces the disturbance of changing the diving depth on the fluid and improves experimental accuracy.
2. The variation characteristics of the load on the square column have been obtained, and the forces are mainly determined by the ISW-induced flow field. As the wave amplitude increases, the force increases, while the trend of force remains unchanged. As the depth of the square column changes, the magnitude and direction of the force change.
3. The motion trend of square column during diving is analyzed. During the process of square column diving, both roll and pitch increase first and then decrease, with the maximum values near the pycnocline. This indicates that the square column is most threatened by ISWs when passing through the pycnocline.

References

1. Cai, S., Xie, J., He, J.: An overview of internal solitary waves in the South China Sea. Surv. Geophys. **33**, 927–943 (2012)
2. Lien, R., Henyey, F., Ma, B.: Large-amplitude internal solitary waves observed in the northern South China Sea: properties and energetics. J. Phys. Oceanogr. **44**, 1095–1115 (2014)
3. Peng, P., Du, H., Wei, G., Wang, S., Xuan, P., Cai, S., Xie, J.: Experimental investigation on the vertical structure characteristics of internal solitary waves. J. Mar. Sci. Eng. **10**(8) (2022)
4. Alford, M., Peacock, T., MacKinnon, J.: Corrigenvdum: the formation and fate of internal waves in the South China Sea. Nature, 528–152 (2015)
5. Wang, T., Huang, X., Zhao, W., Zheng, S., Yang, Y., Tian, J.: Internal solitary wave activities near the Indonesian submarine wreck site inferred from satellite images. J. Mar. Sci. Eng. **10**, 197 (2022)
6. Osborne, A., Burch, T.: Internal solitons in the Andaman Sea. Science. **208**(4443), 451–460 (1980)

7. Du, H., Wei, G., Gu, M.: Experimental investigation of the load exerted by nonstationary internal solitary waves on a submerged slender body over a slope. Appl. Ocean Res. **59**, 216–223 (2016)
8. Gong, Y., Xie, J., Xu, J., Chen, Z., He, Y., Cai, S.: Oceanic internal solitary waves at the Indonesian submarine wreckage site. Acta Oceanol. Sin. **41**, 109–113 (2022)
9. Morison, J., Johnson, J., Schaaf, S.: The force exerted by surface waves on piles. J. Pet. Technol. **2**(05), 149–154 (1950)
10. Du, T., Sun, L.: An estimation of internal soliton forces on a pile in the ocean. J. Ocean U. China. **6**(2), 101–106 (2007)
11. Cai, S., Long, X., Wang, S.: Forces and torques exerted by internal solitons in shear flows on cylindrical piles. Appl. Ocean Res. **30**(1), 72–77 (2008)
12. Lü, H., Xie, J., Xu, J.: Force and torque exerted by internal solitary waves in background parabolic current on cylindrical tendon leg by numerical simulation. Ocean Eng. **114**, 250–258 (2016)
13. Zhang, J., Chen, K., You, Y.: A prediction method of internal solitary wave loads on the semi-submersible platform. China Ocean Eng. **36**(3), 464–473 (2022)
14. Chen, M., Chen, K., You, Y.: Experimental investigation of internal solitary wave forces on a semi-submersible. Ocean Eng. **141**, 205–214 (2017)
15. Ding, W., Ai, C., Jin, S., Lin, J.: Numerical investigation of an internal solitary wave interaction with horizontal cylinders. Ocean Eng. **208**, 107430 (2020)
16. Wang, C., Du, W., Li, G., Du, P., Zhao, S., Li, Z., Chen, X., Hu, H.: Numerical simulation of influence of ocean internal waves on hydrodynamic characteristics of underwater vehicles. Chin. J. Ship Res. **17**(3), 102–111 (2022)
17. Li, Z., Hu, H., Wang, C.: Hydrodynamics and stability of oblique water entry in waves. Ocean Eng. **292**, 116506 (2024)
18. Xie, Z., Jiao, J., Zhao, B.: Theoretical and experimental research on the effect of bi-directional misalignment on the static and dynamic characteristics of a novel bearing. Mech. Syst. Signal Process. **208**, 111041 (2024)
19. Ji, M., Chen, K., You, Y., Zhang, R.: Experimental study of wave loading by internal solitary waves on a submerged slender body. In: International Conference on Offshore Mechanics and Arctic Engineering, 84386. American Society of Mechanical Engineers (2020) V06BT06A059
20. Xuan, P., Du, H., Wang, S., Peng, P., Wei, G.: Experimental and theoretical study of internal solitary wave loads on a submerged slender body. Ocean Eng. **272**, 113936 (2023)
21. Tian, Z., Jia, L., Xiang, J., Yuan, G., Yang, K., Wei, J., Zhang, M., Shen, H., Yue, J.: Excess pore water pressure and seepage in slopes induced by breaking internal solitary waves. Ocean Eng. **267**, 0 (2023)
22. Wang, S., Du, H., Wei, G., Peng, P., Xuan, P.: Effects of the inhomogeneous vertical structure of an internal solitary wave on the force exerted on a horizontal transverse cylinder. Phys. Fluids. **35**, 062117 (2023)
23. Kao, T.W., Pan, F.-S., Renouard, D.: Internal solitons on the pycnocline: generation, propagation, and shoaling and breaking over a slope. J. Fluid Mech. **159**, 19–53 (1985)

Differences in Flow Fields Around Smooth and Coarse Semi-cylindrical Structures

Guangsheng Wang[1], Zishun Yao[2(✉)], and Lidi Shi[2]

[1] China Harbor Engineering Company Ltd., Beijing 100027, China
[2] Key Laboratory of Ministry of Education for Coastal Disaster and Protection, Hohai University, Nanjing 210024, China
zyao132@aucklanduni.ac.nz

Abstract. In ocean engineering projects, submerged semi-cylindrical structures are used as sea crossing tunnels. Previous studies have extensively examined the flow field around cylindrical structures, but research on flow around semi-cylinders is relatively limited. This paper presents a physical experiment on flow features around smooth and coarse semi-cylindrical structures under current conditions. The experiment results demonstrate that the surface roughnesses of the semi-cylinders significantly influences the features of the flow field around the structure, including turbulence intensities and the turbulence kinetic energy of the wake flow. Specifically, a coarser surface on the semi-cylinder could decrease the turbulence intensity of the wake, which could effectively minimize potential local scour downstream of it.

Keywords: Submerged semi-cylinder · Flow field · Surface roughnesses

1 Introduction

Cylindrical structures play a crucial role in various ocean engineering projects. The flow characteristics around these structures, encompassing macroscopic flow patterns and turbulence structures, have been extensively investigated over the past decades [1–5]. Previous studies concluded that intense turbulence or vortex shedding may impair the structural safety or scour problems [6]. Some countermeasures have been developed to address the potential problems in passive or active manners [7]. Among the passive countermeasures, increasing roughnesses of cylindrical structures' surface has been identified as an effective means to alleviate vortex shedding and turbulence problems [8, 9]. Notably, existing studies primarily focus on the impact of surface roughnesses on hydrodynamic conditions around structures [10–12], with limited attention given to the specific case of horizontal semi-cylindrical structures commonly found in maritime applications. Investigating the influence of surface roughnesses on the flow field around semi-cylindrical structures is both academically and practically significant. It is essential to highlight that the height of semi-cylindrical structures in ocean engineering, such as undersea pipelines and tunnels, is significantly less than the depth of the sea. Consequently, the influence of waves on these submerged semi-cylindrical structures is

D. Jeng and B. Cai (Eds.): ICCOE 2024, LNCE 423, pp. 76–82, 2025.
https://doi.org/10.1007/978-981-96-0181-3_8

minimal compared to ocean currents. Therefore, this study focuses exclusively on the current-only condition. Given the study's emphasis on understanding flow patterns and features around semi-cylindrical structures considering two different surface roughness conditions through physical experiments, semi-cylindrical models were installed at the bottom of a current flume. Utilizing Particle Imaging Velocimetry (PIV) techniques, the study aimed to measure flow velocity and investigate flow patterns, vortices, turbulence structures, and related aspects around the structure.

2 Methodology

2.1 Experimental Setup

A current flume was utilized for this study. The flume dimensions, as depicted in Fig. 1, are 11.0 m in length and 0.6 m in both width and height. At the water outlet of the flume, a honeycomb structure was installed to straighten the flow. As illustrated in Fig. 1, an aluminum semi-cylindrical structure model with the same width as the flume was positioned at the bottom. The diameter of the model is 0.05 m. Two different surface roughnesses (denoted as n) were considered in this experiment. Therefore, two models with different surface roughnesses were employed. One model had a smooth surface (n=: 0 mm), while the other had a coarse surface (n = 0.5 mm).

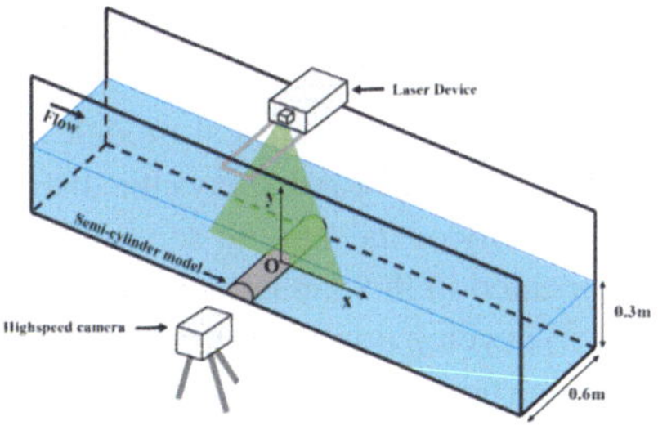

Fig. 1. Experiment setup

Fig. 2. Laser light plane

To obtain the flow characteristics around the semi-cylindrical model in the experiment, a photographic system, including a high-speed camera and a laser device, is installed (see Fig. 1). The laser device is installed above the model. The laser device can project a laser plane, which is parallel to the flow direction. The camera is positioned on one side of the flume to record the images of tracer particles passing through the laser plane during the experiment (see Fig. 2). Based on the images, PIV technique was then used to calculate the flow field features. The detailed method and calculation process are described in detail in [13].

2.2 Experimental Method

Two tests with smooth and coarse semi-cylindrical models were conducted in this study, each maintaining a constant flow discharge ($Q = 28.8 \times 10^{-3}$ m^3/s) and water depth (h = 0.3 m). This ensured uniformity in the flow mean velocity (0.16 m/s) and the flow Reynolds number (47952) in both tests. Only surface roughnesses are different in these two tests. Specifically, test L1 featured a surface roughness of 0 mm, while surface roughness in test L2 is 0.5 mm.

Characteristic parameters such as the average kinetic energy (AKE) and the normalized turbulence kinetic energy (TKE_N) is performed using the following equations, respectively:

$$AKE = \frac{1}{2}\left(u^2 + v^2\right) \tag{1}$$

Where u represents the flow velocity in x direction, while v represents the flow velocity in y direction.

$$TKE_N = \frac{1}{2}\left(\frac{u'^2 + v'^2}{u_0{}^2}\right) \tag{2}$$

Here, u' and v' represent the fluctuating velocity in x and y directions, respectively, while u_0 is the mean velocity of the approaching flow.

3 Experimental Results

3.1 Time-Averaged Wake Flow

To analyse characteristics of the wake flow passing the semi-cylindrical model, time-averaged flow parameters are calculated and presented. Figure 3(a) displays the distribution of streamlines. Three vortices develop in the vicinity of the smooth semi-cylinder: one vortex (v1) develops upstream of the model, located in the gap between the model and the bed; while the other two vortices (v2 and v3) develop downstream of the model. However, in the test with the coarse model, only two vortices are present in the vicinity of the model. Furthermore, compared to the smooth model, v2 and v3 develop closer to the coarser model. This suggests that increased surface roughness alters the boundary layer upstream, diminishing approaching flow velocity and leading to disappearance of v1. The rough surface also influences the position of v2 and v3, advancing vortex shedding.

In Fig. 3(b), the distributions of average kinetic energy (AKE) around the semi-cylinders reveal that the isoline of 0.05 is lower around the coarse model, which indicates that the coarse surface inhibits vortex development downstream, aligning with the observed streamline patterns.

Furthermore, Fig. 3(c) illustrates time-averaged vorticities at various surface roughness levels. The distributions are consistent between different surface roughness conditions. The maximum vorticities, vor1 and vor2, are located downstream of the semi-cylinder, with vor1 greater than vor2. When the n = 0 mm, vor1 and vor2 attain maximum values of 34.32 and −5.151, respectively. With the n = 0.5 mm, vor1 decreases to 31.03, and vor2 decreases to −4.4152. This implies that the peak vorticity around the submerged semi-cylinder decreases by 9.58% to 13.68% as surface roughness increases from 0 mm to 0.5 mm, indicating that the rough surface mitigates vortices behind the semi-cylinder.

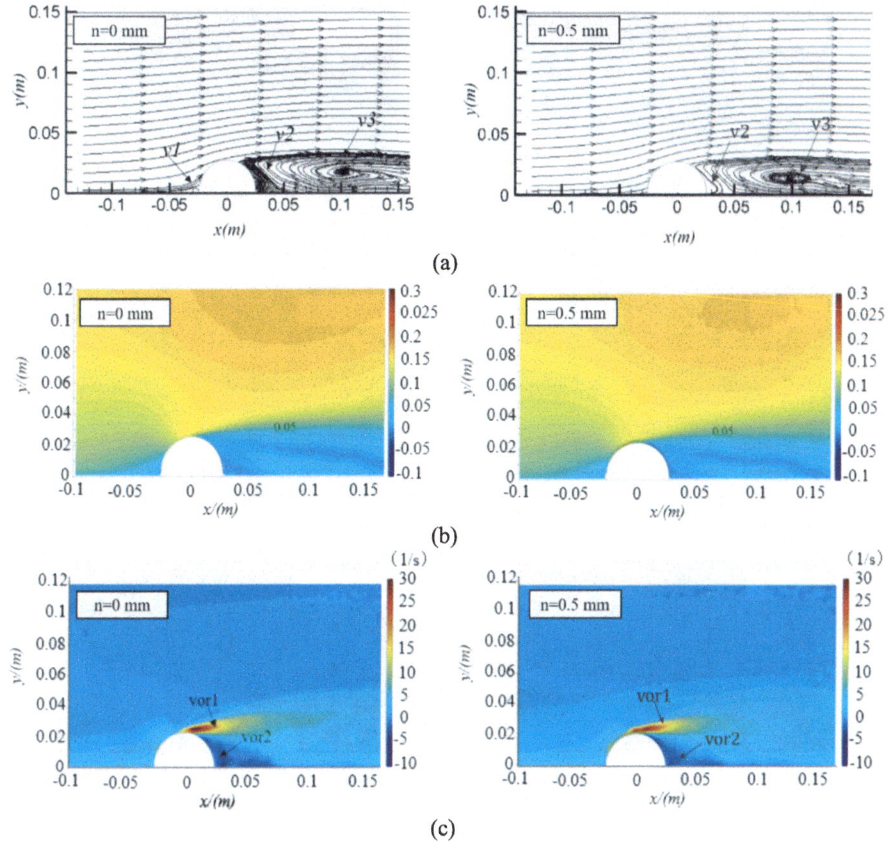

Fig. 3. Time-averaged flow field features between two tests: (a) Streamlines; (b) AKE; (c) Vorticities.

3.2 Wake Turbulence Features

To mitigate scale effects, the turbulence intensity and kinetic energy of both tests were normalized, revealing the characteristics of wake turbulence.

Figures 4 depicts the distributions of normalized turbulence intensities in x (TNx) and y directions (TNy). TNx and TNy downstream of the model are relatively high, with the TNx zone being narrower than the TNy zone. Specifically, TNy downstream of the coarse semi-cylindrical model is decreased significantly, compared to that of the smooth one. It may suggest that the turbulence intensity in the vertical direction is more susceptible to the surface roughness of the model. Furthermore, compared to the model with a smooth surface, larger areas of low TNx and TNy occur downstream of the semi-cylindrical model. This suggests that the semi-cylindrical model with a coarser surface could decrease wake turbulence intensity, especially in the area above the bed downstream of the model. It may also minimize the potential for local scour at the rear side of the structure.

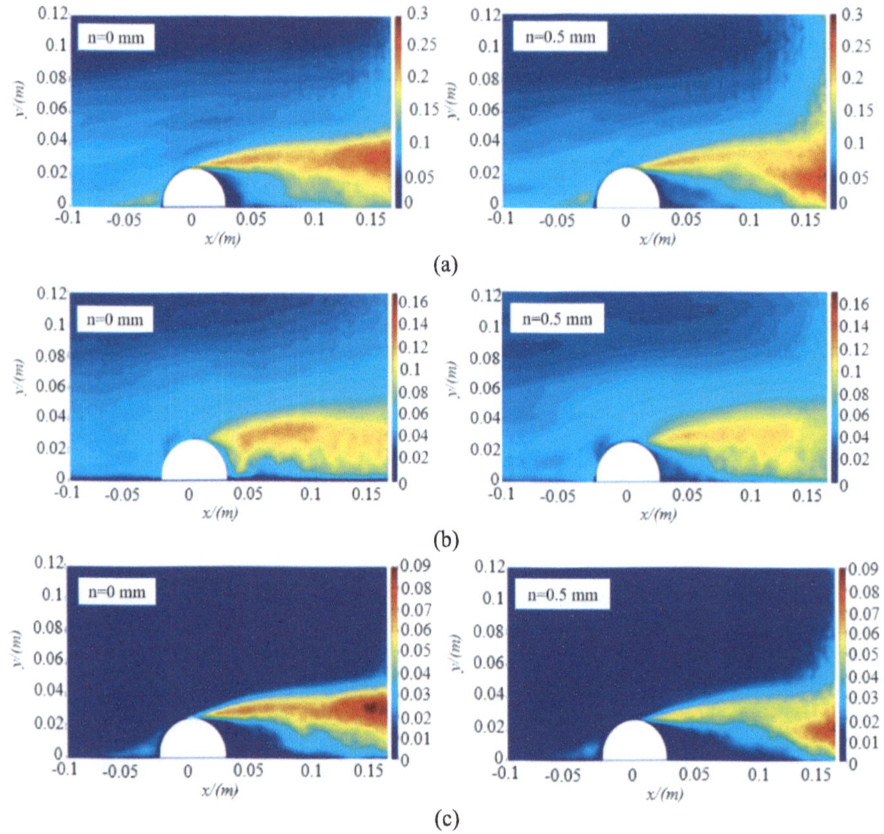

Fig. 4. Comparison of turbulence structure between two tests: (a) TNx; (b) TNy; (c) TKE_N.

The distribution patterns of normalized turbulence kinetic energy (TKEN) in Fig. 4(c) mirror those of horizontal turbulence intensities. Notably, TKEN of wake of the coarser semi-cylindrical model is markedly smaller than that of the smooth one, which may due to the significantly decreased TNy. This suggests that the coarser cylindrical model could absorb more turbulence energy from the wake around it, with increasing surface roughnesses correlating with reduced flow turbulence past the semi-cylinder.

The experimental findings suggest that the surface roughness of semi-cylinders at different physical scales may similarly affect the surrounding flow field. Additionally, semi-cylinder's surface roughnesses may also affect turbulence structures under varying flow discharge conditions. Still, it needs further verification by future experiments.

4 Conclusion

The key conclusions of this study are:

(1) Transitioning from a smooth to a rough semi-cylinder surface led to a reduction in AKE, vorticities, turbulence intensities of the wake flow. Notably, the maximum vorticity around the semi-cylinder decreased by 9.58% to 13.68% with an increase in surface roughness from 0 mm to 0.5 mm.
(2) Surface roughnesses influenced the development of wake vortices. The number of vortices decreases, and their positions move closer to the semi-cylinder when the surface of the model becomes coarser.
(3) The Enhanced roughnesses of the semi-cylinder surface demonstrated potential in mitigating wake turbulence and improving wake flow conditions.

This study contributes to understanding the impact of surface roughnesses on the flow field surrounding submerged semi-cylindrical structures. Future experiments incorporating additional surface roughness parameters are anticipated to further enrich our understanding of these dynamics.

Acknowledgement. This work was supported by National Key R&D Program of China (2023YFE0126300).

References

1. Bearman, P.W., Trueman, D.M.: An investigation of the flow around rectangular cylinders. Aeronaut. Q. **23**(3), 229–237 (2016). https://doi.org/10.1017/s0001925900006119
2. Fredsøe, J., Hansen, E.A.: Lift forces on pipelines in steady flow. J. Waterw. Port Coast. Ocean Eng. **113**(2), 139–155 (1987). https://doi.org/10.1061/(asce)0733-950x(1987)113:2(139)
3. Jenssen, U., Manhart, M.: Flow around a scoured bridge pier: a stereoscopic PIV analysis. Exp. Fluids. **61**(10) (2020). https://doi.org/10.1007/s00348-020-03044-z
4. Lin, W.-J., Lin, C., Hsieh, S.-C., Dey, S.: Flow characteristics around a circular cylinder placed horizontally above a plane boundary. J. Eng. Mech. **135**(7), 697–716 (2009). https://doi.org/10.1061/(asce)0733-9399(2009)135:7(697)
5. Okajima, A.: Numerical simulation of flow around rectangular cylinders. J. Wind Eng. Ind. Aerodyn. **33**(1–2), 171–180 (1990). https://doi.org/10.1016/0167-6105(90)90033-9
6. Dey, S., Singh, N.P.: Clear-water scour below underwater pipelines under steady flow. J. Hydraul. Eng. **134**(5), 588–600 (2008). https://doi.org/10.1061/(asce)0733-9429(2008)134:5(588)
7. Akilli, H., Sahin, B., Filiz Tumen, N.: Suppression of vortex shedding of circular cylinder in shallow water by a splitter plate. Flow Meas. Instrum. **16**(4), 211–219 (2005). https://doi.org/10.1016/j.flowmeasinst.2005.04.004
8. Achenbach, E.: Influence of surface roughness on the cross-flow around a circular cylinder. J. Fluid Mech. **46**(2), 321–335 (1971). https://doi.org/10.1017/s0022112071000569
9. Nakamura, Y., Tomonari, Y.: The effects of surface roughness on the flow past circular cylinders at high Reynolds numbers. J. Fluid Mech. **123**, 363–378 (1982). https://doi.org/10.1017/s0022112082003103

10. Güven, O., Farell, C., Patel, V.C.: Surface-roughness effects on the mean flow past circular cylinders. J. Fluid Mech. **98**(4), 673–701 (2006). https://doi.org/10.1017/s0022112080000341
11. Park, H., Bernitsas, M.M., Kim, E.S.: Selective surface roughness to suppress flow-induced motion of two circular cylinders at 30,000 < Re < 120,000. J. Offshore Mech. Arct. Eng. **136**(4) (2014). https://doi.org/10.1115/1.4028061
12. Zhou, B., Wang, X., Gho, W.M., Tan, S.K.: Force and flow characteristics of a circular cylinder with uniform surface roughness at subcritical Reynolds numbers. Appl. Ocean Res. **49**, 20–26 (2015). https://doi.org/10.1016/j.apor.2014.06.002
13. Guan, D., Chiew, Y.M., Wei, M., Hsieh, S.C.: Visualization of flow field around a vibrating pipeline within an equilibrium scour hole. J. Vis. Exp. (150) (Aug 26 2019). https://doi.org/10.3791/59745

Ship-Borne Wave Simulation by a Boussinesq-Type Model: Extension and Validation

Mingliang Guan[1], Xiaoyu Hu[1], Lei Wang[2], Ping Wang[3], Long Liang[2], Kezhao Fang[2(✉)], and Yu Wang[1]

[1] Changjiang Nanjing Waterway Engineering Bureau, Nanjing 210013, China
[2] The State Key Laboratory of Coastal and Offshore Engineering, Dalian University of Technology, Dalian 116024, China
kfang@dlut.edu.cn
[3] State Environmental Protection Key Laboratory of Marine Ecosystem Restoration, National Marine Environmental Monitoring Center, Dalian 116023, China

Abstract. The Boussinesq-type model developed by Fang et al. (2022) have shown satisfactory performance thanks to the usage of hybrid finite-volume and finite-difference numerical scheme, well balanced wet-dry interface and GPU acceleration. This study attempts to extend the model to simulate ship-borne waves. To this end, the pressure effect caused by the ship motion is introduced as a moving source term into the right hand sided of the Boussinesq momentum equations. This term is discretized on uniform rectangular grid and approximated by finite difference formulation. The simulation results of the extended Boussinesq model are compared against four sets of dataset from the physical experiments and numerical simulations from other models. Agreements and discrepancies are discussed in detail.

Keywords: Boussinesq model · Ship-borne waves · Numerical simulation

1 Introduction

Ship-borne waves refer to the pattern of ripples formed behind a vessel as it moves over the water surface. This pattern generally consists of two parts: waves directly generated by the propulsion of the ship and ripples that spread out due to the ship's movement. Typically, the shape of ship waves is V-shaped, with the angle of the V depending on the ship's speed—the faster the speed, the narrower the opening angle of the V. These ripples are not only visually striking but also have significant implications for marine engineering and ship design.

In narrow canals, solitary waves generated ahead of the ship can be observed, and a series of experiments were conducted by varying the ship's speed and the canal's shape to study this phenomenon. Research has shown that the speed of the ship determines the type and characteristics of the ship waves (Russel [1, 2]). Havelock [3] analyzed the wave patterns produced by ship motion and introduced the concept of the Kelvin angle,

D. Jeng and B. Cai (Eds.): ICCOE 2024, LNCE 423, pp. 83–95, 2025.
https://doi.org/10.1007/978-981-96-0181-3_9

which is the angle between the line of divergence, where the transverse and divergent waves interfere, and the ship's path. He also developed a mathematical framework to describe this angle using the nondimensional depth Froude number: $Fh = U/\sqrt{gh}$ (where U is the ship's speed, h is the local water depth and g is the gravity acceleration). When $Fh < 1$, the Kelvin angle is approximately 19.28°. As the Froude number approaches 1, the Kelvin angle quickly approaches 90°, and as Fh exceeds 1, the Kelvin angle gradually decreases.

Some scholars attempt to simulate ship waves by introducing source terms into the shallow water wave equation (Chen and Uliczka [4]). However, the shallow water equation lacks dispersion and cannot simulate short waves, while the Boussinesq equation combines dispersion and nonlinearity, making it capable of simulating various complex wave propagation and evolution phenomena in the time domain, thus serving as an alternative tool for simulating ship waves. Nascimento et al. [5] modified FUNWAVE model by incorporating a moving pressure source term and used a single pressure distribution to simulate the effects of a moving ship body. Li and Sclavounos [6] represented the ship with an elliptical pressure perturbation and conducted simulations. Subsequently, Nascimento et al. [5] expanded FUNWAVE model with multiple moving pressure sources, demonstrating the general applicability of this approach. Bayraktar and Beji [7] were the first to describe the pressure terms in detail, implementing two different shapes of two-dimensional pressure fields (hemispherical and slender). David et al. [8] added Bayraktar and Beji's [7] slender pressure term to the BOSZ model [9] to test the distribution and propagation speed of the pressure term and its effects on wave characteristics, comparing real-world data from the River Elbe with corresponding numerical data. Shi et al. [10] incorporated Ertekin's [11] wide-body ship pressure term into FUNWAVE-TVD model, reproducing the bow solitary wave observed in Gourlay's [12] experiment using three wave breaking standards. Similarly, Samaras et al. [13] developed a two-dimensional Boussinesq model for weakly nonlinear fully dispersive water waves based on the work of Karambas and Memos [14], in which Ertekin's ship pressure term was introduced into the momentum equation. These scholars have all integrated the advantages of dispersion and nonlinearity in the Boussinesq equation, and have added moving pressure terms to the momentum equation, enabling a more accurate simulation of the complex wave propagation and evolution phenomena associated with ship motion.

In this study, we employ the Boussinesq-type model COWS (Coastal and Offshore Wave Simulator) developed by Fang et al. [15], incorporating a surface pressure term into the momentum equation to simulate ship-borne waves. The paper is organized as follows: Sect. 2 outlines the mathematical background of the applied Boussinesq model and the expression of the moving pressure term; Sect. 3 compares model results with datasets from four experimental cases and computed results from other simulations; conclusions drawn from this work are presented in the final Sect. 4.

2 Model Description

2.1 Boussinesq Equations with Moving Pressure Term

The conservative form of the two-dimensional weakly dispersive nonlinear Boussinesq equation suitable for rapidly varying bathymetry (Fang et al. [12], Kim et al., 2009a [16]) is:

$$\boldsymbol{U}_t + \boldsymbol{F}(\boldsymbol{U})_x + \boldsymbol{G}(\boldsymbol{U})_y = \boldsymbol{S}(\boldsymbol{U}) \tag{1}$$

in which $\boldsymbol{U}$ is the velocity vector, $\boldsymbol{F}$ and $\boldsymbol{G}$ are the flux vectors along x and y directions respectively:

$$\boldsymbol{U} = \begin{pmatrix} \eta \\ U \\ V \end{pmatrix}, \boldsymbol{F} = \begin{pmatrix} P \\ P^2/d + g(\eta^2 + 2h\eta)/2 \\ PQ/d \end{pmatrix}, \boldsymbol{G} = \begin{pmatrix} Q \\ PQ/d \\ Q^2/d + g(\eta^2 + 2h\eta)/2 \end{pmatrix} \tag{2}$$

with

$$\begin{aligned} U &= P - (B + \tfrac{1}{3})h^2 P_{xx} - hh_x \tfrac{1}{3} P_x - hh_{xx} \tfrac{1}{6} P + \tfrac{1}{3} h_x^2 P \\ &- (B + \tfrac{1}{3})h^2 Q_{xy} - hh_x \tfrac{1}{6} Q_y - hh_y \tfrac{1}{6} Q_x - hh_{xy} \tfrac{1}{6} Q + \tfrac{1}{3} h_x h_y Q \end{aligned} \tag{3}$$

$$\begin{aligned} V &= Q - (B + \tfrac{1}{3})h^2 Q_{yy} - hh_y \tfrac{1}{3} Q_y - hh_{yy} \tfrac{1}{6} Q + \tfrac{1}{3} h_y^2 Q \\ &- (B + \tfrac{1}{3})h^2 P_{xy} - hh_y \tfrac{1}{6} P_x - hh_x (\tfrac{1}{6} P_y) - hh_{xy} (\tfrac{1}{6} P) + \tfrac{1}{3} h_x h_y P \end{aligned} \tag{4}$$

where h is the still water depth, η is the free wave surface, $d = h + \eta$ is the total water depth, $P = du$ and $Q = dv$ represent the velocity flux in x and y directions respectively(P, Q, u and v represent the water depth average velocity in the x and y directions respectively. $\boldsymbol{S}$ in Eq. (1) is the source term vector, which is composed of the bottom slope term ($\boldsymbol{S_b}$), moving source term($\boldsymbol{S_{ves}}$) and dispersion term ($\boldsymbol{S_d}$):

$$\boldsymbol{S}(\boldsymbol{U}) = \boldsymbol{S}_b + \boldsymbol{S}_{ves} + \boldsymbol{S}_d = \begin{pmatrix} 0 \\ g\eta h_x \\ g\eta h_y \end{pmatrix} + \begin{pmatrix} 0 \\ \tau_{vesx} \\ \tau_{vesy} \end{pmatrix} + \begin{pmatrix} 0 \\ \varphi_x \\ \varphi_y \end{pmatrix} \tag{5}$$

with dispersion terms φ_x and φ_y are:

$$\varphi_x = Bgh^3(\eta_{xxx} + \eta_{xyy}) + Bgh^2(\eta_{yy} + 2\eta_{xx} + h_y\eta_{xy} + h_{xx}\eta_x + h_{xy}\eta_y) \tag{6}$$

$$\varphi_y = Bgh^3(\eta_{yyy} + \eta_{xxy}) + Bgh^2(\eta_{xx} + 2\eta_{yy} + h_x\eta_{xy} + h_{yy}\eta_y + h_{xy}\eta_x) \tag{7}$$

The work in this paper utilizes a moving pressure field to mimic waves generated due to a moving vessel. The moving source term ($\boldsymbol{\tau_{ves}}$) in Eq. (5) is defined as:

$$\tau_{vesx} = \frac{\partial D(x, y, t)}{\partial x}, \tau_{vesy} = \frac{\partial D(x, y, t)}{\partial y} \tag{8}$$

Three vessel types are provided here. First, the dimensionless cosine pressure distribution used by Wu [17] in the one dimensional calculation is:

$$D(x,t) = \begin{cases} \frac{1}{2}\rho g D_s\{1+\cos[\frac{2\pi}{L}(x-x_s(t))]\}, & |x-x_s(t)| \le \frac{L}{2} \\ 0, & |x-x_s(t)| > \frac{L}{2} \end{cases} \tag{9}$$

In Eq. (9), ρ is the water density, L is the length of the vessel, D_s is the draft at the geometric center x_s of the vessel, and the draft on both sides of the geometric center decreases according to the cosine law until the draft at the bow and stern drops to 0. The pressure value of the moving pressure field is approximately the hydrostatic pressure at the bottom of the vessel. As shown in Fig. 1, the pressure value at the geometric center is large, the pressure value at the bow and stern is zero, and the middle area decreases according to the cosine law.

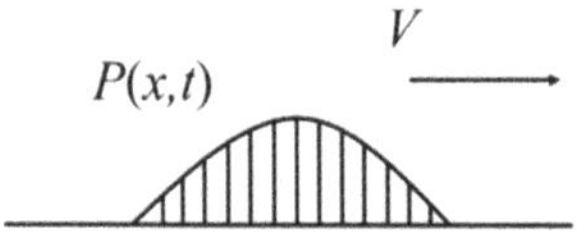

Fig. 1. One-dimensional ship-shaped moving pressure field.

The first type of two-dimensional vessel type adopts the wide-body vessel type proposed by Ertekin et al. [11], and the pressure distribution is expressed as:

$$D(x,y,t) = \rho g D_s f(x')g(y') \tag{10}$$

$$f(x') = \begin{cases} \cos^2\left[\pi(\frac{x'-0.5\alpha_1}{1-\alpha_1})\right], & 0.5\alpha_1 < x' \le 0.5 \\ \cos^2\left[\pi(\frac{x'+0.5\alpha_2}{\alpha_2-1})\right], & -0.5 \le x' < -0.5\alpha_2 \\ 1, & -0.5\alpha_2 \le x' \le 0.5\alpha_1 \end{cases} \tag{11}$$

$$g(y') = \begin{cases} \cos^2\left[\pi(\frac{y'-0.5\beta_1}{1-\beta_1})\right], & 0.5\beta_1 < y' \le 0.5 \\ \cos^2\left[\pi(\frac{y'+0.5\beta_1}{\beta_1-1})\right], & -0.5 \le y' < -0.5\beta_1 \\ 1, & -0.5\beta_1 \le y' \le 0.5\beta_1 \end{cases} \tag{12}$$

$$\begin{pmatrix} x' \\ y' \end{pmatrix} = \begin{bmatrix} \frac{1}{L_s} & 0 \\ 0 & \frac{1}{B_s} \end{bmatrix} \begin{bmatrix} \cos(\theta_s(t)) & \sin(\theta_s(t)) \\ -\sin(\theta_s(t)) & \cos(\theta_s(t)) \end{bmatrix} \begin{pmatrix} x - x_s(t) \\ y - y_s(t) \end{pmatrix} \tag{13}$$

In Eqs. (11)–(12), L_s and B_s represent the length and width of the vessel respectively, α_1, α_2 and β_1 are parameters representing the shape of the draft region, $0 < \alpha_1, \alpha_2 < 1$ are used to control the shape of the bow and stern, $0 < \beta_1 < 1$ is used to Control the shape of the vessel's cross-section.

Figure 2 is schematic diagram of ship pressure field, in Eq. (13), (x, y) are coordinates relative to the geometric center of the pressure field, while $(x_s(t), y_s(t))$ are the coordinates of the moving geometric center of the pressure field. The pressure field is defined within a rectangular area as $-L/2 < x - x_s(t) < L/2$ and $-B/2 < y - y_s(t) < B/2$, with the pressure outside this region being zero, and it adjusts the heading angle $\theta_s(t)$ as needed over time. Thus, this model allows the ship to perform oblique or rotational navigation. To more realistically simulate the actual navigation of ships, the model also uses cubic spline interpolation for specified positions of the ship.

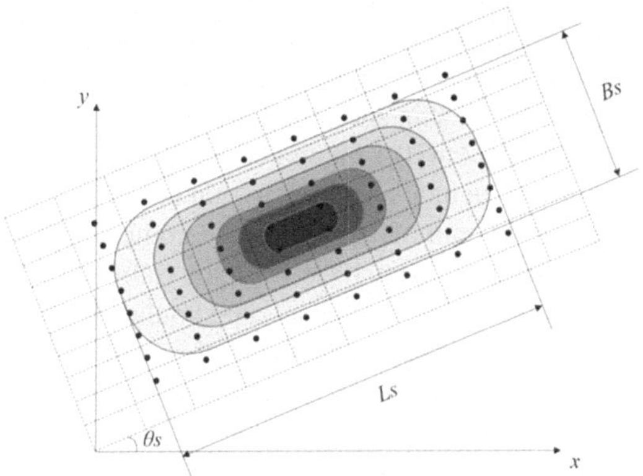

Fig. 2. Schematic diagram of ship pressure field.

2.2 Numerical Scheme

The implementation of the hybrid FV/FD shock-capturing scheme primarily follows our previous work (Fang et al., 2013 [18], 2016 [19]). The main procedures are summarized herein for clarity and completeness. A Godunov-type finite volume method is used on a rectangular cell system to deal with the convective parts while the rest terms are discretized using the finite difference method. The fourth-order Monotone Upstream-Centred Schemes for Conservation Laws construction and central upwind scheme are used for flux calculation. The third-order Runge Kutta scheme with the Strong Stability Preserving property and the adaptive time step is used for time marching. For numerical stability requirement, the time increment must be restricted by the Courant number *CFL*. The velocity can then be obtained by solving the tridiagonal system, resulting from discretizing Eqs. (3) and (4) using the second-order finite difference formula.

2.3 Boundary Conditions

A minim value of water depth ($d_{\min}$) is specified to distinguish the dry ($d \le d_{\min}$) and wet cell ($d > d_{\min}$). The hydrostatic construction technique (Wang et al., 2011 [20]) is then used during variable construction to capture the moving shoreline in an accurate and efficient manner. The model captures breaking wave as a shock by deactivating dispersive terms once local wave height exceeds 0.8 h or wave front slope is steep than 30°. For the reflective wall, tangential velocity and normal velocity on ghost cells are determined from the inner domain by imposing symmetric and anti-symmetric conditions with respect to the solid wall.

3 Model Validations

3.1 Case 1: Moving Pressures in 1D Channels

As the first step in numerical model validation, this study examined a one-dimensional example of wave generation in a channel with a rectangular cross-section. Here, the perturbed pressure will be calculated using Eq. (9) and the model results will be compared with the fKdV equation (Wu, 1987 [17]) and BIEM model (Cao et al., 1993 [21]). This section simulates the numerical model with water depth $h = 1$ m, ship length $L_s = 2$ m, depth Froude number $Fr = 1$, and ship speed $u = 3.13$ m/s. The ship's geometric center draft D_s will be simulated at 0.02 m, 0.1 m, and 0.15 m.

The simulation conditions include a computational domain length of 200 m, spatial step size $dx = 0.195$ m, a grid count of 1024, Courant number $CFL = 0.2$, and the ship moving in the negative x-direction from 10 m from the right edge. A 3 m sponge layer is set on the left side for wave damping, with a total simulation time of 60 s.

Figures 3, 4 and 5 present comparisons between the results of this model and other models, with the ship's bow serving as the coordinate origin and the x-axis defined by a moving coordinate system, where ut denotes the ship's travel distance. Due to the BIEM's accurate resolution of potential flow, its solutions serve as benchmarks for validating other approximate models. According to the results depicted in the figures, this model shows good agreement with the BIEM model when the ship's draft is 0.02 m and 0.15 m. Notably, at a draft of 0.1 m, prominent solitary waves appear, which are characteristic features of shallow water ship waves. To examine the nonlinear effects of the model, it was also compared with the fKdV results provided by Wu. The comparison reveals that the fKdV model fails at drafts of 0.1 m and 0.15 m, exhibiting lower phase velocities and wave heights. As the draft increases, the waves generated by this model become unstable, potentially breaking as the ship moves. Similar observations were made with Cao's BIEM model, and Lee et al. [22] also observed wave breaking in their experiments when the draft exceeded 0.15 m.

3.2 Case 2: Experimental Data of Gourlay (2001)

Gourlay [12] conducted a series of experiments at the Australian Maritime College to study various flow states generated by ships traveling at supercritical speeds in channels. The experimental flume is 60 m long and 3.5 m wide, with two different water depths (0.114 m and 0.205 m). The ship used in the experiments has a length (L_s) of 1.6 m, a width (B_s) of 0.4 m, and a draft (D_s) of 0.1 m, and is a monohull design. The experiments involve adjusting the ship's speed to study changes in the bow solitary waves as the Froude number (Fr) increases gradually from 1.05 to higher values. Shi et al. [10] also observed similar phenomena while studying different breaking patterns of bow solitary waves. Gourlay [12] found that when Fr exceeded 1.12, the bow solitary waves began to break, and as Fr increased, the breaking of solitary waves continued, with the maximum wave height remaining constant while the amplitude of the bow solitary waves gradually decreased.

In the numerical simulation, Eq. (10) is used to model conditions at a water depth(h) of 0.114 m. The computational domain is set to 60 m in length and 3.5 m in width, with

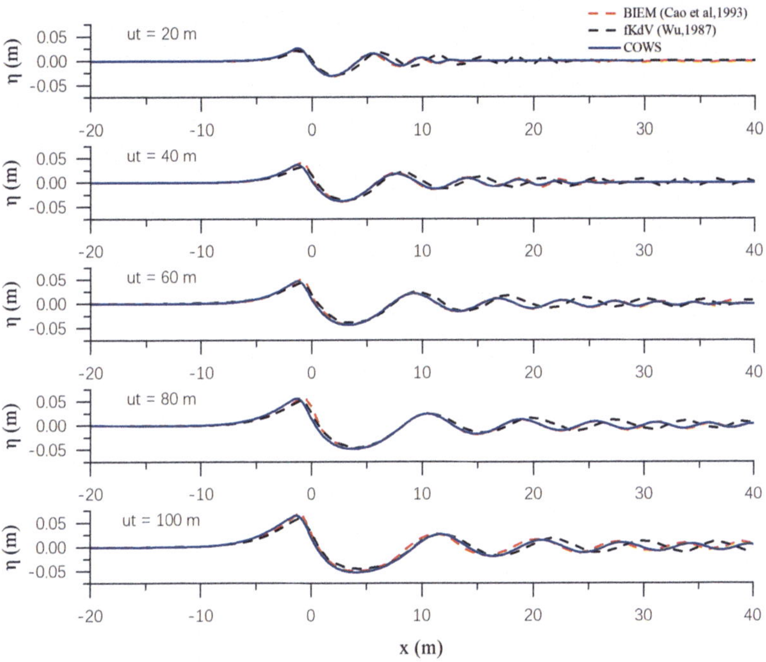

Fig. 3. Ship traveling wave in 1D channel when $D_s = 0.02$ m.

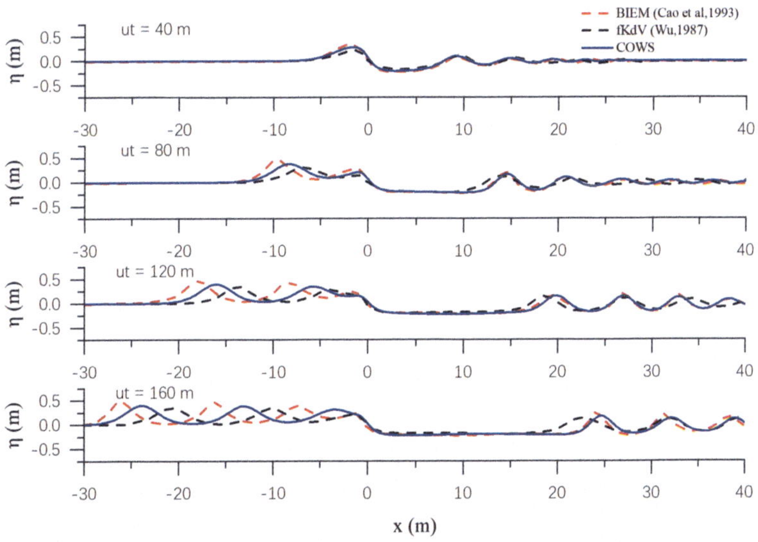

Fig. 4. Ship traveling wave in 1D channel when $D_s = 0.10$ m.

ship parameters consistent with the experimental setup. The parameters α_1 and α_2 are both set to 0.5, and β is set to 0.8. The ship travels at a constant speed from $x = 5$ m

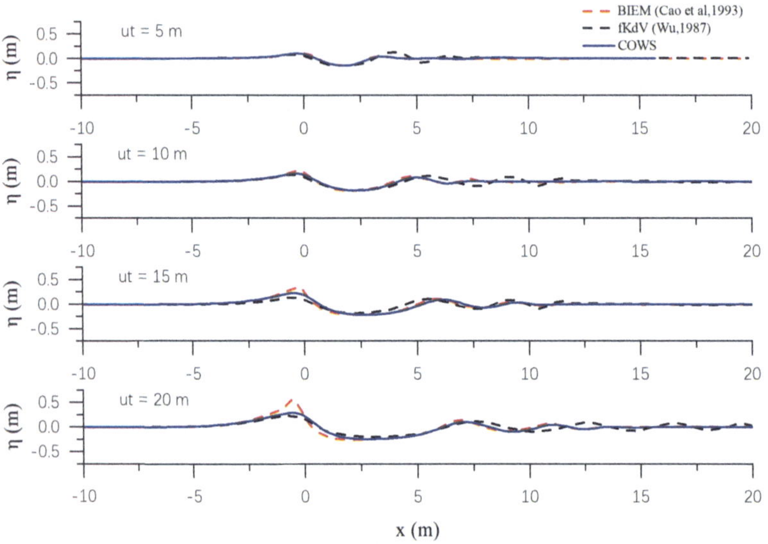

Fig. 5. Ship traveling wave in 1D channel when $D_s = 0.15$ m.

along the central line at $y = 1.75$ m. A 3 m sponge layer is positioned on the left side to absorb the ship-borne waves. The computation domain is discretized into 1024 grids in x-direction and 32 girds in y-direction, with a grid size of $dx = 0.05859$ m and $dy = 0.05469$ m.

Figure 6 presents the wave front conditions at times $t = 4$ s, 16 s, and 30 s for COWS model with a Froude number (Fr) of 1.15, and these results are compared with simulations by Shi et al. [10]. The results show that at $t = 4$ s, curved wave crests begin to appear at the bow, marking the gradual formation of solitary waves. Over time, it is observed that at $t = 16$ s and 30 s, the initially curved wave crests quickly straighten, a phenomenon considered to be Mach reflection [23], and the number of solitary waves at the bow of both models increases, with an equal count of solitary waves. Throughout the process, the wave propagation speed in COWS model is slightly faster than that in the Shi model, which may be due to differences in the wave breaking methods used in the two models.

Figure 7 presents the comparison between model simulation results and experimental data at position ($x = 33.125$ m, $y = 0.72$ m) for $Fr = 1.15$. The vertical axis represents $(h + \eta)/h$, while the horizontal axis represents time t. It can be observed from the figure that due to the significant energy dissipation caused by the breaking of the first solitary wave, the subsequent solitary waves are suppressed in amplitude, consistent with wave breaking theory. Therefore, the maximum value on the vertical axis appears only on the first solitary wave with a value of 1.6, which is very close to the experimental maximum value of 1.61. The computational results are in close agreement with the experimental results. Figure 8 shows the variation of the average wave amplitude a of the bow solitary wave with Fr. As Fr increases, the number of bow solitary waves gradually decreases. Both COWS model and Shi simulations predict the overall trend of amplitude decay.

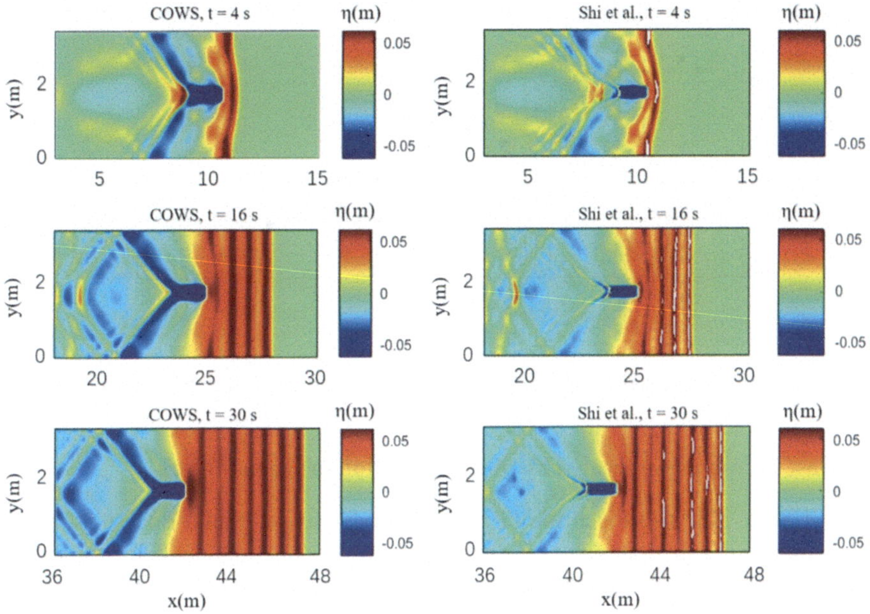

Fig. 6. Comparison of ship wave pattern at $t = 4, 16, 30$ s.

However, Shi's simulation predicts a faster decay rate in amplitude. Shi's simulation predicts the occurrence of a pure bore at $Fr = 1.375$, while COWS model predicts it at $Fr = 1.41$, which is more consistent with the experimental data.

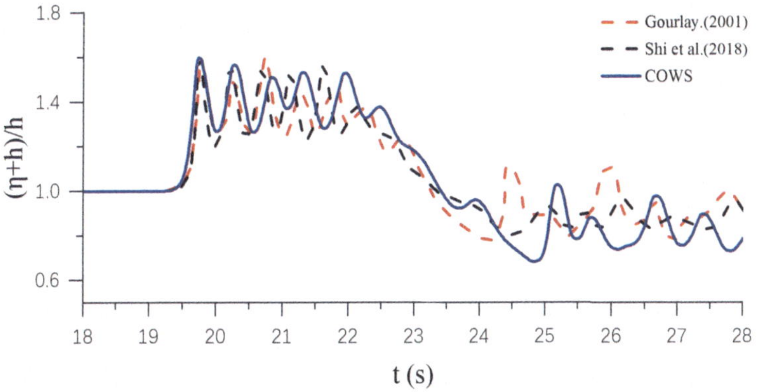

Fig. 7. Model results from COWS and Shi et al. are compared with Gourlay measured data, for $Fr = 1.15$.

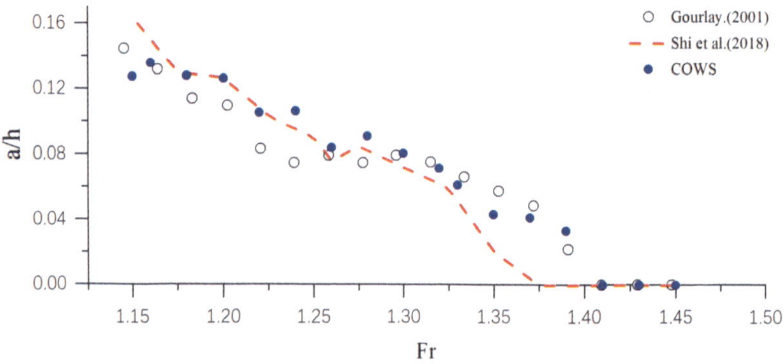

Fig. 8. Comparison of wave amplitude a/h (left axis) with increasing *Fr* between model data

3.3 Case 3: Waves Generated by Ship Navigation in Open Waters

Agarwal et al. [24] developed the FEBOUSS model based on the Boussinesq equations. This section simulates the wave generation by the ship pressure field described in Agarwal's document and compare it with the results from FUNWAVE-TVD simulation provided by Agarwal.

The numerical model covers a domain of 840 m in length and 408 m in width, with 1024 grids in x-direction and 512 girds in y-direction, and a grid size of $dx = 0.8203$ m and $dy = 0.7969$ m. The still water depth is $h = 5$ m, the Courant number is 0.5, and the total simulation time is 120 s. The ship is 36 m long, 6 m wide, and has a draft of 1 m. The perturbed pressure is calculated using Eq. (8), with coefficients $\alpha_1 = 16$, $\beta_1 = 2$, $\beta_2 = 16$. The ship characteristics are similar to those in experiments conducted by David et al. [8]. The ship starts moving from the coordinates (51.0 m, 204.0 m) and travels in the positive x-direction at a constant speed of 5.95 m/s, resulting in a Froude number $Fr = 0.85$. Sponge layers of 30 m are set up on all four boundaries of the computational domain to absorb the wake generated by the ship's navigation, simulating the effect of the ship traveling in open waters. Measurements are taken at seven cross-sections located at distances of 0B, 0.5B, 2.5B, 4.5B, 6.5B, 8.5B, and 10.5B from the ship's travel line. All other numerical model settings are consistent with those of the FEBOUSS model.

Figure 9 shows the wavefront changes measured between 350 m and 750 m at cross-sections A, B, C, D, F, and G at a ship navigation time of $t = 100$ s. The differences between the two models at cross-section A are primarily evident in the ship's draft between 630 m and 680 m, where the draft depth is 1 m. COWS model predicts a lower maximum draft depth, and the wave peaks generated by the converging wake at the stern are also lower than those predicted by FUNWAVE-TVD model. The same issue is observed in Agarwal's FEBOUSS model. At cross-sections B to D, the simulation results of the two models align very well, with similar transverse wave frequencies and amplitudes. At cross-sections F and G, the transverse wave frequencies are also similar, but COWS model estimates higher wave heights for the wake compared to FUNWAVE-TVD model.

The comparison of the Kelvin wedge angle, θK, formed between the wave crest and the ship's path with analytical results is frequently used as a validation method for

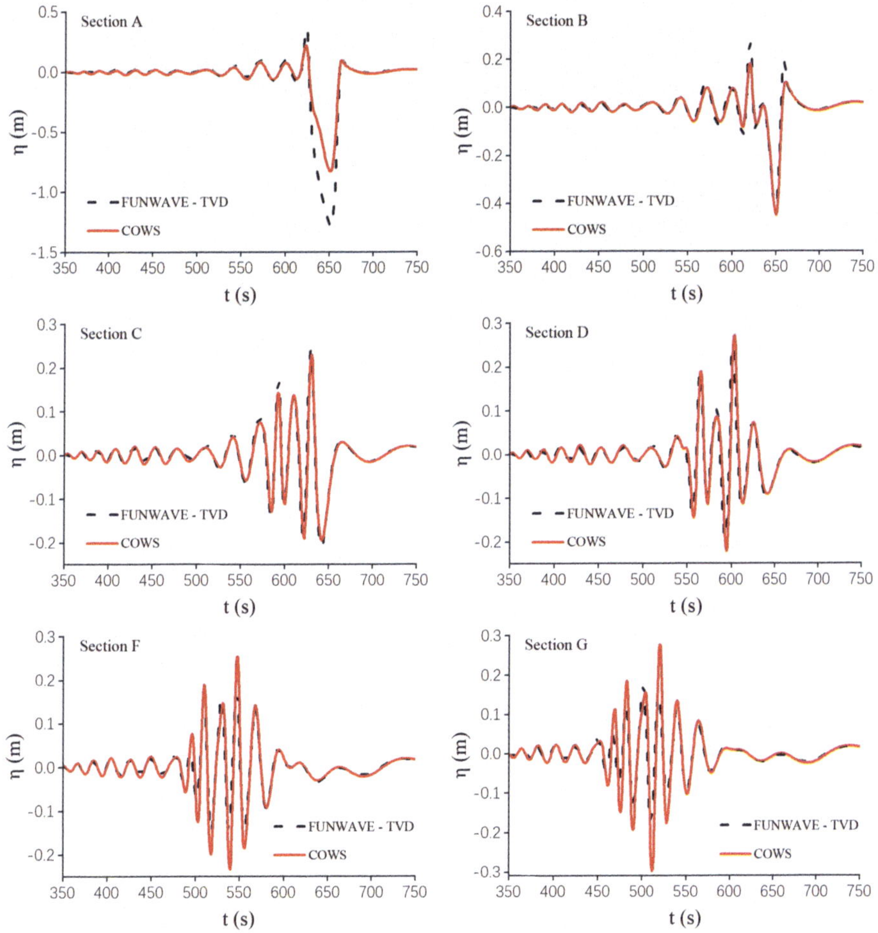

Fig. 9. Comparison of results obtained from FUNWAVE-TVD and COWS for surface elevation at $t = 100$ s.

models of moving ships generating wakes. This has been investigated by researchers such as Bayraktar (2013) [7], David (2017) [8], and Samaras (2021) [13]. With a Froude number, $Fr = 0.85$, the analytical solution gives a Kelvin angle, θK, of 24.76°. Figure 10 in this model illustrates the angle at t = 100 s between the divergent wave crest and the path, which is measured to be $\theta K = 23.85°$. This results in a deviation of 3.82% from the analytical solution, which is within an acceptable range.

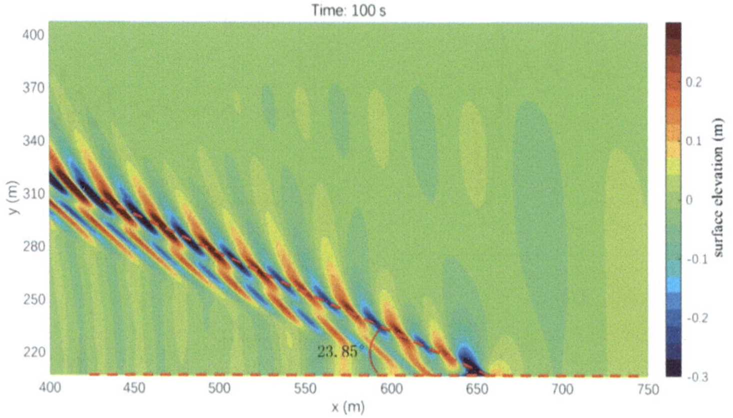

Fig. 10. The wave surface contours and the Kelvin half-wedge angle θK formed by the wave crest at $t = 100$ s

4 Conclusions

In this study, Boussinesq type wave model COWS is applied to simulate ship-borne waves. In the validation within a two-dimensional channel, the simulation results are compared with those from the BIEM model and the fKdV model, and the results of this study are found to be very close to those computed by the BIEM model. In validating the model against bow solitary wave data, comparisons are made with experimental data from Gourlay and simulation results from Shi. It is observed that the bow solitary waves began to break at $Fr = 1.12$, and at $Fr = 1.41$, a clear hydraulic jump occurs in COWS model, consistent with experimental results, showing better alignment with experimental observations. Moreover, comparisons with FUNWAVE-TVD model for ship-generated waves in open water show that in the A-A cross-section, COWS model predicts a lower maximum draft depth and the wave peaks formed by the converging stern flows are also lower than those predicted by FUNWAVE-TVD model. However, the transverse wave frequency and amplitude in the remaining sections are similar to those predicted by FUNWAVE-TVD model, leading to overall satisfactory results. This demonstrates the effectiveness of COWS model in predicting the waves generated by ship navigation.

References

1. Scott Russell, J.: In: BA Reports, British Association for the Advancement of Science (1837)
2. Russell, J.S.: Experimental researches into the laws of certain hydrodynamical phenomena that accompany the motion of floating bodies, and have not previously been reduced into conformity with the known laws of the resistance of fluids. Trans. Roy. Soc. Edinburgh **14**(1), 47–109 (2013)
3. Havelock, T.H., Larmor, J.: The propagation of groups of waves in dispersive media, with application to waves on water produced by a travelling disturbance. In: Proceedings of the Royal Society of London Series A, Containing Papers of a Mathematical and Physical Character, vol. 81, no. 549, pp. 398–430 (1908)

4. Chen, X.-N.: On ships in natural waterways: The Royal Institution of Naval Architects, RINA, International Conference on Coastal Ships and Inland Waterways, London, UK, 1999, ISBN: 0 903055 50 3 Paper: P1999–3 Proceedings. (1999)
5. Nascimento, M.F., Neves, C.F., Maciel, G.D.F.: Waves generated by two or more ships in a channel. Coastal Eng. Proc. **1**(32), 60 (2011)
6. Li, Y., Sclavounos, P.D.: Three-dimensional nonlinear solitary waves in shallow water generated by an advancing disturbance. J. Fluid Mech. **470**, 383–410 (2002)
7. Bayraktar, E.D., Beji, S.: Numerical simulation of waves generated by a moving pressure field. Ocean Eng. **59**, 231–239 (2013)
8. David, C.G., Roeber, V., Goseberg, N., et al.: Generation and propagation of ship-borne waves - Solutions from a Boussinesq-type model. Coast. Eng. **127**, 170–187 (2017)
9. Roeber, V., Cheung, K.F.: Boussinesq-type model for energetic breaking waves in fringing reef environments. Coast. Eng. **70**, 1–20 (2012)
10. Shi, F., Malej, M., Smith, J.M., et al.: Breaking of ship bores in a Boussinesq-type ship-wake model. Coast. Eng. **132**, 1–12 (2018)
11. Ertekin, R.C., Webster, W.C., Wehausen, J.V.: Waves caused by a moving disturbance in a shallow channel of finite width. J. Fluid Mech. **169** (2006)
12. Gourlay, T.P.: The supercritical bore produced by a high-speed ship in a channel. J. Fluid Mech. **434**, 399–409 (2001)
13. Samaras, A.G., Karambas, T.V.: Numerical simulation of ship-borne waves using a 2DH post-Boussinesq model. Appl. Math. Model. **89**, 1547–1556 (2021)
14. Karambas, T.V., Memos, C.D.: Boussinesq model for weakly nonlinear fully dispersive water waves. J. Waterw. Port Coast. Ocean Eng. **135**(5), 187–199 (2009)
15. Fang, K., Sun, J., Song, G., et al.: A GPU accelerated Boussinesq-type model for coastal waves. Acta Oceanol. Sin. **41**(9), 158–168 (2022)
16. Kim, G., Lee, C., Suh, K.-D.: Extended Boussinesq equations for rapidly varying topography. Ocean Eng. **36**(11), 842–851 (2009)
17. Wu, T.Y.-T.: Generation of upstream advancing solitons by moving disturbances. J. Fluid Mech. **184**, 75–99 (1987)
18. Fang, K., Zou, Z., Dong, P., et al.: An efficient shock capturing algorithm to the extended Boussinesq wave equations. Appl. Ocean Res. **43**, 11–20 (2013)
19. Fang, K., Liu, Z., Zou, Z.: Fully nonlinear modeling wave transformation over fringing reefs using shock-capturing boussinesq model. J. Coastal Res. **32**(1), 164–171 (2016)
20. Wang, Y., Liang, Q., Kesserwani, G., et al.: A 2D shallow flow model for practical dam-break simulations. J. Hydraul. Res. **49**(3), 307–316 (2011)
21. Cao, Y., Beck, R.F., Schultz, W.W.: Numerical computations of two-dimensional solitary waves generated by moving disturbances. Int. J. Numer. Meth. Fluids **17**(10), 905–920 (1993)
22. Lee, S.-J., Yates, G.T., Wu, T.Y.: Experiments and analyses of upstream-advancing solitary waves generated by moving disturbances. J. Fluid Mech. **199**, 569–593 (1989)
23. Katsis, C., Akylas, T.R.: On the excitation of long nonlinear water waves by a moving pressure distribution. Part 2. Three-dimensional effects. J. Fluid Mech. **177**, 49–65 (1987)
24. Agarwal, S., Sriram, V., Liu, P.L.F., et al.: Waves in waterways generated by moving pressure field in Boussinesq equations using unstructured finite element model. Ocean Eng. **262** (2022)

Identification of Bluff Body Wake Characteristics Based on a Soap Film Water Tunnel

Hongzhuang Wei, Xiangbo Du, Peng Du(✉), and Haibao Hu

School of Marine Science and Technology, Northwestern Polytechnical University, Xi'an, China
dupeng@nwpu.edu.cn

Abstract. Flow around structures is a common phenomenon in practical engineering. When faced with complex flow conditions, the wake structures formed by structures falling off are relatively complex and diverse. To analyze the flow field structure and characteristics of the flow around the rotating structure, this paper utilizes an experimental device called a soap film water tunnel to investigate the hydrodynamic flow characteristics around the rotating cylinder. First, a vertical soap film experimental device was constructed, and the optical interference method was employed to visualize the wake of the structure. The experiment mainly studies the wake structure of the flow around a rotating cylinder at various relative rotation speeds. It was found that as the relative speed α gradually increases, the shedding wake structure of the rotating cylinder will exhibit three types sequentially: vortex shedding zone, stable zone, and unilateral vortex zone (secondary unstable zone). Finally, the flow field structure generated by the rotating cylinder was analyzed and reconstructed using the Proper Orthogonal Decomposition (POD) method. This analysis revealed a trend where small-scale structures in the wake are progressively captured as the mode number increases. It shows that the POD method can effectively analyze the evolving pattern of the wake structure.

Keywords: flow visualization · hydrodynamics · soap film water tunnel · wake recognition

1 Introduction

Flow around structures is a common phenomenon in practical engineering [1–3]. When fluid flows through a structure, a viscous boundary layer will form on the surface of the structure. Within a specific range of Reynolds numbers, the boundary layer will detach and form a periodic wake vortex known as the Kármán vortex street [4, 5]. Flow around a cylinder is the most common phenomenon of flow around a bluff body. Although the structure of a cylinder is simple, it involves complex fluid mechanics issues such as boundary layer separation, vortex shedding and interaction, and turbulence. Therefore, the flow around a cylinder has always been a topic in fluid mechanics theory. Compared to the flow around a stationary cylinder, the flow around a rotating cylinder will significantly alter the flow pattern around the cylinder. It is a typical unsteady separation flow [6]

D. Jeng and B. Cai (Eds.): ICCOE 2024, LNCE 423, pp. 96–107, 2025.
https://doi.org/10.1007/978-981-96-0181-3_10

and has important applications in flow control research [7]. Studying the flow around a rotating cylinder is of significant theoretical importance for mitigating the vortex-induced vibration of the flow around the cylinder. The research results are of great significance for studying issues in general industrial fields, especially in ocean engineering, aerospace engineering, and related areas [8, 9].

In recent years, many scholars at home and abroad have conducted extensive research on analyzing the flow field characteristics of uniform rotating cylinder flow. He et al. [10] numerically simulated the flow around a rotating cylinder under the influence of uniform incoming flow using the Large Eddy Simulation (LES) method. The research results show that rotating the cylinder can effectively suppress vortex shedding, and the lift-to-drag ratio can be greatly improved with an increase in rotation speed. Sun et al. [11] conducted an experimental study on the flow around a rotating cylinder with different specific speeds using Particle Image Velocimetry (PIV) under the condition of Re = 100. Xu et al. [12] experimentally measured the flow field behind a rotating cylinder with a Reynolds number of Re = 20,000 ~ 90,000, and analyzed the velocity distribution and turbulence distribution at different sections behind the rotating cylinder. Liu et al. [13] utilized wind tunnel tests and Fluent numerical simulation methods to investigate the wake and surface pressure characteristics of a rotating cylinder at various rotation speeds. They also examined the impact of dimensionless rotation speed on the intensity and pattern of vortex shedding. Ezadi et al. [14] studied the flow characteristics in a rotating cylinder with Reynolds numbers ranging from 5900 to 11,800 through experiments and numerical simulations. They summarized the variation of the drag coefficient and lift coefficient with the rotation speed.

The rotation of the cylinder will obviously change the vortex shape in the wake, leading to flow field characteristics that are significantly different from those of a stationary cylinder. Compared to the flow around a stationary cylinder, the movement of the cylinder wall disrupts the symmetry of the flow, causing distortion and asymmetric separation of the boundary layer. This, in turn, alters the shedding pattern of the tail vortex [7]. The flow field structure around a rotating cylinder is more complex due to the interaction between vortex shedding and cylinder rotation. To analyze the wake of the flow more effectively, it is necessary to use a dimensionality reduction method to decrease the complexity of the flow field. This involves extracting the characteristic structures from a vast amount of data. Proper Orthogonal Decomposition (POD) is a powerful tool for extracting and analyzing low-dimensional approximate structures from massive, high-dimensional data. It linearly reduces the high-dimensional nonlinear system by considering the energy contribution rate of each mode. Using a small amount of modal data, the large-scale characteristic structure of the flow field can be obtained and the flow field can be reconstructed [15].

In recent years, the POD technique has been widely used in the analysis and processing of flow field characteristics [16]. For instance, Jiang et al. [17] used Particle Imaging Velocimetry (PIV) technology to investigate the impact of the gap on the cylindrical flow field structure in a water tank. When the POD method was used to analyze the coherent structure of the flow field around a slotted cylinder, it was found that the first six modal forms were the most important coherent structures. Deep et al. [18] employed the POD technique on the vorticity field to study the dominant modes and their corresponding enstrophies, along with the time coefficients. It has been found that the first six modes contribute to more than 95% of the total enstrophy in all the cases studied. Xia et al. [19] studied the velocity field in the near wake of a cylinder coated with and without porous media using PIV measurements. The Snapshot POD technique was applied to analyze the PIV data measured in the horizontal planes, aiming to detect the effect of the porous layer on the dominant flow features. Bai et al. [20] employed the 3D POD method to study the near wake dynamics of smooth and wavy cylinders. The results showed that the first two POD modes account for the majority of the kinetic energy in the near wake. Therefore, they represent the energetically significant large-scale Coherent Structures (CSs) for all the wavy cylinders with varying wavelengths, as well as the smooth cylinder.

In summary, extensive research on the wake characteristics of rotating cylinders has been conducted by numerous scholars both domestically and internationally through simulations and experiments. The wake flow field structure of a rotating cylinder is highly complex, involving phenomena such as vortex shedding, flow separation, and reattachment. These phenomena interact to form a nonlinear and intricate flow field structure, making both analytical and numerical simulations challenging. To simplify the complexity of the flow field and gain a deeper understanding of its characteristics, this study utilizes a soap film water tunnel experimental system to investigate the flow characteristics of a rotating cylinder at a Reynolds number of $Re = 200$. Additionally, the evolution of the cylinder's wake at various speed ratios is analyzed. The Proper Orthogonal Decomposition (POD) dimensionality reduction method is employed to process the flow field, extracting large-scale characteristic structures with a limited amount of modal data and reconstructing the flow field around the rotating cylinder. The analysis and reconstruction of the rotating cylinder's flow characteristics are of significant importance for practical engineering applications.

2 Experimental Platforms

2.1 Vertical Soap Film Water Tunnel Experimental Apparatus

The structure of the vertical soap film device is shown in Fig. 1. The soap film water tunnel is mainly composed of an upper water tank, a nozzle, a flow control valve, a guide wire, a lower water tank and a water pump. The supporting structure is made of 4040C aluminum profile. The overall device is about 2.5 m high and 0.6 m wide.

The soap liquid was prepared with industrial detergent and water in a volume ratio of 1:4. The retention time of soap film formed in the soap film experimental device was compared with soap liquids with different volume fraction ratios. A solution with a volume ratio of industrial detergent and water of 1:4 was selected. The soap film formed by this solution can be maintained for a long time in a relatively stable experimental environment and last for several hours, providing good working conditions for long-term experiments. After being fully stirred and allowed to stand for a period of time, the soap liquid was poured into the lower water tank at the bottom of the device and transported to the upper water tank by the water pump.

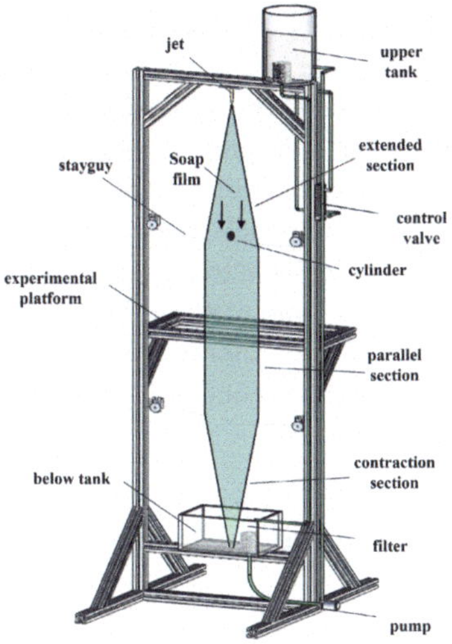

Fig. 1. Structure of the vertical soap film experimental setup

2.2 Measurement of Soap Film Parameters

The flow rate of the soap film is a key flow parameter. The Particle Image Velocimetry (PIV) principle is used in the experiment to measure the flow rate of the soap liquid. The PIV method is to mix the tracer particles with the soap liquid and then add them to the soap film flow field, capture two consecutive frames of particle images, and use the correlation of the two frames of photos to calculate the flow rate of the soap film. In order to reduce the measurement error, three repeated measurements are performed here and the average value is taken (the relative error of the speed is $<2\%$).

When solving the soap film viscosity coefficient and Reynolds number, the empirical formula of the *Re-St* fitting relationship of the single cylinder flow experiment in the literature [21] was used to solve it.

$$Re = \frac{Ud}{v} \tag{1}$$

$$St = \frac{fd}{U} \tag{2}$$

$$St = 0.2731 - \frac{1.1129}{\sqrt{Re}} + \frac{0.4821}{Re}(Re < 1000) \tag{3}$$

Where f represents the cylindrical vortex shedding frequency, d is the cylinder diameter, U is the incoming flow velocity, and v is the kinematic viscosity coefficient of the fluid.

2.3 Flow Field Visualization System

This paper adopts the flow field interferometry method, which uses a high-pressure sodium lamp to illuminate the soap film. The thin soap film interferes under the illumination of sodium light, which can clearly show the wake structure of the flow around the structure. The flow field visualization system mainly consists of two cameras, a high-pressure sodium lamp, a flexible board, and a computer acquisition workbench. The experimental system selected a high-pressure sodium lamp with a power of 800 W as the interference light source. The high-pressure sodium lamp is monochromatic and has high brightness, so the interference effect is better and the overall background color of the soap film is consistent. In order to make the captured trail image clear and reduce the reflection of the floor and other walls onto the soap film, a soft light board is placed in front of the high-pressure sodium lamp to make the sodium lamp light more uniform and the trail image clearer and brighter. After the brightness of the sodium lamp stabilizes, the experimental shooting is carried out, and the shooting devices are high-speed cameras and high-definition cameras.

3 Experiment on Flow Around a Uniformly Rotating Cylinder

3.1 Parameter Setting

During the experiment, the flow rate of the soap film experimental flow valve was Q = 16 mL/min, and the flow rate at this time was $U = 0.954$ m/s. The kinematic viscosity coefficient of the soap liquid is $v \approx 4.77 \times 10^{-5}$ m^2/s, the diameter of the cylinder is d = 10 mm, and Re ≈ 200 is calculated. A large number of literatures have conducted simulations and experiments under this Reynolds number, so this experimental study chooses this Reynolds number for experimental research in order to compare the experimental results and further prove that rotation control is effective in soap film experiments. In order to effectively compare with the experimental results of previous researchers, the dimensionless parameter relative speed α is set:

$$V = 2\pi nr \tag{4}$$

$$\alpha = \frac{2\pi nr}{U} \tag{5}$$

Where n is the motor speed, the radius of the cylinder $r = 5$ mm, the motor rotation drives the cylinder's tangential velocity V, U is the soap film flow velocity, and α is the ratio of the cylinder's rotation tangential velocity to the soap film flow velocity.

The experimental conditions are set for the rotational experimental conditions from 0 (stationary) to 4.5. When the experimental device is at $\alpha > 4.5$, the device cannot reach the speed due to the high speed, so the experiment will not be carried out. When $\alpha < 2$, the dimensionless parameter α of the experiment is set at an interval of 0.25 to set a working condition. After $\alpha > 2$, the experiment sets a working condition every 1, so as to quickly find different types of variable tails and compare them with the simulation experiment.

3.2 Experimental Results

As the relative speed increases, the experimental wake image gradually reveals three distinct levels: the vortex shedding zone, the stable zone, and the unilateral shedding zone (secondary instability zone). When the relative speed range is $0 \leq \alpha < 2$, the vortex shedding zone appears first. When the relative speed exceeds the critical speed ratio ($\alpha = 2$), the wake of the cylindrical flow starts to diminish, the shedding of vortices is inhibited, and a stable zone is established. In this experiment, the range of the stable zone is $2 \leq \alpha < 4.5$. When the relative speed increases to 4.5, the vortex reappears and enters the secondary instability zone. The experimental results correspond to the simulation results in the literature [22]. At Re = 200, at different rotation speeds, the types of wakes are rich and varied, which also reflects that fluid motion is a nonlinear and complex flow. The experimental results prove that the soap film experiment can be used for the study of the wake of the rotating cylinder flow, laying an experimental foundation for subsequent scholars to carry out the study of the wake type of the cylinder flow.

To further analyze the wake shedding process, an experimental condition was selected in the wake vortex shedding zone and the secondary instability zone for time series analysis. As shown in Fig. 2, it is a wake image sequence of the cylindrical flow experiment with Re = 200 and $\alpha = 1$. The wake shedding period is 0.036 s, the time interval is 0.006 s, and a total of 8 wake images were collected. The recirculation zone is elongated during the flow process, and the tail of the recirculation zone swings left and right with the flow. The wake vortex is peeled off from the recirculation zone during the flow process, and gradually becomes larger, stretched, and developed with the flow, and finally falls off to form an independent wake vortex, as shown in vortex A in Fig. 2. Vortex A and vortex B are a pair of tail vortices, which fall off in sequence on both sides of the tail of the recirculation zone. As the tail vortices fall off, they eventually develop into a pair of tail vortices. The difference between the two tail vortices is that the rotation directions of the vortices are opposite, with vortex A rotating counterclockwise and B rotating clockwise. This is related to the rotation direction of the cylinder, which is counterclockwise. Therefore, the flow velocity on the left wall of the cylinder is fast, and the flow velocity on the right wall of the cylinder is slow, resulting in the counterclockwise rotation shearing of the tail vortex on the left side of the recirculation zone and the clockwise rotation shearing of the tail vortex on the right side. As the tail vortex A-B falls off in order, a new vortex pair C-D will be formed in the recirculation zone. Similarly, the C-D vortex pair falls off in sequence on both sides of the recirculation zone, and the rotation directions of the vortices are still opposite. Vortex A and Vortex C have the same rotation direction, while Vortex B and Vortex D have the same rotation direction. The rotation direction of the tail vortex is mainly related to the rotation direction of the cylinder. The rotation directions of each pair of tail vortices are opposite, and the tail vortices on the same side have the same rotation direction. In the range of relative rotation speed $0 \leq \alpha < 2$, the wake rotation law is consistent with the above description.

When the relative rotation is $\alpha = 4.5$, the wake type is a secondary unstable state. Figure 3 shows the unilateral vortex shedding process at different times, and 8 consecutive wake images are selected with a time interval of 0.006 s. Due to the high rotation speed of the cylinder, the recirculation zone is not obvious. The tail vortex is sheared off from the near wall by high speed to form an orderly clockwise single vortex, which gradually develops with the flow rate to form a large-scale tail vortex. The single vortex is clockwise, which is also related to the rotation direction of the cylinder. The high-intensity counterclockwise rotation of the cylinder suppresses the shedding of the

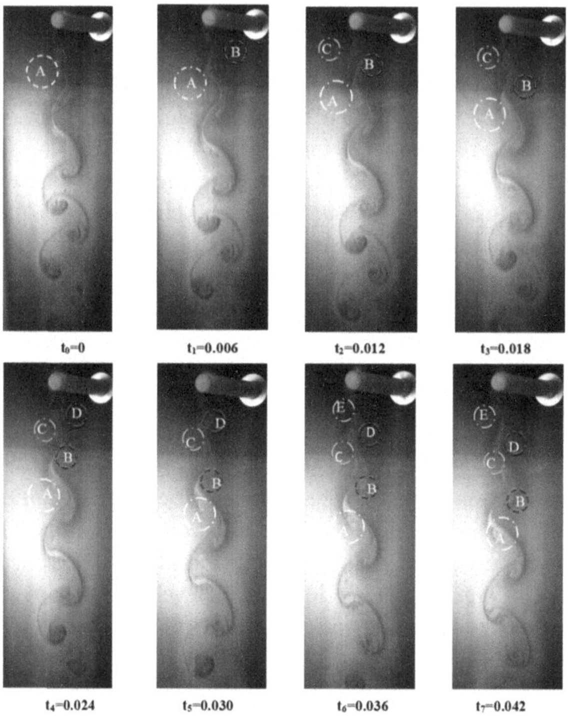

Fig. 2. The process of wake shedding at different times when Re $\approx$ 200 and $\alpha = 1$

counterclockwise tail vortex, leaving only the clockwise tail vortex on the right. When the cylinder rotates at high speed, the unilateral vortex is "thrown out" of the recirculation zone to form a separate clockwise tail vortex. The irregularity of the unilateral vortex also indicates the instability of the flow in the second unstable zone. In the soap film experiment, it is easily disturbed by the external environment. Compared with the literature [22], the vortex structure of the unilateral vortex is consistent, but the vortex spacing of the tail vortex is different. The vortex spacing in the simulation experiment is 3–4 vortex widths, while the vortex spacing of the unilateral vortex in the soap film experiment is relatively dense. It may be that there are certain experimental errors in the soap film experiment under the condition of high-speed rotating cylinders, such as insufficient smoothness of the cylinder, certain differences between soap film flow and real two-dimensional flow, or the soap film experiment is sensitive to the environment, etc., which have a certain impact on the accuracy of the experiment. However, the type of wake remains the same, and the wake is generated on one side of the cylinder, and the direction of the wake rotation is consistent in clockwise rotation. It shows that the soap film experiment still has a certain performance ability for the flow around the high-speed rotating cylinder.

The experiment explores the wake types of the flow around a cylinder with different relative speeds under the condition of Re number 200. The experiment can be divided into three types according to the shedding wake structure: vortex shedding zone, stable

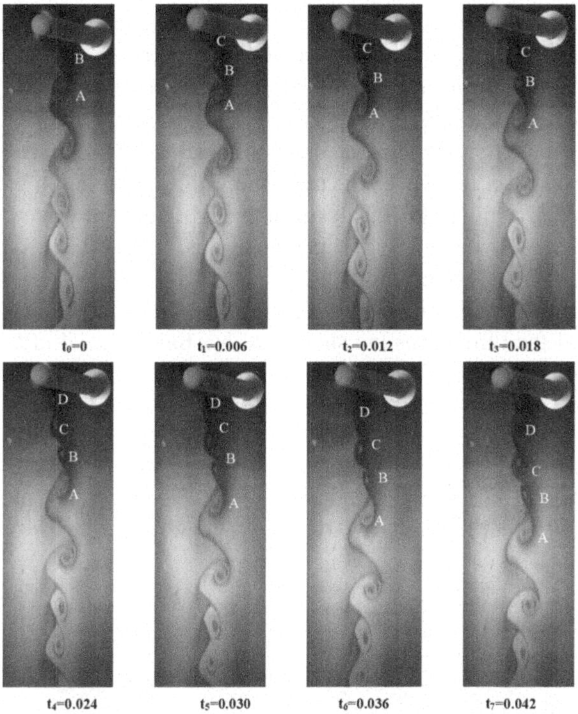

Fig. 3. The process of wake shedding at different times when Re $\approx$ 200 and $\alpha = 4.5$

zone, and unilateral vortex zone. The experimental results correspond one to one with the simulation results, which fully demonstrates the feasibility of the soap film experiment for the flow around a rotating cylinder. In the experiment, the soap film will not be sheared and broken by the high-speed rotating cylinder. The experiment found that under the rotating cylinder, the soap film is better maintained and can last for several hours without breaking. The reason may be that the rotation accelerates the energy mixing between the soap films, better and more fully ensures that the soap liquid is spread on the soap film, and effectively maintains the stability of the soap film.

3.3 POD Analysis of the Wake of a Rotating Cylinder

POD is a method for processing large amounts of data. There have been a lot of studies on flow field modal decomposition and flow field order reduction [23, 24]. POD decomposes the flow field into different time-varying flow structures, that is, different modes. Different modes are the product of orthogonal basis functions and time functions. The essence of POD is to project the original signal (flow field) that varies with time into a superposition of a set of time-varying, mutually orthogonal spatial signals (flow fields).

The acquired wake vortex image sequence is subjected to orthogonal mode decomposition. The large-scale structure in the wake field can be obtained through POD decomposition. The POD analysis is carried out at Re = 200 and the relative rotation speed

$\alpha = 1$. Using the periodicity of soap film wake shedding, the code can solve that the time period of wake shedding under this condition is 0.030 s, and the frequency of wake image acquisition is 0.002 s, so 16 wake images should be collected in one wake shedding cycle. The complete cycle of wake shedding is the prerequisite for POD analysis. The POD analysis under simulation data generally includes 5–8 wake shedding cycles. In the experimental case, the wake structure is unstable, resulting in weak periodicity. Therefore, when performing POD analysis on the experimental wake image, it is necessary to take more wake images with shedding cycles. Here, 10 wake shedding cycles, that is, 160 wake images, are taken for POD analysis. All 160 wake images are stored in the same directory and preprocessed. As shown in Fig. 4(a), (b) and (c), part of the original images of the time series are shown, with a time interval of 0.006 s. In order to make the image better under POD decomposition, the wake image is binarized, that is, the background is represented by black and the wake trace is represented by white. The orthogonal decomposition method is used to extract the features of the wake image structure. The main process includes 4 parts: 1. wake image preprocessing, 2. wake field averaging, 3. removing the average value of each wake field image, and 4. Constructing the correlation coefficient matrix and solving the eigenvector.

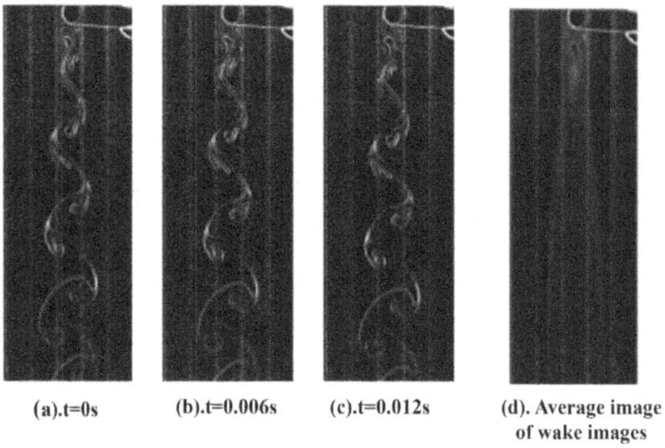

Fig. 4. (a) (b) (c) The first three tail images of the POD time series (d) Average of the tail images

Fig. 4(d) shows the average wake field under the experimental condition. The wake shedding of the wake field shows periodic changes, so the average image of the wake field shows a uniform color. The vertical stripes in the average wake image are caused by image shooting and belong to the noise in the wake field structure. In order to remove the noise in the wake, the wake image needs to be further processed, that is, the wake image is de-averaged, which is also the main process of POD analysis. As shown in Fig. 5, the wake image after de-averaging, the noise of the wake image is removed, making the image clearer.

To solve the principal modes, the code constructs a correlation coefficient matrix and first determines the eigenvalues and eigenvectors of the matrix. The classical orthogonal decomposition method is used to construct the correlation matrix C and solve for the principal modes. Figure 6 shows the first eight principal modes. The first and second modes reveal a large-scale wake structure, highlighting differences in the wake vortex

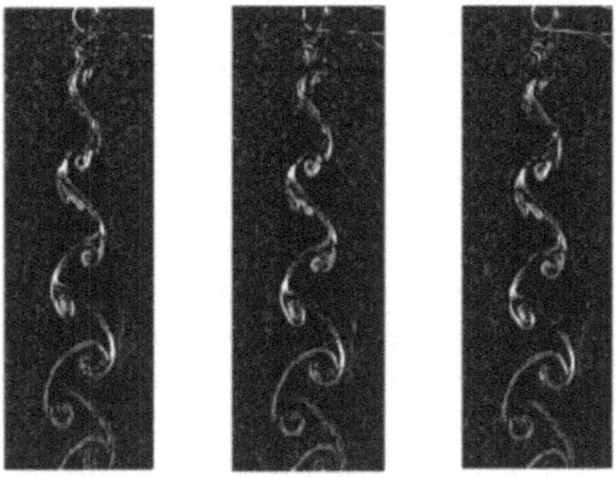

(a).t=0s (b).t=0.006s (c).t=0.012s

Fig. 5. Trail image after removing the mean.

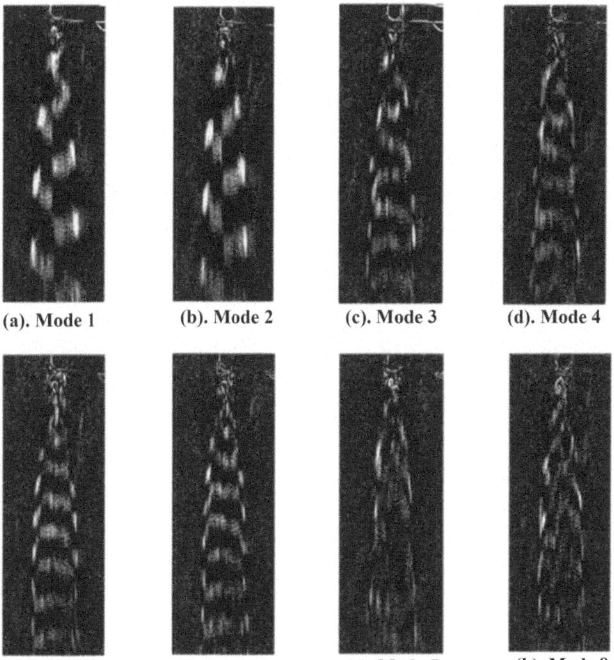

(a). Mode 1 (b). Mode 2 (c). Mode 3 (d). Mode 4

(e). Mode 5 (f). Mode 6 (g). Mode 7 (h). Mode 8

Fig. 6. The first 8 orders of major modes

within the wake structure. Both modes exhibit pulsation phenomena in the flow field. As the modes increase, the smaller-scale structures of the wake are progressively captured. In other words, different POD modes are decomposed by POD. The higher-order modes capture smaller-scale structures or high-frequency changes in the flow field, representing the detailed structure of the flow, while the lower-order modes capture the large-scale structure and main characteristics of the flow field. Therefore, the first and second modes of the flow field around the cylinder represent the periodic motion in the flow field.

4 Conclusions

This paper describes the construction of a vertical soap film water tunnel experimental device designed to observe the wake flow characteristics of a uniformly rotating cylinder at various speed ratios. Proper Orthogonal Decomposition is utilized to decompose the wake field and extract the primary modes for analyzing the flow field structure. The main conclusions are as follows:

(1) By constructing a soap film water tunnel experimental apparatus and employing the flow field interferometry method, the wake patterns of the flow around a cylinder at various relative speeds under the Reynolds number condition of $Re = 200$ were investigated. The experimental results indicate that a vortex shedding zone appears when the relative speed is $0 \leq \alpha < 2$, and the wake remains stable without vortex shedding when the relative speed is $2 \leq \alpha < 4.5$. When the relative speed increases to 4.5, the vortex reappears, indicating a secondary instability phenomenon known as a unilateral shedding zone. The experimental results are consistent with previous simulation results, fully verifying the feasibility of the soap film experiment in studying the flow around a rotating cylinder.

(2) The acquired wake vortex image sequence is subjected to POD to decompose the flow field into different time-varying flow structures, i.e., different modes, to extract the large-scale structure in the rotating cylindrical wake field. In the final solved modes, the first and second modes display the large-scale wake structure and highlight differences in the wake vortex within the wake structure. Both modes exhibit the pulsation phenomenon in the flow field. As the number of modes increases, the small-scale structure of the wake is gradually captured, leading to a comprehensive reconstruction of the rotating cylindrical flow field structure.

The research on rotating cylinders in this paper still remains in the laminar state. In the next step, to align more closely with practical engineering applications, we will investigate and analyze the wake of blunt bodies in turbulent conditions. In turbulent flow, the wake of a rotating cylinder exhibits characteristics such as irregularity and strong pulsation. By analyzing the flow around a rotating cylinder in turbulent conditions, we can gain a more comprehensive understanding of its wake characteristics.

Acknowledgement. This research was supported by the National Natural Science Foundation of China (No. 52201380), Fundamental Research Funds for the Central Universities (No. D5000230080).

References

1. Muddada, S., Patnaik, B.S.V.: Active flow control of vortex induced vibrations of a circular cylinder subjected to non-harmonic forcing[J]. Ocean Eng. **142**, 62–77 (2017)

2. An, H., Cheng, L., Zhao, M.: Steady streaming around a circular cylinder in an oscillatory flow[J]. Ocean Eng. **36**(14), 1089–1097 (2009)
3. King, R.: A review of vortex shedding research and its application[J]. Ocean Eng. **4**(3), 141–171 (1977)
4. Wanderley, J.B., Levi, C.: Vortex induced loads on marine risers[J]. Ocean Eng. **32**(11–12), 1281–1295 (2005)
5. Badr, H.M., Dennis, S.C.R., Kocabiyik, S., et al.: Viscous oscillatory flow about a circular cylinder at small to moderate Strouhal number[J]. J. Fluid Mech. **303**, 215–232 (1995)
6. Numerical study on vortex patterns and critical state characters in the wake of a rotating circular cylinder[J]. Chinese J. Theor. Appl. Mech. **29**(1), 8–16 (1997). https://doi.org/10.6052/0459-1879-1997-1-1995-190
7. Wang Dan, Y., Yong.: Numerical simulation about the secondary instability in the two-dimensional laminar flow past a rotating circular cylinder. J. Beijing Univ. Technol. **35**(11), 1101–1107 (2015)
8. Padrino, J.C., Joseph, D.D.: Numerical study of the steady-state uniform flow past a rotating cylinder[J]. J. Fluid Mech. **557**, 191–223 (2006)
9. Bin, Z., Zhanqi, T., Jiao, S., et al.: Dynamic mode decomposition of flow around rotating cylinder. J. Hebei Univ. Technol. **44**(4), 63–67 (2015)
10. Ying, H., Xinmin, Y., Zhihua, C., et al.: Flow field characteristics of flow past a rotating cylinder[J]. J. Ship Mech. **19**(05), 501–508 (2015). https://doi.org/10.3969/j.issn.1007-7294.2015.05.004
11. Jiao, S., Bin, Z., Zhanqi, T., et al.: Experimental study on the flow past a rotating cylinder with PIV[J]. J. Exp. Fluid Mech. **30**(01), 81–90 (2016)
12. Yihang, X., Shaosong, C.: Analysis of flow characteristics around a rotating cylinder. Chinese J. Theor. Appl. Mech. **53**(7), 1900–1911 (2021). https://doi.org/10.6052/0459-1879-21-153
13. Liu, J., Ma, W., Jin, L., et al.: Experimental and numerical investigation on the wake flow and vortex shedding of a rotating circular cylinder[J]. Phys. Fluids. **35**(7) (2023)
14. Ezadi Yazdi, M.J., Rad, A.S., Khoshnevis, A.B.: Features of the flow over a rotating circular cylinder at different spin ratios and Reynolds numbers: experimental and numerical study[J]. Eur. Phys. J. Plus. **134**, 1–21 (2019)
15. Qiang, L., Zhenbin, L., Xiong, D., et al.: Modal analysis of dual syntehtic jets based on POD method[J]. Acta Aerodynam. Sin. **38**(06), 1027–1033+1046 (2020)
16. Wang, H.F., Cao, H.L., Zhou, Y.: POD analysis of a finite-length cylinder near wake[J]. Exp. Fluids. **55**, 1–15 (2014)
17. Jianhua, J., Feng, B.: POD analysis of a slit circular cylinder near wake[J]. Phys. Gases. **2**(02), 28–36 (2017). https://doi.org/10.19527/j.cnki.2096-1642.2017.02.004
18. Deep, D., Sahasranaman, A., Senthilkumar, S.: POD analysis of the wake behind a circular cylinder with splitter plate[J]. Eur. J. Mech. B/Fluids. **93**, 1–12 (2022)
19. Xia, C., Wei, Z., Yuan, H., et al.: POD analysis of the wake behind a circular cylinder coated with porous media[J]. J. Vis. **21**, 965–985 (2018)
20. Bai, H.L., Alam, M.M., Gao, N., et al.: The near wake of sinusoidal wavy cylinders: three-dimensional POD analyses[J]. Int. J. Heat Fluid Flow. **75**, 256–277 (2019)
21. Leweke, T., Williamson, C.H.K.: Three-dimensional instabilities in wake transition[J]. Eur. J. Mech. B/Fluids. **17**(4), 571–586 (1998)
22. Weiguo, T., Wei, C., Yigang, W., et al.: Numerical simulation of turbulent flow in a rotating cylinder under different flow conditions[J]. J. Wuhan Univ. Technol. (Transp. Sci. Eng.). **46**(02), 247–253 (2022)
23. Taira, K., Brunton, S.L., Dawson, S.T.M., et al.: Modal analysis of fluid flows: An overview[J]. AIAA J. **55**(12), 4013–4041 (2017)
24. Cazemier, W., Verstappen, R., Veldman, A.E.P.: Proper orthogonal decomposition and low-dimensional models for driven cavity flows[J]. Phys. Fluids. **10**(7), 1685–1699 (1998)

Wave-Induced Response of Shell-Sand Mixed Seabed Around Slope Breakwater

Hang Zhu[1](✉), Guangsheng Wang[2], Lifei Gong[3], Ping Sun[3], Richard Asumadu[4], and Titi Sui[1]

[1] Key Laboratory of Coastal Disaster and Protection (Hohai University), Ministry of Education, Nanjing, Jiangsu, China
211303020055@hhu.edu.cn

[2] China Harbour Engineering Company Limited, Beijing, China

[3] Nanjing Hydraulic Research Institute, Nanjing, Jiangsu, China

[4] Centre for Coastal Management, Africa Center of Excellence in Coastal, University of Cape Coast, Cape Coast, Ghana

Abstract. Shell-sand mixed seabed have been observed to reduce pore water pressure amplitudes within the seabed. This characteristic is advantageous for maintaining the stability of foundations supporting marine structures. However, previous research has predominantly focused on idealized conditions, such as wave-induced responses in homogeneous seabed, overlooking the natural heterogeneity of real seabed. This heterogeneity arises from marine biological activities and sedimentation processes, ensuring that natural seabed is inherently spatially heterogeneous. In this study, a series of experiments were conducted using physical models in a wave flume, testing under numerous wave conditions and seabed parameters. The objective was to analyze how varying proportions of shell mixtures affect wave-induced responses in seabed. Key aspects investigated included changes in pore pressure amplitudes. The findings offer valuable insights into how shell content influences the dynamic behavior of seabed under wave action.

Keywords: wave · seabed response · shell-sand mixed seabed · slope breakwater

1 Introduction

The stability of foundations is a critical consideration in the design of marine engineering structures. Unlike the structural strength of the buildings themselves, foundation stability is influenced by a variety of factors, including seabed properties in the structure's location, wave parameters, and other environmental conditions. Pore pressure is a key indicator that significantly affects seabed foundation stability.

Previous studies have extensively investigated the wave-induced responses of seabed based on several assumptions. Yamamoto [1] and Madsen [2] were among the first to explore the dynamic responses of seabed, basing their studies on Biot's three-dimensional consolidation theory [3]. They examined the dynamic response of homogeneous, porous elastic seabed soils and derived theoretical solutions for seabed soil

D. Jeng and B. Cai (Eds.): ICCOE 2024, LNCE 423, pp. 108–117, 2025.
https://doi.org/10.1007/978-981-96-0181-3_11

stresses. Rahman [4] built a one-dimensional finite element model to simulate the buildup of seabed pore pressure under waves. With the advancement of computer technology, numerical simulations have become more prevalent. Sumer et al. [5] conducted a series of experiments to investigate wave-induced pore pressure accumulation and verify their mathematical model. Jeng and Zhao [6] presented a two-dimensional numerical model using the time-phased shear stress obtained from the oscillatory response numerical model. Chen et al. [7] systematically analyzed the applicable conditions of the partially dynamic and fully dynamic models under seismic loads. Chen et al. [8] analyzed the dynamic responses of a transversely isotropic poroelastic seabed subjected to nonlinear random waves with emphasis on liquefaction depth. Zhu et al. [9] probabilistic analyzed of wave-induced dynamic response in a poroelastic sloping seabed using random finite element method. Wang et al. [10] did the experimental study on the dynamic response of a silty seabed under waves and the experimental results clarify that liquefaction and scour promote each other.

In reality, natural seabed is always spatially non-homogeneous, it has been found by previous studies that the non-homogeneity would affect the pore pressure. Yamamoto [11] was the first to analyze heterogeneous seabed, deriving a semi-analytical solution and validating it with empirical data. Gatemiri [12, 13] was the first to use the finite element method to study the effects of various seabed parameters on pore water pressure. Jeng and Lin [14] utilized numerical simulations to study the effects of heterogeneous permeability on seabed responses. Zhang et al. [15] investigated the effect of the horizontally non-homogeneous soil property and three-dimensional non-homogeneous seabed response around a mono-pile. Li et al. [16] analyzed wave-induced dynamic response in a transversely isotropic and multilayered poroelastic seabed. Sui et al. [17] investigated the wave-induced response characteristics of shell and silt-mixed seabed through a physical model experiment. Zhu et al. [18] made a stochastic numerical model for wave-induced dynamic response in a spatially heterogeneous seabed around structures. With this numerical model, Zhu et al. [19] also did investigation on wave-induced responses of heterogeneous seabed around a buried pipeline based on nonstationary random field model. Senjuntichai et al. [20] analyzed the dynamic impedances of multiple strips on multi-layered transversely isotropic poroelastic soils.

In this study, the model was subjected to a series of experiments in a wave flume, testing under dozens of different wave conditions and seabed parameters. The objective was to analyze the differences in wave-induced responses of the seabed with varying proportions of shell content.

2 Physical Model Experiment

2.1 Flume Setup

The experiments were conducted in a wave flume at the Harbor, Waterway and Coastal Engineering Experimental Center of Hohai University. The wave flume has a length of 50 m, a width of 1.0 m, and a height of 1.5 m. A sloping breakwater model was installed at the rear of the flume, with a sand trench excavated in front of the breakwater. The sand trench is located 35 m from the wave generator, with dimensions of 2.3 m in length, 1.0 m in width, and 0.3 m in depth. The layout of the wave flume is illustrated in Fig. 1.

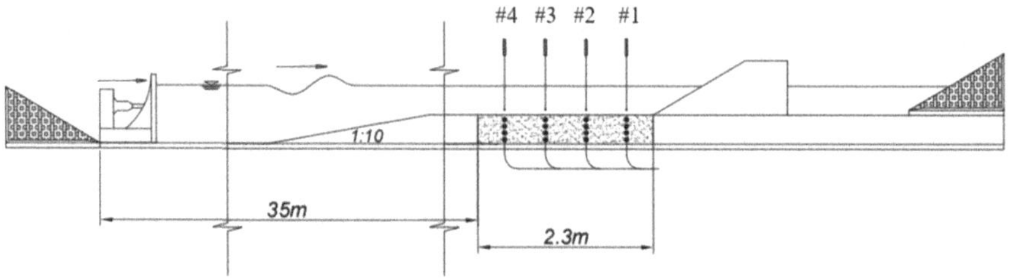

Fig. 1. Flume setup

Table 1. Wave and seabed parameters setup

Cases	H(cm)	T(s)	SC(%)
1–3	6	1.3/1.6/1.9	0
4–6	8	1.3/1.6/1.9	0
7–9	10	1.3/1.6/1.9	0
10–12	6	1.3/1.6/1.9	3.75
13–15	8	1.3/1.6/1.9	3.75
16–18	10	1.3/1.6/1.9	3.75
19–21	6	1.3/1.6/1.9	7.5
22–24	8	1.3/1.6/1.9	7.5
25–27	10	1.3/1.6/1.9	7.5
28–30	6	1.3/1.6/1.9	15
31–33	8	1.3/1.6/1.9	15
34–36	10	1.3/1.6/1.9	15
37–39	6	1.3/1.6/1.9	30
40–42	8	1.3/1.6/1.9	30
43–45	10	1.3/1.6/1.9	30

2.2 Model Setup

The experimental model was a simplified sloping breakwater with a crest elevation of 0.48 m and a slope gradient of 1:1.5. The sand used in the experiments included quartz sand with a median grain diameter of 0.18 mm and shell sand with a median grain diameter of 3 mm, which were manually mixed to create seabed with different proportions of heterogeneous sand.

Four cross-sections were evenly spaced within the sand trench, labeled from distance to nearby as #4, #3, #2, and #1, with each section spaced 0.6 m apart. Section #1 was positioned close to the toe of the breakwater, at a distance of 0.05 m from its front edge. Each cross-section was equipped with one wave gauge to monitor wave surface changes, and four pore pressure transducers were evenly distributed within each section to measure the changes in pore pressure within the seabed.

2.3 Cases Setup

The experiments utilized five different proportions of mixed seabed, specifically: 0%, 3.75%, 7.5%, 15%, and 30%. The water depth in the test section was 29.2 cm, with three different wave heights ranging from 6 to 10 cm, and three different periods ranging from 1.3 to 1.9 s. To minimize the impact of secondary reflected waves, readings were taken from the most stable wave generation stage for each set of conditions. The data acquisition frequency was 50 Hz. The arrangement of the test conditions is summarized in the table below.

3 Experiment Results

3.1 Wave Reflection

During the wave propagation process, three distinct stages were observed:

Wave Incidence Stage: In this initial phase, the wave generator initiates waves over a deep, flat seabed area. As these waves propagate, they encounter a gentle slope with a gradient of 1:10, gradually shoaling until they reach the test section with a water depth of 29.2 cm. The wave height of the incident waves is measured in the test section. Throughout this stage, the wave height increases progressively as the wave generator begins its operation.

First Reflection Stage: As waves approach the sloping breakwater, they reflect off its surface. The reflected waves interact with the incoming waves, forming a pattern resembling partial standing waves. This results in alternating wave crests and troughs in front of the breakwater. Wave forms remain relatively stable during this stage.

Second Reflection Stage: Here, due to the presence of the wave generator, initially reflected waves travel back towards its push plate, causing a secondary reflection. These secondary waves then combine with the original incident waves, resulting in a wave pattern that diverges from real-world conditions. During this stage, wave forms become disordered because the wave generator's push plate is in motion.

The three stages of wave propagation are illustrated in Fig. 2. In this study, the measurements taken during the first reflection stage were chosen as the valid data for the various test conditions.

3.2 Pore Pressure in Seabed

Given the influence of wave reflection discussed earlier, analyzing seabed pore pressure during the first reflection stage provides a more representative measure. This stage typically initiates about seven to eight wave cycles after significant fluctuations are observed. For instance, in cases with a wave period of 1.6 seconds, the 15th and 16th waves exhibit notably stable forms. Figure 3 illustrates the pore pressure at a depth of 8 cm in section #4 under varying wave heights.

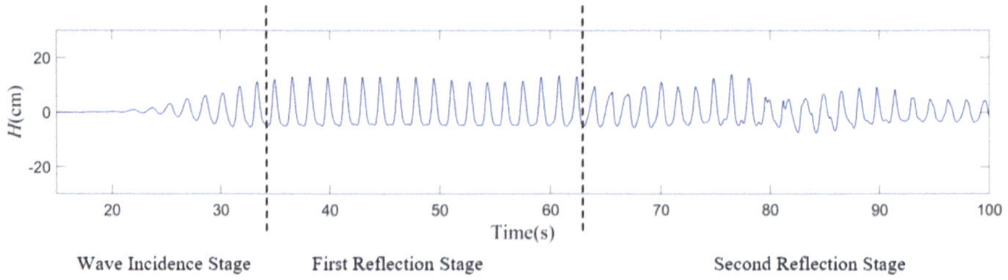

Fig. 2. Stages of incidence and reflection

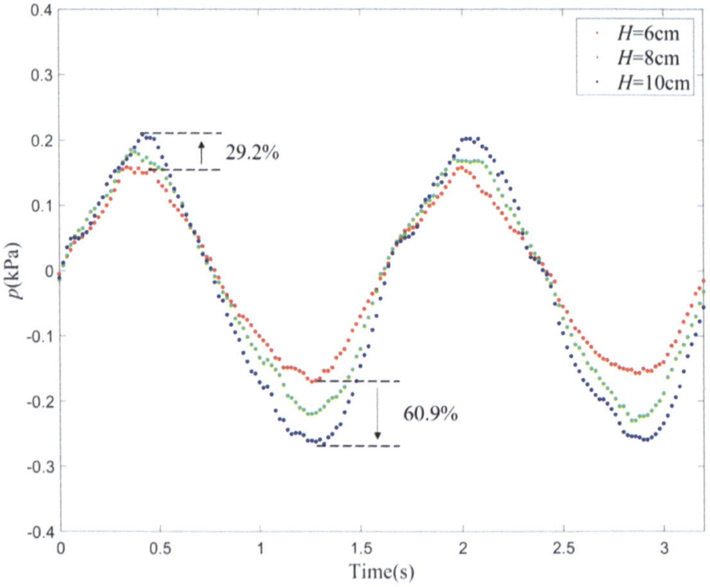

Fig. 3. Difference between positive and negative pressure

Variations in wave height directly influence pore pressure. As wave height increases from 6 cm to 10 cm, the amplitude of pore water pressure also notably rises. When distinguishing between positive and negative pore pressures, it becomes clear that the increase in positive pressure is less pronounced compared to the increase in negative pressure. Specifically, in this scenario, positive pressure rises by 29.2%, whereas negative pressure increases by 60.9%, roughly double the increment observed in positive pressure.

4 Influence of Shell Content

4.1 Impact on Pore Pressure Amplitude

Typically, alterations in seabed composition affect its permeability, thereby influencing fluctuations in pore pressure. A seabed with higher permeability experiences less attenuation of wave pressure as it propagates downwards from the seabed surface, whereas a less permeable seabed shows more pronounced attenuation. With constant wave parameters, the pore pressure amplitude in a highly permeable seabed exceeds that in a less permeable one. Specifically, under wave conditions with a height of H = 8 cm and a period of T = 1.6 s, the study compared changes in wave-induced pore pressure among seabed with different shell contents (SC) (see Fig. 4).

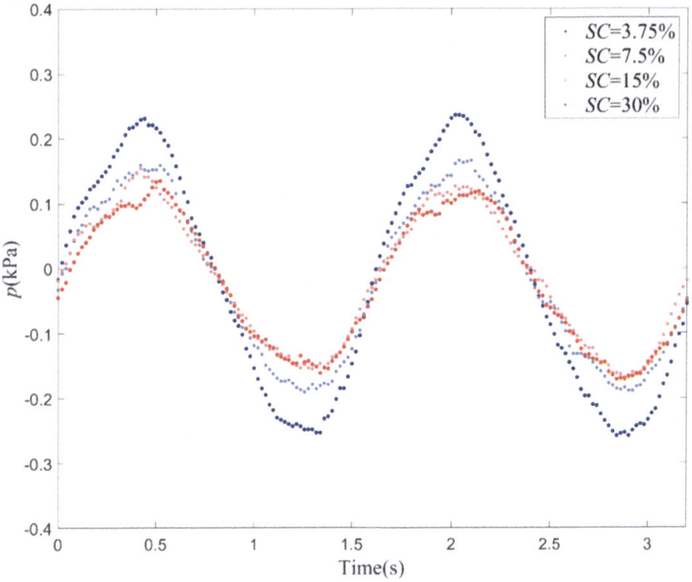

Fig. 4. Impact of SC on pore-pressure

As shell content increases, the amplitude of pore pressure in the seabed decreases, which is a notable observation. Typically, adding larger shell particles to sand would increase the seabed's permeability, resulting in larger pore pressure amplitudes. However, the experiments suggest that the seabed's permeability may actually decrease. Subsequent studies will aim to measure the seabed's permeability coefficient to verify this hypothesis. If confirmed, one possible explanation could be attributed to the unique shape of shell particles, which cannot simply be considered as larger spherical components. The mixture of smaller sand particles and irregularly shaped shell particles likely forms a different microstructure, potentially explaining the observed experimental phenomenon.

As shell content increases from SC = 3.75% to SC = 30%, the amplitude of pore pressure fluctuations in the seabed gradually decreases. This reduction is most pronounced when the shell content increases from SC = 3.75% to SC = 7.5%. Figure 5 illustrates the trend of pore pressure amplitude changes in the seabed across four different shell contents. It shows that both positive and negative pressures behave consistently under the influence of shell content. Specifically, the reduction in pore pressure amplitude is 25.4% from SC = 3.75% to SC = 7.5%, 16.9% from SC = 7.5% to SC = 15%, and 3.3% from SC = 15% to SC = 30%. Despite each increment in shell content being doubled, the effect on reducing pore pressure amplitude diminishes gradually, indicating a trend of diminishing marginal returns. This observation aligns with findings reported by Sui [17].

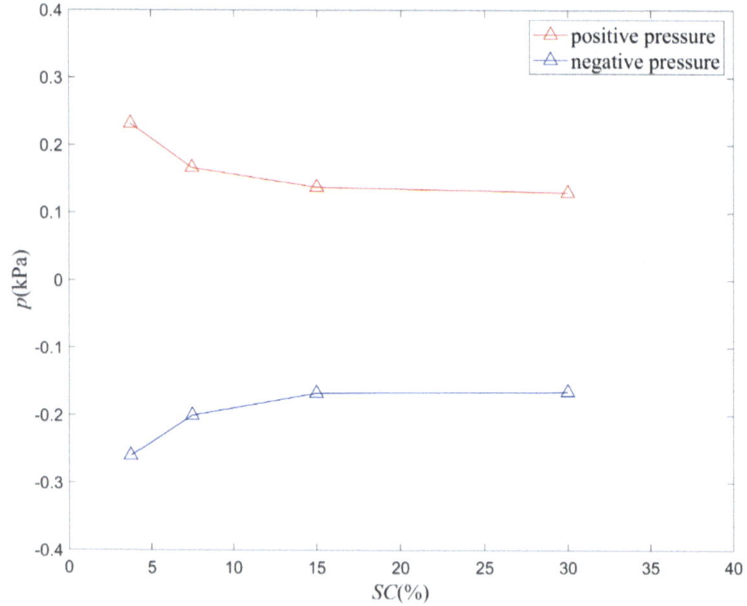

Fig. 5. Impact of SC on amplitude of pore-pressure

4.2 Impact on Pore Pressure Attenuation

Figure 4 compares pore pressure fluctuations at the same point under varying shell content. Additionally, we calculate the amplitude of pore pressure fluctuations at a specific point and conduct a comparative analysis of measurements at different depths and with different shell contents. Figure 6 illustrates the pore pressure amplitude at three different burial depths in section #4 under varying shell content. The lines are color-coded: dark yellow for 3.75%, light yellow for 7.5%, light green for 15%, and dark green for 30% shell content.

Given the data corresponds to different burial depths, normalization is necessary to facilitate comparisons on a consistent scale. In practical terms, since seabed depth is theoretically infinite, water depth serves as the spatial scale for normalization. Pore pressure measurements originate from wave pressure, which travels from the water surface to the seabed surface and through the seabed to the measurement point. Therefore, the pressure measurement value at the seabed surface is chosen as the normalization reference for pore pressure.

After normalizing pore pressure and burial depth, the horizontal axis represents the ratio of pore pressure amplitude at a point to the surface pressure of the seabed,

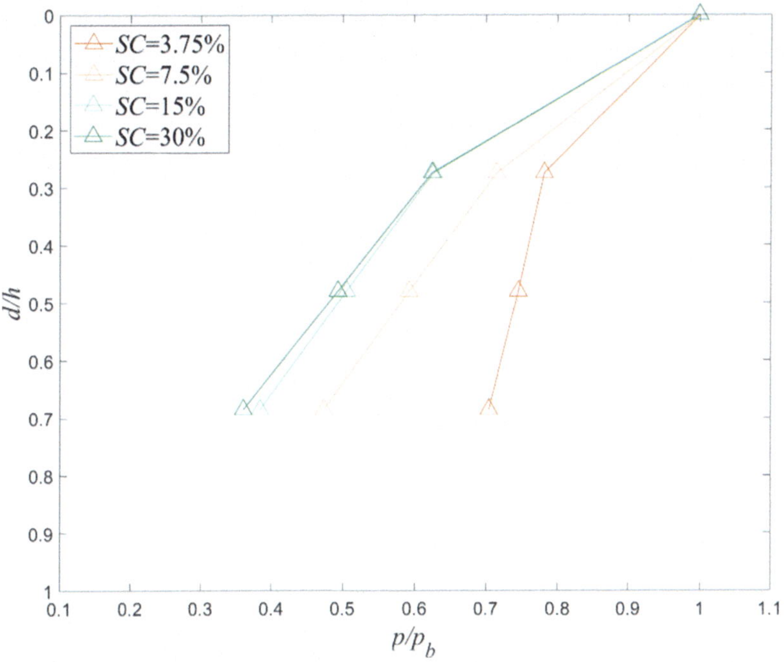

Fig. 6. Impact of SC on attenuation of pore pressure

while the vertical axis represents the ratio of burial depth to water depth. The slope of the graph reflects the seabed's ability, with varying shell content, to transmit pore pressure. A steeper slope indicates more efficient transmission of pore pressure to greater depths, implying minimal attenuation. Conversely, a gentler slope suggests significant impedance in pore pressure transmission by the seabed.

From the figure, it can be observed that under different shell contents, the pore pressure amplitude always decreases with increasing burial depth, but the degree of this attenuation varies with different shell contents. The attenuation of pore pressure amplitude is most significant from SC = 3.75% to SC = 7.5%, and a marginal effect appears at SC = 30%. This spatial characteristic is consistent with the results shown in Fig. 4 and Fig. 5.

5 Conclusion

A series of physical experiments were conducted to study the wave-induced response characteristics of a shell-sand mixed seabed in front of a sloping breakwater. The study analyzed how varying shell content influences the seabed's response to waves.

1. As the shell content increases, the amplitude of pore pressure in the seabed decreases. This phenomenon may be attributed to the unique shape of the shell particles, which could alter the seabed's microstructure and reduce its permeability.

2. Spatially, the pore pressure amplitude in the seabed exhibits a negative correlation with depth. Moreover, an increase in shell content accentuates the vertical attenuation of pore pressure amplitude.

Acknowledgements. We would like to acknowledge the financial support from National Key R&D Program of China (2023YFE0126300).

References

1. Yamamoto, T., Koning, H.L., Sellmeijer, H., et al.: On the response of a poro-elastic bed to water waves[J]. J. Fluid Mech. **87**(1), 193–206 (1978)
2. Madsen, O.S.: Wave-induced pore pressure and effective stresses in a porous bed[J]. Géotechnique. **28**(4), 377–393 (1978)
3. Biot, M.A.: General theory of three - dimensional consolidation. J. Appl. Phys. **12**(2), 155–164 (1941)
4. Seed, H., Rahman, M.S.: Wave-induced pore pressure in relation to ocean floor stability of cohesionless soils. Mar. Georesour. Geotechnol. **3**, 123–150 (1978)
5. Sumer, B.M., Ozgur Kirca, V.S., Fredsoe, J.: Experimental validation of a mathematical model for seabed liquefaction under waves. Int. J. Of Shore Polar Eng. **22**(2), 133–141 (2012)
6. Ye, J., Jeng, D.S., Wang, R., et al.: Numerical simulation of the waveinduced dynamic response of poro-elastoplastic seabed foundations and a composite breakwater. Appl. Math. Model. **39**, 322–347 (2015)
7. Chen, W.Y., Jeng, D.S., Chen, W., Chen, G.X., Zhao, H.Y.: Seismic-induced dynamic responses in a poro-elastic seabed: solutions of different formulations[J]. Soil Dyn. Earthq. Eng. **131** (2020)
8. Chen, X., Lei, Y., Zheng, X.Y.: Dynamic responses of a transversely isotropic poroelastic seabed subjected to nonlinear random waves with emphasis on liquefaction depth[J]. Ocean Eng. **286** (2023)
9. Zhu, B., Hirashi, T., Mase, H., Pei, H., Yang, Q.: Probabilistic analysis of wave-induced dynamic response in a poroelastic sloping seabed using random finite element method[J]. Ocean Eng. **252** (2022)
10. Wang, L., Zhang, J., Jeng, D., Zhang, Q., Chen, T.: Experimental study on the dynamic response of a silty seabed under waves[J]. Ocean Eng. **269** (2023)
11. Yamamoto, T.: Wave-induced pore pressures and effective stresses in homogenous seabed foundations[J]. Ocean Eng. **8**, 1–16 (1981)
12. Gatmiri, B.: A simplified finite element analysis wave-induced effective stresses and pore pressure in permeable seabed[J]. Géotechnique. **40**(1), 15–30 (1990)
13. Gatmiri, B.: Response of cross-anisotropic seabed to ocean waves[J]. J. Geotech. Eng. **118**(9), 1295–1314 (1992)
14. Jeng, D.S., Lin, Y.S.: Finite element modeling for water waves-soil interaction[J]. Soil Dyn. Earthq. Eng. **15**(5), 283–300 (1996)
15. Zhang, C., Sui, T., Zheng, J., et al.: Modelling wave-induced 3D non-homogeneous seabed response. Appl. Ocean Res. **61**, 101–114 (2016)
16. Li, X., Zhang, Z., Pan, E.: Wave-induced dynamic response in a transversely isotropic and multilayered poroelastic seabed[J]. Soil Dyn. Earthq. Eng. **139** (2020)
17. Sui, T., Kirca, O., Sumer, B.M., Carstensen, S., Fuhrman, D.R.: Wave induced seabed liquefaction in a mixture of silt and seashelss [J]. Coast. Eng. **178**, 104215 (2022)

18. Zhu, B., Yang, Q., Hiraishi, T., Mase, H., Pei, H., Wang, T., Lu, M.: A stochastic numerical model for wave-induced dynamic response in a spatially heterogeneous seabed around structures[J]. Ocean Eng. **288** (2023)
19. Zhu, B., Yang, Q., Wang, T., Lu, M., Mo, P., Hiraishi, T., Mase, H.: Investigation on wave-induced responses of heterogeneous seabed around a buried pipeline based on nonstationary random field model[J]. Comput. Geotech. **167** (2024)
20. Senjuntichai, T., Phulsawat, B., Keawsawasvong, S., Kaewjuea, W.: Dynamic impedances of multiple strips on multi-layered transversely isotropic poroelastic soils[J]. Forces Mech. **14** (2024)

High-Performance Dynamic Analysis Method of Top Tensioned Riser with Space-Time Adaptive Scheme

Hu Pengji and Liu Xiuquan(✉)

Centre for Offshore Engineering and Safety Technology, China University of Petroleum, Qingdao 266580, China
lxqmcae@163.com

Abstract. The finite element method (FEM), with consistent meshing and fixed time step, is commonly used to analyze the dynamic response of the top tensioned riser (TTR). The meshing and time step sizes in the dynamic response are often determined with empirical summary, which results in a loss of the riser element accuracy and analysis efficiency. A high-performance space-time adaptive (FEM) of the TTR is proposed to gain satisfactory meshing and time step. The dynamic analysis is divided with the adaptive method into space mesh adaptive and time step adaptive stage. The space mesh adaptive scheme is established considering riser dynamic characteristics to calculate the space discretization error and obtain a new meshing distribution in the space mesh adaptive stage. The time step adaptive scheme is established to evaluate the time discretization error at each time step with time error estimator and acquire the modified time step in the time step adaptive stage. Several cases are given verify the efficiency and accuracy of the proposed method. The proposed method can appropriately resolve the conflict between calculation accuracy and cost, and efficiently improve the dynamic analysis performance of the TTR.

Keywords: top tensioned riser · discretization error estimation · space-time adaptivity

1 Introduction

The top tensioned riser (TTR) is a crucial equipment connecting offshore floating structures with subsea wellhead in offshore oil and gas exploration and development [1, 2]. As a typical slender flexible structure, the TTR is vulnerable to various dynamic loads caused by waves and platform motions in the marine environment [3–4]. The dynamic response analysis is an indispensable step of the TTR structural design and safety evaluation [5].

Several numerical methods are developed for dynamic analysis such as finite element method and finite difference method. FEM is one of the most widely used methods to analyze the dynamic response of the TTR with its outstanding ability to handle complex riser profile and boundary conditions [6]. Extensive research has been carried out by scholars on the dynamic response of the TTR with FEM [7–10]. In these investigations,

D. Jeng and B. Cai (Eds.): ICCOE 2024, LNCE 423, pp. 118–128, 2025.
https://doi.org/10.1007/978-981-96-0181-3_12

a consistent, conservative mesh and time step are selected by experience in almost all simulations, which leads to a compromise of computational accuracy and efficiency in the dynamic analysis of the TTR [11, 12]. At present, a scientific, practical method to determine meshing and time step to maintain the accuracy of dynamic analysis and improve its efficiency is still lacking [13, 14]. The space mesh and time step adaptive method can optimize element size and adjust time step reasonably by space and time error estimation, which compensates for the shortcoming of the FEM caused by discretization [15, 16]. The space mesh and time step adaptive methods are the satisfactory choice to solve the above problem.

A new high-performance dynamic analysis method of the TTR based on space-time adaptive theory is studied in the paper. The main objective of the method is to maintain the dynamic analysis accuracy of the TTR and improve the analysis efficiency by selecting meshing and time step scientific.

2 Numerical Model of the TTR

The dynamic analysis program with the Newmark-β method is illustrated in Fig. 1.

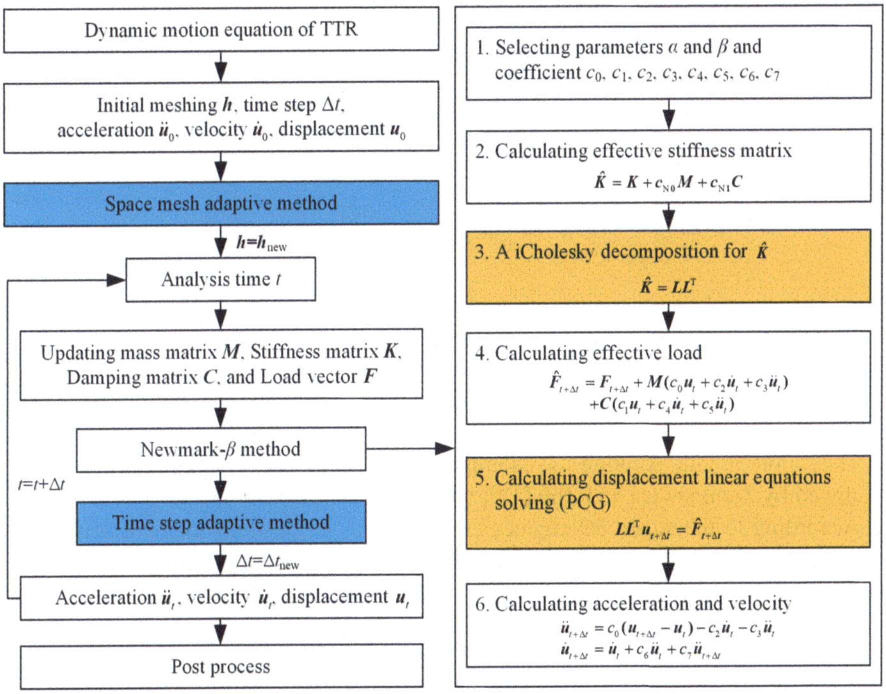

Fig. 1. Dynamic analysis program.

The numerical approximate solution obtained by FEM is widely accepted and applied to analyze the deformation and vibration of the TTR. The governing motion of the TTR can be described as

$$\mathbf{M\ddot{u}} + \mathbf{C\dot{u}} + \mathbf{Ku} = \mathbf{F} \tag{1}$$

where $\boldsymbol{M}$, $\boldsymbol{C}$, and $\boldsymbol{K}$ are the mass, damping, and stiffness matrix of the TTR, respectively; $\ddot{\mathbf{u}}$, $\dot{\mathbf{u}}$ and $\boldsymbol{u}$ are the acceleration, velocity, and displacement vectors of the TTR, respectively; $\boldsymbol{F}$ is the load vector of the TTR.

The space mesh and time step adaptive methods are introduced into the program to optimize the meshing and time step. These two methods are elaborated in the next section.

3 Space-Time Adaptive Method

3.1 Space Mesh Adaptive Method

The space mesh adaptive method includes reliable estimator of the space discretization error and the corresponding adaptive scheme. Space discretization error (SDE) is caused by the element discretization of the TTR model. The Z^2 method is utilized to establish the SDE estimator of the TTR [17]. The energy norm is used as the index of the SDE estimation in the Z^2 method, which is defined as

$$\|\mathbf{e}\|_{\mathrm{s}} = \left[\int_{\Omega} (\mathbf{e}_{\sigma})^{\mathrm{T}} \mathbf{D}^{-1} \mathbf{e}_{\sigma} \mathrm{d}\Omega\right]^{1/2} \tag{2}$$

$$\mathbf{e}_{\sigma} = \boldsymbol{\sigma} - \boldsymbol{\sigma}^{\mathrm{h}} \approx \boldsymbol{\sigma}^{*} - \boldsymbol{\sigma}^{\mathrm{h}} \tag{3}$$

$$\boldsymbol{\sigma}^{\mathrm{h}} = \mathbf{DBu}_{\mathrm{e}} \tag{4}$$

where Ω is the solving domain; σ and σ^{h} are the stress exact solution and numerical solution of the TTR, respectively; σ^{*} is the substitute value of the stress exact solution σ; $\boldsymbol{e}_{\sigma}$ is the error between the exact solution and the numerical solution; $\boldsymbol{D}$ is the riser elasticity matrix; $\boldsymbol{B}$ is the strain matrix operator; $\boldsymbol{u}_{\mathrm{e}}$ is the displacement vectors of the riser element.

The approximate solution is calculated with superconvergence curve method developed by Zienkiewicz and Zhu [18], which is illustrated in Fig. 2.

According to the superconvergence characteristic of the finite element solution, a smooth curve is near the real stress field of the riser element. The smooth stress curve in each riser element can be written as

$$\sigma^{\mathrm{s}} = ax^2 + bx + c \tag{5}$$

where x is the node value; a, b, and c are the curve coefficients.

The curve coefficients are determined by minimizing the objective function F_{m}.

$$F_{\mathrm{m}} = \sum_{\mathrm{i}=1}^{4} \left(\sigma^{\mathrm{h}}(x_{\mathrm{i}}) - \left(ax_{\mathrm{i}}^2 + bx_{\mathrm{i}} + c\right)\right)^2 \tag{6}$$

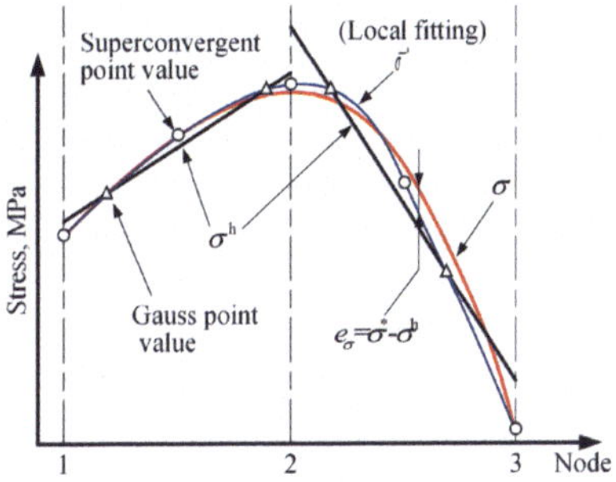

Fig. 2. Superconvergence characteristic and local fitting.

$$(a, b, c) = \min\{F_m\} \tag{7}$$

where x_i is the Gauss point location.

The stress approximate solution σ^* is then gained by interpolation method, and the approximate error e_σ of the TTR is further obtained, which are expressed as

$$\boldsymbol{\sigma}^* = \mathbf{P}\boldsymbol{\sigma}^s \tag{8}$$

where $\boldsymbol{P}$ is the interpolation function.

The space mesh adaptive scheme is employed to adjust the element size of the TTR after the energy norm is calculated. The h-adaptive method is used to adjust the riser element size and optimize the results of the dynamic response. Relative error η_s is generally used as the precision index in the riser element size regeneration, which is defined as

$$\eta_s = \|\mathbf{e}\|_s / \|\mathbf{u}\|_s \tag{9}$$

where $\|\boldsymbol{u}\|_s$ and $\|\boldsymbol{e}\|_s$ is the energy norm for the displacement and its error of the TTR, respectively.

The permissible upper limit of the energy norm $\|\boldsymbol{e}\|_{s\_max}$ each riser element is expressed as

$$\|\mathbf{e}\|_{s\_max} \le \eta_{sp} \left(\frac{\|\mathbf{u}\|_s^2 + \|\mathbf{e}\|_s^2}{M} \right)^{1/2} \tag{10}$$

where η_{sp} is the specified SDE permissible value; M is the number of the riser element.

Mesh refinement should be performed if the current relative error of the riser element η_s is greater than η_{sp}. The current riser elements need to be refined into several equal segments in the mesh refinement strategy. The mesh refinement coefficient ξ of the TTR and the new element size can be written as [19]

$$\boldsymbol{\xi}_i = \|\mathbf{e}\|_{s\_i} / \|\mathbf{e}\|_{s\_max}, \ i = 1, 2, \ldots, M \tag{11}$$

$$\mathbf{h}_{new} = \mathbf{h}_c / \boldsymbol{\xi}^{1/p} \tag{12}$$

where $\|e\|_{s\_i}$ is the energy norm of the ith riser element; $\boldsymbol{h}_c$ is the current riser element size distribution, and p is the order of the polynomials used in the shape functions.

The mesh coarse strategy of the TTR is that two adjacent riser elements are merged if the energy norm of the new element is less than the permissible minimum. The permissible lower limit and riser element coarsening of energy norm can be defined as [20]

$$\|\mathbf{e}\|_{s\_min} = \lambda\|\mathbf{e}\|_{s\_max}, \lambda \in (0, 1) \tag{13}$$

$$\|\mathbf{e}\|^2_{s\_min} \leq \|\mathbf{e}\|^2_{s\_i\_new} = \|\mathbf{e}\|^2_{s\_i} + \|\mathbf{e}\|^2_{s\_i+1} \leq \|\mathbf{e}\|^2_{s\_max} \tag{14}$$

Meanwhile, the limiting refinement coefficient ξ_{refine} and limiting coarse coefficient ξ_{coarse} are defined to prevent unlimited mesh refinement or coarsening.

3.2 Time Step Adaptive Method

Time discretization error (TDE) is derived from the approximate of the numerical integration method. The Newmark-β method is utilized to analyze the dynamic response of the TTR in this paper, and its specific error estimator is used to obtain the TDE of the TTR [21]. The estimator takes local truncation error as the estimation index of TDE, which is illustrated in Fig. 3.

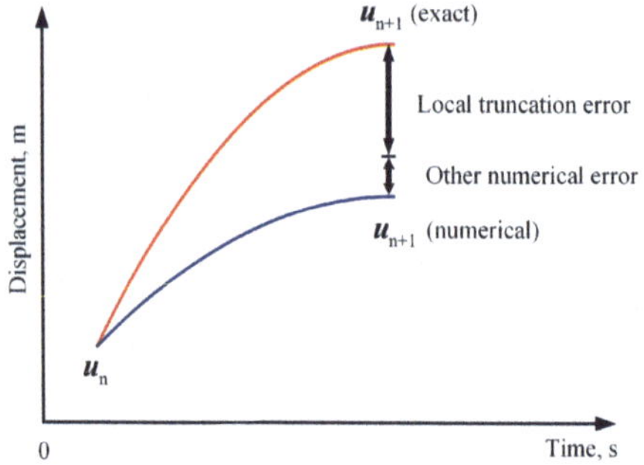

Fig. 3. Local truncation error.

The local truncation error of the Newmark-β method is written as

$$\mathbf{e}_t = \frac{1}{2}\Delta t^2\left(2\beta - \frac{1}{3}\right)(\ddot{\mathbf{u}}_{n+1} - \ddot{\mathbf{u}}_n) \tag{15}$$

where Δt is the time step, β is a coefficient of the Newmark-β method, $\ddot{\mathbf{u}}_{n+1}$ is the acceleration at the $(n + 1)$th time step, and $\ddot{\mathbf{u}}_n$ is the acceleration at the nth time step.

The relative TDE is commonly used as an estimation index, which can be defined as

$$\eta_t = \frac{\|\mathbf{e}_t\|_{L_2}}{\|\mathbf{u}\|_{L_2}} = \frac{\frac{1}{2}\Delta t^2\left(2\beta - \frac{1}{3}\right)\|\ddot{\mathbf{u}}_{n+1} - \ddot{\mathbf{u}}_n\|_{L_2}}{\|\mathbf{u}\|_{L_2}} \tag{16}$$

where $\|\boldsymbol{u}\|_{L2}$ is the L_2 norm of displacement $\boldsymbol{u}$.

The new time step is written as [22]

$$\Delta t_{\text{new}} = \alpha \left(\frac{\eta_{\text{tp}}}{\eta_{\text{t}}} \right)^{\frac{1}{3}} \Delta t_{\text{old}} \quad (17)$$

where α is a safety coefficient, $\alpha \in (0,1)$; η_{tp} is the permissible upper limit of TDE; Δt_{old} is the time step before changing.

The frequent change of time step has a negative effect on dynamic response analysis. Consequently, the time step control needs to be relaxed, which is shown as

$$\gamma_1 \eta_{\text{tp}} \leq \eta_{\text{t}} \leq \gamma_2 \eta_{\text{tp}} \quad (18)$$

where γ_1, γ_2 is the time step control coefficient.

Similar to the space mesh adaptive scheme, an allowable range is defined to prevent an extreme time step size. Consequently, the new time step can neither be more than twice nor less than half the original time step, which can be expressed as

$$\Delta t_{\text{new}} = \min\left\{ \max\left\{ \frac{1}{2} \Delta t_{\text{old}}, \Delta t_{\text{new}} \right\}, 2\Delta t_{\text{old}} \right\} \quad (19)$$

4 Results and Discussion

A TTR numerical model is established to study the effect of STA-FEM on the accuracy and efficiency of dynamic analysis. The parameters of TTR are listed in Table 1. The parameters of STA-FEM are determined in Table 2.

Table 1. The parameters of TTR.

Parameters	Value	Parameters	Value
Length of riser system, m	1508.72	External diameter of risers, m	0.5334
Length of each riser joint, m	22.86	Thickness of risers, m	0.0254
Top tension coefficient of TTR	1.2	Density of risers, kg/m^3	7850
Elasticity modulus of TTR, GPa	210	Dry weight of riser joint, kg	24,869
Density of seawater, kg/m^3	1025	Wet weight of riser joint, kg	2563

4.1 Accuracy Evaluation of STA-FEM

The dynamic analysis with STA-FEM is subsequently conducted. The accuracy of STA-FEM is expressed by SDE and TDE. The element size distribution with STA-FEM is depicted in Fig. 4. Mesh refinement is carried out at the top and bottom of the TTR after

Table 2. The parameters of STA-FEM.

Parameters	Value	Parameters	Value
Initial elements number per riser	20		
Upper limit of SDE η_{sp}	0.001	Initial time step Δt	0.02
Mesh refinement coefficient ξ_{refine}	4	Upper limit of TDE η_{tp}	$2 \times 10 - 8$
Mesh coarse coefficient ξ_{coarse}	2	Time step control coefficient γ_1, γ_2	0.8, 1.2
SDE coefficient λ	0.98	Time adaptive safety coefficient α	1

adaptive meshing, whereas mesh coarsening is performed at the middle. A large stress gradient is at the top and bottom of the TTR due to the boundary conditions defined, which results in a large SDE. The smaller element sizes are used in these locations to keep SDE within the allowable range. By contrast, larger element sizes can be used to meet the requirement on SDE at the middle part of the TTR with low stress gradients. The SDE of the TTR with STA-FEM is shown in Fig. 5. The SDE of each riser element is more reasonable than those of the fixed mesh and within the permissible range after applying STA-FEM. However, the SDE of some elements at the top and bottom of the TTR still exceeds its allowable upper limit because the definition of refinement coefficient ξ_{refine} prevents the riser element size from becoming smaller.

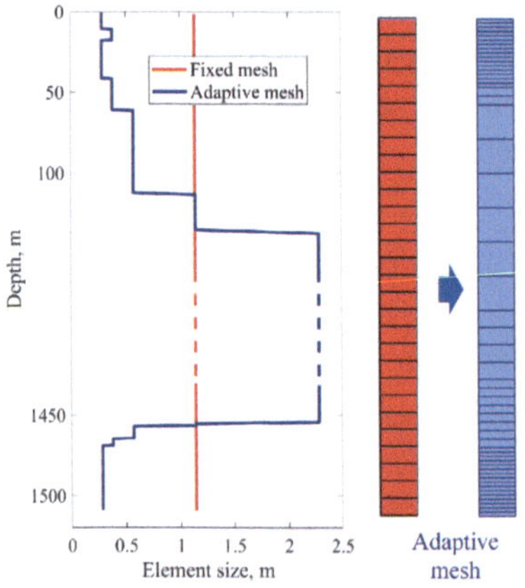

Fig. 4. Element size distribution with STA-FEM.

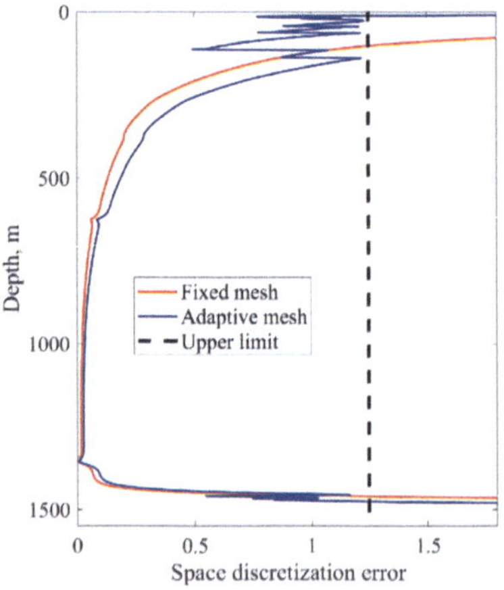

Fig. 5. Space discretization error with STA-FEM.

The TDE with traditional FEM is depicted in Fig. 6. The TDE is positively correlated with the time step. The TDE is obtained by calculating the L_2 norm of the difference between two acceleration vectors of the adjacent time step. The acceleration history of the TTR has evident periodicity due to the influence of cyclic loads such as wave and platform movement. Therefore, the TDE of the riser system in dynamic analysis has the same periodic characteristics. The TDE with STA-FEM is shown in Fig. 7. The time step is adjusted according to time step adaptive scheme once TDE exceeds the defined permissible range. The TDE corresponding to each time step is within the allowable range after adaptive iteration optimization.

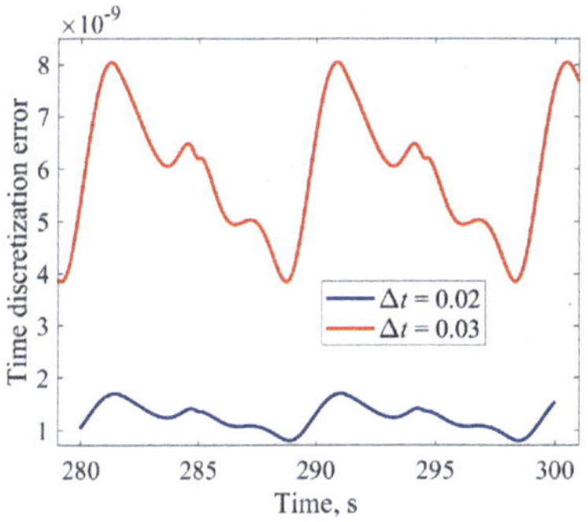

Fig. 6. Time discretization error with traditional FEM.

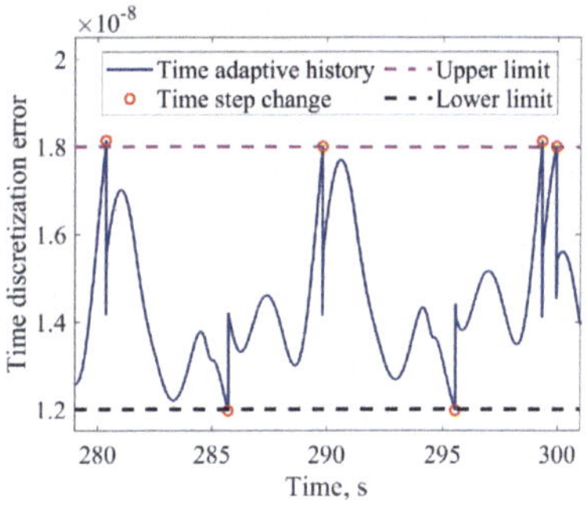

Fig. 7. Time discretization error with STA-FEM.

4.2 Efficiency Evaluation of STA-FEM

Several dynamic analysis cases are given to evaluate the time consuming of STA-FEM, using a computer system with the configuration of 8GB RAM, Intel(R) Core™ i5-10,400 CPU. The simulation time is defined as 1200 s, and Case0 is a reference case with traditional FEM. The parameters of the STA-FEM are listed in Table 3. The CPU time and proportion of the STA-FEM in each case are listed in Table 4. The proportion of time saved by STA-FEM are shown in Fig. 8. The cases show that STA-FEM can remarkably improve the efficiency of the dynamic response analysis of the TTR, whereas the space-time adaptive method takes up less than 3% of total time in dynamic analysis. The space and time adaptive parameters listed in Table 3 can be used as the reference value of the dynamic response analysis of the TTR with STA-FEM.

Table 3. The parameters of the STA-FEM for analysis cases.

Cases	η_{sp}	λ	η_{tp}	γ_1, γ_2
Case 0	–	–	–	–
Case 1	0.001	0.98	1×10^{-8}	0.8, 1.2
Case 2	0.003	0.98	1×10^{-8}	0.8, 1.2
Case 3	0.005	0.98	1×10^{-8}	0.8, 1.2
Case 4	0.007	0.98	1×10^{-8}	0.8, 1.2
Case 5	0.001	0.90	1×10^{-8}	0.8, 1.2
Case 6	0.001	0.92	1×10^{-8}	0.8, 1.2
Case 7	0.001	0.95	1×10^{-8}	0.8, 1.2

(continued)

Table 3. (*continued*)

Cases	η_{sp}	λ	η_{tp}	γ_1, γ_2
Case 8	0.001	0.98	2×10^{-8}	0.8, 1.2
Case 9	0.001	0.98	3×10^{-8}	0.8, 1.2
Case 10	0.001	0.98	4×10^{-8}	0.8, 1.2
Case 11	0.001	0.95	1×10^{-8}	0.85, 1.15
Case 12	0.001	0.95	1×10^{-8}	0.9, 1.1
Case 13	0.001	0.95	1×10^{-8}	0.95, 1.05

Table 4. CPU time and proportion of STA-FEM.

Cases	Space mesh adaptive (%)	Time step adaptive (%)	CPU time (s)	Improved efficiency (%)
Case 0	–	–	307.285	–
Case 1	1.122%	1.168%	165.466	46.152%
Case 2	1.320%	1.135%	121.384	60.498%
Case 3	1.399%	1.176%	110.976	63.885%
Case 4	1.411%	1.094%	107.898	64.887%
Case 5	1.402%	0.047%	132.514	56.876%
Case 6	1.386%	0.043%	132.598	56.848%
Case 7	1.132%	1.204%	165.968	45.989%
Case 8	1.705%	0.040%	106.284	65.412%
Case 9	1.952%	0.041%	93.591	69.542%
Case 10	2.084%	0.042%	87.689	71.463%
Case 11	1.360%	0.038%	136.437	55.599%
Case 12	1.401%	0.039%	130.879	57.408%
Case 13	1.451%	0.040%	126.880	58.709%

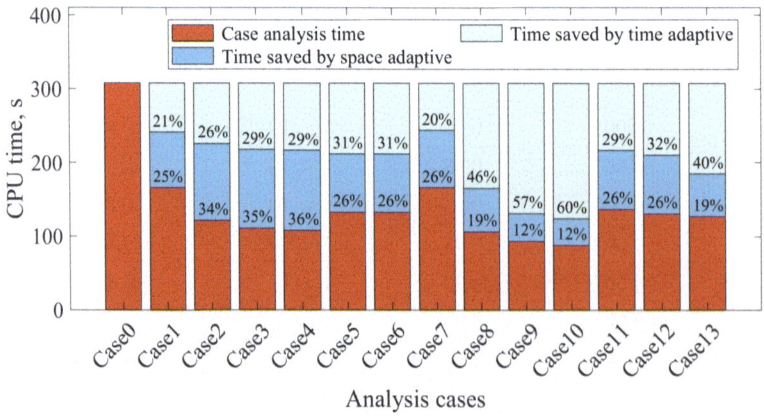

Fig. 8. Proportion of time saved by STA-FEM.

5 Conclusions

In this paper, a high-performance dynamic analysis method of the TTR is proposed with space-time adaptive (STA) theory. The governing equation of the TTR is established by FEM and solved by Newmark-β method. The SDE and TDE estimators are established to obtain the space mesh and time step error during the dynamic analysis. The space mesh adaptive scheme is developed based on the space error estimator to obtain an optimal element size distribution. While the time step adaptive scheme is developed based on the time error estimator to obtain a suitable time step. The STA-FEM analysis program is then proposed base on the space mesh and time step adaptive schemes. The proposed method has definite mechanism, which is fundamentally different from the empirical method that uses consistent element size distribution and fixed time step size.

The performance of the proposed STA-FEM is further explored by case analyses. Dynamic analyses of the TTR with FEM and with STA-FEM are performed respectively. The analysis results obtained by the two methods are compared to comprehensively evaluate the accuracy of STA-FEM. The results indicate that the element size distribution and time step are adjusted after space-time adaptive optimization while SDE and TDE are constrained within the allowable range. Several cases are given to evaluate the CPU time of STA-FEM. The developed method can substantially improve the efficiency of the dynamic analysis of the TTR. The developed method can maintain the analysis accuracy of the TTR and reduce the computational cost properly.

References

1. Chang, Y., Zhang, C., Wu, X., et al.: A Bayesian network model for risk analysis of deepwater drilling riser fracture failure. Ocean Eng. **181**, 1–12 (2019)
2. Qiu, N., Liu, X., Liu, Z., et al.: Mechanical model and characteristics of deep-water drilling riser-wellhead system under internal solitary waves. Mar. Struct. **90**, 103439 (2023)
3. Liu, X., Wang, X., Jiang, Y., et al.: An optimization method for the suppression device configuration of risers. Ocean Eng. **194**, 106619 (2019)
4. Ragheb, H.A., Goodridge, M., Pham, D.C., et al.: Extreme response based reliability analysis of composite risers for applications in deepwater. Mar. Struct. **78**, 103015 (2021)
5. Tian, D., Fan, H., Leira, B.J., et al.: Study on the static behavior of installing a deep-water drilling riser on a production platform. J. Pet. Sci. Eng. **185**, 106652 (2020)
6. Yu, Y., Xu, S., Yu, J., et al.: Dynamic analysis of steel catenary riser on the nonlinear seabed using vector form intrinsic finite element method. Ocean Eng. **241**, 109982 (2021)
7. Ni, P., Li, J., Han, H., et al.: Stochastic dynamic analysis of marine risers considering fluid-structure interaction and system uncertainties. Eng. Struct. **198**, 109507 (2019)
8. Liu, X., Li, C., Zhang, N., et al.: Improved axial dynamic analysis of risers based on finite element method and data-driven models. Ocean Eng. **214**, 107782 (2020)
9. Wang, X., Liu, X., Zhang, N., et al.: Improved recoil dynamic analysis of the deepwater riser system after emergency disconnection. Appl. Ocean Res. **113**, 102719 (2021)
10. Zhou, X., Li, Y., Chen, J., et al.: Numerical evaluation of the coupled/uncoupled effects of lower marine riser package and/or blowout preventer with a lower flex joint in a hang-off drilling riser. Ocean Eng. **266**, 113004 (2022)
11. Monsalve-Giraldo, J.S., Cortina, J.P.R., Sousa, F.J.M., et al.: Hybrid Parabolic Interpolation-Artificial Neural Network Method (HPI-ANNM) for long-term extreme response estimation of steel risers. Appl. Ocean Res. **76**, 221–234 (2018)

12. Chen, R., Low, Y.M.: Efficient long-term fatigue analysis of deepwater risers in the time domain including wave directionality. Mar. Struct. **78**, 103002 (2021)
13. Kim, J.D., Jang, B., Yun, R.: Application of analytical model in the prediction of dynamic responses and fatigue damage of flexible risers: Part I - Improvement of analytical model considering shear deformation and varying tension effects. Mar. Struct. **79**, 103044 (2021)
14. Hejazi, R., Grime, A., Randolph, M., et al.: An efficient probabilistic framework for the long-term fatigue assessment of large diameter steel risers. Appl. Ocean Res. **118**, 102941 (2022)
15. Ahmed, N., John, V.: Adaptive time step control for higher order variational time discretizations applied to convection-diffusion-reaction equations. Comput. Methods Appl. Mech. Eng. **285**, 83–101 (2015)
16. Ngo, L.C., Choi, H.G.: A multi-level adaptive mesh refinement for an integrated finite element/level set formulation to simulate multiphase flows with surface tension. Comput. Math. Appl. **79**(4), 908–933 (2020)
17. Zienkiewicz, O.C., Zhu, J.Z.: A simple error estimator and adaptive procedure for practical engineering analysis. Int. J. Numer. Methods Eng. **24**(2), 337–357 (1987)
18. Zienkiewicz, O.C., Zhu, J.Z.: Adaptivity and mesh generation. Int. J. Numer. Methods Eng. **32**(4), 783–810 (1991)
19. Dutta, A.: Adaptive finite element analysis of structures subjected to transient dynamic loads using time marching scheme. Comput. Struct. **80**(27), 2313–2319 (2002)
20. Ray, T., Sinha, R.K.: An adaptive finite element method for parabolic interface problems with nonzero flux jumps. Comput. Math. Appl. **82**, 97–112 (2021)
21. Zienkiewicz, O.C., Xie, Y.M.: A simple error estimator and adaptive time stepping procedure for dynamic analysis. Earthq. Eng. Struct. Dyn. **20**(9), 871–887 (1991)
22. Wang, Y., Xue, T., Tamma, K.K., et al.: A three-time-level a posteriori error estimator for GS4-2 framework: adaptive time stepping for second-order transient systems. Comput. Methods Appl. Mech. Eng. **384**, 113920 (2021)

Research Progress on Wave Transmission Coefficient of Porous Rubble Mound Breakwater

Zanqiang Liu(✉)

China Institute of Water Resources and Hydropower Research, Beijing 100038, China
liuzanqiang@163.com

Abstract. Rubble mound breakwaters, typically made of core rock and armor blocks, are porous structures through which a portion of waves and flow can penetrate. These breakwaters can serve as enclosure structures for the lower reservoir of seawater pumped-storage power stations, and it's important to limit the transmitted wave height to ensure safe operation. The wave transmission coefficient of rubble mound breakwaters is a critical research area with both theoretical and practical significance. This paper describes research findings on wave transmission coefficients, summarizes empirical formulas for the transmission coefficient, and analyzes various factors that influence the wave transmission coefficient. These insights can be valuable for future reference in engineering design and scientific research.

Keywords: Rubble mound breakwater · Seawater pumped-storage power station · Transmission coefficient · Empirical formula · Influence factors

1 Introduction

Waves are a crucial environmental factor for maritime structures as they have a significant impact. A breakwater is a man-made structure designed to reduce wave energy and protect harbors, anchorages, marinas, coastlines, and other areas. Conventional breakwaters, like rubble mound breakwaters, offer advantages such as simplicity, easy construction, and wide applicability. However, these structures are impermeable, which can lead to water pollution and harm the ecological environment (Agarwala and Saengsupavanich 2023). Therefore, breakwaters should be designed to be permeable for environmental protection or to meet specific flow requirements, such as for the lower reservoir of a seawater pumped-storage power station.

Seawater pumped-storage power stations, located in coastal areas with the lower reservoir being the ocean, need to limit wave height when the power station units are running to ensure water intake safety. In Japan, the only seawater pumped-storage power station used a rubble mound breakwater as the enclosure structure of the lower reservoir (OMOTO 1994; TANAKA 2004). This structure is both the intake and outlet, and its porosity allows for water exchange.

D. Jeng and B. Cai (Eds.): ICCOE 2024, LNCE 423, pp. 129–134, 2025.
https://doi.org/10.1007/978-981-96-0181-3_13

Rubble mound breakwaters, typically made of core stones and armor blocks, are commonly used in port and coastal engineering. However, if soil bags or small gravels are used as core stones, the water exchange rate is limited. Increasing the size of core stones is an effective way to improve water quality and increase the water exchange rate, but it can also generate transmitted waves that may disrupt the tranquility of a harbor.

The transmission coefficient K_t is defined as:

$$K_t = H_t/H_i \tag{1}$$

The incident and transmitted wave heights are denoted as H_i and H_t, respectively. Research indicates that the wave transmission coefficient (K_t) is influenced by factors such as H_i, water depth, structural scale, and the size of core stones. Understanding the wave transmission coefficient of rubble mound breakwaters is valuable for engineering applications and theoretical research.

Previous studies have primarily used percolation theory and empirical formulas derived from physical model tests and numerical simulations. These formulas are convenient for practical engineering use. This paper introduces several empirical formulas and analyzes the factors influencing wave transmission.

2 Empirical Formulas of Wave Transmission

Atsushi Numata (1975) conducted a study on the transmission of waves in breakwaters under regular wave conditions without overtopping. This study was carried out using a 55 m long, 1.5 m high and 1m wide laboratory wave flume. The model breakwater used quadripod armor units. An empirical formula was developed to calculate the transformation coefficient based on the analysis of test data.

$$K_t = \frac{1}{\left[1 + 1.135(B_{swl}/D_s)^{0.66}(H_i/L)^{0.5}\right]^2} \tag{2}$$

$$D_s = 2 \times \sqrt[3]{\frac{3M}{4\pi\rho_s}} \tag{3}$$

where:

B_{swl} = width of breakwater at SWL (still water level)
L = wave length
ρ_s = mass density of armour rock
D_s = nominal diameter of rock
M = average mass

Wave attenuation in rubble mound breakwater was studied by Stephenson (1979). He considered the flow to be in the square region of flow resistance. The wave transmission coefficient formulas without overtopping were then provided.

$$K_t = \frac{1}{\sqrt{1 + \mu B H_i^2/2D_s d^2}}, \tag{4}$$

where:

μ = resistance coefficient of water flow in breakwater
B = mean breadth of breakwater body under SWL
d = water depth

According to the experimental result of Stephenson (1979), μ was defined by:

$$\mu = \frac{800}{\mathrm{Re}} + \mu_t, \tag{5}$$

where Re is the Reynolds number, and μ_t takes on different values according to the shape of the stone: $\mu_t = 1$ for smooth pebbles, $\mu_t = 2$ for gravel, and $\mu_t = 3$ for crushed stone.

In a study conducted by Ahrens J. P. in (1987), model tests were carried out using a laboratory wave flume to investigate stone stability, wave transmission, reflection, and wave energy dissipation of a rubble mound breakwater under irregular waves without overtopping. The model breakwater consisted of representative and uniform stones. The wave transmission coefficient formulas were as follows:

$$K_t = \frac{1}{1 + \left(\frac{H_i A_t}{L D_s^2}\right)^{0.592}}, \tag{6}$$

where A_t = cross section area of breakwater under SWL.

The formula for calculating the wave transmission coefficient was derived by Xi (1988) based on the relationship between the energy loss and the rate of change of the wave energy current. This formula demonstrates a theoretical connection between the wave transmissibility coefficient and various factors, including the original wave height (H_i), relative depth (d/L), wave steepness (H_i/L), stone diameter (D_s), width of the breakwater body (B), and more. The calculated values from the formula align closely with experimental results. The formula is as follows:

$$K_t = \frac{1}{1 + \frac{\mu m_d H_i B}{L D_s}}, \tag{7}$$

where:

$$m_d = \frac{\pi}{5\sqrt{2}} \frac{\tanh kd (\sinh 2kd)^{2.5}}{\sinh^3 kd \left(1 + \frac{2kd}{\sinh 2kd}\right)\left(1 - e^{-2kd}\right)} \tag{8}$$

Wang (2008) conducted an experimental study on the impact of core material within rubble-mound breakwaters on wave transmission, according to typical section structure of rubble-mound breakwater. The study discussed the relationship between wave height, permeable core material, incident wave element and breakwater width. Wang also proposed an empirical formula of the transmitted coefficient K_t in terms of $H_i/(gT^2)$ and B_{swl}/D_s under different core material.

$$K_t = 0.018\left(100\frac{H_i}{gT^2}\right)^{-0.58}\left[3.9 - \log\left(\frac{B_{swl}}{D_s}\right)\right], \tag{9}$$

where T = incident wave period.

In line with related research, it has been found that long period waves can more easily pass through a rubble mound breakwater. The transmission of long period waves in a rubble mound breakwater was investigated in a laboratory wave flume by Ge (2014), who derived an empirical formula.

$$K_t = \frac{1}{1 + \left(100\frac{H_i}{gT^2}\right)^{0.75}\left(\frac{B}{D_s}\right)^{0.77}(kd)^{-0.59}} \tag{10}$$

Feng (2014) conducted a study on the influence of the core stone properties on wave propagation for rubble mound breakwaters in deep water using regular waves in physical experiments. The study discussed the relations between the transmission coefficient of the breakwater, the period and height of the incident wave, and the weight distribution of the core stone, without overtopping. Additionally, an empirical formula for the transmission coefficient as a function of dimensionless parameters H_i/d, gT^2/d and B_{swl}/D_{50} was derived.

$$K_t = 7 \times 10^3\left(\frac{H_i}{d}\right)^{-0.209}\left(\frac{gT^2}{d}\right)^{1.315}\left(\frac{B_{swl}}{D_{50}}\right)^{-0.255} \tag{11}$$

where D_{50} = median diameter of core stones.

The influence of the core stone properties on the propagation of irregular waves in deep water was studied by Cao in (2015) through physical model tests. Statistical analysis was used to examine the regularities of irregular waves. By establishing relationships between the transmission coefficient of the breakwater, the period and height of the incident wave, and the weight distribution of the core stone, an empirical formula for the transmission coefficient of irregular waves was proposed.

$$K_t = 7.4 \times 10^3\left(\frac{H_i}{d}\right)^{-0.089}\left(\frac{gT^2}{d}\right)^{0.904}\left(\frac{B_{swl}}{D_{50}}\right)^{-0.268} \tag{12}$$

The values of H_i/d, gT^2/d and B_{swl}/D_{50} were proposed to be $0.25 \leq H_i/d \leq 0.5$, $39.516 \leq gT^2/d \leq 172.051$ and $73.9 \leq B_{swl}/D_{50} \leq 85.0$.

The incident wave height H_i and period T, corresponded to the significant wave height and period of irregular waves in Eq. (11).

Sun (2018) analyzed the relationship between the ratio of block diameter of core material to the equivalent width of dike D_s/B_{swl} and the dimensionless wave steepness $H_i/(gT^2)$ through a series of sectional model tests under regular waves. An improved empirical formula of the transmission coefficient of permeable breakwater was proposed based on previous researches. The formula to be used to calculate the transmission coefficient of the permeable breakwater was given by:

$$K_t = 0.042 \times \left(\frac{H_i}{gT^2}\right)^{-0.47}\left(\frac{D_s}{B_{swl}}\right)^{0.63} \tag{13}$$

The values of the $H_i/(gT^2)$ and D_s/B_{swl} were proposed to be $0.041 \leq 100H_i/(gT^2)$ ≤ 0.421, $0.04 \leq D_s/B_{swl} \leq 0.086$ in Eq. (12).

3 Applicability and Comparison of the Empirical Formulas

In the study conducted by Ge in (2014), an experimental investigation of the transmission of long period waves by a rubble mound breakwater was performed. The results were compared with calculations from Eqs. (2), (3), (5), (6), and (8). The findings revealed that the results of Eq. (2) underestimated the experimental values due to differences in the type of core stones used. Additionally, the head loss caused by quadripod armor units was found to be greater than that caused by crushed stones used in the study by Ge (2014). Equation (3) yielded results approximately 3–4 times higher than the experimental values by Xi (1988), Ge (2014), and Sun (2018). This discrepancy was attributed to the assumption of a circular motion track of water particles by Stephenson (1979), which was found to be quasi-ellipsoid in shallow coastal areas. Furthermore, Stephenson (1979) only considered wave energy in half of the water depth, leading to deviations between the calculation results of Eq. (5) and the experimental values of Ge (2014) and Sun (2018). The discrepancies may be related to the cross-section area of the breakwater under SWL. On the other hand, Sun (2018) demonstrated that the results from Eqs. (8) and (9) aligned well with the experimental findings, indicating the applicability of Eq. (12).

4 Influence Factors of Transmission Coefficient of Rubble Mound Breakwater

The researcher mentioned earlier proposed several formulas for wave transmission coefficients based on theoretical analysis and model tests. The factors influencing the transmission coefficients of rubble mound breakwaters were analyzed.

1. The wave transmission coefficient of rubble mound breakwaters is related to the incident wave height, wave period, size of core stones, and width of the breakwater at SWL. In other words, the wave transmission coefficient is linked to dimensionless parameters $H_i/(gT^2)$ and D/B_{swl}.
2. The wave transmission coefficient (K_t) shows a decreasing trend with an increase in $H_i/(gT^2)$ and an increasing trend with an increase in D/B_{swl}.

The transmission of waves through rubble mound breakwaters can be compared to water flow through pores, and the attenuation of waves can be seen as a loss of energy. The loss of energy increases with the velocity of water particles or the incident wave height (H_i), leading to an increase in wave attenuation loss and a decrease in K_t. Additionally, larger core stone diameters result in decreased wave attenuation loss or increased K_t. Furthermore, with an increase in wave period, the decrease in water particles leads to a decrease in energy loss, translating to an increase in K_t with increasing wave length or period under constant water depth.

5 Conclusion

In this paper, various empirical formulas for calculating the transformation coefficient of rubble mound breakwaters have been presented. These formulas can be utilized as a reference for engineering design or for further investigation. Due to the complex interactions between waves and breakwaters, it is suggested that the wave transformation coefficient be thoroughly investigated through model tests conducted in a laboratory.

References

Agarwala, N., Saengsupavanich, C.: Oceanic environmental Impact in seaports. Oceans **4**(4), 360–380 (2023)

Omoto, W.: Visit on Okinawa sea water pumped storage power station. Jpn. Civ. Infrastruct. **42**(10), 47–50 (1994)

Tanaka, K.: Okinawa seawater pumped storage power station. J. Jpn. Electr. Soc. **124**(9), 583–586 (2004)

Numata, A.: Experimental study on wave-dissipating effect of block embankments. In: Proceedings of Coastal Engineering Lectures, vol. 22, pp. 501–505 (1975)

Stephenson, D.J.: Rockfill in Hydraulic Engineering. Elsevier, Amsterdam (1979)

Ahrens, J.P.: Characteristics of reef breakwaters. Coastal Engineering Research Center TR87-17: US Army Corps of Engineers (1987)

Xi, H.Z., Pan, J.N.: A study on the wave transmissibility of breakwater. J. Hydrodyn. **3**(4), 57–66 (1988)

Wang, D.T., Zuo, Q.H.: The experimental research of the rubble mound breakwater core material's influence on wave transmission. Chin. J. Hydrodyn. (2008)

Ge, X.D.: Experimental Research of Wave Transmission Characteristics of Rubble Mound Breakwaters. Dalian University of Technology (2014)

Feng, W.B., Tang, Z.S., Yu, D.J.: Experimental study on influence of wave propagation by different weight distributions of core stone. Port Waterw. Eng. **509**(11), 42–46 (2015)

Cao, R.Y., Feng, W.B., Yu, D.J.: Experimental study on influence of irregular wave propagation by different weight distributions of core stone. Port Waterw. Eng. **527**(4), 89–93 (2017)

Sun, T.T., Liu, Q.J., Wang, D.T.: Influence of relative block size of core material on the effect of wave transmission. Port Waterw. Eng. **547**(10), 56–61+66 (2018)

Physical Modeling of a Water Intake Tunnel of Sanmen Nuclear Power Station

Minghui Jin[1,2], Yuanzhe Zhi[1,3], Yue Yang[1,2], Wanwei Zhang[1,3], Zhiwei Li[1], and Shichang Yan[1,2(✉)]

[1] Hohai University, Nanjing 210024, China
yanshichang@hhu.edu.cn
[2] Key Laboratory of Ministry of Education for Coastal Disaster and Protection, Nanjing 210024, China
[3] Institute of Water Science and Technology, Hohai University, Nanjing 210098, China

Abstract. To ensure the wave stability of the water intake facilities for Units 5 and 6 of the Sanmen Nuclear Power Project and to measure the wave dissipation characteristics of the combined pump house, a wave flow test was conducted in the harbor basin using a full-terrain simulation. This test established a 1:30 scale normal wave local physical model to analyze the wave and velocity distribution in each compartment of the pump house under various combinations of water levels, waves, and flows. The test results indicate that within the effective propagation length, the influence of wave reflection is significant in the compartments affected by wave radiation, forming stable standing waves and directly governing the wave height distribution in each compartment. Additionally, the steadying effect of the water flow on the waves is notable. When the tidal level drops, the water velocity inside the compartments increases to a range related to the position, and changes in the tidal level can alter the relative velocity distribution of different flow channels at specific locations.

Keywords: Water Intake Pump House · physical modeling · wave reflect · flow velocity · combined wave-current action

1 Introduction

The site for Units 5 and 6 of the Sanmen Nuclear Power Project is located on the eastern coast of Zhejiang Province, adjacent to Sanmen Bay. The project area is influenced by waves from the southeast, at an angle 12° east of south (azimuth 123°). The site is planned to host six pressurized water reactor nuclear power units. Units 5 and 6 of the Sanmen Nuclear Power Project will share a combined pump house, which draws water through intake heads and tunnels. The geographical location and general layout of the project are shown in Fig. 1.

To ensure an optimal operating environment for the cooling water pumps, the intake channels within the combined pump house must exhibit a stable flow state, and the wave fluctuation amplitude in the forebay should not exceed the permissible limit. Additionally, it is crucial to prevent false alarms in the differential level meters of the rotating

D. Jeng and B. Cai (Eds.): ICCOE 2024, LNCE 423, pp. 135–150, 2025.
https://doi.org/10.1007/978-981-96-0181-3_14

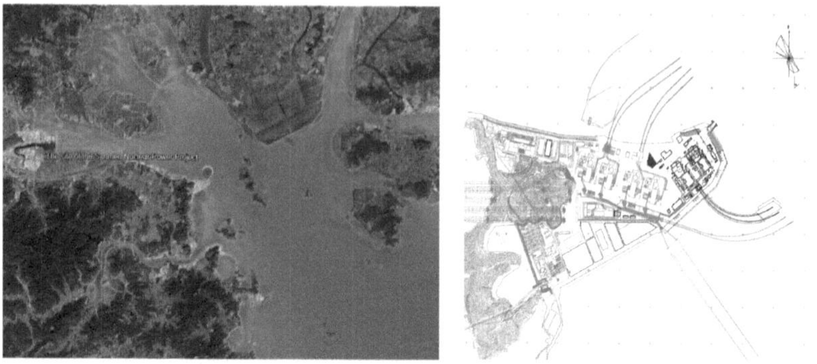

Fig. 1. Geographical location of the project and overall layout

screens caused by wave fluctuations, which could lead to the shutdown of the circulating water pumps. This necessitates model tests to verify the wave dissipation capability of the intake head, intake tunnel, and the current layout of the combined pump house. Furthermore, physical model tests are required to determine how different wave elements and tidal levels under still water and constant flow conditions impact the dynamic and steady-state balance of the water flow. The objectives of this study are as follows:

1. To measure the water surface fluctuations in each channel compartment of the combined pump house intake tunnel under various flow and wave conditions, analyze the wave intrusion and final behavior under the set conditions, and identify the wave dispersion patterns.
2. To measure the velocities at different depths and positions in the intake tunnel of the combined pump house under different flow and wave conditions, and to analyze the characteristics of the velocity field and the relationships of the relevant variables based on changes in position and tidal levels.

To achieve these objectives, an undistorted 1:30 scale model of the combined pump house intake tunnel, adhering to the Froude similarity law, was constructed. This model includes the intake head and the forefront of the revetment within the boundaries of the local topography. It was used to replicate the wave conditions and the interaction process with the study object in a wave basin. The model tests were conducted in the wave basin at Hohai University, where wave and velocity data in each compartment under various wave and flow combinations were measured.

The physical model testing adhered to the principles and methods proposed by Frostick et al. [1] (2011) and the Chinese national standards: NB/T 25046-2015 [2], SL 155-2012 [3], SL 156–165-95 [4], and JTS/T 231-2021 [5]. Additionally, it referred to Code 6349-2 of BSI Group, BS [6], and Engineer Manual 1110-2-1100 of the U.S. Army Corps of Engineers [7]. The study also referenced the measurement and analysis methods for flow velocity in channels by F. G. Carollo et al. [8] and M. Salih Kirkgoz [9], the measurement and theoretical summary of wave reflection principles by Michael Isaacson et al. [10], and the handling and calculation methods of reflected waves in irregular waves by Goda et al. [11]. This paper summarizes the physical model testing and data collection process, presenting the results and analysis of the interaction between water flow and waves under multiple tidal levels and wave elements, as well as the interaction with chamber-type tunnel structures. The influence of structural changes on the flow state of the water body is also discussed.

2 Experimental Conditions

2.1 Hydrological Conditions

Water level. The experiment used the China National 85 Elevation Standard and set the experimental water levels as follows: 1. 100-year return period high tide level: +5.23 m; 2. Average low tide level: −1.86 m; 3. 100-year return period low tide level: −3.82 m.

Waves. Wave direction is southeast by east 12°, with experimental elements detailed in Table 1.

Table 1. Wave elements at a depth of −20 m outside Sanmen nuclear plant intake

water level & wave return period	$H_{1\%}$ [m]	$H_{4\%}$ [m]	$H_{5\%}$ [m]	$H_{13\%}$ [m]	H_m [m]	T_m [s]	L [m]	Dir. [°]
100-year return level of high tide + 100-year return level of waves	5.66	4.81	4.66	3.92	2.51	10.4	137.9	112
Mean low tide + 100-year return level of waves	4.31	3.66	3.55	2.98	1.90	9.9	120.2	126
100-year return level of low tide + 100-year return level of waves	3.15	2.67	2.59	2.17	1.38	9.6	106.4	129

Water Flow. Under different combinations of water level and wave conditions, the operational configurations of pumps are as follows: 1 Circulating Water Pump: 36.2 m^3/s; 1 Essential Plant Service Water Pump: 3800 m^3/h; 2 Rotary Screen Backwash Pumps: 600 m^3/h each; 1 Seawater Pump: 400 m^3/h. The pump flow rates are based on the steady-state conditions and normal operating conditions of each system under different tidal levels. Under these conditions, the flow rate per single flow channel is 18.5 m^3/s.

2.2 Structures

Intake Tunnel and Head Section. For a single unit, the intake tunnel is a square pipe with dimensions of 4900 mm × 4900 mm, totaling 60.00 m in length. The intake employs a single-sided rectangular bar screen, set at the same angle as the original rock face. The intake tunnel connects to the forebay of the pump house. The arrangement of the intake is shown as cyan arrange in Fig. 2, with particular focus on the northwest flow channel and pump house as the primary simulation and measurement targets for this experiment.

Combined Pump House. The interior of the combined pump house consists of straight chambers divided into four channels by three straight dividing piers. The circulating water pump, serving as the primary hydraulic power source within the combined pump house, is installed together with all other functional water pumps located behind chambers. In both experimental and engineering setups, dual pipelines providing constant water flow serve as boundary conditions for the experimental environment of the combined pump house channels, shown as red range in Fig. 2. The plan layout of the intake tunnel is also depicted in Fig. 2.

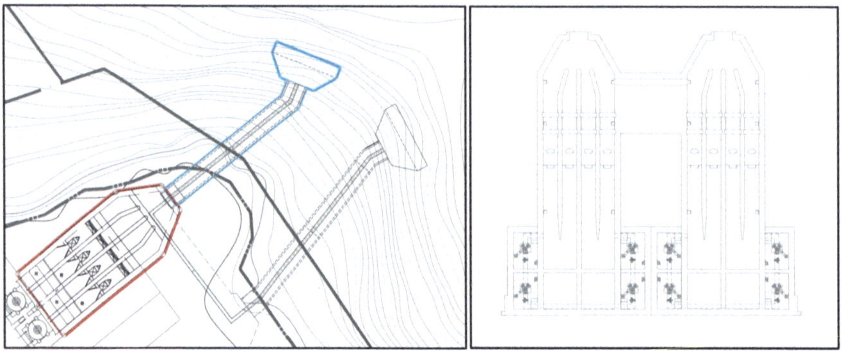

Fig. 2. Layout diagram of intake head (Left) & intake tunnel and combined pump house (Right)

3 Experimental Setup

3.1 Test Facility, Measurement Method

Physical modeling study was carried out in the shallow water wave basin at Hohai University, and the dimensions of the basin is 40 m long, 30 m wide, and 1.2 m deep. The wave basin is equipped with a combined paddles multi-directional piston-type wave generator. Figure 3 presents a photograph of the shallow water wave basin and the wave generator, along with the listed modeling scope of the test. This experiment will measure wave height, water level, and flow velocity, with the accuracy of the measuring instruments listed in Table 2.

Wave height measurements utilize the DJ800 wave height acquisition system. For measuring wave height distribution inside the combined pump house, each set of channels

Fig. 3. Shallow water wave basin at Hohai University and layout of modeling

Table 2. Measurement instruments and accuracy

measuring instruments	measurements	range and accuracy (R&A)
DJ800 wave height gauge	wave height/period	R: 60 cm A: ±0.3 mm (<2% F.S)
photoelectric current meter	flow velocity	R: 0–120 cm/s A: ±1.2 cm/s (<1% F.S)

is equipped with 6 wave height measurement points located as follows: near the forebay wave attenuation baffle (A), between the channel gate well and fine baffle (B), in front of the rotary screen chamber dividing pier (C), at the exit of the rotary screen chamber (D), at 1.5 times the pump outlet diameter in front of the centerline of the circulating water pump room (E), and at 3 times the pump outlet diameter (F). Additionally, outside the combined pump house, at the −20 m depth contour outside the channels (near the intake head of Unit 5), 1 control point is set. The planar positions of the wave height gauges inside the combined pump house are shown in Fig. 5.

Flow velocity measurements are conducted using a photoelectric flow velocity meter, placed at various locations within the channels: Section B, C, D, E, F. Each flow measurement section is horizontally arranged with 1–3 rows of measurement points to measure the horizontal distribution of flow velocity, and vertically arranged with 3–5 rows of measurement points to measure the vertical distribution of flow velocity.

Specifically: Sections B, C, and D are horizontally arranged with 1 row of measurement points and vertically arranged with 3 rows of measurement points (at depths of 1.5 m, 4.5 m, and 7.5 m from the bottom). Sections E and F are horizontally arranged with 3 rows of measurement points (−10 m, 0 m, +10 m) and vertically arranged with 4 rows of measurement points (at depths of 1.5 m, 4.5 m, 7.5 m, and 10.5 m from the bottom). The planar positions of the flow velocity meters inside the combined pump house are shown in Fig. 4.

3.2 Simulation Method

Similarity Criteria. To simultaneously validate the flow regime and hydraulic characteristics at different water levels and wave combinations within the pump house inlet

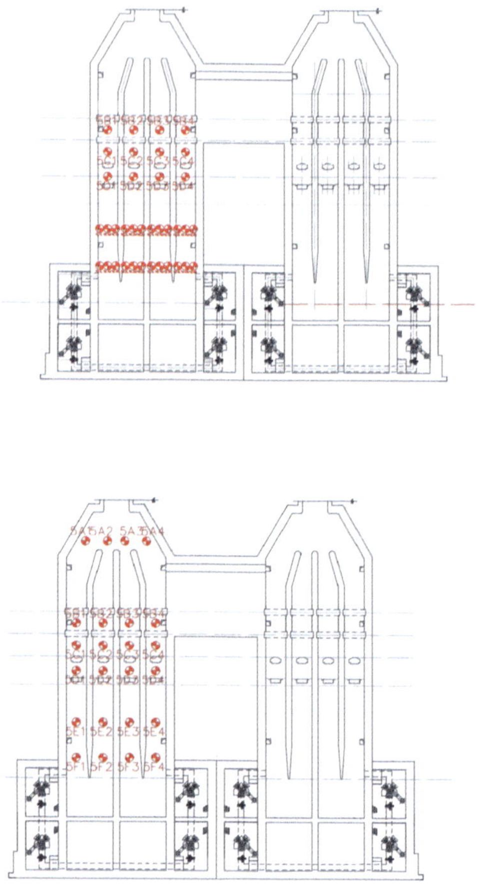

Fig. 4. Flow velocity meter locations schema (Left) and Wave height gauge station layout schema (Right)

channels, at the intake of each functional water pump, and along the intake channels, it is essential to study the surface wave fluctuations, velocity distributions, and the generation and dissipation of eddies within both the combined pump house and the channels. This research necessitates adherence to at least two fundamental principles: Gravity Similarity Criterion (Froude Number Similarity); Viscosity Similarity Criterion (Reynolds Number Similarity). By satisfying these similarity criteria, the experimental setup can reliably replicate the hydraulic conditions experienced by the actual system, enabling comprehensive analysis of flow patterns, water surface dynamics, and the interaction of flows with structural elements within the pump house and intake channels.

Scale. The model experiments primarily adhere to the gravity similarity criterion, employing a regular wave model to simulate wave behavior. The geometric scale of the model is primarily set at 1:30, based on considerations such as building structure dimensions, water depth and wave conditions, experimental site, and equipment capabilities. The scaling of key physical quantities is detailed in Table 3.

Table 3. Scale of the main physical quantities

Scales	Expressing Equation	Values
Geometric length	$\lambda_L = L_p/L_m$	30
Time	$\lambda_T = \lambda_L^{1/2}$	5.48
Weight	$\lambda_W = \lambda_L^3$	27000
Flow rate (unit width)	$\lambda_Q = \lambda_L^{3/2}$	164.32

Similarity of Waves. The irregular wave modeling adheres to the gravity similarity criterion, using a JONSWAP spectrum for wave generation. The spectral density function is represented by Eq. 1.

$$
\begin{aligned}
S(f) &= \alpha H_s^2 T_p^{-4} f^{-5} exp\left[-\tfrac{5}{4}\left(T_p f\right)^{-4}\right]\gamma^{exp\left[-(T_p f-1)^2/2\sigma^2\right]} \\
\alpha &= \frac{0.0624}{0.230+0.0336\gamma-0.185(1.9+\gamma)^{-1}} \\
\sigma &= \begin{cases} 0.07, f \le f_P \\ 0.09, f \ge f_P \end{cases}
\end{aligned}
\tag{1}
$$

In the equation: **S(f)** represents spectral density [$m^2 s$]; $\mathbf{H_s}$ represents significant wave height [m]; $\mathbf{T_p}$ represents peak period [s]; $\mathbf{f_p}$ represents peak frequency [Hz]; **f** represents frequency [Hz]; $\boldsymbol{\gamma}$ spectral peak parameter, set to 3.3.

In each test run, irregular waves exceed 1000 waves, and each test run is repeated 4–5 times to ensure the reliability of the experimental data and results. The deviation of significant wave height and wave period in the model test is less than ±5%. The results and the deviation of calibration were listed in Table 4.

Table 4. The results of wave calibration

Water level [m]	Target $H_{13\%}$ [m]	Target T_m [s]	Calibrated $H_{13\%}$ [m]	Calibrated T_m [s]	δH [–]	δT [–]
5.23	3.92	10.4	3.94	10.6	1%	2%
−1.86	2.98	9.9	3.03	10.0	2%	1%
−3.82	2.17	9.6	2.24	9.8	3%	2%

Similarity of Flow. The water flow is simulated using constant flow conditions. Given that under normal operational conditions, the flow rate of the circulation pump within a single pump chamber is significantly higher than the combined flow rates of the essential plant water pump, rotary screen backwash pump, and chlorine production water pump. The latter collectively amount to only 2% of the former (0.8 m^3/s compared to 36.2 m^3/s). In the experiment, each pump chamber is equipped with one pump to simulate the corresponding flow rate (37.0 m^3/s) of the pump operation combination under site-normal conditions.

Similarity of Structures. The intake headworks, intake tunnel box culvert, and combined pump house: fabricated from acrylic based on scaled dimensions from design drawings, with model errors within ±1 mm. The overall model configuration is illustrated in Fig. 5.

Fig. 5. Aerial view of the model and test site

3.3 Experimental Procedure

Experiment 1: Flow Test. The hydraulic experiments measured and analyzed flow velocities at various locations within the pump house under conditions where water flow acted independently. The experimental setup is detailed in Table 5.

Table 5. Cases of Experiment 1: hydrodynamic test

Tide level	Wave	Flow velocity
100-year return level of high tide +5.23 m	/	37.0 m^3/s
100-year return level of low tide −3.82 m	/	37.0 m^3/s

Experiment 2: Wave Test. Wave tests measured and analyzed the fluctuation of liquid levels at various locations within the pump house under the separate action of waves and the combined action of waves and water flow. Statistical analysis included effective wave height, transmission coefficient (where the relative wave height equals the ratio of measured wave height to incident wave height), and average period. Experimental configurations for the test groups are shown in Table 6.

Table 6. Cases of Experiment 2: wave test

Tide level	Wave	Flow velocity
100-year return level of high tide +5.23 m	$H_s = 3.92$ m, $T_m = 10.4$ s (100-year)	/
Mean low tide −1.86 m	$H_s = 2.98$ m, $T_m = 9.9$ s (100-year)	/
100-year return level of low tide −3.82 m	$H_s = 2.17$ m, $T_m = 9.6$ s (100-year)	/
100-year return level of high tide +5.23 m	$H_s = 3.92$ m, $T_m = 10.4$ s (100-year)	37.0 m^3/s
Mean low tide −1.86 m	$H_s = 2.98$ m, $T_m = 9.9$ s (100-year)	37.0 m^3/s
100-year return level of low tide −3.82 m	$H_s = 2.17$ m, $T_m = 9.6$ s (100-year)	37.0 m^3/s

4 Results and Discussion

4.1 Flow Velocity Effect and Distribution

Distribution of flow velocity results were demonstrated as below Fig. 6 and Fig. 7, for 2 cases of tide level with 4 flow channels (E and F sections omit symmetrical flow channels and average statistics):

Overall Analysis of Distribution Characteristics. In section B, the flow velocity is higher in the deep water layer compared to the surface layer. Specifically, in the 5-2/5-3 channels located at depths of 1.5 m to 4.5 m from the bottom, the water velocity is significantly higher than in the 5-1/5-4 channels. This characteristic gradually diminishes towards the rear along the water intake tunnel of the pump house. In contrast, section D exhibits an opposite pattern: the flow velocity is slower in the deep water layer while it increases in the surface layer. In the surface layer, the water velocity in the 5-2/5-3 channels is higher than in the 5-1/5-4 channels, whereas in the deep water layer, the water velocity in the 5-1/5-4 channels is higher than in the 5-2/5-3 channels. Sections E and F show consistent characteristics where the water velocity remains stable, with velocities in the surface layer being higher than those in the bottom layer.

Individual Analysis of Distribution Characteristics. In section B, the horizontal distribution of upper layer water flow is relatively uniform, being closer to the vertical wall base of the gate well and overall smaller. However, there are deviations in the horizontal flow velocity distribution of the middle layer water flow: during high tide, channels 5-2/5-4 have relatively higher velocities compared to channels 5-1/5-3; the velocities in channels 5-1/5-4 are relatively higher on both sides, while channels 5-2/5-3 have relatively lower velocities in the middle. In section C, the horizontal flow velocity distribution of lower layer water flow shows that channels 5-2/5-3 have relatively higher velocities while channels 5-1/5-4 have relatively lower velocities on both sides; the middle layer water flow in channels 5-2/5-4 has relatively higher velocities while channels 5-1/5-3 have relatively lower velocities; the upper layer velocities show that channels 5-1/5-4 have relatively higher velocities on both sides while channels 5-2/5-3 have relatively lower velocities in the middle.

4.2 Wave Test Result and Wave Dispersion

Wave test results were demonstrated as below Fig. 8, including 3 cases of wave-tide level with each two sets of symmetrical data for 4 flow channels.

Pure Wave Action. As expressed by the blue bars, under wave action alone, significant surface fluctuations are observed within the pump house, exhibiting an overall 'larger at

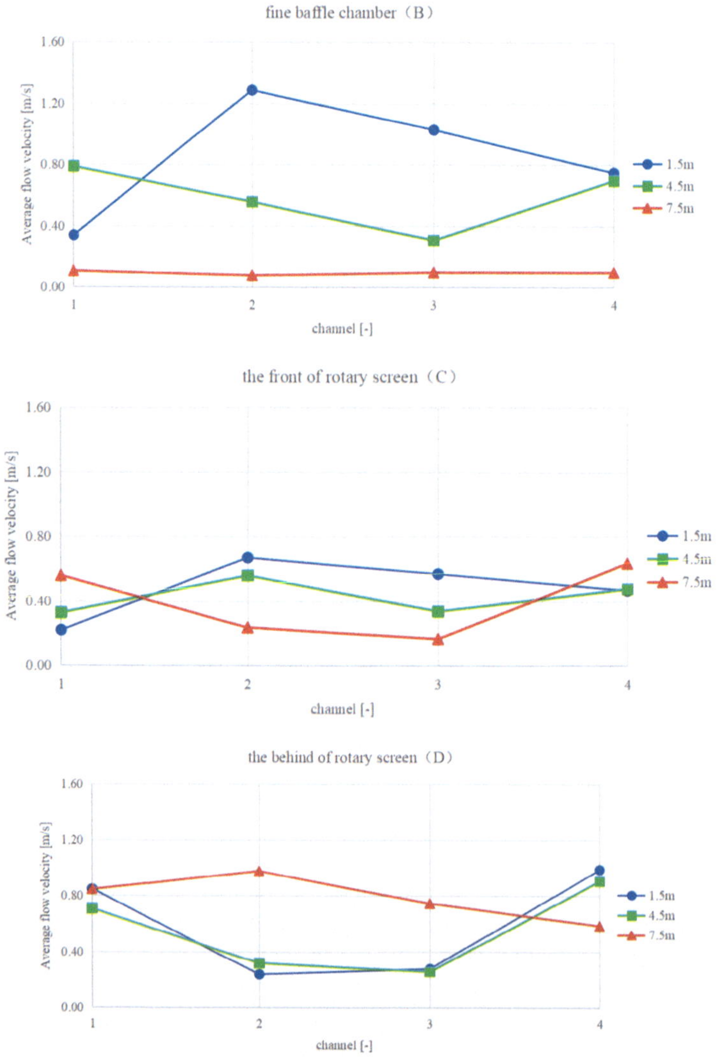

Fig. 6. Velocity Distribution of Each Channel in Each Compartment of Case 1

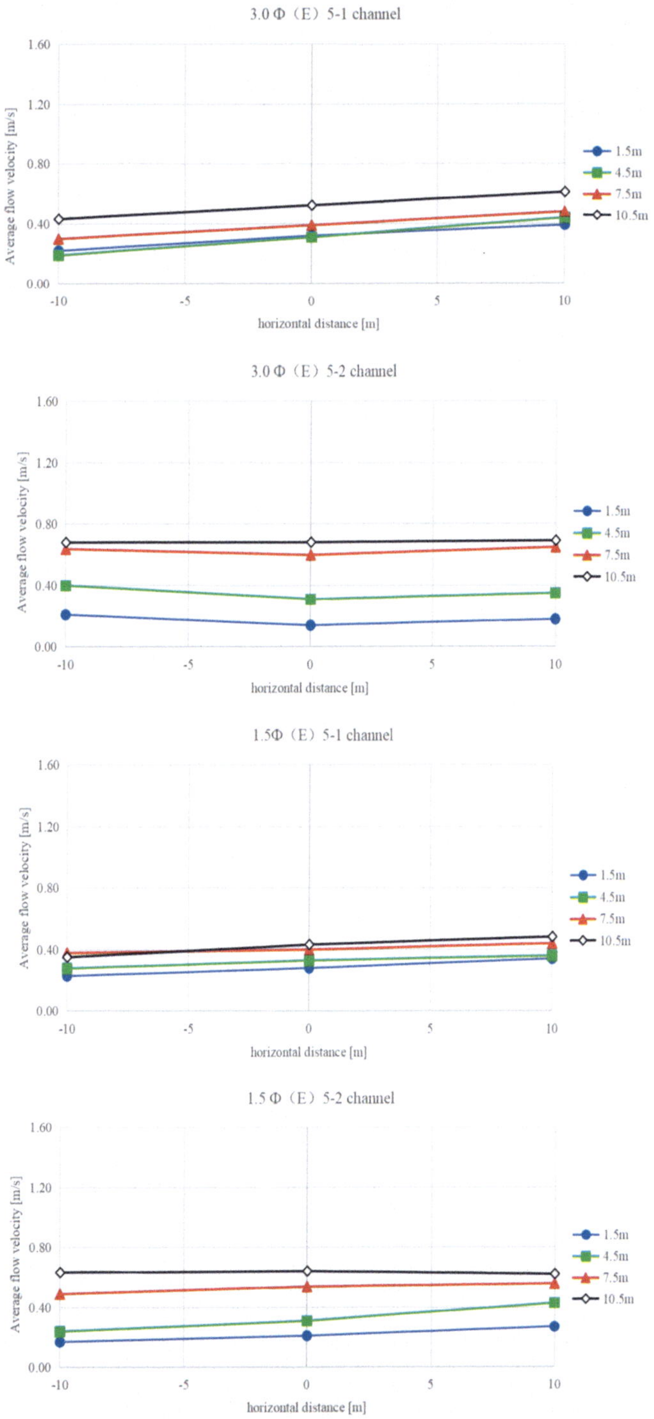

Fig. 6. (*continued*)

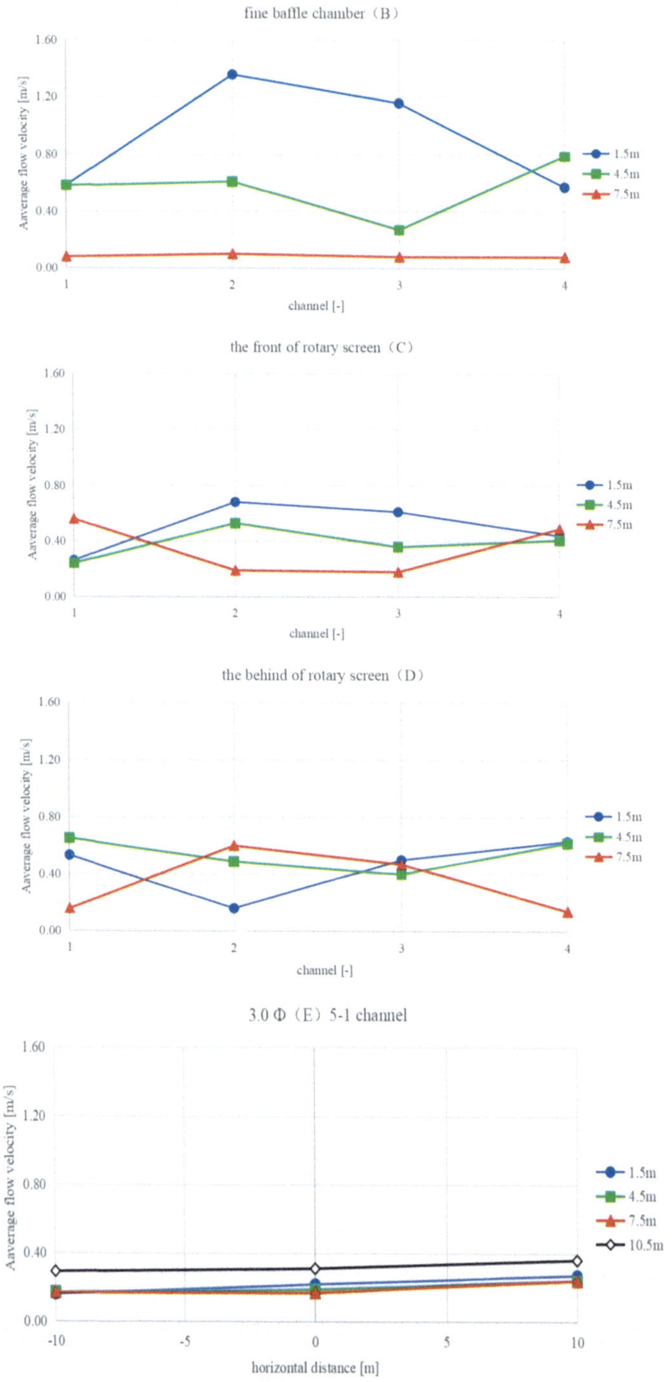

Fig. 7. Velocity Distribution of Each Channel in Each Compartment of Case 2

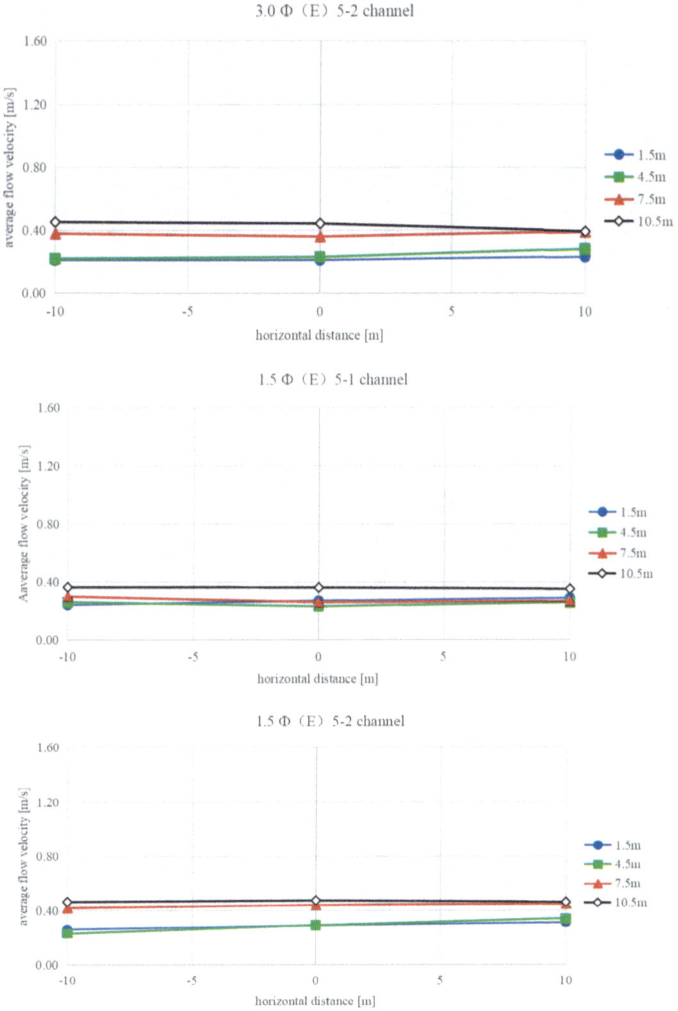

Fig. 7. (*continued*)

ends, smaller in middle' pattern. This phenomenon is attributed to wave reflection off the vertical walls of the gate well in section A and the suction chamber in section F, where wave heights are greater compared to other areas within the pump house. Specifically, the front pool of the pump house experiences the highest wave heights due to direct wave influence from the tunnel, with a maximum transmission coefficient reaching 0.20 and a maximum significant wave height ($H_{1/3}$) of 0.59 m. In contrast, the suction chambers exhibit transmission coefficients not exceeding 0.12, with a maximum significant wave height of 0.40 m.

In sections B and C, despite relatively smaller wave heights, there are long-period wave oscillations with wave periods ranging from 2 to 3 times the incident wave period. In the remaining parts of the pump house, wave periods are slightly longer than the

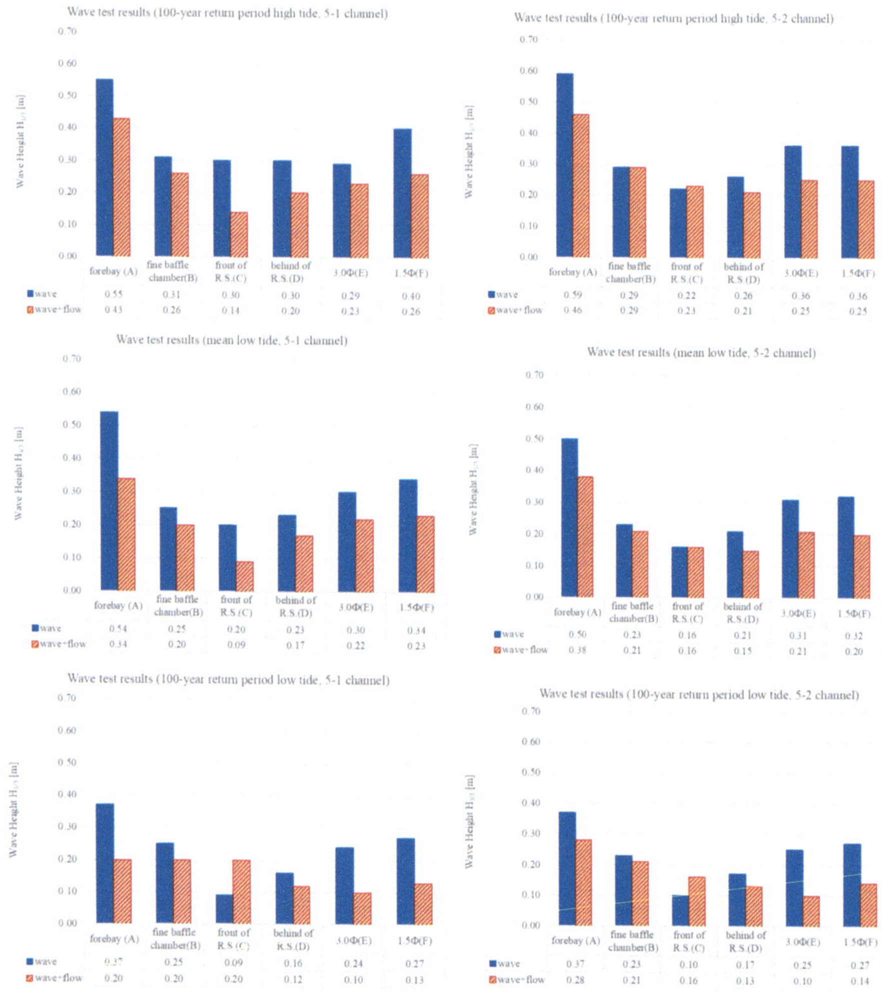

Fig. 8. Wave height distribution in 6 segments of chambers in flow channels 5-1/5-2

incident wave periods, with the former being less than 1.5 times the latter. The transmission coefficients at various locations within the pump house increase as the water level decreases. However, due to the significant decrease in incident wave heights from the open sea as the water level lowers, the overall wave heights within the pump house decrease with decreasing water levels.

Wave-Current Combined Action. As expressed by the red bars, under the combined action of waves and currents, the surface fluctuations within the pump house continue to exhibit an overall "larger at ends, smaller in middle" pattern. However, due to the smoothing effect of the currents, wave heights overall decrease compared to when waves act alone. The stabilizing effect notably affects locations with higher wave heights, while its impact on sections C and D, where wave heights are smaller, is less pronounced.

Sections B and C still experience long-period wave oscillations with wave periods ranging from 1.5 to 2 times the incident wave period. Additionally, the characteristics observed are consistent with those under wave action alone.

5 Conclusions

5.1 Hydrodynamic Test

1. Starting from the intake head, the flow velocity of water in the middle channels of the deep water layer is significantly higher than in the side channels. This characteristic gradually diminishes along the pump house intake tunnel chambers towards the rear until it reverses: the flow velocity in the deep water layer becomes slower, while the velocity in the surface layer increases. In the surface layer, the flow velocity in the middle channel is higher than in the side channels, whereas in the deep water layer, the flow velocity in the side channels is higher than in the middle channel. Near the constant flow pump outlet, the characteristics of water flow velocity stabilize, with velocities in the surface layer being higher than those in the bottom layer.
2. The anticipated changes in tidal levels have a systematic impact on the overall flow velocities in the channels, generally causing them to increase as the tide lowers. However, these changes do not significantly alter the distribution characteristics and patterns of flow velocities within different sections of the structure under the same operating conditions. The impact of tidal changes varies in degree across different segmented intervals, but overall, it has minimal effect on the flow distribution ratios among the channels.

5.2 Wave Test

1. Within the effective propagation length, chambers exposed to wave radiation exhibit significant wave reflection effects. Throughout the entire chamber segment, stable standing waves are formed due to prominent wave reflection, directly influencing the distribution of wave heights in each chamber. The reflection of standing waves is a dominant factor in the formation of wave modes within the model length, surpassing the energy dissipation effects of perforated layers on wave energy.
2. When water flow has a certain velocity, it exerts a stabilizing effect on waves persistently acting on the water surface. However, its effect is uncertain for locations with small significant wave heights or standing wave nodes.

Acknowledgement. The authors were appreciate to China Nuclear Power Engineering Co., Ltd. through Sanmen Nuclear Power Plants 5 and 6 Engineering Construction Project and Physical Model Verification Project. The authors also thank the College of Harbour, Coastal and Offshore Engineering, and Hohai University offering full laboratory resources and technical assistance.

References

1. Frostick, L.E., Mclelland, S.J., Mercer, T.G.: Users Guide to Physical Modelling and Experiment. CRC Press, Boca Raton (2011)
2. National Energy Administration of the People's Republic of China, NB/T 25046-2015: Code for hydraulic design of nuclear power plants. China Planning Press, Beijing (2015)
3. Ministry of Water Resources of the People's Republic of China, SL 155-2012: Specification for normal hydraulic model test. China Water & Power Press, Beijing (2012)
4. Ministry of Water Resources of the People's Republic of China, SL 156–165-95: Test regulation for special hydraulic model. China Water & Power Press, Beijing (1995)
5. Ministry of Water Resources of the People's Republic of China, JTS/T 231-2021: Technical specification for simulation test of water transportation engineering. China Communications Press, Beijing (2021)
6. BSI Group. BS 6349-2: Maritime Works. Code of Practice for the Design of Quay Walls, Jetties and Dolphins. BSI Group, London (2019)
7. U.S. Army Corps of Engineers. Engineer Manual 1110-2-1100: Coastal Engineering Manual U.S. Army Corps of Engineers, Washington, D.C. (2002)
8. Carollo, F.G., Ferro, V.I.T.O., Termini, D.: Flow velocity measurements in vegetated channels. J. Hydraul. Eng. **128**(7), 664–673 (2002)
9. Kirkgöz, M.S.: Turbulent velocity profiles for smooth and rough open channel flow. J. Hydraul. Eng. **115**(11), 1543–1561 (1989)
10. Isaacson, M.: Measurement of regular wave reflection. J. Waterw. Port Coast. Ocean Eng. **117**(6), 553–569 (1991)
11. Goda, Y., Suzuki, Y.: Estimation of incident and reflected waves in random wave experiments. Coast. Eng. **1976**, 828–845 (1976)

Numerical Simulation Study of Dam-Break Based on SPH Method

Ye Huo[1], Yuguang Zhong[1](✉), HuanYu Zhao[2], Songcheng Li[3], and Yuanbo Jiang[3]

[1] College of Mechanical and Electrical Engineering, Harbin Engineering University, Harbin 150001, China
zhongyuguang@126.com

[2] Liaoning Provincial Key Laboratory of Aircraft Ice Protection, AVIC Aerodynamics Research Institute, Shenyang 110034, China

[3] Liao Shen Industries Group Co., Ltd., Shenyang 110045, China

Abstract. The smooth particle hydrodynamics (SPH) method is a meshless method in a pure Lagrangian framework, which is suitable for simulating wave overturning and breaking, high-speed jets and other dynamic boundary and large deformation problems. This manuscript introduces the basic principles and improvement methods of the SPH method, and uses the SPH method to establish different types of numerical flume models, and simulates the dam-breaking motions of different types of flumes in detail. Comparison of the computational results of the SPH simulation and the experimental results from the literature reveals that the computational results of the adopted SPH method match the experimental results from the literature to a high degree, and are able to accurately capture the droplet splashing, free surface deformation, etc., which validates the stability of the numerical model. By comparing the simulation of dam failure flow of two different forms of energy dissipating cans, the energy dissipation effect of the two types of barriers is analyzed and compared.

Keywords: The smooth particle hydrodynamics · Dam-break · Obstacle

1 Introduction

Dam-break is a sudden and catastrophic natural phenomenon that has received widespread attention. When a water storage structure is damaged, the huge dam-break flow slopes downward, causing a huge impact on downstream buildings and organisms. The dam-break flow problem not only needs to be focused on in practical engineering applications, but also occupies an important position in hydrodynamics, of which it is a very classical model. Therefore, it is necessary to study the numerical simulation of the coupled flow-solidity of the dam failure process.

D. Jeng and B. Cai (Eds.): ICCOE 2024, LNCE 423, pp. 151–161, 2025.
https://doi.org/10.1007/978-981-96-0181-3_15

The accuracy of dam failure predictions relies heavily on the appropriateness of mathematical formulations and the accuracy of numerical simulations. The dam failure flow is a typical free surface flow with kinematic boundaries and complex large deformation problems. Commonly used solution methods for the dam failure flood evolution problem include finite difference method (FDM), finite element method (FEM), finite volume method (FVM) and smooth particle hydrodynamics (SPH) method, etc. The SPH method, as a kind of particle method, has complete Lagrangian particle characteristics, which can deal with the problems of free-surface water and deformable boundaries efficiently, and it is very suitable for simulating the dam failure water flow [1]. The SPH method is very suitable for simulating dam-break water flow. Using the SPH method in simulating fluid motion, the fluid is discretized into particles which have their own physical properties such as density, pressure, velocity and acceleration [2]. The fluid particles are subjected to Lagrangian motion by a combination of internal forces (viscous forces, pressure gradients, and surface tension) and external forces (gravity and solid-to-fluid forces). Since the meshless method requires only node information and no connection information between nodes, the mesh is completely or partially eliminated, leaving behind the initial delineation of the mesh and mesh reconstruction [3]. During the particle motion, the pressure and velocity of the particle are updated in the time domain based on the control equations. At each moment, the properties of the particle itself represent the physical properties at that location within the flow field.

Various experts and scholars have carried out extensive in-depth analysis and research on the dam failure problem based on the SPH method. Monaghan [4], firstly, applied the SPH method to the free surface flow problem and pointed out that the SPH method is well adapted to the simulation of free-surface flow with large deformation, and then used the dam failure flow to form an isolated wave, and combined with the laboratory experiments to study the breakup process of the isolated wave on a gentle slope in detail. Wang et al. [5] applied the SPH method to the Shallow Water Equations (SWE) for the first time for the non-viscous dam burst flow, and captured and accurately modeled the sharp discontinuities in the flow. ZHANG Yongxiang uses the FVM method combined with the VOF format to track the free surface to simulate the two-dimensional dam failure problem [6]. Savic and Holly [7] solved the one-dimensional equations and simulated the dam failure wave using two improved Godunov formats, and the results were in good agreement with the theoretical solutions.

2 The SPH Method and Its Improvement

SPH is a meshless method, its basic idea is not to use a predefined grid structure to discretize the solution domain, but a series of node arrangements to replace the grid structure, using a kind of approximation related to the weight function (kernel function), using the weight function to characterize the physical information on the nodes, a node on the solution domain can affect the mechanical properties of any point on the whole solution domain, so that in the process of numerical computation The concept of unit or grid is eliminated from the numerical calculation process, which provides the possibility of solving various discontinuous problems [8].

The SPH method consists of two key steps, the first step is the integral representation, also known as the kernel approximation method, and the second step is the particle approximation method. The kernel approximation method is a stepwise integration of arbitrary and smooth functions. The function integral expression is then approximated by cumulative summation of the values of the nearest neighboring particles. In the SPH method, the function integral expression is defined as,

$$f(x) = \int_{\Omega} f(x')\mathrm{W}(x - x', h)dx' \tag{1}$$

where, Ω is the entire solution domain; $f(x)$ is a function of the coordinate vector x; $\mathrm{W}(x - x', h)$ is the kernel function; h is the smooth length; and x' is the particle in the domain of the particle x support.

In the SPH method, the value of the function at any point i and its derivative can be expressed as,

$$f(\boldsymbol{x}_i) = \sum\nolimits_j f(\boldsymbol{x}_j)W(\boldsymbol{x}_i - \boldsymbol{x}_j, h)V_j \tag{2}$$

$$\nabla_x f(\boldsymbol{x}_i) = \sum\nolimits_j f(\boldsymbol{x}_j)\nabla_x W(\boldsymbol{x}_i - \boldsymbol{x}_j, h)V_j \tag{3}$$

where, V_j is the volume of particle j.

The smooth function (or called kernel function) is very important in the SPH method because it not only determines the form of the function approximation, defines the size of the particle tight-branching domain, but also determines the accuracy of the SPH approximation. Compared with previous kernel functions, the Wendland kernel function has a more uniform distribution of particles, so the Wendland kernel function is chosen as the kernel function in this paper [9].

$$W(\boldsymbol{x} - \boldsymbol{x}', h) = \begin{cases} (1 + 2q)(2 - q)^4, & q < 2 \\ 0, & q \geq 2 \end{cases} \tag{4}$$

where, $q = (\boldsymbol{x} - \boldsymbol{x}')/h$, , and $(\boldsymbol{x} - \boldsymbol{x}')$ is the distance between particles.

The dam-break hydrodynamic model can be represented by the N-S equation, which in the Lagrangian framework can be described as,

$$\frac{1}{\rho}\frac{\mathrm{D}\rho}{\mathrm{D}t} + \nabla \cdot \boldsymbol{u} = 0 \tag{5}$$

$$\frac{\mathrm{D}\boldsymbol{u}}{\mathrm{D}t} = -\frac{1}{\rho}\nabla p + \upsilon\nabla^2\boldsymbol{u} + \boldsymbol{g} \tag{6}$$

where, ρ is the density of the material; p is the pressure of the fluid; υ is the dynamic viscosity; $\boldsymbol{u}$ is the velocity vector, and $\boldsymbol{g}$ is the volumetric force.

In terms of SPH methods, many different improved algorithms have been proposed in recent years, among which, δ-SPH is the widely influenced one. δ-SPH method can minimize the pseudo-high-frequency numerical oscillations in the pressure field, and thus improve the accuracy and precision of the SPH method. δ-SPH framework of the equations of mass conservation and momentum conservation for the particle i is expressed as,

$$\frac{D\rho_i}{Dt} = -\rho_i \sum_j (\boldsymbol{u}_j - \boldsymbol{u}_i) \cdot \nabla_i W_{ij} V_j + \delta h c_0 D_i \tag{7}$$

$$\frac{D\boldsymbol{u}_i}{Dt} = -\frac{1}{\rho} \sum_j (p_i + p_j) \nabla_i W_{ij} V_j + \boldsymbol{g} + \alpha h c_0 \frac{\rho_0}{\rho_i} \sum_j \Pi_{ij} \nabla_i W_{ij} V_j \tag{8}$$

The SPH particle approximation method is similar to the center-difference format in the finite-difference method, which is prone to unphysical oscillations in the flow field. When modeling the motion of a non-viscous fluid, it is often necessary to impose some artificial viscosity to ensure the stability of the numerical simulation. In the paper, the artificial viscosity term and density dissipation term proposed by Monaghan are chosen, and the formulas are as follows,

$$\pi_{ij} = \frac{(\boldsymbol{u}_j - \boldsymbol{u}_i) \cdot (\boldsymbol{r}_j - \boldsymbol{r}_i)}{(\boldsymbol{r}_j - \boldsymbol{r}_i)^2} \tag{9}$$

$$D_i = 2 \sum_j \psi_{ji} \frac{(\boldsymbol{r}_j - \boldsymbol{r}_i) \cdot \nabla_i W_{ij}}{(\boldsymbol{r}_j - \boldsymbol{r}_i)^2 + (0.01h)^2} V_j \tag{10}$$

$$\psi_{ji} = (\rho_j - \rho_i) - \frac{1}{2} \left(\langle \nabla \rho \rangle_i^L + \langle \nabla \rho \rangle_j^L \right) \cdot (\boldsymbol{r}_j - \boldsymbol{r}_i) \tag{11}$$

In the ψ_{ji} term, $\langle \nabla \rho \rangle_i^L$ and $\langle \nabla \rho \rangle_j^L$ are calculated using the modified particle approximation [24, 26].

3 Algorithm Simulation

3.1 Dam-Break Simulation and Validation

In order to verify the validity of the SPH method for simulating dam failure, which is a problem of large deformation of free surface fluid, a flume model was developed [10]. Model size is 0.584 × 0.35 m^2 and rectangular water body size is 0.146 × 0.293 m^2. The total number of model particles is 5899, of which the number of fluid particles is 2628. The smooth kernel function uses the Wendland function. Figure 1 shows the particle pressure distribution at the initial moment of the dam failure model.

The simulation results under the working condition are processed and visualized, and the scatter plots of flow pattern and flow velocity distribution are made for different moments in 1.0 s time (t = 0.2 s, t = 0.4 s, t = 0.6 s, t = 0.8 s, and t = 1.0 s and other special moments are taken), and the results of the SPH calculations are compared with the experimental results of the model in the literature, and the detailed results are shown in Fig. 2.

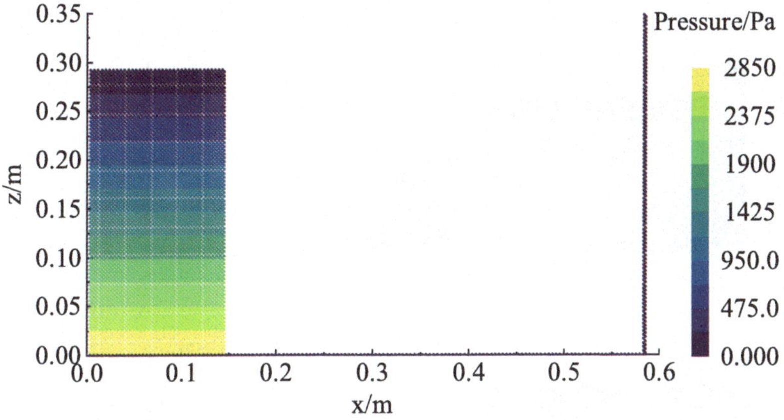

Fig. 1. Particle pressure distribution at the initial moment of the dam failure model

The overall movement of the dam burst water body process is: in T = 0.2 s time the overall flow pattern is streamlined, the dam burst to the front of the rapid flow. At T = 0.4 s, the front end of the water flow reaches the right side wall, after being restrained by the right wall climbed up the wall. At this moment, the kinetic energy of the dam break is gradually converted into potential energy. At 0.4–0.6 s, the water particle reaches the highest position on the right side wall, and the water body falls down. In the process, the liquid breaks up and splashes, and at 1.0 s, the whole body of water flows back to the left side wall, where it is restrained by the wall, and the particles meet and merge with each other.

During the whole process in 1 s time, it can be seen from the analysis of the flow pattern that the simulation results of the SPH method corresponding to each moment are basically consistent with the model experiments, and the SPH method shows good linear fluid characteristics, and demonstrates the morphology of droplet splashing and fragmentation, which shows that the SPH method is more detailed in the depiction of the fluid, and it has a good simulation result.

For the presentation of the simulation results of the dam-break water flow, one of the evaluation methods is to compare the free surface height curve of the fluid, in order to further explore the differences between the SPH numerical simulation and the experimental results, based on the method, t = 0.2 s, t = 0.4 s and t = 0.8 s are selected to make the free surface height curve of the two and compare the two, and the results show that the simulation obtained with the experimental match is better. The results show that the simulation is in good agreement with the experiment (Fig. 3).

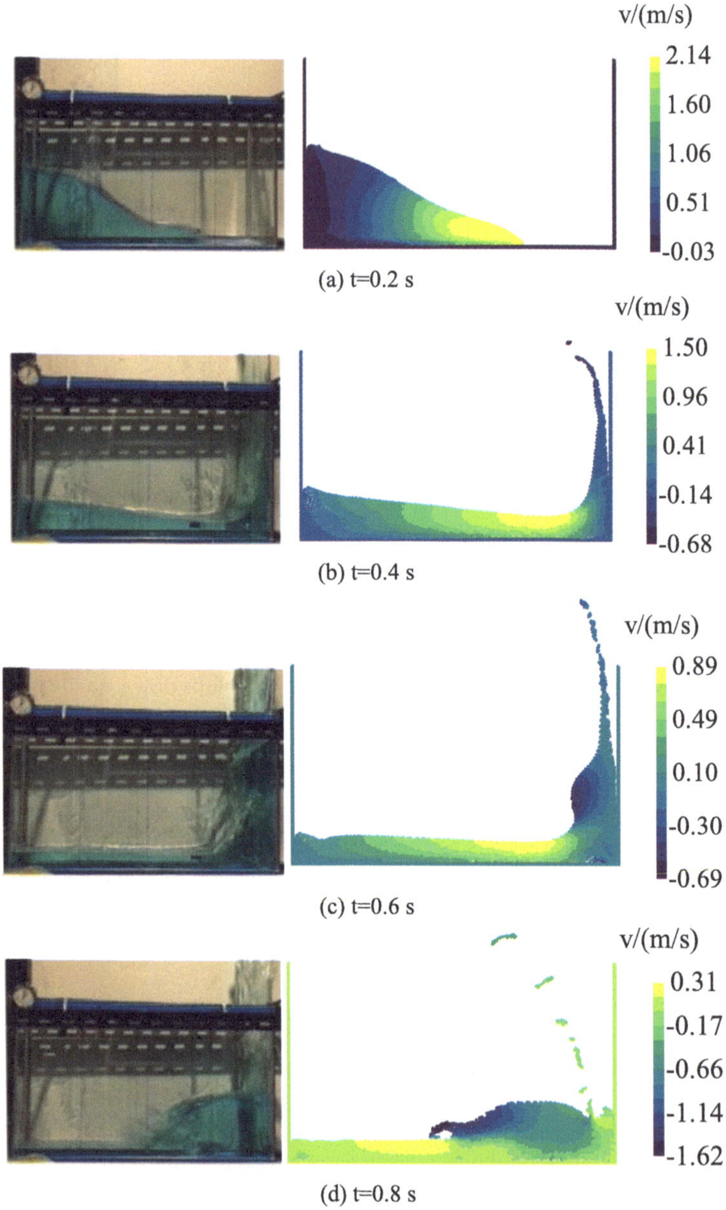

Fig. 2. Comparison of the experimental results with the simulation results of the SPH

(e) t=1.0 s

Fig. 2. *(continued)*

3.2 Simulation of Dam-Break Flow for Two Energy Dissipating Cans

In this paper, the classical flume model is used, as shown in Fig. 4, with a flume size of 1.2 × 0.6 m. The obstacles are divided into two forms, namely, a 0.06 × 0.2 m obstacle with no holes at 0.6 m and a 0.06 × 0.2 m obstacle with holes of 0.06 m size at 0.6 m. The obstacles are also divided into two forms, namely, a 0.06 × 0.2 m obstacle without holes at 0.6 m and a 0.06 × 0.2 m obstacle with holes of 0.06 m size.

From the left side of the 0.4 m to set up a gate, the initial water level upstream is 0.5 m, the density of the water body is 1000 kg/m^3. The water body is in hydrostatic equilibrium at the initial moment, and the situation when the water column breaks the dam is simulated by quickly lifting the gate. The water body is in hydrostatic equilibrium at the initial moment.

Figure 5 shows the comparison of particle distributions in several moments after the dam failure calculated by numerical simulation, from which it can be seen that the water level decreases after the dam failure, the water body moves to the right at high speed, and the water body flows in an orderly manner before overtopping the obstacle.

As shown in Fig. 6, the comparison of the pressure of the water body in impacting the right wall surface is calculated in the text, from which it can be seen that under the obstruction of the perforated obstacle, the pressure of the water body in impacting the right wall surface is less than the situation when there is no perforated obstacle, so that the perforated obstacle has a better dissipation of energy effect within a certain range.

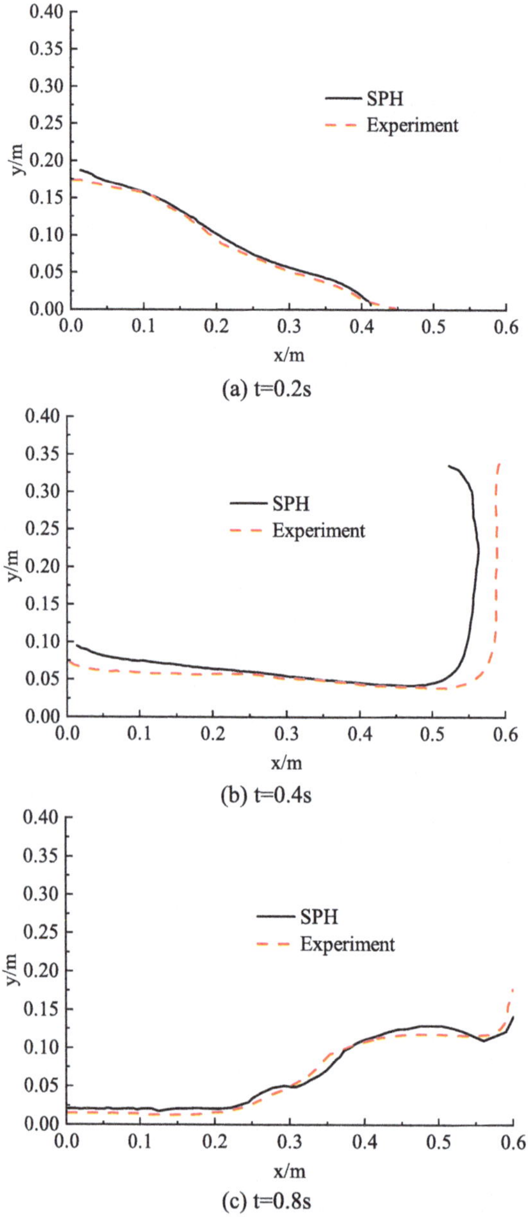

Fig. 3. Comparison between SPH simulation results and experimental free liquid surface height curve

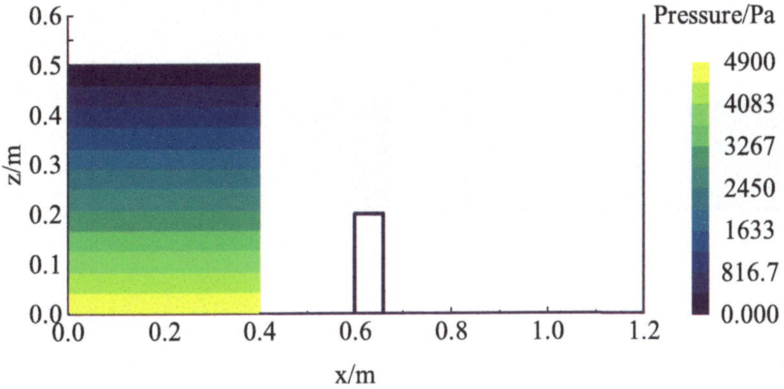

(a) Modeling of non-porous obstacles

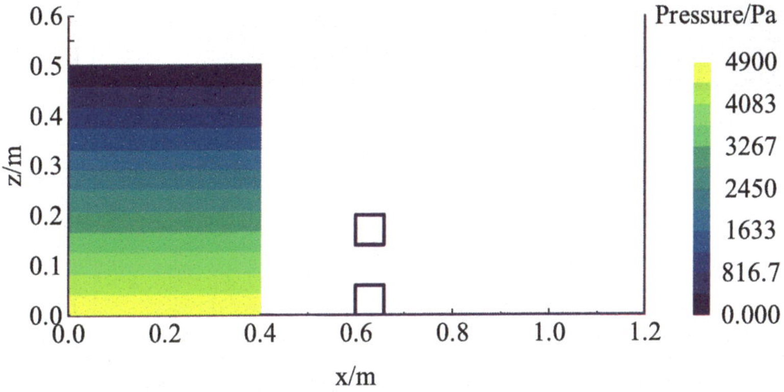

(b) Modeling of perforated obstacles

Fig. 4. Initial pressure distribution for different flume models

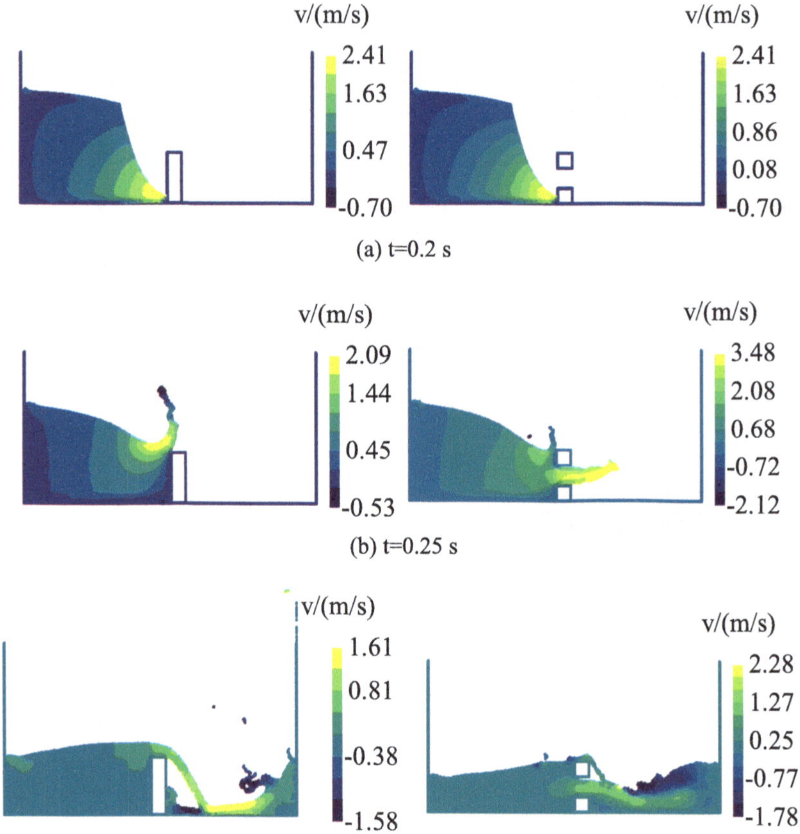

(a) t=0.2 s

(b) t=0.25 s

(c) t=0.9 s

Fig. 5. The comparision of freesurface flow variation in the two different model sat the same time

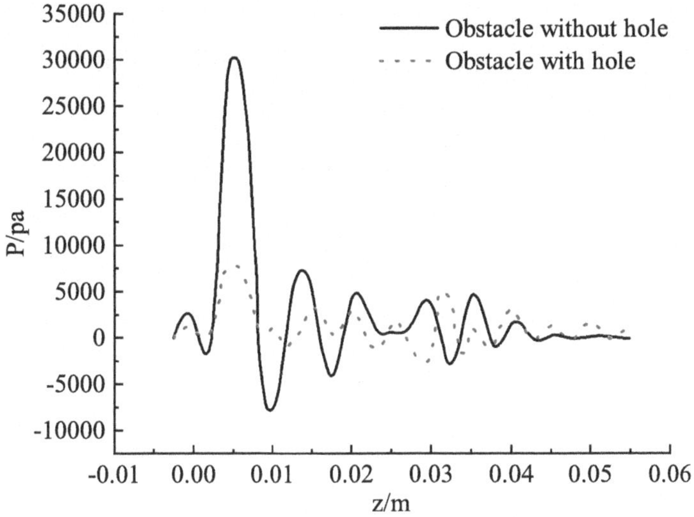

Fig. 6. The comparision of pressure when water hit the right wall at 0.9 s

References

1. Zhang, A., Sun, P., Ming, F., Colagrossi, A.: Smoothed particle hydrodynamics and its applications in fluid-structure interactions. J. Hydrodyn. **29**(2), 187–216 (2017)
2. Huo, Y., Zhong, Y., Ning, J., Liu, G., Duan, J., Cao, D.: Restoring particle consistency in discontinuous smoothed particle hydrodynamics. Eur. J. Mech. B. Fluids **100**, 223–238 (2023)
3. Dong, X., Huang, X., Liu, J.: Modeling and simulation of droplet impact on elastic beams based on SPH. Eur. J. Mech. A. Solids **75**, 237–257 (2019)
4. Monaghan, J.J., Kajtar, J.B.: SPH particle boundary forces for arbitrary boundaries. Comput. Phys. Commun., 1811–1820 (2009)
5. Wang, Z.L., Shen, H.T.: Lagrangian simulation of one-dimensional dam-break flow. J. Hydraul. Eng. **125**(11), 1217–1220 (1999)
6. Zhang, Y., Chen, J.: The conservation of element and share win method for numerical simulation of two dimensional dam bursting flood wave. J. Water Conserv. **36**(10), 1224–1229 (2005)
7. Savic, L.J., Holly, F.M.: Dambreak flood waves computed by modified Godunov method. J. Hydraul. Res. **31**(2), 187–204 (1993)
8. Huang, C., Lei, J.M., Liu, M.B., Peng, X.Y.: An improved KGF-SPH with a novel discrete scheme of Laplacian operator for viscous incompressible fluid flows. Int. J. Numer. Meth. Fluids **81**, 377–396 (2016)
9. Xu, F., Zhao, Y., Yan, R.: Multi-dimensional meshfree particle formulation for simulation shock waves. Shock Wave **93**, 1125–1146 (2013)
10. Liu, W., Zhang, Y.: Application of SPH method to numerical simulation of dam-break. J. Chongqing Univ. **40**(11), 11–19 (2017)

Preliminary Discussion on the Turbulent Behaviors of the Flow Field During the Resonance Among Wave, Current and Rippled Bottoms

Jun Fan[1,2], Aifeng Tao[1,2](✉), and Jinhai Zheng[1,2]

[1] Key Laboratory of Ministry of Education for Coastal Disaster and Protection, Hohai University, Nanjing 210024, China
aftao@hhu.edu.cn

[2] College of Harbour, Coastal and Offshore Engineering, Hohai University, Nanjing 210024, China

Abstract. Submarine continuous sand waves and complex tidal currents in the estuary and coastal areas have gradually drawn attention to the hydrodynamic interaction processes between ambient flow and continuous rippled bottoms. Focusing on the situation for steady free-surface flow over continuous undulating topography, the new wave components (as upstream-propagating waves) are generated on the free surface of the steady flow under specific conditions. The flow field during the wave excitation process is quantitatively observed based on the further exploration of observational data measured using Acoustic Doppler Velocimetry from the flume experiments performed. The turbulence intensity and spatial distribution characteristics above the rippled bottoms are analyzed. This aims to elucidate the turbulent dynamic characteristics of the flow field during wave excitation and its relationship with the wave excitation process.

Keywords: Wave field turbulent behavior · upstream-propagating waves · wave-current-bottom resonance · flume experiment · turbulent intensity

1 Introduction

Considering the existence of sand waves in estuarial and coastal regions, the free-surface responses for steady flow over rippled bottoms have drawn various attentions as the classical hydrodynamic problem, including the stationary free-surface undulations (Kennedy 1963, Mei 1969, Mizumura 1995) and the corresponding instability behaviors (Yih 1976, McHugh 1988, 1992). Especially, when steady flow passes over continuous undulating bottom topography, it triggers new wave components on the free surface under specific water depth and flow velocity conditions. This phenomenon is also known as the upstream-propagating waves (Kyotoh and Fukushima 1997, Fan et al. 2016, 2021, 2022) and is attributed to the dynamic interaction between ambient flow and undulating bottom

D. Jeng and B. Cai (Eds.): ICCOE 2024, LNCE 423, pp. 162–169, 2025.
https://doi.org/10.1007/978-981-96-0181-3_16

topography, especially in areas with extensive underwater continuous sand wave topography affected by tidal currents. Previous studies have shown that this phenomenon originates from resonant interactions (Liu et al. 1998, Raj and Guha 2019, Choi et al. 2021, Peng et al. 2022) and the simultaneous satisfaction of critical energy transfer conditions among waves, ambient flow, and rippled bottoms (Fan et al. 2021). This leads to the significant amplification of minor perturbation wave components participating in specific resonant dynamic interactions on the free water surface, continuously gaining energy and increasing their wave amplitudes. The dynamic factors that trigger this phenomenon primarily include flow velocity, water depth, wavelength, amplitude, and length of the undulating bottom topography. Despite some research on the principles and mechanisms behind the occurrence of the upstream-propagating waves as well as the related hydrodynamic problems (Fan et al. 2021, 2023a, 2023b), there is still limited investigation on the turbulent characteristics of the flow field during the excitation of upstream-propagating waves. Understanding the turbulent dynamic characteristics of the underwater flow field during the triggering of this phenomenon remains unknown and requires further analysis.

In this study, further analysis of flow field observations from flume experiments is carried out to elucidate the spatial and temporal distribution of turbulent characteristics of the flow field above the undulating topography during the excitation process of upstream-propagating waves. In detail, the turbulence intensity σ and the turbulent kinetic energy (TKE) are calculated and analyzed from the instantaneous flow velocity data measured by Acoustic Doppler Velocimetry.

The remainder of this paper is organized as follows. The description of flume experiment setup, apparatus, and case parameters are stated in Sect. 2. The turbulent behaviors of the flow field under the excitation process of the upstream-propagating waves are examined in Sect. 3. The Sect. 4 gives the conclusions.

2 Experimental Setup

The experiment was conducted in a wave flume with effective dimensions of 51.0 m × 1.0 m × 1.5 m (length × width × height). The side walls are made of glass, and beneath the flume, water pipes and pumps are buried to generate a constant water flow with adjustable velocity (see Fig. 1). The wavelength L_b of the fixed undulating bottom topography is 0.24 m, with an amplitude A_b of 0.04 m (see Fig. 2a). Its average height position coincides with the flat bottom of the flume (partially embedded in the flume bottom). Eight sinusoidal undulating bottom waves are arranged, with a total length of 1.80 m for the undulating bottom section. The midpoint of the bottom section is 33.0 m away from the flow inlet of the flume at upstream side, and flow stabilization facilities are installed. By analyzing typical test cases of flume experiments conducted for investigating the upstream-propagating waves, with a focus on phenomena and data characteristics of underwater flow fields, additional analysis of the flow field characteristics of wave-flow resonance on continuous undulating topography is supplemented.

The flow field measurements were concentrated between the first and second undulating bottom waves on the upstream side. The Acoustic Doppler Velocimetry (SonTek MicroADV) was used for flow field measurements, with the sampling frequency of 50 Hz

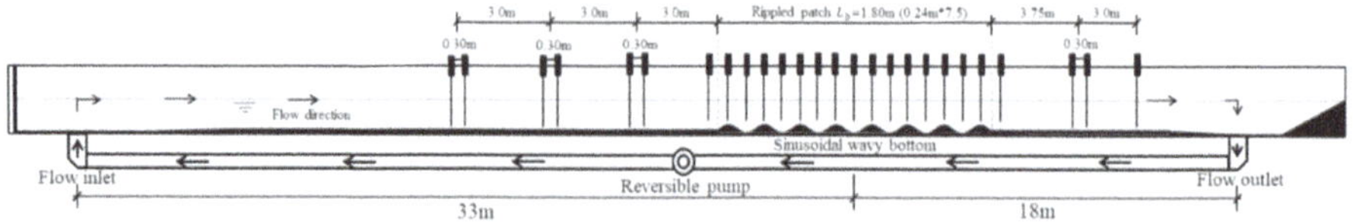

Fig. 1. Sketch of the flume experiments for flow over rippled bottoms.

and the sampling time of 60 s. The collected data included instantaneous flow velocity variations in three vertical directions over time. Spatial sampling points were located at the center section of the flume, with horizontal and vertical spatial intervals of 2 cm (except one point near the bottom trough). The spatial distribution of these points is illustrated below (see Fig. 2b).

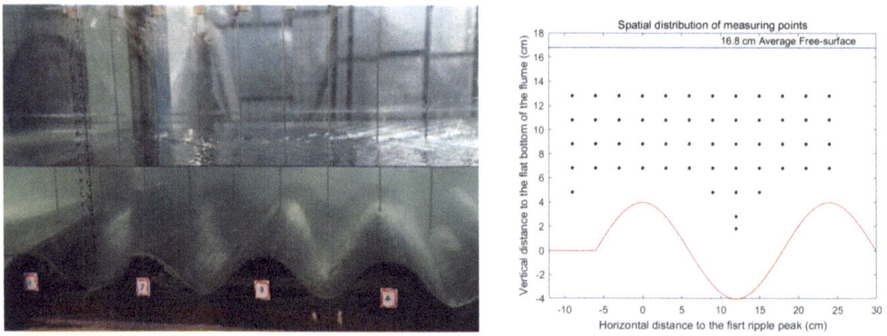

Fig. 2. (a-Left) Rippled bottom in flume experiment; (b-Right) Spatial distribution of the measuring point above the rippled bottoms.

For the cases analysis of the turbulent behaviors under the generation process of upstream-propagating waves for steady flow over rippled bottoms, the representative cases are selected with the distinguished generation phenomenon of these waves (see Table 1).

Table 1. Representative cases for flow field turbulent behavior analysis.

Water depth h (m)	Relative water depth h/L_b	Flow velocity U ($m{\cdot}s^{-1}$)	Froude Number
0.168	0.7	0.347	0.27

Note: The flow velocity U in this table represents the cross-sectional average velocity at the flat-bottom position in the flume. There are spatial differences in the flow velocities spatial distribution above the rippled bottoms due to variations of cross-sectional flow area.

Under the condition of the representative case, the upstream-propagating waves are generated above the rippled bottom. The corresponding free-surface elevation and spectrum are presented in Fig. 3.

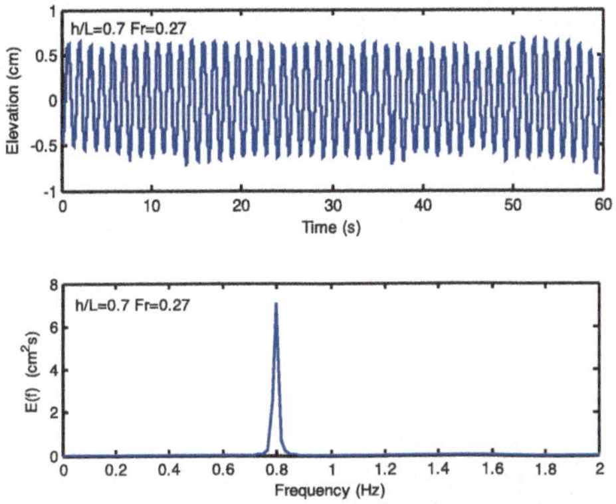

Fig. 3. (a-Above) Time-series of the free-surface elevation under the generation condition of upstream-propagating waves; (b-Below) Corresponding spectrum of the free-surface elevation.

3 Analysis of the Experimental Results

In this section, the characteristics of the spatial distribution of the turbulence intensity and turbulent kinetic energy (TKE) under the excitation process of upstream-advancing waves for the representative case are discussed in detail and compared with the qualitative flow field observation in flume experiments.

3.1 Spatial Distribution of Turbulence Intensity

The instantaneous flow velocity $u(t)$ in the x direction is expressed as Eq. 1.

$$u(t) = \overline{u} + u'(t) \tag{1}$$

in which, u' is the pulsatile flow velocity, and $\overline{u}$ is the time-averaged flow velocity, which is calculated by Eq. 2.

$$\overline{u} = \frac{1}{T}\int_0^T u'(t)dt \tag{2}$$

In Eq. 2, T is measuring time. Based on the measured flow velocity data by ADV, the time-averaged flow velocity is the arithmetic mean of the time series of instantaneous flow velocities.

Then, the turbulence intensity could be calculated as below,

$$\sigma_x = \sqrt{\frac{\sum_{i=1}^{N}(u_i - \overline{u})^2}{N}}, \sigma_y = \sqrt{\frac{\sum_{i=1}^{N}(v_i - \overline{v})^2}{N}}.\sigma_z = \sqrt{\frac{\sum_{i=1}^{N}(w_i - \overline{w})^2}{N}} \tag{3}$$

in which, σ_x, σ_y, and σ_z are the turbulence intensity of the x, y, and z directions respectively. And u_i, v_i, and w_i are the instantaneous velocity of the x, y, and z directions respectively. N is the total amount of the sample size, which is 3000 for the cases of the flume experiments in this study.

The spatial distribution of the turbulence intensity σ_x of x direction, which is horizontal along the flow direction presents distinctive spatial difference for both horizontal and vertical directions (see Fig. 4). Both the regions with large and small turbulence intensity are above the first trough of the rippled bottoms with the closed form, in which, the regions with large turbulence intensity σ_x are below that region with the smaller value. And the turbulence intensity σ_x is relatively larger in the downstream side than the upstream side.

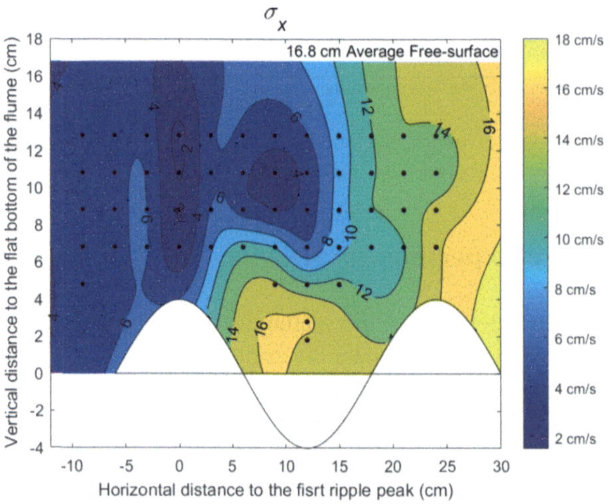

Fig. 4. Spatial distribution of the turbulence intensity of x direction (horizontal along the flow direction) above rippled bottoms for the case $h/L_b = 0.7$, $Fr = 0.27$.

Compared to the primary flow direction x, the turbulence intensity σ_y of y direction (horizontal in lateral direction) above rippled bottoms are relatively smaller than σ_x (see Fig. 5). The values of σ_y present the lateral non-uniformity of the flow field in the experimental flume, which are relatively smaller in the upstream side and upwards. The regions with the maximum values are still located in the trough between two ripples' crest.

As the turbulence intensity of the vertical direction z (see Fig. 6), the spatial distribution of σ_z has the similar pattern of σ_x. The regions with relatively large value of σ_z are concentrated in the location of the topographic trough with the circular-shaped enclosed area.

3.2 Spatial Distribution of Turbulent Kinetic Energy

The turbulent kinetic energy TKE (simplified as T below, with the unit of Joule, J) could comprehensively consider the turbulence intensity in three directions of the spatial domain. It is calculated on the basis of the turbulence intensity as below,

$$T = \frac{1}{2}(\sigma_x^2 + \sigma_y^2 + \sigma_z^2) \tag{4}$$

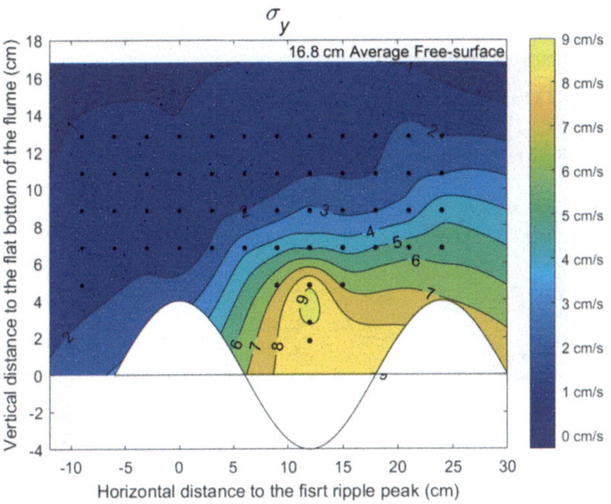

Fig. 5. Spatial distribution of the turbulence intensity of y direction (horizontal in lateral direction) above rippled bottoms for the case $h/L_b = 0.7$, $Fr = 0.27$.

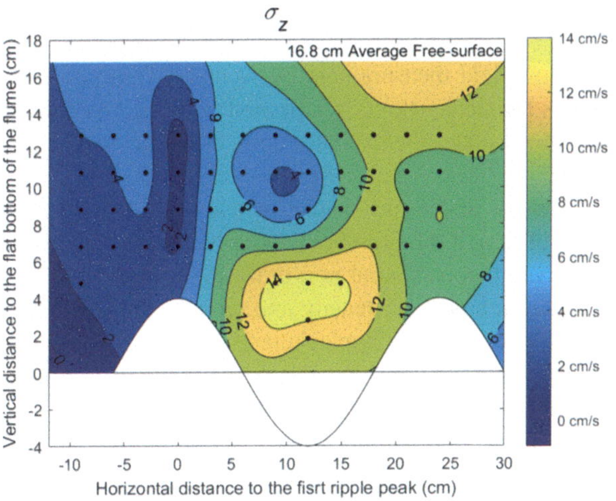

Fig. 6. Spatial distribution of the turbulence intensity of z direction (vertical direction) above rippled bottoms.

in which, σ_x, σ_y, and σ_z are the turbulence intensity of the x, y, and z directions respectively.

The spatial distribution of the turbulent kinetic energy integrates the turbulence intensity characteristics across all three directions. The circular-shaped enclosed region with larger energy values still located at the trough of the rippled bottoms. Conversely, the similar circular-shaped enclosed region with smaller energy values just located above it. And the general turbulent kinetic energy increases towards the downstream side (Fig. 7).

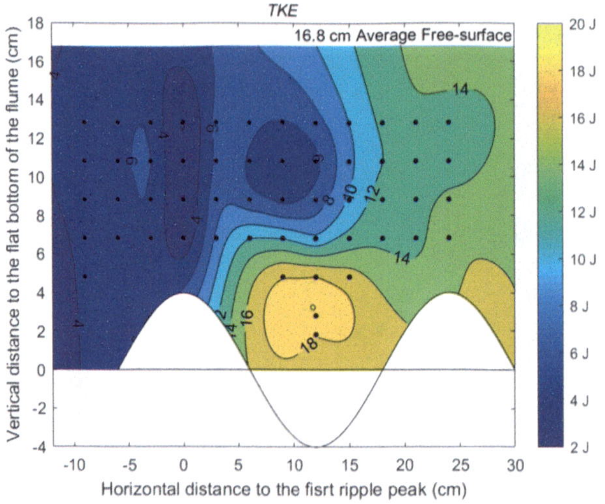

Fig. 7. Spatial distribution of the turbulent kinetic energy (TKE) above rippled bottoms.

4 Conclusions

On the basis of the flow field measurement by Acoustic Doppler Velocimetry during the flume experiments for steady flow over rippled bottoms, the quantitative turbulent behaviors of the flow field under the excitation process of upstream-propagating waves are revealed.

The spatial distribution of the turbulence intensity (σ_x, σ_y, σ_z), as well as the turbulent kinetic energy (TKE) of the flow field under the excitation process of upstream-propagating waves have obvious spatial non-uniformity behavior. It exists the regions with large values of the turbulence intensity and circular-shaped enclosed form in the trough area between the neighboring ripples' crests, which reveals the vortex behaviors from the quantitative perspective. Besides, the region with smaller turbulence intensity (limited to σ_x and σ_z) is observed and located just above the region with larger values, which needs further analysis in the future work.

Acknowledgements. This research was supported by the National Natural Science Foundation of China (Grant No. 52101308), the Fundamental Research Funds for the Central Universities (Grant No. B240201012), and the National Natural Science Foundation of China (Grant No. 52271271).

References

Kennedy, J.F.: The mechanics of dunes and antidunes in erodible-bed channels. J. Fluid Mech. **16**(4), 521–544 (1963)
Mei, C.C.: Steady free surface flow over wavy bed. J. Eng. Mech. Div. **95**(6), 1393–1402 (1969)
Yih, C.S.: Instability of surface and internal waves. Adv. Appl. Mech., 369–419 (1976)
McHugh, J.P.: The stability of stationary waves in a wavy-walled channel. J. Fluid Mech. **189**, 491–508 (1988)

McHugh, J.P.: The stability of capillary-gravity waves on flow over a wavy bottom. Wave Motion **16**(1), 23–31 (1992)

Mizumura, K.: Free-surface profile of open-channel flow with wavy boundary. J. Hydraul. Eng. **121**(7), 533–539 (1995)

Kyotoh, H., Fukushima, M.: Upstream-advancing waves generated by a current over a sinusoidal bed. Fluid Dyn. Res. **21**(1), 1–28 (1997)

Liu, Y.M., Yue, D.K.P.: On generalized Bragg scattering of surface waves by bottom ripples. J. Fluid Mech. **356**, 297–326 (1998)

Fan, J., Zheng, J.H., Tao, A.F., Yu, H.F., Wang, Y.: Experimental study on upstream-advancing waves induced by currents. J. Coast. Res. **SI75**, 846–850 (2016)

Raj, R., Guha, A.: On Bragg resonances and wave triad interactions in two-layered shear flows. J. Fluid Mech. **867**, 482–515 (2019)

Choi, W., Chabane, M., Taklo, T.M.A.: Two-dimensional resonant triad interactions in a two-layer system. J. Fluid Mech. **907**, A5 (2021)

Fan, J., Zheng, J.H., Tao, A.F., Liu, Y.: Upstream-propagating waves induced by steady current over a rippled bottom: theory and experimental observation. J. Fluid Mech. **910**, A49 (2021)

Peng, J., Tao, A.F., Fan, J., Zheng, J.H., Liu, Y.M.: On the downshift of wave frequency for Bragg resonance. China Ocean Eng. **36**(1), 76–85 (2022)

Fan, J., Tao, A.F., Zheng, J.H., Peng, J.: Numerical investigation on temporal evolution behavior for triad resonant interaction induced by steady free-surface flow over rippled bottoms. J. Mar. Sci. Eng. **10**, 1372 (2022)

Fan, J., Tao, A.F., Xie, S.Y., Wu, D., Wang, G.: Numerical investigation on nonlinear evolution behavior and water particle velocity of wave crests for narrow-band wave field with Gaussian spectrum. Ocean Eng. **268**, 113518 (2023a)

Fan, J., Tao, A.F., Shi, M.Q., Li, Y., Peng, J.: Flume experiment investigation on propagation characteristics of tidal bore in a curved channel. China Ocean Eng. **37**, 131–144 (2023b)

Offshore Wind Power Generation and Structural Mechanics of Foundation Platform

Real-Time Monitoring and Analysis for Digital Twin of Floating Wind Turbine

Yi Liu, Hou-Chun Ning, and Zhen-Zhong Hu(✉)

Institute for Ocean Engineering, Shenzhen International Graduate School, Tsinghua University, Shenzhen 518055, China
huzhenzhong@tsinghua.edu.cn

Abstract. The increasing demand for clean energy has made floating wind turbines (FWTs) a viable option. However, the operation and maintenance of FWTs face numerous challenges. The emergence of digital twins provides a potential solution. This study focuses on real-time monitoring and analysis techniques for FWTs, aiming to address key challenges in building the digital twin of such systems. We propose a monitoring and analysis framework for the digital twin of FWTs, covering three main aspects: ocean environment, FWT's pose, and structural state. For the tower structure of FWTs, we present a state inversion method based on limited sensors, enabling rapid and comprehensive assessment of structural health. Based on this, we construct a scaled model of the FWT and apply the proposed methods to the monitoring and analysis of an actual FWT, realizing a complete process from data perception, storage, integration, and fusion to feedback control, creating an information closed-loop between the physical and virtual worlds. Furthermore, the application of the constructed digital twin platform validates the feasibility and effectiveness of the methods. By deeply integrating digital models, ocean environment data, and FWT state data, we create a comprehensive digital twin of the FWT, providing crucial support for transparency of state information, safety risk management, intelligent parameter adjustments, and informed decision-making.

Keywords: Floating Wind Turbine · Digital Twin · Real-Time Monitoring · Data Fusion · State Inversion

1 Introduction

With the increasing global demand for renewable energy, offshore wind turbines have gained significant attention as a clean energy generation technology. Floating wind turbines (FWTs), with their ability to overcome geographical limitations, have become an important approach for offshore wind power generation. However, FWTs face significant challenges in terms of stability and reliability due to their long-term exposure to complex ocean conditions [1]. Real-time monitoring and analysis of FWT state are of great significance for ensuring their safe operation and optimizing performance [2].

The state monitoring of FWTs involves multiple aspects, including environmental conditions, pose changes, and structural health, which are crucial for ensuring the

D. Jeng and B. Cai (Eds.): ICCOE 2024, LNCE 423, pp. 173–185, 2025.
https://doi.org/10.1007/978-981-96-0181-3_17

safe and reliable operation. For example, Kou et al. discussed environmental monitoring technologies and equipment for offshore wind farms, including meteorological radars, anemometers, observation stations, buoys, and earthquake monitoring [3]. Ciuriuc et al. analyzed various sensors and monitoring techniques for FWTs, including vibration-based damage detection, temperature measurement, strain measurement, and photogrammetry [4]. Jaksic et al. monitored the dynamic response of a tension leg platform in a wave tank using load cells, a camera-based motion recognition system, and a laser Doppler vibrometer, analyzing the nonlinear characteristics of the platform response [5]. Wang et al. proposed an identification method for the tower response of FWTs using artificial intelligence and optimized sensor deployment to obtain more tower response information [6].

However, traditional monitoring methods are often constrained by factors such as time, space, and cost, making it difficult to meet the real-time monitoring requirements of FWTs. In recent years, digital twin (DT) technology has been introduced as an innovative means in the field of ocean engineering, providing a new approach for real-time monitoring and analysis. DT is a technology that connects physical entities with their digital models, synchronizing, comparing, and validating the real-world operational conditions with the digital models through real-time data acquisition and analysis, thereby simulating, predicting, and optimizing the behavior of the physical entity [7]. DT has four main characteristics: (1) real-time synchronization, where the virtual model is updated as quickly as possible to stay synchronized with the physical entity [8]; (2) bidirectional interaction, enabling information exchange between the physical entity and the virtual model in both directions [9]; (3) fidelity, where the virtual model accurately reflects the characteristics, state, and performance of the physical entity [10]; and (4) simulatability, where the behavior and future states of the physical entity can be simulated in a virtual environment [11]. The introduction of DT technology allows for the comprehensive and accurate monitoring of FWTs, utilizing various monitoring data, providing real-time data analysis and decision support, and ultimately improving the efficiency, reliability, and safety of FWTs.

In recent years, there has been a rapid growth in research applying DT technology to the field of ocean engineering. Some studies have focused on applying DTs to FWTs for real-time monitoring and prediction of turbine states. For example, Moghadam et al. proposed a stochastic physics-based model combined with DT technology to predict the remaining useful life of transmission system components [12]. Walker et al. developed a DT for the mooring system of FWTs to accurately detect long-term drifts and predicting near-term axial tension [13]. Wang et al. proposed a DT-based monitoring framework for offshore wind turbines to support fatigue crack prediction in supporting structures [14]. Haghshenas et al. developed a DT platform using Unity3D for visualization, which can detect possible failures and present them using augmented reality [15]. However, the application of DT in the field of FWTs still faces challenges such as complex sensor deployment, data fusion and real-time analysis.

Therefore, this study aims to investigate real-time monitoring and analysis techniques for FWTs, and provide effective solutions for intelligent operation and performance optimization by incorporating DT-related technologies. The focus of this study will be on the environmental, pose, and structural monitoring of FWTs. By constructing accurate digital models and utilizing data acquisition and state description methods, corresponding solutions for data fusion, analysis, and visualization will be proposed. Furthermore, a DT platform for FWTs will be developed, and the application of DTs in intelligent operation will be explored. Through this study, we expect to provide novel insights, methods, practices, and tool support for real-time monitoring and analysis of FWTs, and promote the application of DT technology in the field of ocean engineering.

2 Monitoring and Analysis Methods

2.1 Overall Framework

This study aims to achieve real-time monitoring and analysis of the position, attitude, and structural state of FWTs. Considering the significant influence of environment, we divide the work into three main components: ocean environment, FWT's pose, and structural state. The overall framework is illustrated in Fig. 1.

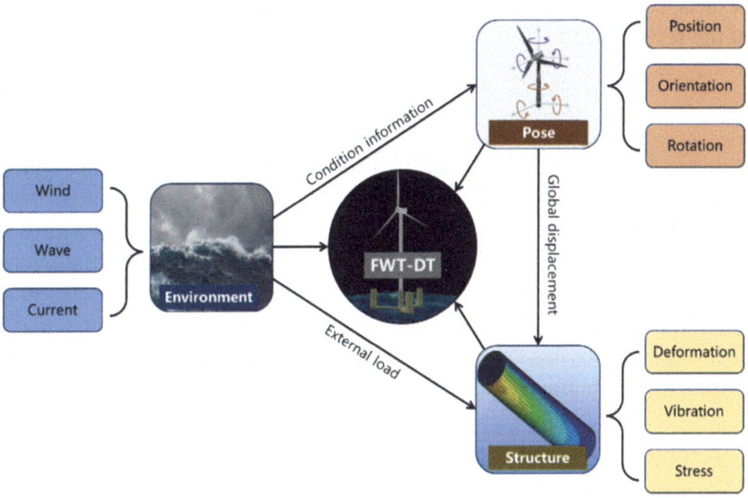

Fig. 1. Overall monitoring and analysis framework for FWTs

2.2 Ocean Environment

The ocean environment where the FWTs are located often differs from the predefined conditions, necessitating the monitoring of environment in the local area around the FWTs. For the purposes of turbine motion and structural monitoring, key environmental factors include sea surface waves, offshore wind fields, and subsea flow fields. The following are descriptions and monitoring methods for these three factors.

Regarding sea surface waves, it is necessary to consider parameters such as wave height H, wave period T, and wave direction $\overrightarrow{n} = (n_x, n_y)$. Monitoring of sea surface waves can be achieved using wave buoys, wave pressure sensors, and remote sensing techniques. For visualizing waves, one approach is to represent them through surface deformations by constructing a time-varying surface height. For instance, in the case of deep-water conditions and second-order Stokes waves, the sea surface height function $h(x, y, t)$ can be expressed as follows (g is the gravitational acceleration):

$$\begin{cases} \xi(x, y, t) = \frac{4\pi^2}{g \cdot T^2} \cdot \left(x \cdot n_x + y \cdot n_y\right) - \frac{2\pi}{T} \cdot t + \phi_0 \\ h(x, y, t) = \frac{H}{2} \cdot \cos(\xi) + \frac{\pi^2 \cdot H^2}{2g \cdot T^2} \cdot \cos(2\xi) \end{cases} \tag{1}$$

Concerning offshore wind fields, the wind speed near the FWT blades is typically of interest, including both wind direction and wind speed magnitude. This information can be monitored using anemometers and wind vanes, and visualized by a wind direction indicator and a wind speed display. Subsea flow velocity also require consideration of both flow direction and flow velocity magnitude. Measurements can be carried out using subsea flow buoys, acoustic Doppler current profilers, electromagnetic flowmeters. The visualization of ocean flow fields can be demonstrated through the horizontal movement of the sea surface, for example, by implementing coordinate translations using a mapped texture of the sea surface.

2.3 FWT's Pose

When monitoring the pose of a FWT, the main consideration is the motion of the turbine foundation and the relative rotation of the upper blades. Therefore, there are 9 degrees of freedom, consisting of 3 translational and 6 rotational degrees. The translational motion can be recorded using a translation vector $\overrightarrow{v} = \left(v_x, v_y, v_z\right)$. For the rotation, special treatment is required due to the different order of rotations leading to different results. Any rotation can be achieved by sequentially rotating around the global x-axis by ϕ_x, then the global y-axis by ϕ_y, and finally the global z-axis by ϕ_z. The direction of the local coordinate system after rotation can be expressed as follows [16]:

$$\begin{cases} \overrightarrow{n_x} = \left(\cos\phi_z \cos\phi_y, \sin\phi_z \cos\phi_y, -\sin\phi_y\right) \\ \overrightarrow{n_y} = \left(\cos\phi_z \sin\phi_y \sin\phi_x - \sin\phi_z \cos\phi_x, \cos\phi_z \cos\phi_x + \sin\phi_z \sin\phi_y \sin\phi_x, \cos\phi_y \sin\phi_x\right) \\ \overrightarrow{n_z} = \left(\cos\phi_z \sin\phi_y \cos\phi_x + \sin\phi_z \sin\phi_x, \sin\phi_z \sin\phi_y \cos\phi_x - \cos\phi_z \sin\phi_x, \cos\phi_y \cos\phi_x\right) \end{cases} \tag{2}$$

It can be deduced that any rotation result can also be represented by a set of ϕ_x, ϕ_y, and ϕ_z. Furthermore, it can be derived that there is a relationship between the direction vector $\overrightarrow{d_0}$ before rotation and the direction vector $\overrightarrow{d}$ after rotation as follows:

$$\overrightarrow{d} = T \cdot \overrightarrow{d_0}, T = \left[\overrightarrow{n_x}\ \overrightarrow{n_y}\ \overrightarrow{n_z}\right] \tag{3}$$

Here, T is referred to as the rotation matrix. It is important to note that in the actual rotation process, these three parameters often change simultaneously.

The aforementioned parameters may not provide an intuitive representation of the FWT's pose. Therefore, we propose three new parameters: (1) The angle ϕ_a between the projection of the local x-axis on the horizontal plane and the global x-axis (azimuth angle). (2) The angle ϕ_b between the local x-axis and the horizontal plane (elevation angle). (3) The angle ϕ_c between the local y-axis and the horizontal plane (roll angle). The new parameters have the following relationships with the previous parameters:

$$\phi_a = \phi_z, \phi_b = -\phi_y, \sin\phi_c = \cos\phi_y \sin\phi_x \tag{4}$$

By substituting Eq. (4) into Eq. (2), the rotation matrix corresponding to the new parameters can be obtained.

For any point $\overrightarrow{p_0}$ on the wind turbine base, the coordinates $\overrightarrow{p}$ after the motion is:

$$\overrightarrow{p} = T \cdot \overrightarrow{p_0} + \overrightarrow{v} \tag{5}$$

The parameters T and $\overrightarrow{v}$ are determined by 3 rotational parameters and 3 translational parameters, respectively. For the rotation of the upper blades, an additional rotation can be incorporated, considering the changes in the FWT foundation:

$$\overrightarrow{p_{up}} = T_{up} \cdot \left(\overrightarrow{p} - \overrightarrow{c}\right) + \overrightarrow{c} = T_{up} \cdot T \cdot \left(\overrightarrow{p_0} - \overrightarrow{c_0}\right) + T \cdot \overrightarrow{c_0} + \overrightarrow{v} \tag{6}$$

where $\overrightarrow{c_0}$ represents the rotation center of the blades before the pose transformation.

Therefore, after obtaining the 9 parameters through monitoring calculations, the FWT's pose can be described and visualized by applying Eqs. (5) and (6). These equations allow us to determine the position of each point on the FWT after the transformations, thereby enabling a comprehensive representation of the FWT's pose.

2.4 Tower Structure

Due to the complex loads, the structural state of a FWT is often a focal point for users. In this study, the FWT's tower is chosen as the main object to develop monitoring, analysis, and visualization methods. Traditional methods may primarily focus on the positions of maximum stress under design conditions. However, the actual structural state under environmental conditions can be significantly different, and examining the local positions alone cannot guarantee the overall structural safety. Therefore, it is necessary to integrate monitoring data with the FWT's structure information to achieve a deeper level of real-time monitoring and synchronization of the structural state.

For the tower structure, considering that its vertical dimension is much larger than the horizontal dimension, the primary concern is the bending deformation. Taking the tower base as a reference, assuming the length of the tower is L and the z-coordinate at the base is 0, the bending stiffness in a certain horizontal direction is denoted as $I(z)$. The elastic modulus of the tower material is E, and its density is ρ. Under the assumption of linear elasticity and small deformation, the axial rotation and deformation function under a concentrated force F at the top can be calculated as follows:

$$\begin{cases} \theta(z) = \int_0^z \frac{F \cdot (L - z_1)}{E \cdot I(z_1)} dz_1 \\ \Delta(z) = \int_0^z \theta(z_1) \cdot dz_1 \end{cases} \tag{7}$$

Apart from the tower base, the upper part of the tower is primarily subjected to the loads transmitted from the blades and nacelle. Therefore, this can be considered as one of the main loading conditions for the tower. Specifically, when the cross-section of the tower remains relatively unchanged, it can be obtained that:

$$\begin{cases} \theta(z) = \frac{F}{EI} \cdot z \cdot \left(L - \frac{z}{2}\right) \\ \Delta(z) = \frac{F}{EI} \cdot z^2 \cdot \left(\frac{L}{2} - \frac{z}{6}\right) \end{cases} \tag{8}$$

For non-static scenarios, it is also necessary to consider the vibration of the tower. The dynamic behavior of the structure is mainly determined by its natural frequencies and mode shapes, with low-frequency modes typically playing a dominant role. Assume that the deformation function of the tower under resonance conditions is given by

$$\Delta(z, t) = \varphi(z) \cdot Q \cdot \sin(\omega \cdot t + \alpha) \tag{9}$$

Where φ represents the mode shape function, Q is the amplitude coefficient, and ω is the vibration frequency corresponding to that mode. Assuming the cross-sectional area of the tower is $A(z)$, based on the equilibrium relationship, we have

$$E \cdot I(z) \cdot \frac{\partial^4 \Delta}{\partial z^4} + \rho \cdot A(z) \cdot \frac{\partial^2 \Delta}{\partial t^2} = 0 \tag{10}$$

Simplifying further, we have:

$$E \cdot I(z) \cdot \frac{d^4 \varphi}{dz^4} - \rho \cdot A(z) \cdot \omega^2 \cdot \varphi = 0 \tag{11}$$

Now consider the boundary conditions. We approximate that the bottom of the tower is fixed, and components such as the nacelle and blades can be represented as a concentrated mass m. Therefore, the boundary conditions can be expressed as

$$\varphi(0) = 0, \varphi'(0) = 0, \varphi''(L) = 0, E \cdot I(z) \cdot \varphi^{(3)}(L) + m \cdot \omega^2 \cdot \varphi(L) = 0 \tag{12}$$

Specifically, when the cross-section of the tower remains unchanged, we have:

$$\begin{cases} 1 + \cosh \lambda \cdot \cos \lambda + \beta \cdot \lambda \cdot (\sinh \lambda \cdot \cos \lambda - \cosh \lambda \cdot \sin \lambda) = 0 \\ \varphi(z) = \cosh\left(\lambda \cdot \frac{z}{L}\right) - \cos\left(\lambda \cdot \frac{z}{L}\right) - \frac{\cosh(\lambda)+\cos(\lambda)}{\sinh(\lambda)+\sin(\lambda)}\left[\sinh\left(\lambda \cdot \frac{z}{L}\right) - \sin\left(\lambda \cdot \frac{z}{L}\right)\right] \end{cases} \tag{13}$$

Where $\beta = m/(\rho A \cdot L)$ represents the ratio of the upper mass to the tower mass, $\lambda = L \cdot \sqrt[4]{\left(\omega^2 \cdot \rho A\right)/EI}$ is related to the natural frequency.

For cases involving cross-sectional variations, obtaining analytical solutions for the above equations may be challenging. In such situations, it is possible to approximate the tower by dividing it into multiple segments of uniform beams for analysis.

During the monitoring of actual tower, the initial step involves selecting the fundamental states of the structure. It is assumed that the true state of the structure is a linear combination of several fundamental states. The overall structural state of the tower is obtained by fitting real-time monitoring data. For instance, if the deformation function $\delta(z)$ under the concentrated force at the top and the first n mode shapes $\varphi_i(z)$ are chosen as the fundamental states, the actual deformation state $\Delta(z)$ can be expressed as:

$$\Delta(z) = a_0 \cdot \delta(z) + a_1 \cdot \varphi_1(z) + a_2 \cdot \varphi_2(z) + \cdots + a_n \cdot \varphi_n(z) \tag{14}$$

By using the actual displacements measured at different heights, the coefficients a_i can be determined through fitting, resulting in the final deformation function of the tower. It is important to have at least as many monitoring points as coefficients. With this method, it is possible to obtain the deformation values at any height of the tower, rather than being limited to the information from the monitoring points alone.

Based on Eq. (14), further information regarding the tower's angular rotation and bending moments can be obtained:

$$\begin{cases} \theta(z) = \frac{d\Delta}{dz} = a_0 \cdot \frac{d\delta}{dz} + a_1 \cdot \frac{d\varphi_1}{dz} + a_2 \cdot \frac{d\varphi_2}{dz} + \cdots + a_n \cdot \frac{d\varphi_n}{dz} \\ M(z) = -EI \cdot \frac{d^2\Delta}{dz^2} = -EI \cdot \left(a_0 \cdot \frac{d^2\delta}{dz^2} + a_1 \cdot \frac{d^2\varphi_1}{dz^2} + a_2 \cdot \frac{d^2\varphi_2}{dz^2} + \cdots + a_n \cdot \frac{d^2\varphi_n}{dz^2} \right) \end{cases} \quad (15)$$

Through Eq. (15), on one hand, fitting the deformation function implies obtaining other information such as the cross-sectional rotation and internal forces of the tower. On the other hand, since the rotations and internal forces for each fundamental state are known, it is also possible to fit actual states using other types of data, such as using angular sensors or stress sensors, or even combining multiple sensors.

3 Case Study

3.1 FWT Model

To validate the feasibility and effectiveness of the proposed method, a scaled model was designed and fabricated based on the DTU 10-MW offshore FWT model [17]. Following the principles of Froude's similarity and considering manufacturing limitations such as thickness, diameter, as well as the installation of control and monitoring systems, the basic parameters of the scaled FWT model were determined as follows: The tower was constructed using aluminum material, with a height of 42 cm. The tower had a uniform cross-section with an outer diameter of 25 mm and a wall thickness of 1.5 mm. The combined mass of the blades and nacelle was 3.2954 kg (Fig. 2).

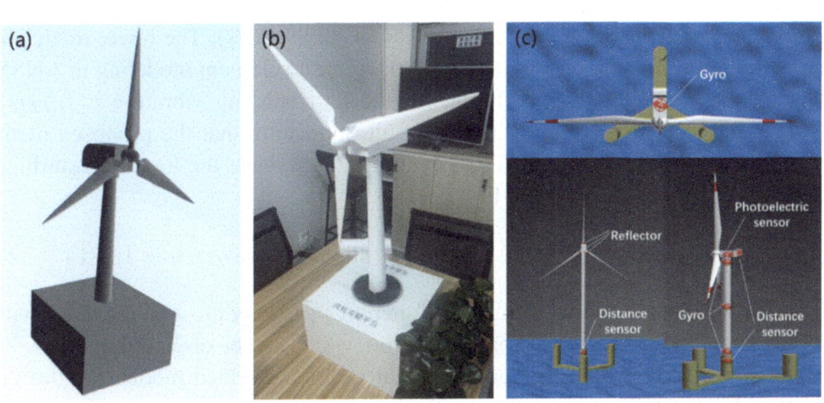

Fig. 2. FWT design model (a), physical model (b), and sensor layout (c)

The sensor layout scheme is as follows, utilizing a total of 10 sensors to monitor 15 data points. Regarding the FWT environment, we employed anemometer sensor and wind vane sensor for wind speed and wind direction monitoring, respectively. For the FWT's pose, we installed laser distance sensors to measure the displacement of the FWT's base in three directions. We also placed gyroscopes at the FWT's base and the nacelle's top to measure the rotation data of them (note that the nacelle does not roll). Reflective markers were placed on the three blades, and rotation speed of the blades was measured using a photoelectric sensor. For structural monitoring, we installed gyroscopes at the middle and top sections of the tower to measure rotational data around two horizontal axes. Finally, a sensor network was established using LoRa technology for remote data transmission, and a database was designed to store the acquired data.

3.2 Tower Structure Analysis

The deformation of the tower is monitored using two angular sensors located at the middle and top sections of the tower, and an additional angular sensor at the base is used for data calibration. In a single direction, a maximum of two fundamental force states can be selected as they cannot exceed the number of monitoring data points.

To verify the accuracy of tower angular measurements, tests were conducted using concentrated forces applied at the top. According to Eq. (8), when a concentrated static force is applied at the tower end, the base angle θ_L, middle section angle θ_M, and top section angle θ_U should satisfy $(\theta_M - \theta_L)/(\theta_U - \theta_L) = 0.75$. After applying three different loads at the top of the tower, the calculated ratios were found to be 0.7311, 0.7447, and 0.7763. Considering the effects of installation and measurement errors, the deviations in the actual results are within an acceptable range, indicating that this method can effectively monitor the deformation of the tower structure.

Furthermore, the first-order free vibration of the FWT is considered. The aluminum material has a density of $\rho = 2700\,\text{kg/m}^3$ and an elastic modulus of $E = 70\,\text{GPa}$. Based on the tower's cross-sectional parameters, the cross-sectional area is calculated as $A = 1.1074 \times 10^{-4}\,\text{m}^2$, and the moment of inertia is $I = 7.6757 \times 10^{-9}\,\text{m}^4$. The ratio of the concentrated mass to the tower mass is $\beta = 3.2954/0.1256 = 26.24$. Substituting this value into Eq. (13) yields the λ value for the first mode as $\lambda_1 = 0.5802$, and the corresponding frequency is $f_1 = 12.87\,\text{Hz}$ ($\omega_1 = 80.89\,\text{rad/s}$). The tower model with concentrated mass at the top is also analyzed using finite element modeling in ANSYS, and the approximate frequency solution for the first bending vibration is $f_{1,FEM} = 12.74\,\text{Hz}$. This result is very close to our result, indicating that the proposed method can rapidly and accurately calculate the bending vibration of the tower. According to Eq. (13), the shape function for the first mode is:

$$\varphi_1(z) = \cosh \eta z - \cos \eta z - 1.7301 \times (\sinh \eta z - \sin \eta z), \eta = 1.3814 \quad (16)$$

It is worth noting that the mode shape function φ_1 is very close to the deformation function δ under the concentrated force at the top, which can be observed in Fig. 3. This is because, in order to satisfy Froude's similarity for the scaled model, the designed concentrated mass at the upper section is much larger than the tower mass. Figure 3 also shows the shapes of several other vibration modes of the FWT structure.

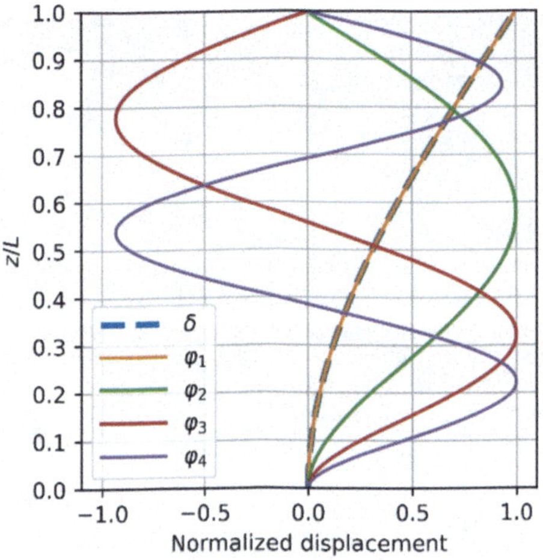

Fig. 3. Deformation under concentrated force at the top of the tower and the shapes of the first four vibration modes (normalized)

3.3 DT Platform Development

Based on the above content, we have designed and constructed a DT platform for monitoring the environmental conditions, pose, and tower structure of an FWT. The overall functional logic of the platform is shown in Fig. 4.

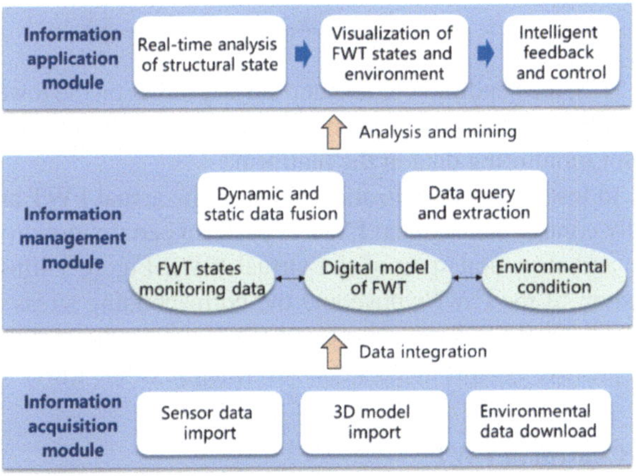

Fig. 4. Overall functional logic of the DT platform

The digital model of FWT includes the geometric shape of FWT, material and composition, and the aforementioned structural calculation model, etc.

Figure 5 provides visual representations of the ocean environment under different wind speed and wave parameters.

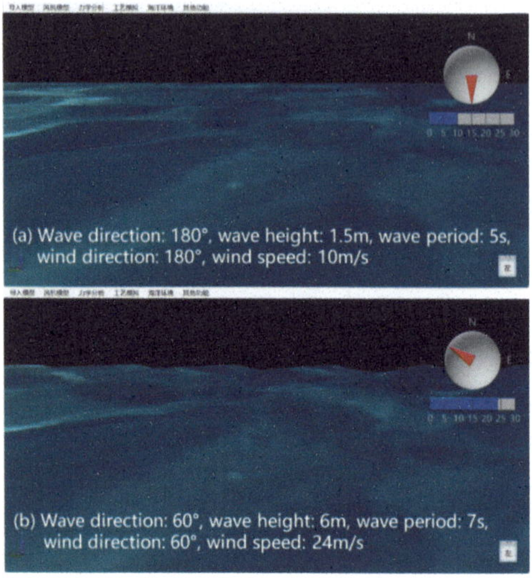

Fig. 5. Visual representation of the ocean environment under different wind fields and wave parameters

Since the tower's deformation under end static load is almost identical to the first-mode vibration, we only consider the deformation under end static load as the fundamental state. By utilizing two measured angular data at the middle and top of the tower, the best estimation of the tower deformation is obtained through least squares fitting:

$$\Delta(z) = \frac{24(\theta_M - \theta_L) + 32(\theta_U - \theta_L)}{25L^2} \cdot z^2 \cdot \left(\frac{L}{2} - \frac{z}{6}\right) \tag{17}$$

Based on Eq. (15), the bending moment and surface stress on the tower cross-section can be further calculated. Figure 6 presents the tower structure analysis information based on the given sensor monitoring data in the platform.

Furthermore, to test the synchronization between the actual FWT and the DT platform, we manually control the physical FWT to perform certain motion commands and observe the pose and structural state of the virtual FWT. Figure 7 illustrates the real-time synchronization of the FWT's pose and the corresponding stress analysis within the platform. The physical FWT and the virtual FWT exhibit the same motion, and the DT platform can provide synchronized stress information of the tower structure.

3.4 Intelligent Control of FWT

The previous experiments validated the information transfer and analysis from the physical space to the virtual space. However, a key feature of DTs is their ability to achieve bidirectional interaction, meaning that the analysis information from the digital space should also be transmitted back to the physical space. To demonstrate that the proposed method in this study can meet the real-time and bidirectional interaction characteristics of the DT, we have added a functionality in the DT platform to adjust the rotor direction based on the monitoring wind data. For convenience, we have formulated the strategy to

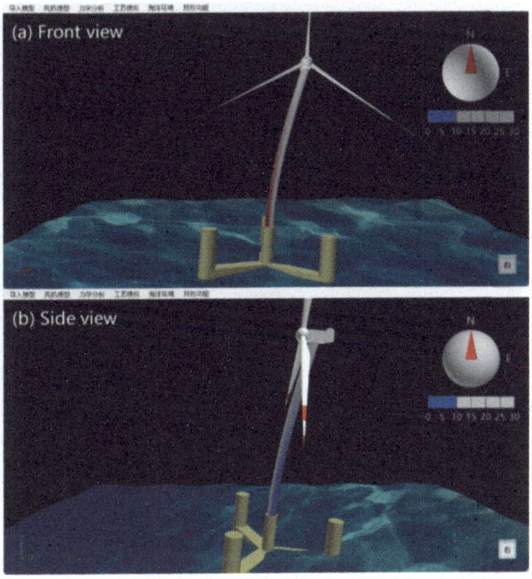

Fig. 6. Structural information under specific conditions (deformation amplified)

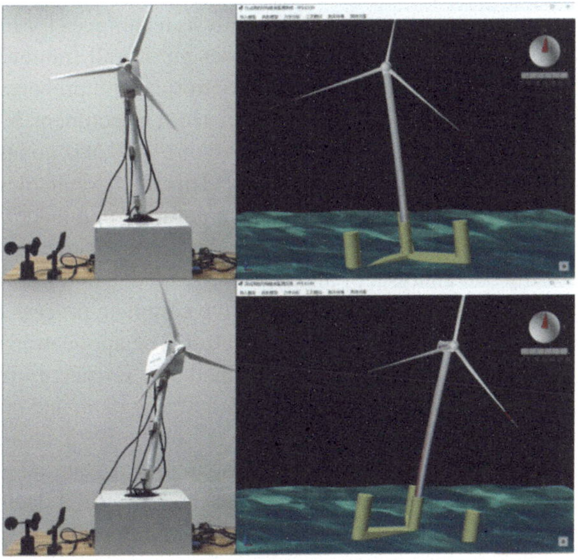

Fig. 7. Real-time synchronization and analysis of the FWT's pose and stress

align the rotor direction as consistent as possible with the wind direction. The specific logic and experimental process are depicted in Fig. 8. It can be observed that the wind direction sensor, virtual wind vane, actual FWT orientation, and virtual FWT orientation are generally consistent, indicating that the DT platform can support a closed-loop between the physical space and the virtual space.

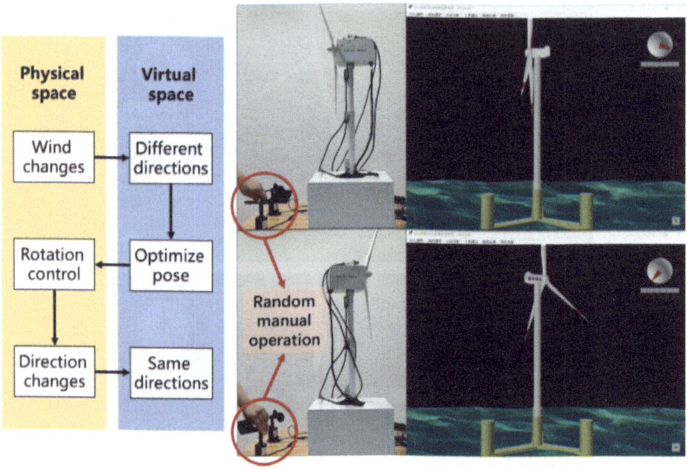

Fig. 8. FWT intelligent control based on the DT

4 Conclusion

This study focuses on real-time monitoring and analysis techniques for FWTs to address key challenges in building DTs. The study proposes an overall framework for the DT of FWTs, providing methods for the systematic description, comprehensive monitoring, unified fusion, and integrated visualization of the marine environment, FWT's pose, and structural status. Specifically, for the tower structure of FWTs, a state inversion method is proposed based on multiple and limited sensors. This method enables the derivation of deformation and stress states beyond localized monitoring points, providing comprehensive structural health information for FWTs. The proposed methods were applied to real-time monitoring and analysis of an actual FWT model, and a corresponding DT platform was constructed. The platform supports the complete process of data perception, storage, integration, fusion, and feedback control. The application results validate the feasibility and effectiveness of the proposed methods. By deeply integrating digital models, marine environment data, and FWT monitoring data, and conducting rapid analysis, visualization, and feedback, a complete DT of the FWT is constructed. This DT plays a key supporting role in achieving transparency of state information, safety risk management, intelligent parameter adjustment, and informed decision-making of FWTs. Further improvements can be made in incorporating real wave and flow monitoring, coupling analysis with environmental conditions, and implementing advanced control strategies based on the FWT's pose and structural status.

References

1. McMorland, J., Collu, M., McMillan, D., et al.: Operation and maintenance for floating wind turbines: a review. Renew. Sustain. Energy Rev. **163**, 112499 (2022)
2. Gao, Z., Odgaard, P.: Real-time monitoring, fault prediction and health management for offshore wind turbine systems. Renew. Energy, 119258 (2023)

3. Kou, L., Li, Y., Zhang, F., et al.: Review on monitoring, operation and maintenance of smart offshore wind farms. Sensors **22**(8), 2822 (2022)
4. Ciuriuc, A., Rapha, J.I., Guanche, R., et al.: Digital tools for floating offshore wind turbines (FOWT): a state of the art. Energy Rep. **8**, 1207–1228 (2022)
5. Jaksic, V., O'Shea, R., Cahill, P., et al.: Dynamic response signatures of a scaled model platform for floating wind turbines in an ocean wave basin. Philos. Trans. R. Soc. A Math. Phys. Eng. Sci. **373**(2035), 20140078 (2015)
6. Wang, Z., Qiao, D., Tang, G., et al.: Monitoring system framework design for floating wind turbine using the deep learning technology and tower response identification considering sensor optimization. Ocean Eng. **299**, 117316 (2024)
7. Tao, F., Zhang, H., Liu, A., et al.: Digital twin in industry: state-of-the-art. IEEE Trans. Ind. Inf. **15**(4), 2405–2415 (2018)
8. Jiang, Y., Li, M., Li, M., et al.: Digital twin-enabled real-time synchronization for planning, scheduling, and execution in precast on-site assembly. Autom. Constr. **141**, 104397 (2022)
9. Jones, D., Snider, C., Nassehi, A., et al.: Characterising the digital twin: a systematic literature review. CIRP J. Manuf. Sci. Technol. **29**, 36–52 (2020)
10. Singh, M., Fuenmayor, E., Hinchy, E.P., et al.: Digital twin: origin to future. Appl. Syst. Innov. **4**(2), 36 (2021)
11. Boschert, S., Rosen, R.: Digital twin—the simulation aspect. In: Hehenberger, P., Bradley, D. (eds.) Mechatronic Futures: Challenges and Solutions for Mechatronic Systems and their Designers, pp. 59–74. Springer, Cham (2016). https://doi.org/10.1007/978-3-319-32156-1_5
12. Moghadam, F.K., Nejad, A.R.: Online condition monitoring of floating wind turbines drivetrain by means of digital twin. Mech. Syst. Signal Process. **162**, 108087 (2022)
13. Walker, J., Coraddu, A., Collu, M., et al.: Digital twins of the mooring line tension for floating offshore wind turbines to improve monitoring, lifespan, and safety. J. Ocean Eng. Mar. Energy, 1–16 (2022)
14. Wang, M., Wang, C., Hnydiuk-Stefan, A., et al.: Recent progress on reliability analysis of offshore wind turbine support structures considering digital twin solutions. Ocean Eng. **232**, 109168 (2021)
15. Haghshenas, A., Hasan, A., Osen, O., et al.: Predictive digital twin for offshore wind farms. Energy Inform. **6**(1), 1 (2023)
16. Liu, Y., Zhang, J.M., Min, Y.T., et al.: A digital twin-based framework for simulation and monitoring analysis of floating wind turbine structures. Ocean Eng. **283**, 115009 (2023)
17. The DTU 10-MW reference wind turbine. https://backend.orbit.dtu.dk/ws/portalfiles/portal/55645274/The_DTU_10MW_Reference_Turbine_Christian_Bak.pdf. Accessed 21 June 2024

Fracture of Mooring Lines of Offshore Floating Wind Turbines Based on the Finite Particle Method

Ying Yu(✉) and Peng Luo

Shantou University, Shantou 515063, Guangdong, China
yuying@stu.edu.cn

Abstract. This study explores the dynamic fracture behavior of mooring lines in offshore floating wind turbines under ocean currents using the Finite Particle Method (FPM). Following the Newton's second law, the FPM models mooring lines as discrete particles connected by massless elements, allowing for the simulation of large deformations and nonlinear behaviors. A fracture model and fracture criteria for mooring lines were established. The study examines the fracture processes under both quasi-static and dynamic ocean current conditions. The results indicate that ocean currents accelerate the fracture process and alter potential fracture locations. Stress distribution and post-fracture motion are significantly different between the two scenarios, emphasizing the need to consider ocean currents in mooring system design. These findings provide valuable insights into the dynamic response and structural integrity of mooring systems, contributing to the safe operation of offshore floating wind turbines in complex marine environments.

Keywords: Mooring Line · Finite Particle Method · Fracture

1 Introduction

Due to the harsh and complex marine environment, the mooring lines of offshore floating wind turbines experience alternating tension-relaxation states, leading to sudden jumps and even abrupt breaks. This results in significant impact tension within the mooring lines, severely threatening the safety of the mooring system and the floating wind turbine.

The mooring force of catenary mooring lines is primarily provided by their own weight. However, since the mooring lines are flexible components with tensile but not compressive strength, they undergo alternating tension and relaxation as the floating platform moves. This causes the dynamic tension within the lines to fluctuate sharply, generating impact tension. The maximum value of this tension far exceeds the normal tension amplitude of the mooring line, reaching several to tens of times the pre-tension of the mooring line [1]. Impact tension can lead to fatigue accumulation damage or even complete breakage of the mooring lines, causing mooring system failure and severely affecting the safe operation of the floating platform.

D. Jeng and B. Cai (Eds.): ICCOE 2024, LNCE 423, pp. 186–196, 2025.
https://doi.org/10.1007/978-981-96-0181-3_18

Current computational models mainly include the catenary model based on catenary theory [2], the lumped mass model based on multibody dynamics [3], and the Finite Element Model based on flexible rod theory [4]. Since mooring line breakage involves the calculation of large deformations, large rotations, fractures, and other strongly nonlinear and discontinuous behaviors, current models require modifications to handle these issues [5]. In numerical analysis methods for mooring lines, a new numerical analysis approach has emerged in recent years. It directly uses discrete "particle values" and particle motion laws to describe structural behavior and employs generalized vector mechanics as the criterion for motion and deformation [6]. Luo et al. first proposed the Finite Particle Method (FPM), a new numerical analysis method for structural engineering problems based on the basic concepts of vector mechanics [7]. This method aims to simulate and analyze various strongly nonlinear problems in structural engineering. It discretizes the analysis region into particles and uses Newton's second law to replace the partial differential equations of continuous media to describe particle motion. The motion and deformation of the structure are characterized by the trajectories of the particle system.

The current FPM and vector mechanics analysis theory have been successfully applied to the nonlinear dynamic analysis of structures [8], wind-induced damage analysis of prestressed cable-net structures [9], vehicle-bridge vibration coupling dynamic analysis [10], and preliminary applications in marine engineering structure analysis such as submarine pipelines [11] and offshore wind turbine pile foundations [12]. These research results demonstrate that the FPM has significant advantages in analyzing strongly nonlinear and discontinuous behaviors of structures, providing new insights and powerful analytical tools for the dynamic analysis of deep-sea floating wind turbine mooring systems.

Considering the deficiencies of current computational models and numerical analysis methods in calculating the strongly nonlinear processes of large deformations and material nonlinearity during mooring line breakage, this paper's main research content is as follows: Establish a nonlinear dynamic model and motion equations for mooring lines based on the Finite Particle Method. On this basis, a mooring line fracture model and fracture criteria are developed to study the fracture process of mooring lines under different ocean current speeds. The results indicate that ocean currents may accelerate line breakage and alter the potential fracture locations of the lines.

2 Finite Particle Method for Mooring Line Analysis

2.1 Particle Motion Equation

The mooring lines of an offshore floating wind turbine can be modeled as particles and cable elements using the FPM, as illustrated in Fig. 1. The motion equation of any particle α follows Newton's second law, with the mooring line's mass assumed to be represented by each particle. The particles are connected by massless elements, whose deformations only represent the force relationships between the connected particles.

$$\mathbf{M}_\alpha \ddot{\mathbf{d}}_\alpha = \mathbf{F}_\alpha^{\text{ext}} + \mathbf{F}_\alpha^{\text{int}} \tag{1}$$

where $\mathbf{M}_\alpha$ represents the mass matrix, $\ddot{\mathbf{d}}_\alpha$ is the acceleration vector, and $\mathbf{F}_\alpha{}^{\text{ext}}$ and $\mathbf{F}_\alpha{}^{\text{int}}$ are the external and internal force vectors of particle α, respectively. The parameter $\mathbf{F}_\alpha{}^{\text{int}}$ equals the sum of the internal nodal forces exerted by the elements connected to particle α. Since each particle in Eq. (1) is considered to be in dynamic equilibrium under the internal and external forces, the static and dynamic analyses can be combined into a single procedure. To avoid an iterative solution process, this study suggests using explicit time integration with a simple central difference method. For detailed information on solving the particle motion equation, refer to Reference [9].

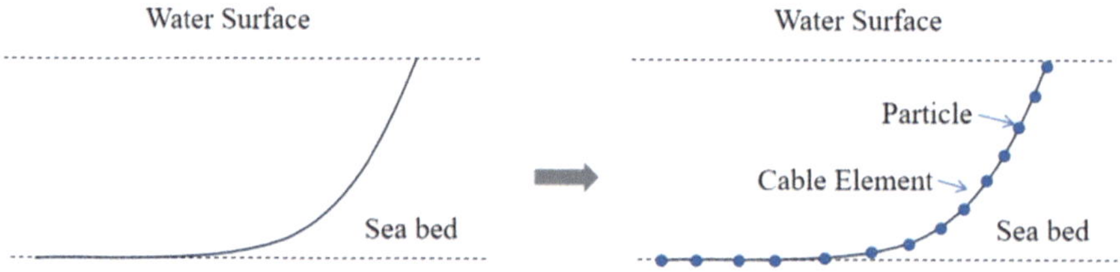

Fig. 1. FPM model of a catenary mooring line

2.2 FPM for Modeling Geometric Nonlinearity

The large deformation of mooring lines due to the combined effects of wind, waves, and currents constitutes a geometrically nonlinear problem. The FFPM utilizes fictitious motions (including both fictitious reverse and forward motions) to handle geometric nonlinearities in the calculation of internal forces of the elements [7]. For an in-depth explanation of fictitious motion, please refer to Reference [8]. Following this approach, the internal force of the cable element is determined as follows:

$$\mathbf{f}_{a'} = \mathbf{f}_a + \Delta\mathbf{f}_a = \begin{cases} (\sigma_a A_a + \frac{E_a A_a \Delta_L}{L_a})\mathbf{e}_{AB} & \text{if } (\sigma_a A_a + \frac{E_a A_a \Delta_L}{L_a}) > 0 \\ 0 & \text{otherwise} \end{cases} \tag{2}$$

where, $\mathbf{f}_a$ represents the internal force of element AB at time t_a, and $\Delta\mathbf{f}_a$ denotes the incremental internal force of element AB at time t_a. The axial stress at time t_a is σ_a, and A_a is the cross-sectional area of element AB. E_a is the Young's modulus, Δ_L is the length variation of element AB between t_a and t_b, L_a is the length of element AB at time t_a, and $\mathbf{e}_{AB}$ is the directional vector of element AB at time t_b. The axial force is then applied to the corresponding particle. The internal force of a particle is determined by summing all the axial forces of the elements connected to it.

2.3 FPM for Modeling Material Nonlinearity

The mooring lines studied in this paper are primarily composed of R4S grade studless chain links. According to the mooring chain standards in China [13], R4S grade studless chain links are mainly manufactured using hot-rolled round steel. For steel materials, the ideal elastoplastic model is shown in Fig. 2.

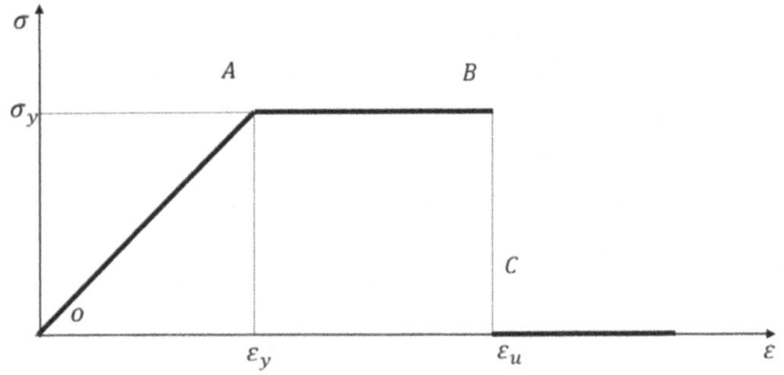

Fig. 2. Elastoplastic Model of the Mooring Line

When the steel specimen is subjected to tension, it initially enters the elastic stage, as depicted by segment OA in the figure. During this stage, strain increases proportionally with stress, and the stress-strain relationship is linear. As the stress continues to increase, the specimen exhibits yielding when the stress reaches the yield stress σ_y, corresponding to the yield strain ε_y. Subsequently, the specimen enters the plastic strengthening stage, represented by segment AB in the figure, where the stress reaches a maximum value and remains constant while the strain continues to increase until it reaches the ultimate strain ε_u, leading to the failure of the specimen. When the cable element is subjected to compression, the element fails.

3 Fracture Model for the Mooring Line

3.1 Fracture Criterion

The mooring lines simulated in this study are primarily composed of R4S grade studless chain links. For these types of chain links, the mooring chain standards [13] typically use breaking tension to determine if a fracture has occurred. Specifically, when the maximum tension in the chain link exceeds the breaking tension specified in the standards, the chain link is considered to have fractured. Since the plastic zone in a studless chain link is not significant throughout the link at the point of fracture, this study adopts the breaking stress corresponding to the breaking tension specified in the standards as the criterion for determining chain link fracture:

$$\rho \geq \rho_u \tag{3}$$

where, ρ represents the axial stress of the rod element, and ρ_u is the stress value corresponding to the breaking load when under tension.

3.2 Fracture Model

When a cable reaches the failure criterion, member fracture becomes inevitable. The mooring line model established in this paper is based on macroscopic analysis, and does not take into account initial defects or the various development processes of material failure. Since the FPM models a cable element using two particles and one element, crack propagation is not considered in this study. A straightforward method for modeling cable failure is introduced.

As illustrated in Fig. 3, if element AB meets the fracture criterion, the forces on both particle A and particle B are examined. If the force on particle B is greater, the fracture occurs at particle B. A new particle is then generated at the same location as particle B, and the element is detached from particle B. If the forces on the two particles are equal, the fracture occurs at both ends.

After the fracture, the fundamental computational procedure of the FPM remains unchanged. The motion of both the newly added particles and the original particles continues to follow the basic motion equations. Only the properties of these particles, such as mass and internal force, need to be updated according to the new topology. For further details, refer to reference [11].

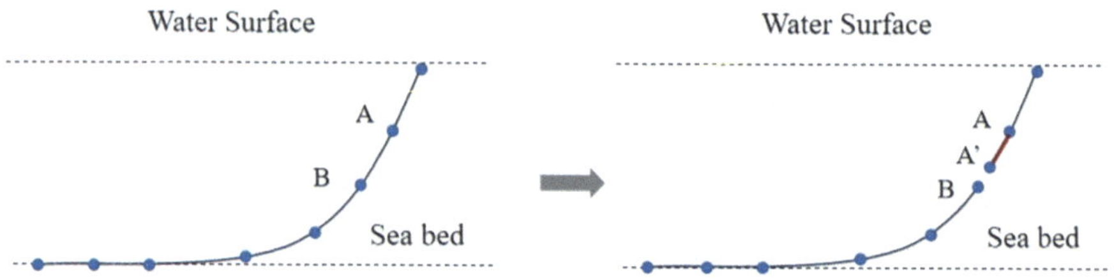

Fig. 3. Structural model after fracture

4 Simulation of Mooring Line Fracture

This section simulates the fracture behavior of the catenary mooring line used in the OC3 Hywind spar platform under ocean current conditions. The initial static configuration of the catenary mooring line is shown in Fig. 4. The parameters of the mooring line are listed in Table 1.

In the analysis, the mooring line is discretized into 50 cable elements. During the simulation process, the damping coefficient is set to $\xi = 2$, with a time step of $\Delta t = 1 \times 10–4$ s. The ocean current is considered quasi-static. According to the mooring chain specifications [13], the minimum breaking load of the mooring line is determined to be 4116 kN, corresponding to a breaking stress of $\rho_u = 647$ MPa. Since the upper platform is not modeled in this study, the displacement of the platform under the combined action of wind, waves, and current is simplified as a sinusoidal displacement excitation applied at the top of the mooring line in the x-direction. The excitation amplitude is 15 m, and the excitation frequency is 0.05 Hz. The fracture process of the mooring line under quasi-static flow is shown in Fig. 5.

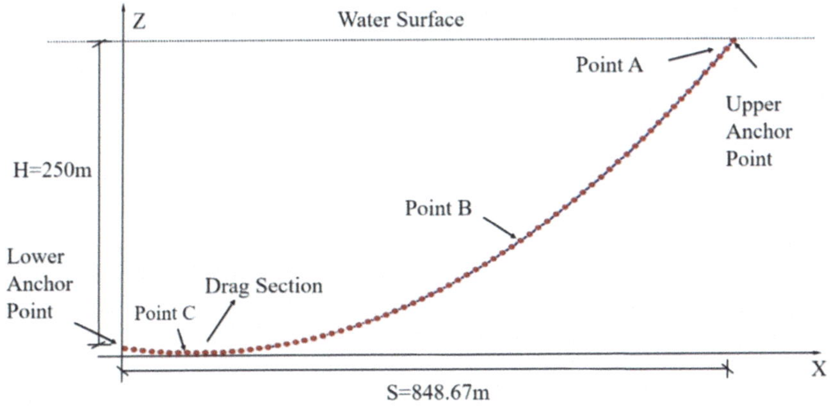

Fig. 4. Initial Configuration of the Catenary Mooring Line

Table 1. Mooring Line and Environmental Parameters

Parameter	Symbol	Unit	Value
Original Length	L_0	m	902.2
Diameter	d	m	0.09
Mass per Unit Length	m	kg/m	77.7066
Tensile Stiffness	EA	N	3.84×10^8
Mooring Depth	h	m	250
Mooring Length	s	m	848.67
Added Mass Coefficient	C_a		1.0
Drag Coefficients	C_{dt}, C_{dn}, C_{db}		0.1,1.6,1.6
Current Speed	U_C	m/s	0.0
Seawater Density	ρ_w	kg/m^3	1029
Underwater Unit Length Weight	w_0	N/m	698.094

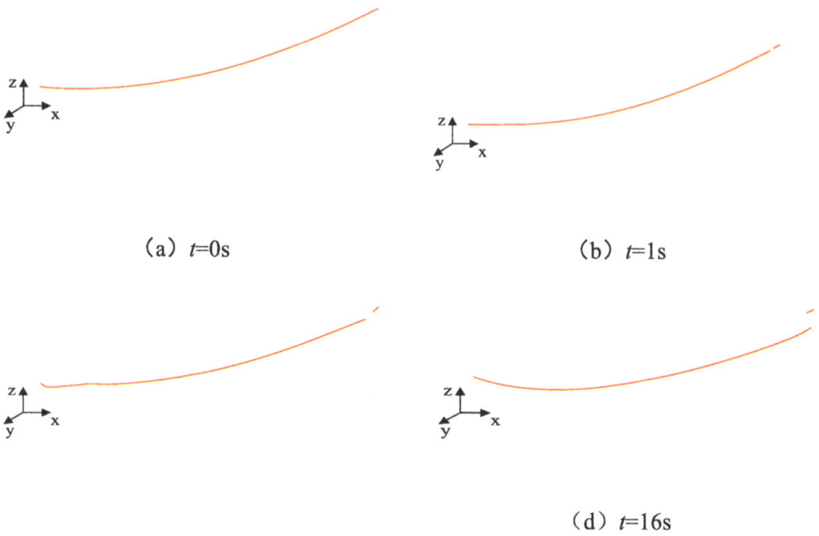

Fig. 5. Simulation of the fracture process of the mooring line with a flow velocity of 0 m/s

From Fig. 5, it can be observed that at $t = 1$ s, the cable fractures at the upper end due to the displacement excitation from the upper platform. After the fracture, the cable does not immediately sink under the influence of gravity; instead, it continues to oscillate, dissipating the residual kinetic energy due to the impact force of the fracture, as shown in Fig. 5(c). By $t = 16$ s, the fractured part of the cable has dissipated all of its residual kinetic energy. Since the mooring line is in a quasi-static flow and the influence of the ocean current on the cable is ignored, the cable begins to sag naturally under the influence of gravity. The stress state of the cable throughout the entire process is shown in Fig. 6.

Figure 6 also reflects the basic motion process of the mooring line after fracture. At $t = 1$ s, the mooring line fractures, causing a rapid decrease in internal stress within the elements. Subsequently, due to the residual kinetic energy and strain energy, the mooring line remains in motion, with minor fluctuations in element stress. By $t = 16$ s, the mooring line has dissipated all its residual kinetic energy and strain energy. At this point, the stress at the upper end (Point A) is zero, while the stress at the middle (Point B) and lower end (Point C) of the cable remains unchanged, and the mooring line naturally sags under the influence of gravity.

To compare the possible different fracture modes of the mooring line under the influence of ocean currents, the fracture simulation conditions in the quasi-static flow were modified. The initial static configuration shown in Fig. 4 was used, and the simulation time step was set to 1×10^{-4} s. The excitation applied by the upper platform's mass point was consistent with the fracture simulation in the quasi-static flow. The only difference was that the cable was no longer in a quasi-static flow; the ocean current speed was set to 0.5 m/s, flowing in the positive x-direction. The fracture process of the mooring line under these conditions is shown in Fig. 7.

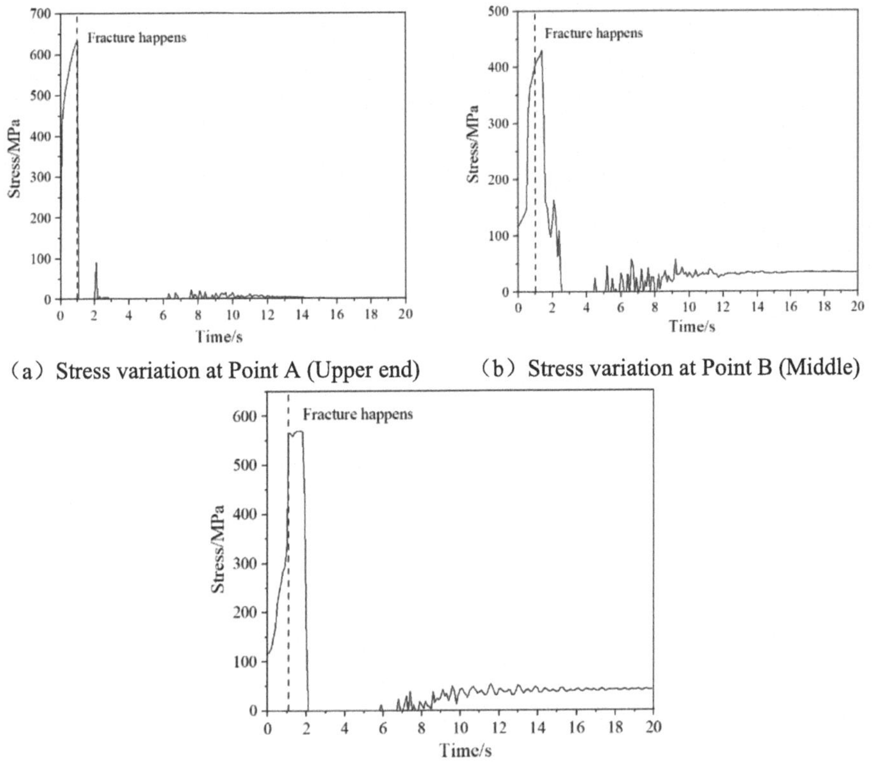

(a) Stress variation at Point A (Upper end)

(b) Stress variation at Point B (Middle)

(c) Stress variation at Point C (Lower End)

Fig. 6. The stress variation during the cable fracture process with a flow velocity of 0 m/s

From Fig. 7, it can be observed that the fracture position of the cable in the non-quasi-static flow is completely different from that in the quasi-static flow. The fracture occurs in the middle of the cable rather than at the upper end. The primary fracture process is as follows: at $t = 1.6$ s, the cable fractures in the middle. From $t = 1.6$ s to 6 s, the cable continues to move under the influence of the ocean current and residual kinetic energy. By $t = 16$ s, the post-fracture motion pattern of the cable is different from that in the quasi-static flow. The upper half of the fractured cable swings due to the combined effects of the ocean current and platform excitation, and the cable near the fracture point curls due to the action of the ocean waves. Figure 8 shows the element stress variation during the cable fracture process.

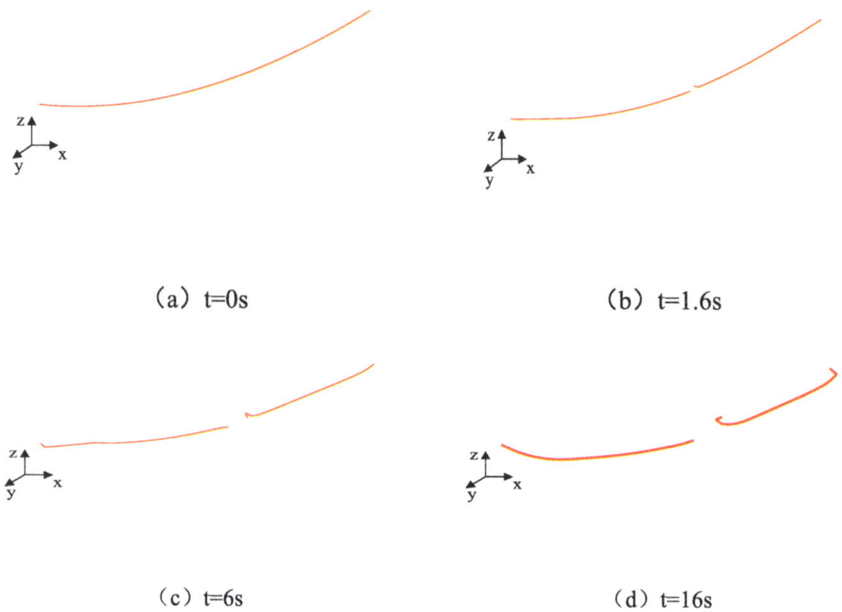

Fig. 7. Simulation of the fracture process of the mooring line with a flow velocity of 5 m/s

From the stress variation diagram of the cable elements at the upper end shown in Fig. 8, this trend can also be observed. At $t = 1.6$ s, the cable fractures in the middle, causing the stress at Point B (middle of the cable) to drop rapidly to zero after the fracture. Since Point A (upper end of the cable) is farther from the fracture location, its stress does not immediately decrease but shows a transitional phase. Subsequently, the stress at Point A drops quickly and begins to fluctuate within a small range. Under the influence of the ocean current and the excitation at the top, the cable continues to swing, causing the stress at Point A to start increasing.

Similarly, for Point C (lower end of the cable), being farther from the fracture location, there is also a transitional phase before the stress starts to decrease rapidly. After the residual kinetic energy is dissipated, the cable sags under the influence of gravity and the ocean current. Consequently, the stress at Point C stabilizes and remains constant during the sagging process.

The observed transitional phases and subsequent stress fluctuations highlight the impact of the cable's motion under combined effects of the ocean current, platform excitation, and gravity. This detailed understanding of stress distribution and motion dynamics is crucial for designing mooring systems that can withstand diverse environmental conditions and ensure structural integrity.

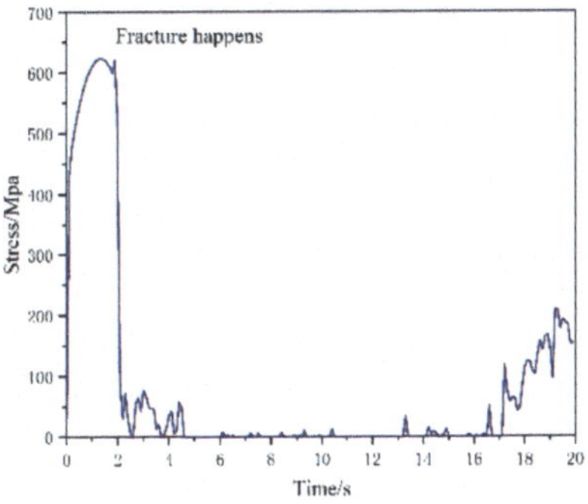

(a) Stress variation at Point A (Upper end)

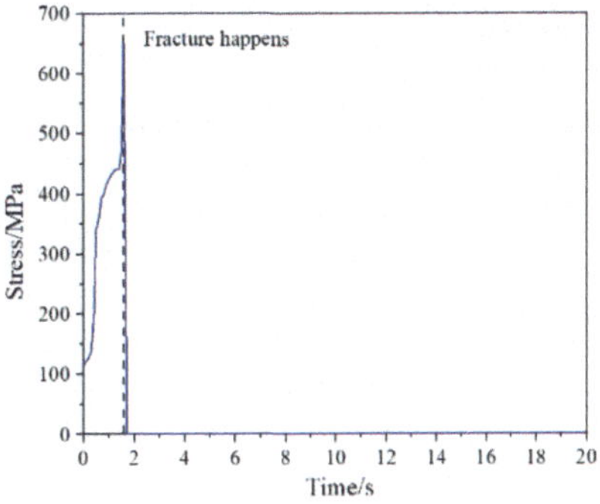

(b) Stress variation at Point B (Middle)

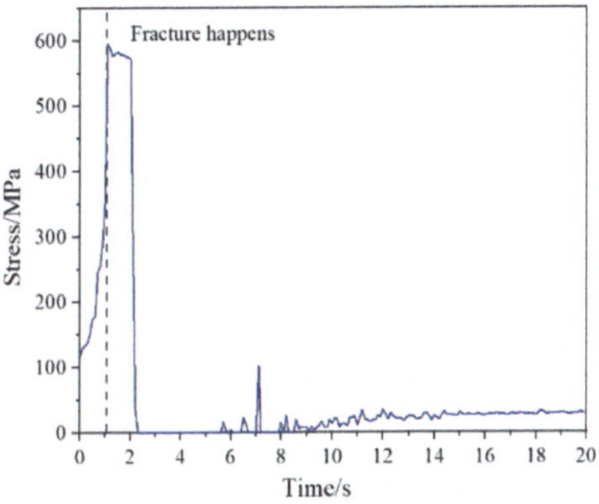

(c) Stress variation at Point C (Lower End)

Fig. 8. The stress variation during the cable fracture process with a flow velocity of 5 m/s

5 Conclusion

This paper presents a novel approach for analyzing the dynamic performance and fracture behavior of mooring lines under combined wind, wave, and current conditions using the FPM. The study reveals that ocean currents significantly influence the fracture location and post-fracture dynamics of mooring lines. Under quasi-static conditions, the mooring line fractures at the upper end due to platform excitation, while in dynamic current conditions, the fracture occurs in the middle, leading to different stress distributions and motion patterns. These findings highlight the importance of considering ocean currents in the design and analysis of mooring systems for offshore floating wind turbines, providing a new theoretical basis and analytical method for ensuring the safe and stable operation of deep-sea floating wind turbines.

Acknowledgments. This work is supported the National Natural Science Foundation of China under grant Nos. 52378165 and 52238001; and the Natural Science Foundation of Guangdong, China (2023A1515240049 and 2024A1515010573), Li Ka Shing Foundation STU-GTIIT Joint Research Grants (2024LKSFG05).

References

1. Li, Y., Guo, S., Chen, W., Yan, D., Song, J.: Analysis on restoring stiffness and its hysteresis behavior of slender catenary mooring-line. Ocean Eng. **209**, 107521 (2020)
2. Touzon, I., Nava, V., Gao, Z., Mendikoa, I., Petuya, V.: Small scale experimental validation of a numerical model of the HarshLab2.0 floating platform coupled with a non-linear lumped mass catenary mooring system. Ocean Eng. **200**, 107036 (2020)
3. Hermawan, Y.A., Furukawa, Y.: Coupled three-dimensional dynamics model of multi-component mooring line for motion analysis of floating offshore structure. Ocean Eng. **200**, 106928 (2020)
4. Zhang, H., Kong, L., Guan, W., Huang, S., Fang, G.: Dynamic response analysis of the equivalent water depth truncated point of the catenary mooring line. China Ocean Eng. **31**, 37–47 (2017)
5. Ramaswamy, B., Kawahara, M.: Arbitrary Lagrangian-Eulerianc finite element method for unsteady, convective, incompressible viscous free surface fluid flow. Int. J. Numer. Methods Fluids **7**, 1053–1075 (2010)
6. Ting, E.C., Shih, C., Wang, Y.K.: Fundamentals of a vector form intrinsic finite element: Part I. Basic procedure and a plane frame element. J. Mech. **20**, 113–22 (2004)
7. Yu, Y., Paulino, G.H., Luo, Y.: Finite particle method for progressive failure simulation of truss structures. J. Struct. Eng. **137**, 1168–1181 (2011)
8. Yu, Y., Zhu, X.: Nonlinear dynamic collapse analysis of semi-rigid steel frames based on the finite particle method. Eng. Struct. **118**, 383–393 (2016)
9. Yu, Y., Xia, P., Yang, C.: Form finding and collapse analysis of cable nets under dynamic loads based on finite particle method. Comput. Model. Eng. Sci. **117**, 73–89 (2018)
10. Wang S.: Dynamic analysis of wind-turbine-track-bridge coupled system based on vector finite element method. Ph.D thesis. Zhejiang University (2018)
11. Xu, L., Lin, M.: Unevenness analysis of long-distance submarine pipelines based on vector finite element method. Oil Gas Storage Transp. **35**, 208–214 (2016)
12. Wang, Q.: Collision analysis of offshore wind turbine pile foundations based on vector finite element method. Ph.D thesis. Harbin Engineering University (2017)
13. National Standard of the People's Republic of China: Mooring Chain (GB/T20848-2017), Beijing (2017)

Aerodynamic Characteristics of Offshore Floating Counter-Rotating Double-Impeller VAWT

Dan Yu[1,2], Yuanbo Ma[3], Yupeng Duan[1](✉), Quanyu Li[3], Keyi Wang[2], and Hengxu Liu[1,3]

[1] College of Shipbuilding Engineering, Harbin Engineering University, Harbin 150001, China
duanyupeng@hrbeu.edu.cn
[2] College of Mechanical and Electrical Engineering, Harbin Engineering University, Harbin 150001, China
[3] Yantai Research Institute of Harbin Engineering University, Yantai 265500, China

Abstract. Wind energy is expected to play a key role in achieving the green energy transition, as offshore wind turbine technology advances quickly, offshore floating VAWTs have drawn a lot of attention. Under the action of wind and waves, the motion and torque of the floating offshore VAWT undergo periodic changes and generate additional aerodynamic loads. In this paper, the aerodynamic characteristics of a Counter-Rotating Double-Impeller VAWT (CR-DI-VAWT) under longitudinal oscillation and longitudinal rocking motions are investigated based on STAR-CCM+. First, key structural parameters such as the number of blades and impeller radius are optimized by the CFD method and the Taguchi method. Then, based on the optimized Single-Impeller VAWT model, the aerodynamic performance of the DI-VAWT under different rotor spacing L conditions is analyzed. Finally, the effects of longitudinal oscillation and longitudinal rocking motions with different amplitudes and periods on the aerodynamic characteristics of the DI-VAWT were investigated using the customized functions in STAR-CCM+. It is found that the power coefficient of wind turbines can be affected to differing degrees by variations in rotor spacing, longitudinal oscillation motion, and longitudinal rocking motion. These findings will provide a reference for the design and optimization of OF-VAWT.

Keywords: Floating Offshore Vertical-Axis Wind Turbine · Double-Impeller · Computational Fluid Dynamics

1 Introduction

Wind energy, as a clean and renewable energy source, plays a crucial role in the global green energy transition and the fight against climate change [1]. Especially in the marine environment, the development and utilization of offshore wind power have great potential, and OF-VAWTs have received extensive attention from both academia and industry due to their unique design and operational characteristics [2].

D. Jeng and B. Cai (Eds.): ICCOE 2024, LNCE 423, pp. 197–207, 2025.
https://doi.org/10.1007/978-981-96-0181-3_19

Compared with traditional horizontal-axis wind turbines, VAWTs have the advantages of lower start-up wind speed requirement, better wind energy capture efficiency, and more stable output power, which make them more adaptable and reliable in the complex and capricious marine environment. However, the operation of OF-VAWTs in marine environments faces complex aerodynamic and structural dynamics challenges. The combined effect of waves and winds can lead to periodic motion and torque variations in the wind turbine, which in turn induce additional aerodynamic loads, all of which may affect the performance and lifetime of the wind turbine. Therefore, an in-depth study of the aerodynamic characteristics of OF-VAWTs in extreme marine environments is important for optimizing their design and improving their operational efficiency and reliability.

In this paper, the aerodynamic characteristics of a CR-DI-VAWT under longitudinal oscillation and longitudinal rocking motion conditions are systematically investigated by the advanced computational fluid dynamics (CFD) software STAR-CCM+. By simulating and analyzing the aerodynamic response of the wind turbine under wind and wave conditions, the potential advantages of the Counter-Rotating Double-Impeller configuration for reducing aerodynamic loads, suppressing longitudinal oscillations, and optimizing the operational performance of the wind turbine are revealed. In addition, this paper discusses the influence of structural design parameters on the aerodynamic performance of wind turbines, which provides an important theoretical basis and guidance for the design and optimization of future OF-VAWTs.

2 Methodology

2.1 Design of Wind Turbine

The VAWT used in this paper for the validation of CFD calculation results is a H-type VAWT modeled with reference to the experimental model of Elkhoury et al. [3]. The relevant geometrical parameters of the wind turbine model are listed in Table 1. Figure 1 shows the schematic diagram of the wind turbine model. Before CFD calculation, in order to improve the efficiency of numerical simulation, the VAWT which has a complex structure, is simplified to some extent. Theoretically, in the power generation process of VAWT, the impeller is the part of the wind turbine that directly obtains wind energy, the influence of other structures on the aerodynamic performance is relatively small. Rezaeiha [4] et al. compared the aerodynamic performance of VAWT with and without support structure through numerical simulation and found that the support structure has little effect on the aerodynamic performance of the wind turbine as well as the flow field of the wind turbine. Therefore, the internal support structure is not considered for wind turbine modeling in this paper.

2.2 Mode of Pitching and Pitching Motion

In order to simulate the motion of a DI-VAWT under wave action, the longitudinal oscillation and longitudinal rocking motion of the wind turbine under wave action is simplified to a periodic simple harmonic motion, and this motion pattern is realized by

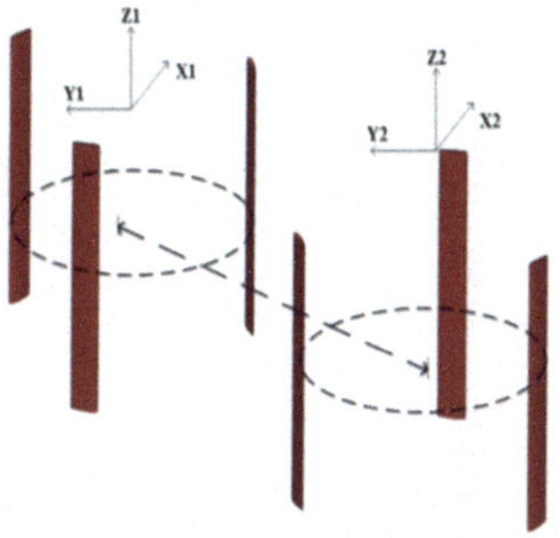

Fig. 1. Schematic diagram of CR-DI-VAWT model

Table 1. Structural parameters of CR-DI-VAWT

Parameters	Numerical value
Number of impeller N	2
Diameter D	5 m
Height h	4 m
Chord c	0.5 m
Number of blade N	3
Phase difference	0°
Airfoil shape	NACA 0018

using the user-defined field function of the Star-CCM+. As shown in Fig. 2 for the DI-VAWT longitudinal oscillation motion side view, the sequence of longitudinal oscillation motion in accordance with the figure shows 1 → 2 → 3 for the periodic reciprocating motion, and when the impeller is in the initial position of longitudinal oscillation, the azimuthal angle of the blades 1-A is 0°. In order to better study the effect of longitudinal oscillation and longitudinal rocking motion on the aerodynamic characteristics of the wind turbine, this section selects the speed inlet to be a constant 8 m/s, and the rotational speed of the DI-VAWT to be a constant 7.68 rad/s (TSR = 2.4), and the motion period to be an integer multiple of the rotational period of the impeller (Fig. 3).

The main contents of the user-defined field function are the displacement function S_{surge} and velocity function V_{surge} of the longitudinal oscillatory motion of the impeller, which can be expressed respectively:

$$S_{surge} = A_{surge} \cdot \sin(2\pi f_{surge} t) \tag{1}$$

$$V_{surge} = 2\pi f_{surge} A_{surge} \cdot \cos(2\pi f_{surge} t) \tag{2}$$

Where A_{surge} is the maximum amplitude of longitudinal oscillation, f_{surge} is the frequency of longitudinal oscillation, and t is the motion time. The rotation axis of longitudinal oscillation is the axis where the diameter of the impeller bottom is parallel to the x-axis.

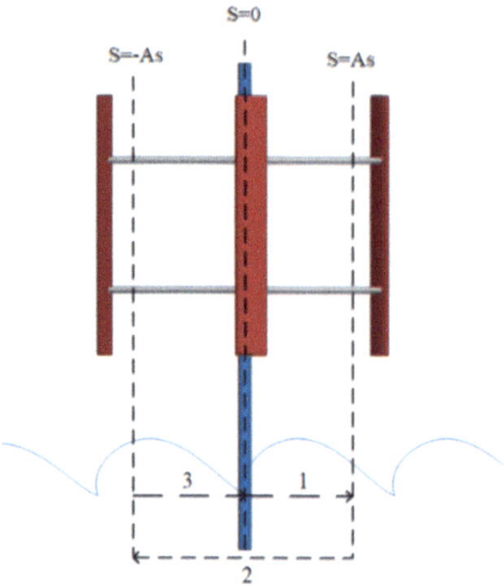

Fig. 2. Side view of longitudinal swinging motion

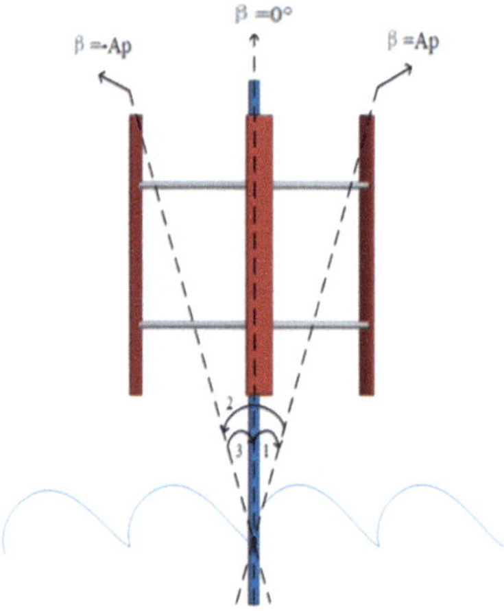

Fig. 3. Side view of longitudinal rocking motion

The longitudinal oscillation motion of wind turbine under wave action is simplified into a periodic simple harmonic motion, and the user-defined field function is utilized to realize the DI-VAWT longitudinal oscillation motion mode.

The angular function β_{pitch} and the angular velocity function ω_{pitch} of the longitudinal motion of the impeller can be expressed as follows, respectively:

$$\beta_{pitch} = A_{pitch} \cdot \sin(2\pi f_{pitch} t) \tag{3}$$

$$\omega_{pitch} = 2\pi f_{pitch} A_{pitch} \cdot \cos(2\pi f_{pitch} t) \tag{4}$$

where ω_{pitch} is the maximum longitudinal amplitude, f_{pitch} is the longitudinal frequency, and t is the motion time.

2.3 Computational Domain

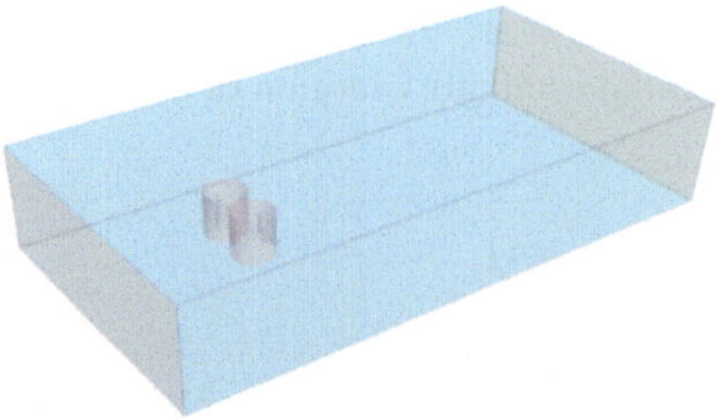

Fig. 4. Computation domain division

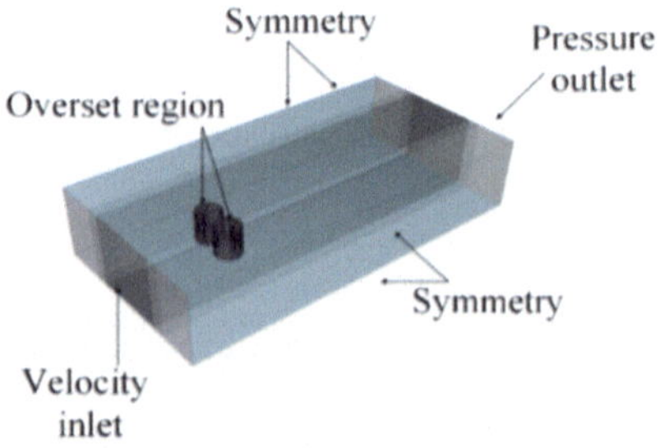

Fig. 5. Computational domain meshing

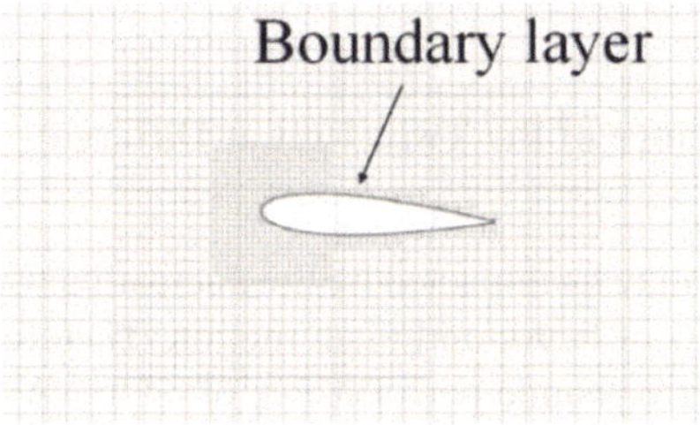

Fig. 6. Airfoil boundary layer delineation

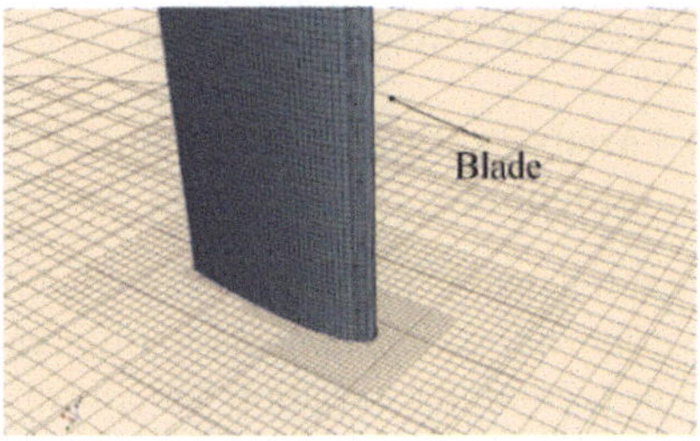

Fig. 7. Mesh delineation of blade surface

In this paper, the schematic diagram of the VAWT three-dimensional computational domain is shown in Fig. 4, which consists of a fixed rectangular region and a cylindrical overlap region. The size of the computational domain is $20D \times 10D \times 5H$, the height of the overlap region is $2H$, and the diameter is $2D$, of which D is the diameter of the impeller of the selected wind turbine model, and H is the height of the blades of the selected wind turbine model. The velocity inlet distance from the wind turbine center is $5D$, the pressure outlet distance from the wind turbine center is $15D$ in order to make the wind turbine wake fully developed, the width of the computational domain is $10D$, the distance of both sides of the boundary from the center of the wind turbine is $5D$, and the height of the computational domain is $5H$ (Figs. 5, 6 and 7).

The computational domain consists of three parts: the rotating domain of the turbine impeller, the turbine motion area, and the outer basin. In this study, the overlapping mesh technique is used for meshing to better capture the flow details. The SST $k-\omega$ turbulence model is used for simulation to obtain more accurate flow information. The boundary layer meshing method is used for the high accuracy of meshing near the wall. In order to meet the requirements of SST $k-\omega$ for the grid resolution of $y^+ < 1$, the number of boundary layers is 10, the velocity growth rate of the boundary layer is 1.25, and the total thickness of the boundary layer is 1.779×10^{-3} m. The SIMPLE algorithm is used to calculate the pressure and velocity of the boundary layer. The SIMPLE algorithm is used to couple the pressure and velocity, and the second-order windward format is used to discretize the other convective and diffusive terms. The velocity inlet of the computational domain is set to the flow rate of $U_0 = 8.0\,\mathrm{m/s}$, and the pressure outlet is set to the constant pressure of $P_0 = 0.0\,\mathrm{Pa}$. The air density in the computational domain is $\rho = 1.29\,\mathrm{kg/m^3}$, and the dynamic viscosity is $1.85508 \times 10^{-5}\,\mathrm{Pa \cdot s}$. The corresponding turbulence velocity scale, turbulence viscosity ratio, and turbulence intensity are set to 1 m/s, 10, and 0.8% respectively, and the time step is selected to be 0.5° for each step. And the total computational time is 8 cycles of the VAWT rotations, and the 8th rotational cycle is taken as the computational result to be analyzed.

3 Results and Discussion

3.1 Analysis of the Aerodynamic Performance of the Twin Unit at Different Amplitudes

The aerodynamic performance of OF-VAWTs may be impacted by the fluctuating wave environment, which can lead to different levels of heave and pitch motion. Therefore, in this section, the variation patterns of the aerodynamic performance of a OF-VAWT under different longitudinal oscillation amplitudes will be investigated and discussed. Here, four different longitudinal amplitudes As = 0.25 m (As = 1/20D), As = 0.5 m (As = 1/10D), As = 1 m (As = 1/5D), As = 1.25 m (As = 1/4D), and the same longitudinal period (Ts = 4T) are used to carry out the study. The angular velocity of the wind turbine remains constant at $\omega = 7.68$ rad/s.

Figure 8 illustrates the instantaneous variation curves of VAWT aerodynamic load with time for four different longitudinal oscillation amplitudes ($\omega = 7.68$ rad/s) during one longitudinal oscillation period. Figures 8(a) and 8(b) depict the torque changes of blade 1 and impeller under four different longitudinal oscillation amplitudes, respectively.

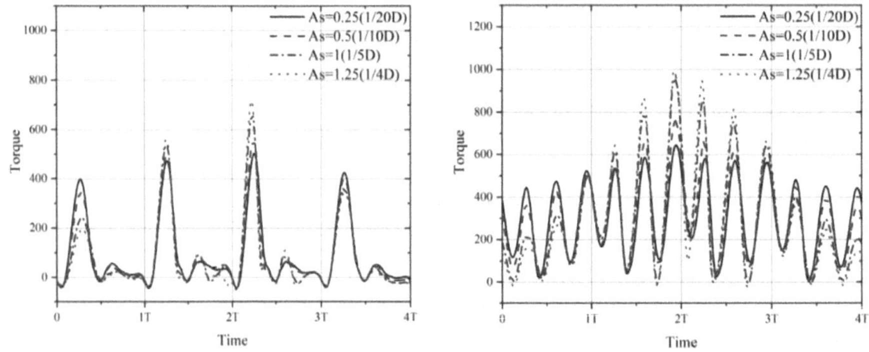

(a) Blade moment diagrams (b) Impeller torque diagrams

Fig. 8. Time-varying wind turbine aerodynamic loads with different longitudinal oscillation amplitudes in one cycle

It can be found that the effect of different longitudinal rocking amplitudes on the blade torque in one cycle have a same pattern. Lower rocking amplitudes cause larger torque values in the time [0, 1T] and [3T, 4T], while higher rocking amplitudes cause larger torque values only in the time [1T, 2T] and [2T, 3T]. It can be inferred that these phenomena can be attributed to the variation of the induced velocity Vs. In the time [0, 1T] and [3T, 4T], the Vs is in the same direction as the free incoming flow V_0, thus the relative velocity Vr decreases while the torque increases. On the contrary, Vr increases in the time [1T, 2T] and [2T, 3T] due to the opposite directions of Vs and V_0.

In order to understand the power output performance of the OF-VAWT under different longitudinal oscillation amplitudes, we investigated the variation of the average power coefficients under the four longitudinal oscillation amplitudes mentioned above.

Table 2. Comparison of power coefficients Cp for non-longitudinal oscillation motion and longitudinal oscillation motion with different amplitudes

Oscillation amplitude	Non- oscillation	As = 0.25	As = 0.50	As = 1.00	As = 1.25
Dual impeller average Cp	0.39874	0.406655	0.406302	0.399717	0.393995

Table 2 lists the average power coefficients of the wind turbine corresponding to these four longitudinal oscillation amplitudes with and non-longitudinal oscillation motion ($\omega = 7.68$ rad/s). It can be found that the longitudinal motion slightly increases the average power coefficient of the wind turbine compared to the no-longitudinal condition, and the value of the average power coefficient Cp gradually increases as the longitudinal amplitude decreases. Compared with the average power coefficient in the no longitudinal condition, the values of the average power coefficient Cp for longitudinal amplitude As = 1, 0.5 and 0.25 increased by 0.245%, 1.896% and 1.985%, respectively. When the longitudinal amplitude As = 1.25, the average power coefficient decreased by 1.19% compared to the no longitudinal motion. Based on the above analysis, the smaller amplitude longitudinal motion can not only reduce the variation amplitude of wind turbine load and blade load, but also increase with different amplitudes of longitudinal rocking angle in one longitudinal rocking cycle the output power of the wind turbine.

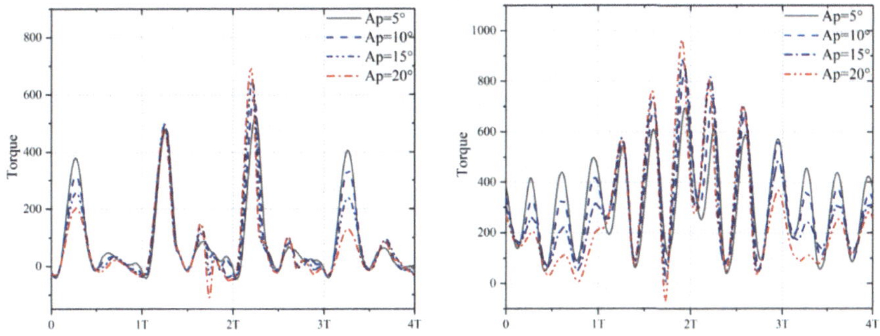

(a) Blade moment diagrams (b) Impeller torque diagrams

Fig. 9. Time-varying aerodynamic loads on wind turbine

Figure 9(a) and Fig. 9(b) shows the blade and impeller torque at four different rocking amplitudes during the same longitudinal rocking cycle, respectively. It can be seen that the larger the rocking amplitude, the larger the range of torque fluctuation.

In order to better compare the power output of the wind turbine at different rocking amplitudes, Table 3 shows the wind energy utilization of the wind turbine at these four angular amplitudes and under non-longitudinal rocking motion. From the table, it is easy to see that proper rocking motion can help improve power generation efficiency Compared with the non-longitudinal rocking motion, the Cp values corresponding to

Table 3. Comparison of power coefficient Cp for longitudinal motion without longitudinal motion and longitudinal motion with different angular amplitudes

Rocking angle amplitude	Non-rocking	Ap = 5°	Ap = 10°	Ap = 15°	Ap = 20°
Dual impeller average Cp	0.39874	0.421186	0.426778	0.406835	0.358091

Ap = 5° (Ap = π/36), Ap = 10° (Ap = π/18), Ap = 15° (Ap = π/12), Ap = 20° (Ap = π/9) motions increase by 5.629%, 7.031%, 2.030%, and −10.194%, respectively. In the actual production design, the wind energy utilization can be improved by controlling the appropriate longitudinal rocking amplitude.

3.2 Analysis of Aerodynamic Performance of Twin Units Under Different Cycles

Figure 10 depicts the aerodynamic loads on the wind turbine for different longitudinal oscillation periods at ω = 7.68 rad/s. It can be seen that a larger oscillation period is more favorable to the stability of the VAWT power output, and on the other hand, a larger longitudinal period also reduces the peak value of the VAWT external load.

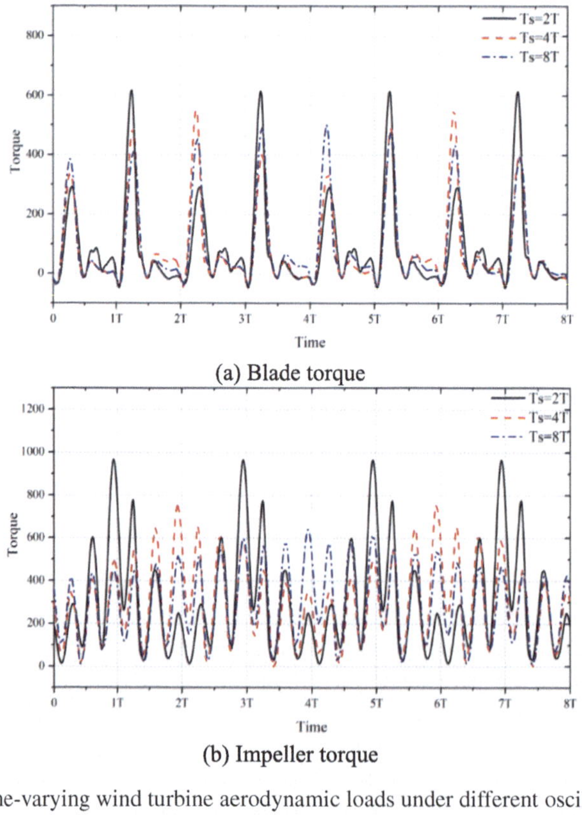

Fig. 10. Time-varying wind turbine aerodynamic loads under different oscillation cycles

Table 4 compares the average power coefficients of the wind turbine at $\omega = 7.68$ rad/s for the VAWT for different period of longitudinal oscillation motion and no longitudinal oscillation motion. It can be found that the value of the average power coefficient Cp increases with the growth of the oscillation period. The average values of wind turbine power coefficient Cp corresponding to longitudinal periods of Ts = 2T, Ts = 4T and Ts = 8T increased by −0.711%, 0.688% and 3.608%, respectively, compared to the non-oscillation condition. It can be concluded that the larger period of longitudinal oscillatory motion is not only beneficial to the stability of the VAWT, but also has a certain enhancement on the power coefficient of the wind turbine.

Table 4. Comparison of power coefficient Cp between no longitudinal oscillation motion and longitudinal oscillation motion with different periods

Longitudinal period	Non-oscillation	Ts = 2T	Ts = 4T	Ts = 8T
Dual impeller average Cp	0.39874	0.395904	0.401485	0.413129

In addition to the influence of different angular amplitudes, the period of pitch motion also significantly affects the aerodynamic performance of wind turbines (Fig. 11).

In order to better compare the power output of the wind turbine under different longitudinal rocking cycles, Table 5 shows the power output coefficients Cp of the wind turbine for these three cases and for the non-rocking motion. It can be found that the longitudinal rocking motion with different period increases the power generation efficiency and the value of Cp increases first and then decreases with the cycle Tp to some extent. Compared with the non-longitudinal oscillatory motion, the corresponding values of Cp under the longitudinal oscillatory motion with Tp = 2T, Tp = 4T and Tp = 8T increased by −2.969%, 6.833% and 6.491%, respectively. Therefore, choosing a suitable longitudinal oscillation period can increase the aerodynamic performance of a DI-VAWT to a certain extent.

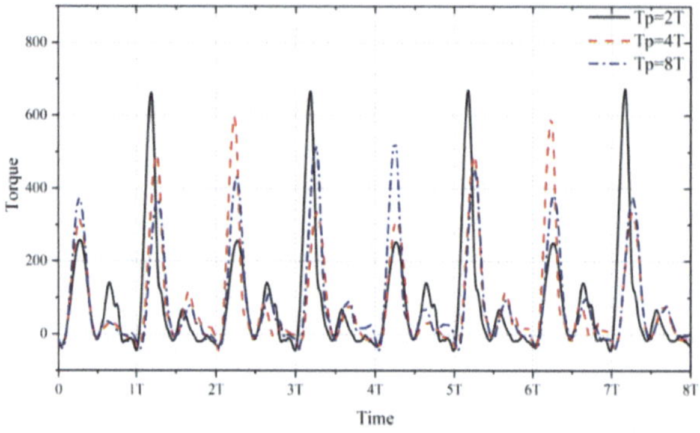

(a) Blade torque

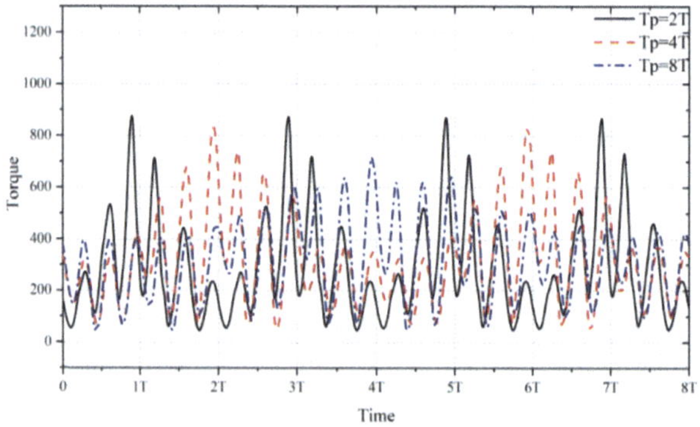

(b) Impeller torque

Fig. 11. Time-varying wind turbine aerodynamic loads under different rocking cycles

Table 5. Comparison of power coefficients Cp for rocking without longitudinal rocking motion and longitudinal rocking motion with different cycles

Longitudinal period	Non-rocking	Ts = 2T	Ts = 4T	Ts = 8T
Dual impeller average Cp	0.39874	0.386903	0.425988	0.424625

4 Conclusion

In this paper, the aerodynamic characteristics of a dual-impeller vertical-axis wind turbine are investigated under single-degree-of-freedom periodic simple harmonic longitudinal oscillation and longitudinal rocking motions, respectively. The aerodynamic performance of the dual impeller vertical axis wind turbine under motion and without motion is compared in detail, and the aerodynamic characteristics of the dual impeller vertical axis wind turbine are investigated under different motion amplitudes and motion periods. The following conclusions are drawn:

(1) The longitudinal oscillatory motion significantly affects the aerodynamic characteristics of wind turbines. It is found that a smaller longitudinal oscillation amplitude and a larger longitudinal period help to increase the power output of the wind turbine and reduce the peak value of the external load of the wind turbine. For example, when the longitudinal oscillation amplitude As = 1 m (As = 1/5D), the average power coefficient Cp of the double impeller is increased to 0.406302, which is 1.896% higher compared to 0.39874 under no longitudinal oscillation condition. However, when the longitudinal oscillation amplitude was increased to As = 1.25 m (As = 1/4D), the average power coefficient decreased to 0.393995, which was 1.19% lower.
(2) The study in the longitudinal rocking motion mode reveals the effects of different longitudinal rocking angle amplitudes and periods on the aerodynamic performance of the wind turbine. The average power coefficient Cp of the double impeller decreases to 0.358091 for the longitudinal oscillation amplitude of Ap = 20° (Ap = π/9), which is 10.194% lower than that in the case of no longitudinal oscillation motion. And when the longitudinal rocking angle amplitude is Ap = 5° (Ap = π/36), Cp increases to 0.421186, which increases by 5.629%. In addition, the Cp is 0.425988 for longitudinal rocking period Tp = 4T, which is 6.833% higher than that in the absence of longitudinal rocking motion, while the Cp is 0.386903 for Tp = 2T, which is 2.969% lower.

References

1. Birol, F.: World Energy Outlook 2023. International Energy Agency, Paris (2023)
2. Hand, B., Cashman, A.: A review on the historical development of the lift-type vertical axis wind turbine: from onshore to offshore floating application. Sustain. Energy Technol. Assess. **38**(2020)
3. Elkhoury, M., Kiwata, T., Aoun, E.: Experimental and numerical investigation of a three-dimensional vertical-axis wind turbine with variable-pitch. J. Wind Eng. Ind. Aerodyn., 111–123 (2015)
4. Rezaeiha, A., Montazeri, H.: Towards optimal layout design of vertical-axis wind-turbine farms: double impeller arrangements. Energy Convers. Manag. **226** (2020)

Numerical Simulation of Hydrodynamic and Topographical Changes Caused by an Offshore Wind Farm on a Muddy Coast

Huarong Shao[1], Dake Chen[2,3](✉), Limin Xiao[2], and Linyun Sun[2]

[1] Taizhou Port and Shipping Development Center, Taizhou 318000, China
[2] Nanjing Hydraulic Research Institute, Nanjing 210029, China
chdake@126.com
[3] The National Key Laboratory of Water Disaster Prevention, Nanjing 210024, China

Abstract. During the recent decades, offshore wind power has been developing rapidly, and the scale of wind farms has become larger and larger. The impact of wind farms on the marine environment and topography should be paid more attention. This study applied a 2HD mathematical model to investigate the hydrodynamic and topographical changes caused by an offshore wind farm on a muddy coast. The simulation results revealed that the installation of the wind turbines did not sufficiently affect the hydrodynamic and geomorphology of the study area from a macroscopic perspective. The hydrodynamic and geomorphological changes (i.e., sediment erosion and siltation) engendered by tidal flow and waves before and after the installation of offshore wind turbines are mainly distributed inside and near the wind farm and pose little impact on the nearby channels.

Keywords: offshore wind farm · wind energy · flow field · topographical change · sediment transport · seabed evolution

1 Introduction

Driven by rapid electricity demand growth and carbon emission reduction strategies, the demand for renewable energy has surged in recent years. Offshore wind energy, with its abundant resources, minimal land footprint, high power generation capacity, and scalability, has become a crucial focus for the renewable energy development of many countries, including China. According to the China Wind Energy Association (CWEA), in 2023, China added 7.183 GW of offshore wind capacity, bringing the cumulative installed capacity to over 37.71 GW - a remarkable 23.5% year-on-year increase.

Zhejiang Province of China has abundant offshore wind energy resources and has been a key region for the construction of offshore wind farms in China. It has an estimated exploitable capacity of approximately 43 GW in nearshore wind power technology. However, the marine economy is well-developed in Zhejiang Province, and its coastline features numerous ports and shipping channels. The construction of offshore wind farms disrupts the existing balance between hydrodynamics, sediment transport,

D. Jeng and B. Cai (Eds.): ICCOE 2024, LNCE 423, pp. 208–222, 2025.
https://doi.org/10.1007/978-981-96-0181-3_20

and topography in the marine environment, leading to changes in hydrodynamics and sediment transport. These alterations may impact the marine environment and ecology. Variations in hydrodynamics and sediment movement can further induce seabed morphological changes, potentially affecting nearby navigation channels. In this study, we employ a mathematical model to investigate the local hydrodynamics and seabed morphological changes induced by the construction of a nearshore wind farm in Zhejiang Province, aiming to assess the impact of wind farm development on local waterways.

2 The Offshore Wind Farm Project and Study Area

2.1 The Xiangshan Offshore Wind Farm Project

The Xiangshan offshore wind power project is situated outside the mouth of Sanmen Bay in Zhejiang Province. The project is being developed in two phases: the first phase covers the northern area, while the second phase encompasses the southern area. Figure 1 illustrates the geographical location of both project phases. In the first phase, located approximately 17 km offshore, the seabed elevation ranges from −7 m to −12 m (based on CD datum). The project area for this phase is an irregular hexagon, spanning a maximum length of 7.5 km from north to south and a maximum width of 4.9 km from east to west, covering approximately 30 km^2. The first phase was completed and operationalized in December 2021.

Moving to the second phase, the center point of the wind power project site is approximately 25 km offshore, with a mud surface elevation of about −12 m. The project area for this phase has an approximately trapezoidal shape, with a maximum length of 14.7 km from north to south and a maximum width of 5.8 km from east to west, totaling a planned area of approximately 58.9 km^2. Construction of the second phase is scheduled to commence in 2023. This research paper focuses on analyzing the impact of the second phase of the wind power project on hydrodynamics and topography within the study area.

The second phase of the project aims to install 56 wind turbines, each with a single unit capacity of 9.0 MW, resulting in a total installed capacity of 504,000 kW. Additionally, the project includes an offshore booster station for electricity transmission. Among the wind turbines, 20 peripheral turbines around the farm and one internal turbine utilize high-pile-supported foundations. The remaining 36 turbines employ single-pile structures. Specifically, the high-pile-supported wind turbine foundation consists of 8 steel pipe piles with diameters ranging from 2.0 to 2.2 m. In contrast, the single pile foundation has a diameter between 7.5 and 10 m. The offshore booster station utilizes a four-pile jacket foundation, with each steel pipe pile having a diameter of 2.3 m.

Xiangshan offshore wind farm is surrounded by several waterways, including the north channel, south channel, and the main channel of Jiantiao Port. These channels accommodate ships weighing 10,000 tons or more. Frequent ship traffic to Jiantiao Port in Taizhou every day contributes to the bustling waterway. Additionally, there are four north-south waterways near the wind farm: the customary route for small coastal ships on the east side, the west route, the east route, and the north route to Taizhou Bay.

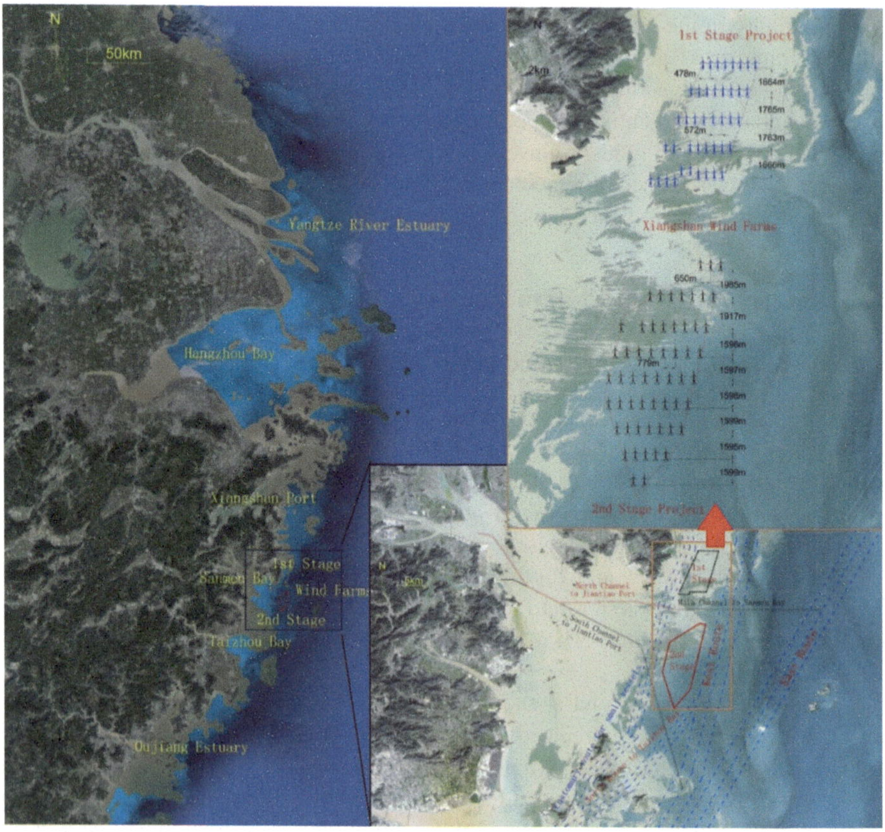

Fig. 1. The location of the Xiangshan offshore wind farms.

2.2 Hydrological and Sediment Conditions

The tidal regime in the wind farm project area exhibits regular semi-diurnal characteristics, with an annual average tidal range of 3.74 m and a maximum tidal range of 6.78 m. The tidal current follows a semi-diurnal pattern, with rotational flow dominating. The tidal-period-average flow velocity of the spring tide in winter is about 0.4–0.5 m/s, and the tidal-period-average flow velocity of the spring tide in summer is about 0.6–0.8 m/s. The waves in the project sea area are strong. The frequency of occurrence of light waves ($0.5 < H_s < 1.25$, H_s is the significant wave height) and medium waves ($1.25 < H_s < 2.5$) are 67.73% and 25.40%, respectively. Seabed sediments in the engineering sea area consist of silt and clayey silt, with a median particle size ranging from approximately 0.01 to 0.014 mm. Under the combined influence of waves and tidal flow, sediment transports primarily in the form of suspended load. The suspended sediment concentration (SSC) in winter is about 0.1 kg/m^3, and in summer it is about 0.2–0.3 kg/m^3.

3 Methodology

A depth-averaged mathematical model has been developed to examine the alterations in the hydrodynamics and topography caused by the second phase of the Xiangshan offshore wind farm project. The model comprises four interconnected modules, namely the hydrodynamic module, wave module, sediment transport module and morphology module. Specifically, the hydrodynamic and wave modules simulate the tidal flow patterns and wave fields, which serve as inputs for the sediment transport module to compute the suspended sediment transport. The resulting topographic changes due to the suspended sediment transport are then quantified by the morphology module.

3.1 Hydrodynamic Module

The governing equations for the hydrodynamic module are the depth-averaged continuity equation and the depth-averaged momentum equations, given as below:

$$\frac{\partial h}{\partial t} + \frac{\partial h\overline{u}}{\partial x} + \frac{\partial h\overline{v}}{\partial y} = 0 \tag{1}$$

$$\begin{aligned}&\frac{\partial h\overline{u}}{\partial t} + \frac{\partial h\overline{u}^2}{\partial x} + \frac{\partial h\overline{uv}}{\partial y} = f\overline{v}h - gh\frac{\partial \eta}{\partial x} - \frac{h}{\rho_0}\frac{\partial p_a}{\partial x} - \frac{gh^2}{2\rho_0}\frac{\partial \rho}{\partial x} + \frac{\tau_{sx}}{\rho_0} - \frac{\tau_{bx}}{\rho_0}\\ &- \frac{1}{\rho_0}\left(\frac{\partial S_{xx}}{\partial x} + \frac{\partial S_{xy}}{\partial y}\right) + \frac{\partial}{\partial x}(hT_{xx}) + \frac{\partial}{\partial y}(hT_{xy})\end{aligned} \tag{2}$$

$$\begin{aligned}&\frac{\partial h\overline{v}}{\partial t} + \frac{\partial h\overline{uv}}{\partial x} + \frac{\partial h\overline{v}^2}{\partial y} = -f\overline{u}h - gh\frac{\partial \eta}{\partial y} - \frac{h}{\rho_0}\frac{\partial p_a}{\partial y} - \frac{gh^2}{2\rho_0}\frac{\partial \rho}{\partial y} + \frac{\tau_{sy}}{\rho_0} - \frac{\tau_{by}}{\rho_0}\\ &- \frac{1}{\rho_0}\left(\frac{\partial S_{yx}}{\partial x} + \frac{\partial S_{yy}}{\partial y}\right) + \frac{\partial}{\partial x}(hT_{xy}) + \frac{\partial}{\partial y}(hT_{yy})\end{aligned} \tag{3}$$

where t is time; x and y are horizontal Cartesian coordinates; h is the water depth; $\overline{u}$ and $\overline{v}$ are the depth averaged horizontal flow velocity in x and y directions, respectively; f is the Coriolis parameter; g is the gravitational acceleration; η is the water surface elevation; ρ is the density of water; ρ_0 is the reference density of water; p_a is the atmospheric pressure; S_{xx}, S_{xy}, S_{yx} and S_{yy} are components of the radiation stress tensor; τ_{sx} and τ_{sy} are the x and y components of the surface wind stresses; τ_{bx} and τ_{by} are the x and y components of the bottom stress; T_{xx}, T_{xy}, T_{yx} and T_{yy} are the horizontal stress terms.

3.2 Wave Module

The dynamics of the gravity waves are described by the transport equation for wave action density. The two wave phase parameters, i.e., the relative angular frequency σ and direction θ are chosen as the independent variables in the model. The action density, $N(\sigma, \theta)$, is related to the energy density $E(\sigma, \theta)$ by $N(\sigma, \theta) = E(\sigma, \theta)/\sigma$. The governing equation, i.e., the wave action balance equation, reads:

$$\frac{\partial N}{\partial t} + \nabla \cdot (\boldsymbol{v}N) = \frac{S}{\sigma} \tag{4}$$

where $\boldsymbol{v} = \left(c_x, c_y, c_\sigma, c_\theta\right)$ is the propagation velocity of a wave group in the four-dimensional phase space x, y, σ and θ with c_x, c_y, c_σ and c_θ being the corresponding velocity components; S is the source term for the energy balance equation. In this study, $S = S_{in} + S_{nl} + S_{ds} + S_{bot} + S_{surf}$, where S_{in} represents the generation of energy by wind, S_{nl} is the wave energy transfer due to non-linear wave-wave interaction, S_{ds} is the dissipation of wave energy due to white-capping, S_{bot} is the dissipation due to bottom friction and S_{surf} is the dissipation of wave energy due to the depth-induced breaking.

3.3 Sediment Transport Module

The sediment transport module solves the depth-averaged advective-dispersion equation, which is written as:

$$\frac{\partial S}{\partial t} + u\frac{\partial S}{\partial x} + v\frac{\partial S}{\partial y} = \frac{1}{h}\frac{\partial}{\partial x}\left(hD_x\frac{\partial S}{\partial x}\right) + \frac{1}{h}\frac{\partial}{\partial y}\left(hD_y\frac{\partial S}{\partial y}\right) + \frac{F_s}{h} \tag{5}$$

where S is the depth-averaged sediment concentration; D_x and D_y are the depth-averaged diffusion coefficients in the x and y directions; F_s is the deposition and erosion flux, which is calculated by:

$$F_s = \begin{cases} \omega_s S\left(\tau_b / \tau_{cr,d} - 1\right) & \tau_b \le \tau_{cr,d} \\ 0 & \tau_{cr,d} < \tau_b < \tau_{cr,e} \\ M\left(\tau_b / \tau_{cr,e} - 1\right) & \tau_b \ge \tau_{cr,e} \end{cases} \tag{6}$$

where τ is the bed shear stress; $\tau_{cr,e}$ and $\tau_{cr,d}$ are the critical shear stress for erosion and deposition, respectively; ω_s is the settling velocity of suspended sediment; M is the erosion coefficient.

3.4 Morphology Module

The erosion and deposition fluxes lead to the changes in bed levels, that is computed by the following equation in the morphology module:

$$\rho_d\frac{\partial z}{\partial t} = -F_s \tag{7}$$

where z is the bed level; ρ_d is dry bulk density of the seabed sediment.

4 Model Setup and Validation

4.1 The Model Scope and Grid

The model encompasses an extensive marine area, extending up to 175 km in length from north to south and 120 km in width from east to west (see Fig. 2). The northern boundary aligns with Putuo Mountain in Zhoushan City, while the southern boundary corresponds to Shitang Peninsula in Wenling City. The western boundary, delineating the transition from sea to land, follows the coastline. To the east, the boundary lies in the open waters approximately 100 km east of Sanmen Bay. The model incorporates the most recent bathymetric data, and adjustments to the channel topography and wind power project area are based on the latest field measurements.

The calculation domain of the model is divided into a triangular mesh. The grids at the wind farm project area and the waterways have been encrypted (Fig. 3). The spatial resolution of the grid near the waterway is about 40 m, and the spatial resolution of the grid near the wind farms is about 1.9 m. The grids where the wind turbine pile foundation is located do not participate in the hydrodynamic calculation.

4.2 The Main Parameters

Manning Coefficient. The model adopts the Manning coefficient to reflect the seabed roughness, which is determined based on the validation of the tidal level and current. The Manning coefficient varies between 100 to 50 $m^{1/3}$/s from the deep waters to the near coast zone.

Friction Coefficient. The JONSWAP spectra are adopted to describe the irregular waves. The friction coefficient is set constant as 0.01 by reference to the study of Gong and Dai (1983).

Settling Velocity. In the estuarine and coastal environments, fine cohesive particles are usually settling in flocs instead of separated particles. The settling velocity of flocs is far larger than that of a single particle. In this study, the average settling velocity of flocs, 0.5 mm/s, is adopted (Liu & Zhang 1992).

Dry Bulk Density of the Seabed Sediment. The dry bulk density of the seabed sediment is calculated by the following empirical formula (Chen et al. 2017):

$$\rho_d = 1750 D_{50}^{0.183} \tag{8}$$

where D_{50} is the median size of the sediment in a unit of mm.

Critical Shear Stress. When the bed shear stress exceeds the critical shear stress for erosion, the erosion of seabed occurs and the seabed sediment is entrained into suspension. When the bed shear stress is lower than the critical shear stress for deposition, the suspended sediment begins to settle down on the seabed. The critical shear stresses for erosion and deposition are related to the physiochemical properties of seabed sediment and the turbulence characteristics of the near-bed flow. The critical shear stress for erosion varies between 1.0–1.25 Pa at the engineering area based on the studies of Chen et al. (2018); Chen et al. (2022); Chen, Zheng, et al. (2021).

Erosion Rate. There are numerous formulas for the erosion rate of non-cohesive sediment (Chen et al. 2021). However, the erosion of cohesive sediment is much more complicated and there is currently no accurate formula available for the erosion rate of cohesive sediment. In this study, the erosion rate of cohesive sediment is calculated by $M\left(\tau_b / \tau_{cr,e} - 1\right)$, where the coefficient M is regarded as a calibration coefficient.

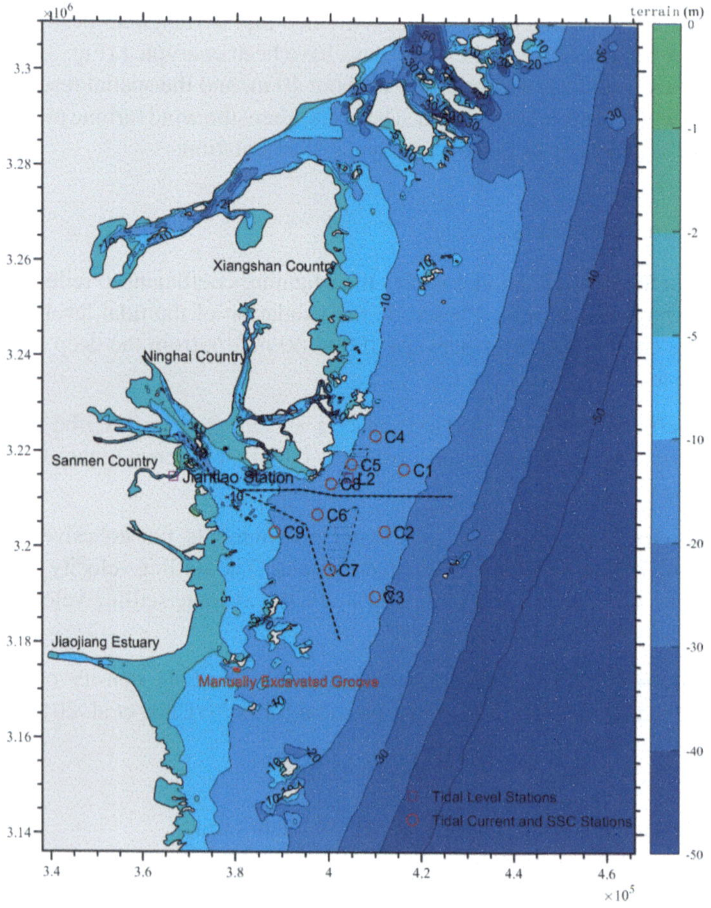

Fig. 2. The scope of the model.

4.3 Model Validation

The simulated tidal level, current velocity and direction, and suspended sediment concentration (SSC) have been successfully validated to demonstrate the ability of the model. The data of tidal level, current velocity and direction, and SSC for validation were measured during the Marth of 2018. Figure 3 shows the locations of the measurement stations of tidal level, flow and SSC. Figures 4, 5 and 6 show part of the validation results. To further demonstrate the accuracy of the model, the topographical change for a manually excavated groove has been also simulated. The groove is located south of Toumen Island and is 600 m long, and 70 m wide (also shown in Fig. 3). According to the terrain monitoring of the groove, the average 1-year siltation thickness inside the groove is about 0.6 m, well agreeing with the calculated value, being 0.64 m.

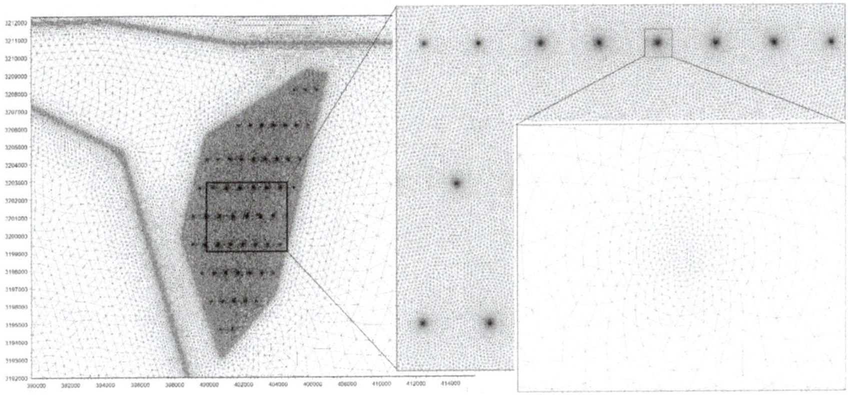

Fig. 3. The grid near the offshore wind farm.

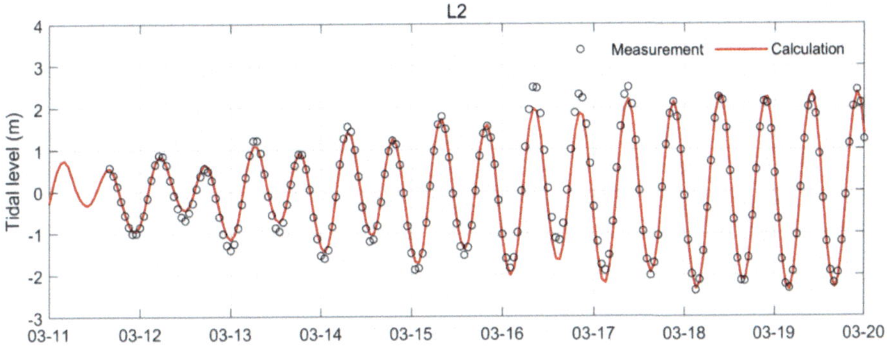

Fig. 4. The simulated tidal level versus the measured data.

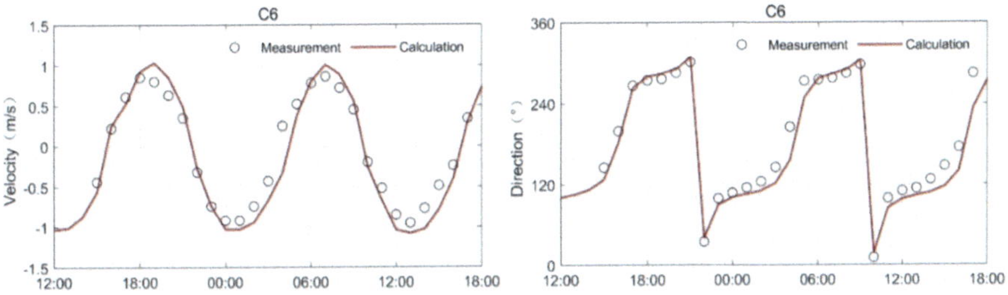

Fig. 5. The simulated tidal current versus the measured data.

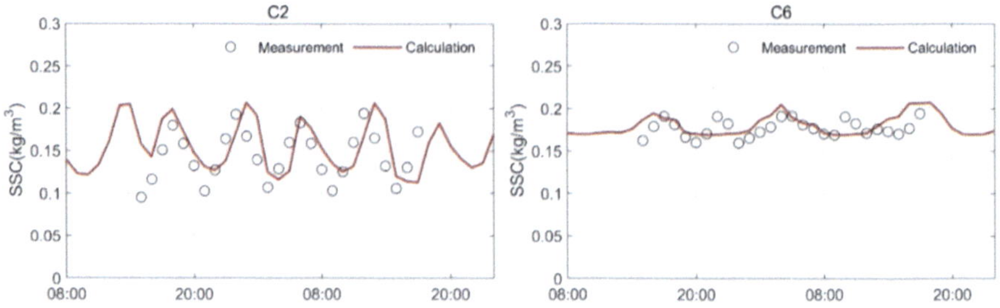

Fig. 6. The simulated suspended sediment concentration versus the measured data.

5 Model Setup and Validation

5.1 The Tidal Current Characteristics

Figures 7 and 8 depict the flow field in the engineering sea area at the peak flooding velocity moment and the maximum ebb velocity moment, respectively. The tidal flow simulation results reveal the tidal current outside the Sanmen Bay mouth exhibits rotational flow, while the tidal current in the bay follows a reciprocating pattern. The flooding flow predominantly enters the bay from the open sea in a southeast direction. Additionally, a minor influx occurs along the coastline near Ma'an Mount and Xiaoji Mount, as well as through the channel between Nantian Island and the mainland coastline. Conversely, the ebb tidal flow exits the bay along a path consistent with the flooding flow but in the opposite direction.

During the flooding phase, the flow traverses the second phase of the wind power project from southeast to northwest. Conversely, the ebb flow descends from northwest to southeast. Notably, near the wind farms, the maximum flooding velocity moment registers velocities ranging from 0.70 to 1.60 m/s. Specifically, within the second phase area, the maximum velocity during flooding ranges from 0.70 to 1.00 m/s, while in the deep trough on the south side of Nantian Island and the southwest and southeast sides of Hua'ao Island, it exceeds 1.2 m/s.

Furthermore, during spring tides, the maximum falling flow velocity in the two wind farm areas lies between 0.70 and 1.20 m/s. Within the second phase area, this velocity ranges from 0.70 to 0.90 m/s. Notably, the deep trough on the south side of Nantian Island and the southwest and southeast sides of Huaao Island experience relatively high velocities, exceeding 1.10 m/s during high tides.

5.2 The Impact on Tidal Current

Figures 9 and 10 show the velocity change after and before the second phase of the Xiangshan offshore wind farm project at the peak flooding velocity moment and the maximum ebb velocity moment, respectively. They show the second phase of the wind farm project has little impact on the flow pattern from a macroscopic perspective. The impact on the tidal flow field is mainly concentrated near the pile foundations.

When the tide rises, the velocity on the south side of the pile foundations tends to increase, and it decreases on the north side. Affected by the overall layout of the

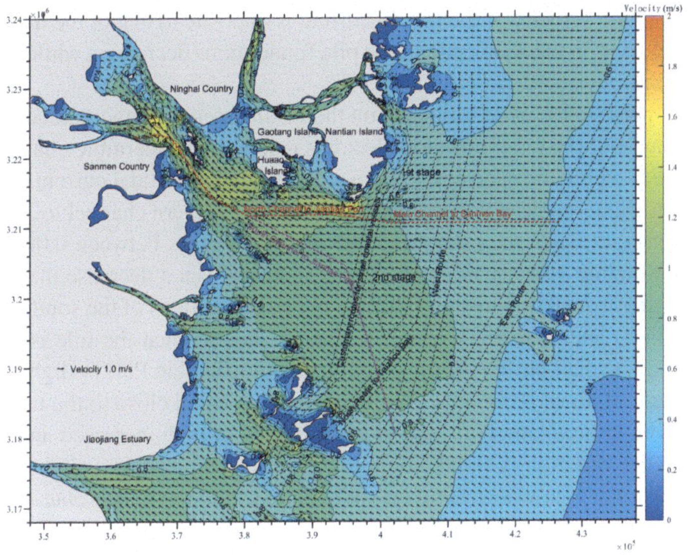

Fig. 7. The simulated tidal current field at the maximum flooding velocity moment.

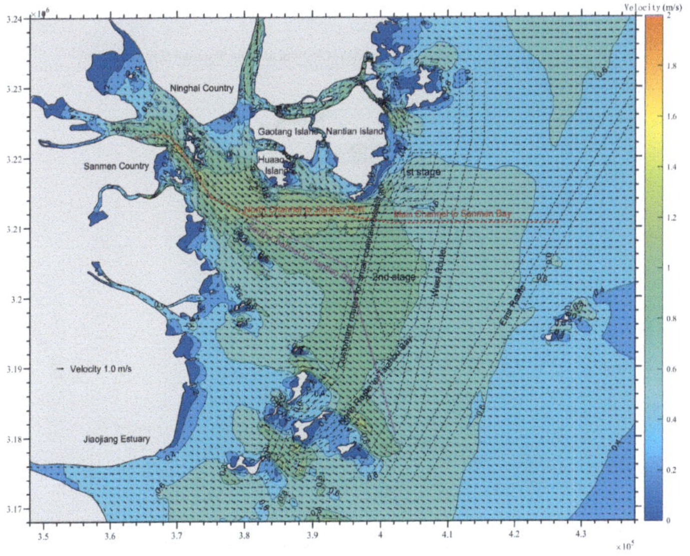

Fig. 8. The simulated tidal current field at the maximum ebb velocity moment.

wind farm and the direction of the rising tide, the intensity and range of the increase in velocity are more obvious in the south of the wind farm, and the intensity and range of the increase decrease as it go north. On the contrary, the intensity and range of the decrease in velocity are smaller in the south of the wind farm, and the intensity and range of the decrease increase as one go north. When the tide fall, the law of increase and decrease in the velocity of the falling current is opposite to that of the flooding flow, that is, the velocity on the south side of the pile foundations decreases, while it increases on the north side.

From the perspective of its influence on the local waterways, the implementation of the second phase of the wind power project has resulted in discernible changes in tidal current velocity within the nearby waterways. Specifically, the tidal current velocity has experienced a slight increase in certain sections of the southern channel to Jiantiao Port at the southern end when the tide rises. The increase ranges between 0.005–0.01 m/s for the peak flooding velocity moment. There also has been a decrease in tidal current velocity, ranging from 0.005 m/s to 0.01 m/s, in certain sections of the southern channel (to Jiantiao Port) and customary routes for small ships. When the tide is falling, the flow velocity in a small portion of the main channel to Jiantiao Port is slightly reduced, with a decrease of about 0.005 m/s. The west route, situated close to the main channel to Jiantiao Port, exhibits intervals and localized areas of both increased and decreased velocity. But the variation is small, about 0.005 m/s. There is also a slight decrease in flow velocity in the area where the west route connects to the north route into Taizhou Bay, with a decrease of about 0.005 m/s.

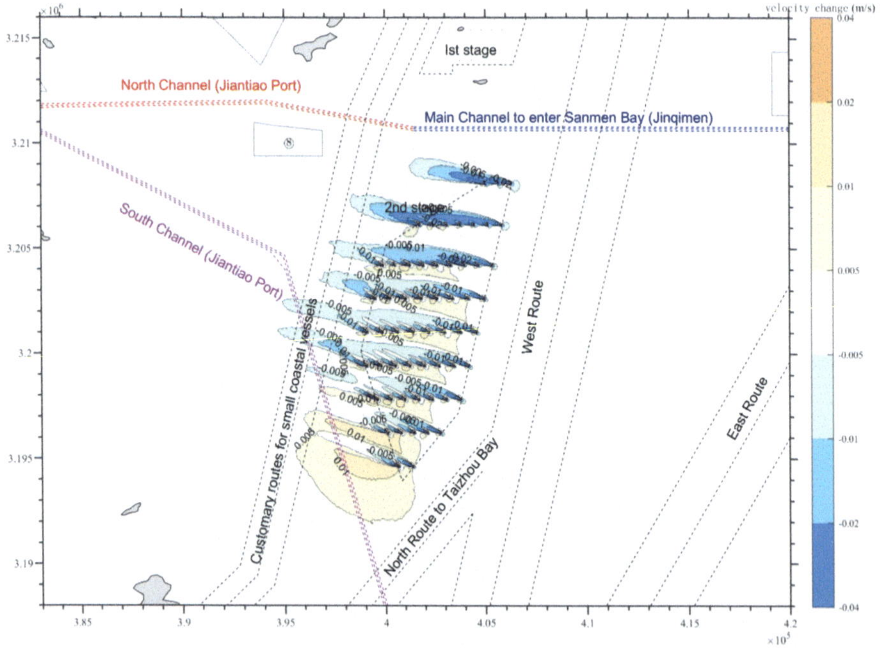

Fig. 9. The velocity change at the maximum flooding velocity moment.

5.3 Prediction of Seabed Evolution Caused by the Offshore Wind Farm

Figure 11 shows 1-year sediment erosion and deposition distribution caused by the second phase of the Xiangshan offshore wind farm project. As depicted in the figure,

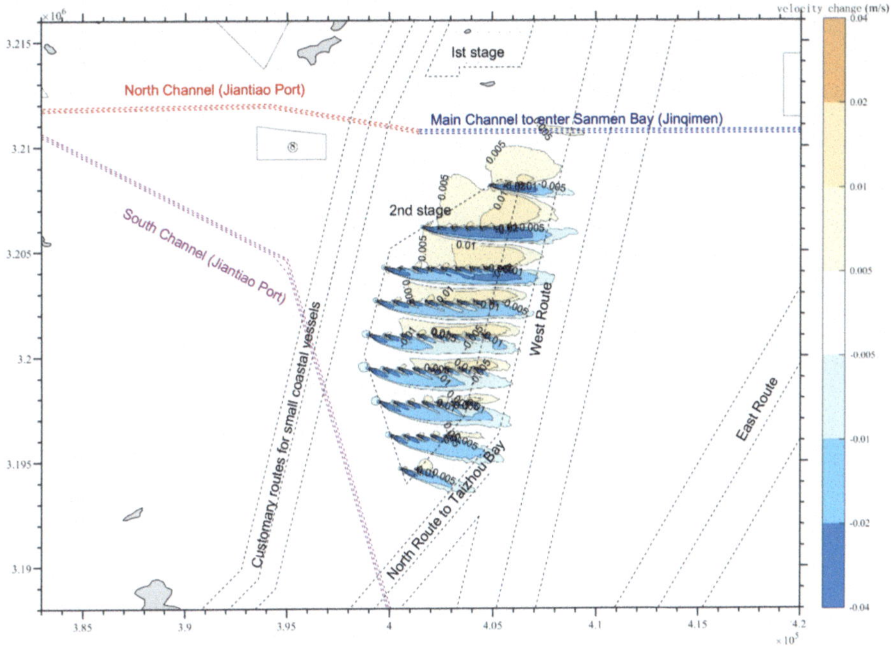

Fig. 10. The velocity change at the maximum ebb velocity moment.

the topographical changes predominantly concentrate in the vicinity of the project site. On the whole, the project exerts minimal influence on the underwater topography of the large-scale sea area. The presence of pile groups in the project zone leads to increased resistance, resulting in a discernible micro-silting trend across the entire project area. The annual siltation thickness ranges from 0.01 to 0.06 m. Notably, a localized micro-scouring region exists in the northeast corner of the wind power project area, with an annual scouring depth ranging from 0.01 to 0.04 m. Similarly, on the western side of the project area, a distinct range of micro-scouring areas is observed, with annual scouring depths between 0.01 and 0.02 m. Additionally, micro-scouring or micro-silting areas, characterized by depths of 0.01 to 0.02 m, occur in nearby waterways.

In general, the temporal evolution of erosion and siltation in the nearshore sea area reveals non-linear patterns following the completion of engineering work that impacts seabed dynamics. Initially, the changes in topography are more pronounced. However, over time, the intensity of the erosion and deposition processes gradually diminishes as the hydrodynamics, sediment transport, and underwater topography adapt to one another. Figures 12 and 13 illustrate the 3-year and 5-year sediment erosion and deposition distribution caused by the second phase of the project.

Comparing the distribution of changes after 1 year, one observe a slight expansion in the range of silting and scouring areas. Specifically, after 3 and 5 years, there is a slight increase in silting near the pile foundations and the northwest corner of the wind power farm. Additionally, minor increases in scouring occur on the western and northeastern corners of the project area. Notably, these changes remain relatively small.

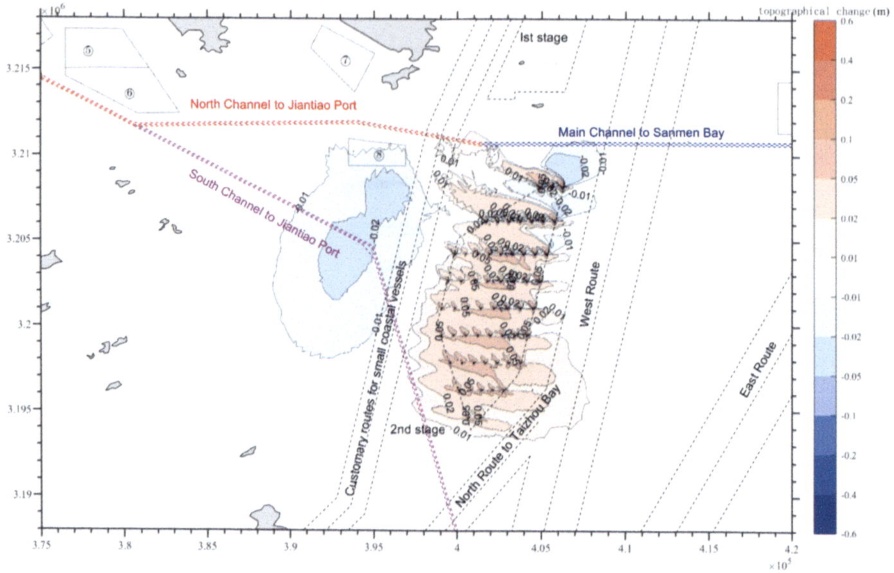

Fig. 11. The 1-year topographical change caused by the offshore wind farm.

The similarity in the distribution of topographical changes after 3 and 5 years suggests that the impact of the second phase of the wind power project has essentially reached a dynamic equilibrium.

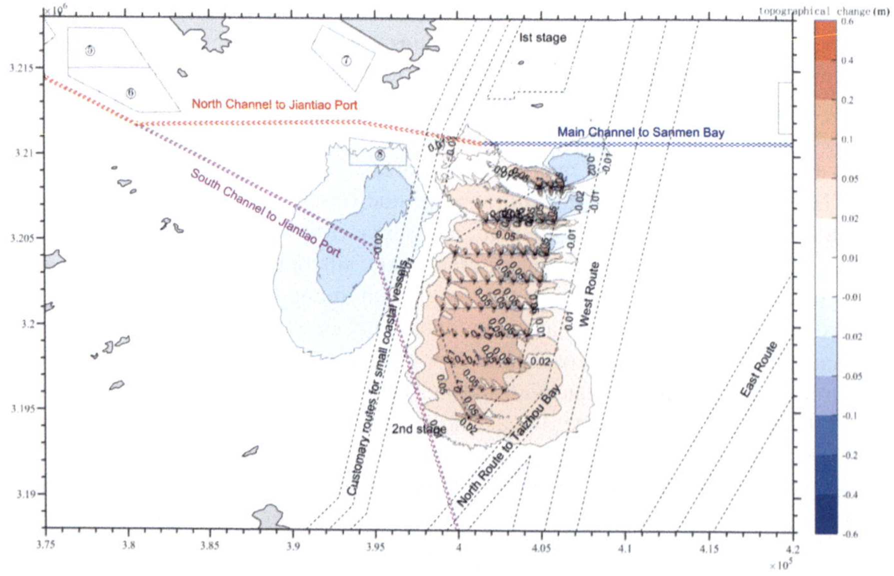

Fig. 12. The 3-year topographical change caused by the offshore wind farm.

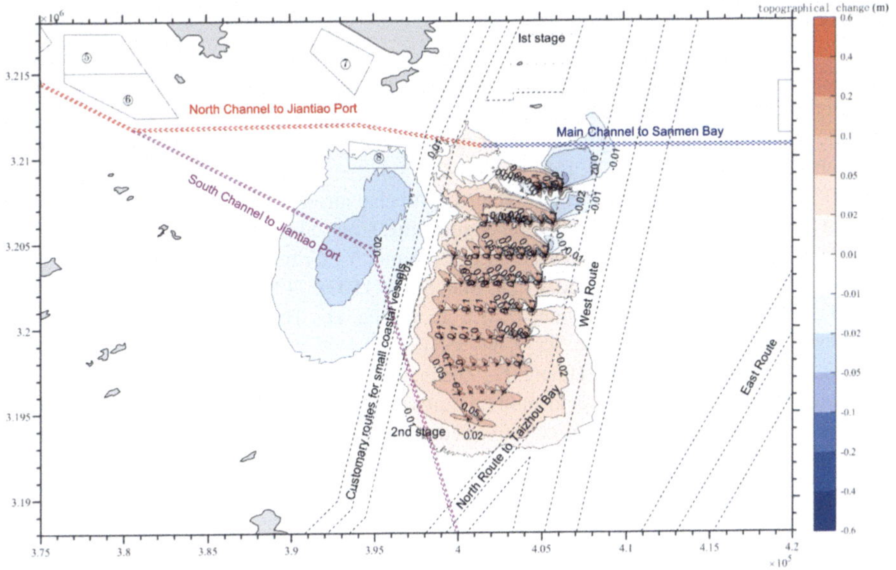

Fig. 13. The 5-year topographical change caused by the offshore wind farm.

6 Conclusion

A two-dimensional (2HD) mathematical model was developed to investigate the impact of the second phase of the Xiangshan offshore wind farm project on hydrodynamics and seabed evolution. The findings reveal that the influence on the flow field primarily concentrates near the pile foundations, with minimal effects on large-scale hydrodynamics and surrounding waterways. Specifically, the flow velocity in local channel experiences slight fluctuations, within a range of ± 0.01 m/s. From a macroscopic perspective, the second phase of the wind power project exerts limited impact on seabed evolution. Topographical changes predominantly occur in proximity to the project site, albeit with small magnitudes. Generally, the project area experiences slight silting, resulting in an annual siltation thickness ranging between 0.01 and 0.06 m. Additionally, localized micro-scouring or micro-silting areas, with depths of 0.01–0.02 m, are observed in nearby waterways. Over time, these topographical changes gradually attenuate, ultimately reaching a dynamic equilibrium within about five years.

Funding. This work was supported by the National Natural Science Foundation of China [52101310]; The Fundamental Research Funds for the Central Research Institutes, NHRI (Y223007).

References

Gong, C., Dai, G.: Mathematical model for wave deformation in shoaling water and determination of bottom friction of silt beach. Ocean Eng. **03**, 21–33 (1983)

Liu, J.J., Zhang, J.C.: Siltation prediction for navigation channels and harbour basins on muddy beach. China Ocean Eng. **6**(2), 157–172 (1992)

Chen, D., Wang, Y., Huang, H., Chen, C., Yuan, C.: A behavior-oriented formula to predict coastal bathymetry evolution caused by coastal engineering. Cont. Shelf Res. **135**, 47–57 (2017)

Chen, D., Zheng, J., Zhang, C., Guan, D., Li, Y., Wang, Y.: Critical shear stress for erosion of sand-mud mixtures and pure mud. Front. Mar. Sci. **8**(1502), 713039 (2021)

Chen, D., Zheng, J., Zhang, C., Guan, D., Li, Y., Huang, H.: Threshold of surface erosion of cohesive sediments. Front. Mar. Sci. **9**, 847985 (2022)

Chen, D., Wang, Y., Melville, B., Huang, H., Zhang, W.: Unified formula for critical shear stress for erosion of sand, mud, and sand-mud mixtures. J. Hydraul. Eng. **144**(8), 04018046 (2018)

Chen, D., et al.: Pickup rate of non-cohesive sediments in low-velocity flows. J. Hydraul. Res. **60**(1), 125–135 (2021)

Design and Analysis of a Semi-submersible Floating Wind Turbine with Active Pump

Ali Azmat Sher Khan[1], Xiangqian Zhu[1](✉), Zhiqiang Yang[1], and Qing'an Li[2,3]

[1] Key Laboratory of High-Efficiency and Clean Mechanical Manufacture of MOE, National Demonstration Center for Experimental Mechanical Engineering, School of Mechanical Engineering, Shandong University, Jinan 250061, China
{aliazmat07,yangy101}@mail.sdu.edu.cn, xqzhu@sdu.edu.cn

[2] Engineering Thermophysics Institute, Chinese Academy of Sciences, 11 North Fourth Ring Road West, Beijing 100190, China
liqingan@iet.cn

[3] CSSC Haizhuang Windpower Co., Ltd., Chongqing 401122, China

Abstract. The wind sector has undergone significant advancements in recent years, transitioning from onshore to offshore sites. In China, most of the offshore wind sites are located in intermediate waters. Typhoons generated in the Pacific Ocean sometimes occur in the South China Sea, which is rich in wind energy reserves. FOWTs face challenges in dealing with harsh environmental conditions, leading to the need for expensive mooring system designs. However, the probability of encountering such extreme conditions is relatively low, resulting in an overly conservative approach to mooring system design for most operational scenarios. To address this issue, this paper proposes the integration of an active pumping system to introduce seawater into the platform's vessel during challenging environmental conditions. While the mooring system of the semi-submersible wind turbine ensures stability under most operational conditions, the active pumping system is essential to enhance safety during challenging environmental conditions. This paper uses passive semi-submersible Model A and the active pump-assisted semisubmersible Model B as its research objectives. Wind loads acting on the system is not considered in this work to evaluate the hydrodynamic responses of the system. Det Norske Veritas (DNV) is utilized for modelling the mooring system, calculating the motions that impact the platform's stability and performance under ordinary and extreme conditions. The FOWT's structural responses, and mooring line tension are determined by frequency domain and time domain simulations, afterward evaluated with statistical procedures. The results demonstrate that incorporating an active pump in a wind turbine enhances platform stability.

Keywords: Floating Offshore Wind Turbines (FOWTs) · Mooring system · Active pump · Hydrodynamic behavior

1 Introduction

FOWTs have been a major focus in ocean engineering for several decades, driven by the growing demand for environmentally friendly and sustainable energy sources, and the exploration of offshore wind power. FOWTs are specifically designed for deployment in

D. Jeng and B. Cai (Eds.): ICCOE 2024, LNCE 423, pp. 223–233, 2025.
https://doi.org/10.1007/978-981-96-0181-3_21

deep ocean areas where wind resources are more reliable and consistent than in onshore or nearshore locations [1]. As a result, FOWTs have become a significant research area in the renewable energy sector. In order to meet its ambitious goals for offshore renewable energy, the European Commission plans to have at least Sixty gigawatts (GW) of offshore wind power by 2030 [2]. Floating wind turbines have gained substantial scientific interest recently and hold great promise for the future of offshore wind energy. Current industry research focuses on coastal conditions and the effects of anomalous waves [3, 4], which can significantly displace floating wind turbines and potentially cause mooring lines to break. Most floater designs are similar to offshore oil and gas platforms. Despite progress, there are unresolved issues that must be addressed for the widespread adoption of floating wind turbines in larger markets. One key aspect is the development of mooring systems suitable for water depths of 50 to 80 m, a range considered transitional between floating and bottom-fixed wind turbines. In shallower waters, the distance among the seabed and the fairlead, crucial for the mooring system's proper function, is significantly reduced, presenting a greater design challenge than in deeper waters [5]. While there is extensive expertise in mooring design within the offshore oil and gas industry, requirements for floating wind turbines are distinct. Rapid progress in China's floating wind turbine industry has highlighted the need for designs suitable for shallow waters, but significant challenges remain in addressing these specific conditions. Floating wind turbines can be used in moderate water depths, typically ranging from 50 to 200 m. It is crucial to guarantee the floating structure's and the mooring system's safety. Analyzing the movement responses of the floating structure and the response of mooring tension is necessary. High-precision computational fluid dynamics (CFD) analysis methods and time-domain procedures based on potential flow theory were employed to estimate motion responses and severe mooring loads of floating wind turbines [6–11]. Previous studies on the hydrodynamic behavior of different platforms have focused on the impact of environmental factors and predicting hydrodynamic responses. For example, studies have investigated how irregular wave spectrum peaks impact the hydrodynamic performance of semi-submersible platforms [12], examined the hydrodynamic response of newly developed V-shaped platforms [13], and investigated the response of barge-type platforms under various wave conditions [14]. Compared to fixed wind turbines, floating wind turbine platforms are more complex, making research on their hydrodynamic response highly significant. Oscillations may result from wave forces acting on the turbine and support structure, particularly in rough conditions, affecting energy generation efficiency. This paper focuses on comparing semi-submersible design concepts, prioritizing representative load cases covering motion responses. While existing research has centered on structural integrity and hydrodynamic loads [18], this study aims to understand better the role of active pumps in mitigating platform motion and improving turbine performance. This research investigates the effects of motion response on the surge, heave, and pitch degrees of freedom in a semi-submersible wind turbine, revealing significant reductions in oscillation amplitudes and improved stability. Additionally, the influence of mooring line tension on the dynamic behavior of FOWTs is examined, providing a reference for designing other floating platforms and their mooring systems. This study utilizes advanced software such as DNV-GeniE [15], DNV-HydroD [16], and WADAM [17] for platform design, hydrostatic analysis, and hydrodynamic analysis, respectively.

These tools are renowned for their reliability and accuracy, facilitating detailed analysis of platform performance in various hydrodynamic environments. The aim is to improve understanding of how the platform behaves with an active pump addition.

2 Methodology

Hydrodynamic analysis is carried out in the frequency domain using the panel model from GeniE by the Wadam program in HydroD [16]. Frequencies ranging from 0.1 to 1.2 Hz, with a step of 0.05 Hz, are taken into account while solving the potential flow model. This system demonstrates simple harmonic motion at the frequency of the incoming waves. Hence, the equation that describes the movement of the platform's center of gravity in the frequency domain is expressed as [20]:

$$M\ddot{x} + C\dot{x} + k(s)x = F \tag{1}$$

In this context, M represents the mass matrix. The symbol C represents the radiation damping matrix, which is also associated with the influence of frequency waves. K(s) indicates the stiffness matrix. The terms x refers as generalized displacement, $\dot{x}$ refer as velocity, and $\ddot{x}$ refer as acceleration, respectively, of the platform within a six degrees of freedom (6-DOF) framework, induced by wave action at frequency. Lastly, F represents the wave loads.

The time-domain response is computed using DNV-Sima, with Simo solving the equations of motion for the simulated body to obtain motion responses in the time domain. This combines the hydrodynamic loads analyzed in Wadam with time-varying loads, such as external forces from moorings. The hydrodynamic coefficients calculated for the platform in the frequency domain are dependent on the frequency of incoming waves. This means that to use these coefficients in time domain analyses, their frequency dependence must be considered. When taking into account the combined effects of waves, current, and mooring forces, the equation of motion for the platform's center of gravity in the time domain is established as follows [20]:

$$\begin{aligned} M\ddot{x}(t) + C\dot{x}(t) + k(s)x(t) + \int_0^t R(t-\tau)\dot{x}(t)d\tau = F_{wave}(t) \\ + F_{current}(t) + F_{mooring}(t) \end{aligned} \tag{2}$$

The symbol M represents the mass matrix. R(t) denotes the matrix of velocity pulse functions; the terms x(ω) refer as generalized displacement, $\dot{x}(\omega)$ refer as velocity, and $\ddot{x}(\omega)$ refer as acceleration, of the platform at time t, respectively. $F_{wave}(t)$, $F_{current}(t)$, $F_{mooring}(t)$ represent the wave load, current load, and mooring restoring force, respectively, acting on the platform with 6 degrees of freedom at time t. The calculation of added mass and potential flow damping using potential flow theory varies with frequency. The floating platform's motion response is then calculated using the fourth-order Runge-Kutta numerical method.

3 Description of Semi-submersible FOWT

3.1 Semi-sub FOWT Model

The OC4 DeepCWind FOWT has been widely utilized in previous research on FOWT. However, in the research we conducted, we used a new semi-submersible structure comprising of columns, pontoons, and heave plates as shown in Fig. 1 and 2. To increase stability, the whole structure has been configured up as an equilateral triangle. In the current study, apart from traditional semi-submersible turbine another semi-submersible turbine with active pump is investigated at a water depth of 67 m. To counter the impact of extreme wave loads active pump semi-submersible platform is analyzed. Lowering the center of gravity has the benefit of improving dynamic stability. To ensure that a floating structure remains level and stable in the water, it is crucial that it displaces the right amount of water to create a balance between the downward force of gravity and the upward buoyant force. This design methodology greatly enhances the structural capacity to sustain loading moments and enhanced the structural stability. The rotor hub and nacelle were modeled as rigid bodies, each incorporating six degrees of freedom (DOF) motion responses. The SIMO package was used to predict the corresponding oscillations. Additionally, the semi-submersible base was created as a rigid object, and all 9 mooring lines is created using bar elements. The characteristics of the models have been described in Table 1.

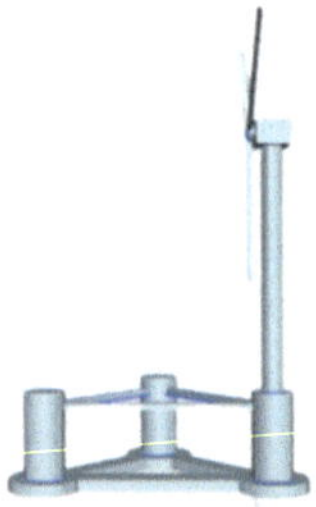

Fig. 1. Sketch of Conceptual Model

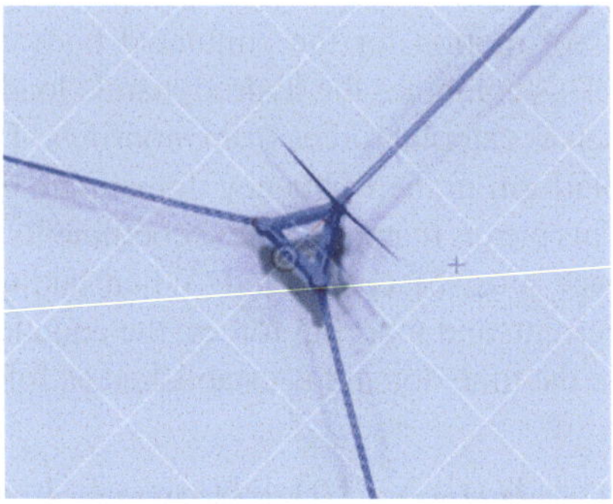

Fig. 2. Analysis in DNV-SIMA

Table 1. Parameters of Models

Main Particular	Model A	Model B
Mass of the whole FOWT structure	1.5529e+07 kg	1.6343e+07 kg
I_{xx}	1.7254e+10 kgm^2	1.4728e+10 kgm^2
I_{yy}	2.7176e+10 kgm^2	2.5377e+10 kgm^2
I_{zz}	2.3336e+10 kgm^2	2.5394e+10 kgm^2
Water depth	67 m	67 m

The design approach for active pump includes adjusting the amount of ballast within the model B columns to maintain proper balance and counterbalance the weight of the tower, rotor nacelle assembly (RNA), and blades. Figure 3 shows the function of model b under different load conditions. This design approach significantly increases the structure's ability to withstand loading moments and ensures greater structural stability.

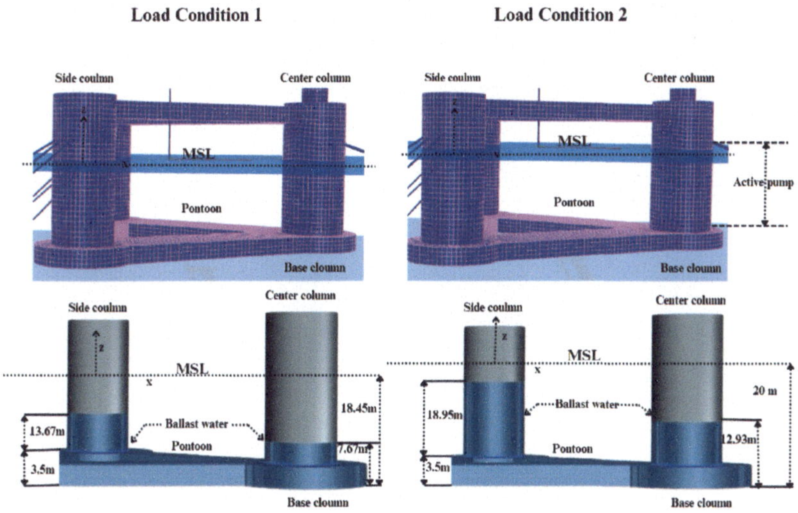

Fig. 3. Illustration of Model B active pump system

3.2 Design Criteria of Mooring System

A comprehensive mooring design assessment, following DNVGL [19] guidelines, incorporates the model motion responses, and structural safety of the mooring lines. The primary design criteria for shallow water mooring in this study are as follows:

1. All mooring lines must withstand load effects under all considered environmental conditions, making sure that the maximum tension in the lines does not surpass its maximum strength while accounting for the safety factor.
2. The length of the catenary mooring line that lies on the seabed must be adequate to keep it from being entirely raised during its lifetime and to stop the drag anchor from supporting vertical loads, even in the most adverse conditions.

4 Result and Discussion

4.1 Impact of Wave Load on Motion Response

This section presents the analysis of the motion responses to the load scenarios. For all load cases, the waves approach from a heading of 0°. This study aimed to analyze the dynamic behavior of a model of a semi-submersible wind turbine located at a water depth of 67 m. The research focused on following oceanic conditions defined as Load Case 1 (LC1) as shown in (Fig. 4 and 5). This study used two separate models, Model A and Model B, which were subjected to a specific set of wave dynamics. Wave heights of 0.67 m and periods of 4.8 s were used in the sea state simulation. Model A exhibited a vertical displacement of 1.0803 m, a vertical movement of 0.68672 m, and a tilt angle of 1.8807°. In contrast, Model B had more controlled motions, with a surge of 0.35011 m, a heave of 0.37639 m, and a pitch of 1.0004°.

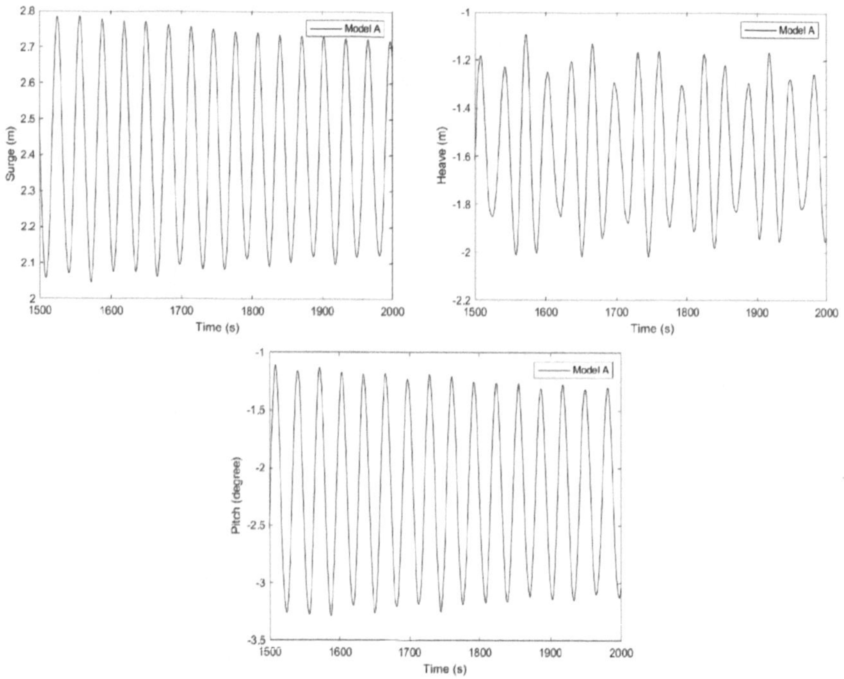

Fig. 4. Time history curve of Model A under LC1

These results reveal differences in the dynamic behavior of the two models under identical wave conditions. This sensitivity is critical for assessing the structural integrity and operating feasibility of the wind turbine in typical sea circumstances.

For load case, the waves approach from a heading of 0 degrees under the extreme wave conditions (H = 15 T = 19.2). As a result, we concentrate on evaluating the platforms' movements in the surge, heave, and pitch degrees of freedom. In order to prevent collisions, floating wind turbines must have their motion responses assessed during the design process, as per the DNV requirements.

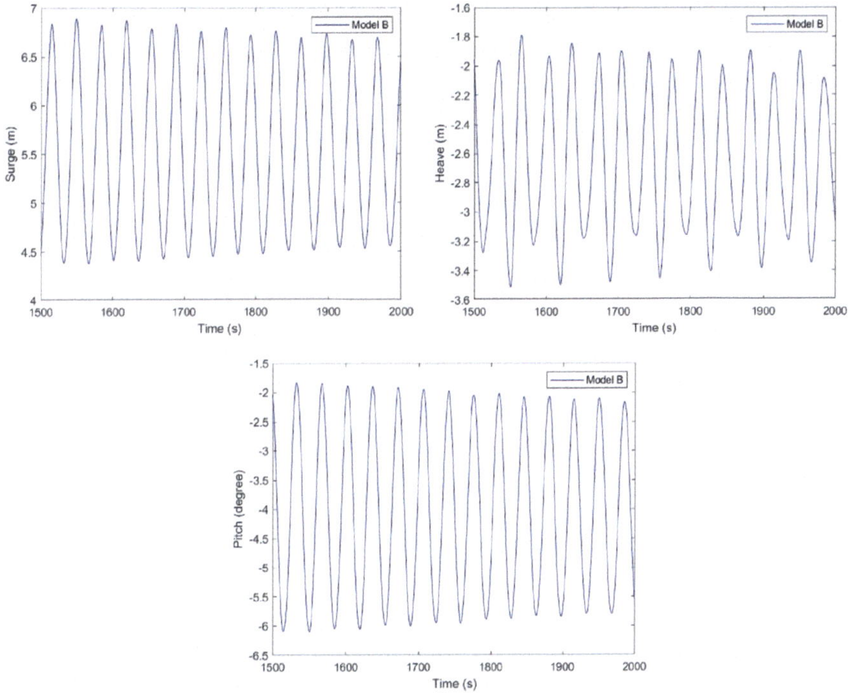

Fig. 5. Time history curve of Model B under LC1

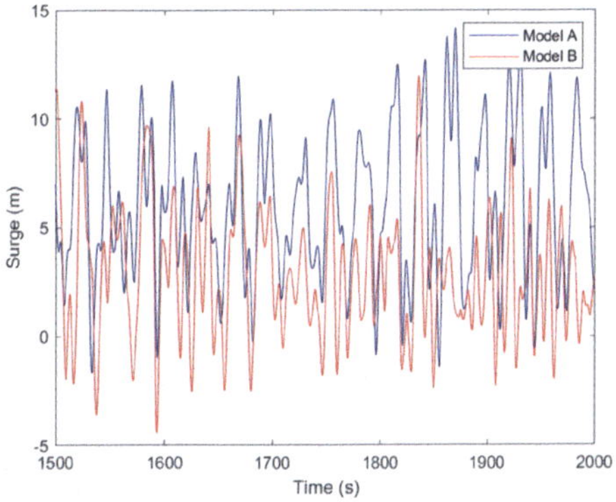

Fig. 6. Time history comparison curve of surge motion under LC2

According to the obtained results of surge time histories under extreme operating conditions from LC2 (Fig. 6) it can be seen that for Model A, the standard deviation is 3.38 m, showing a considerable increase in surge motion due to higher wave height and period. In Model B, the standard deviation drops to 2.95 m, indicating significant mitigation of surge motion and showcasing the pump's ability to stabilize the platform under extreme conditions. The results demonstrated that integrating active pump in

Model B effectively reduced resonant amplitudes in surge motion, showcasing their ability to dampen oscillations. Additionally, the presence of active pump positively affected the surge responses.

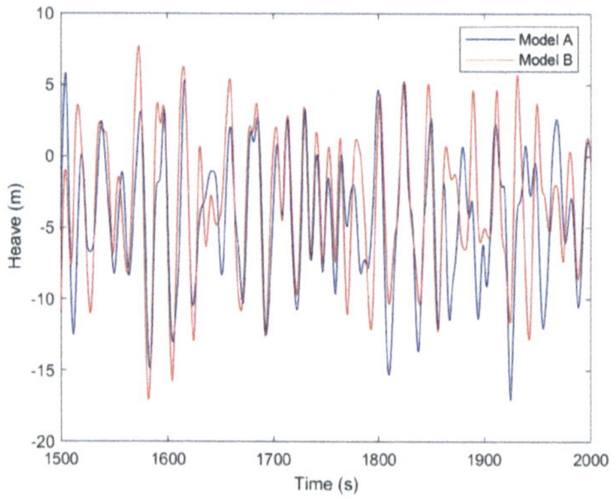

Fig. 7. Time history comparison curve of heave motion under LC2

By analyzing the heave time histories under extreme operating conditions from LC2 (Fig. 7), it is evident that for Model A, the standard deviation is 4.59 m, reflecting more significant heave motion due to more intense wave conditions. In Model B, the standard deviation rises to 4.88 m, indicating controlled heave motion but with increased variability under harsh conditions.

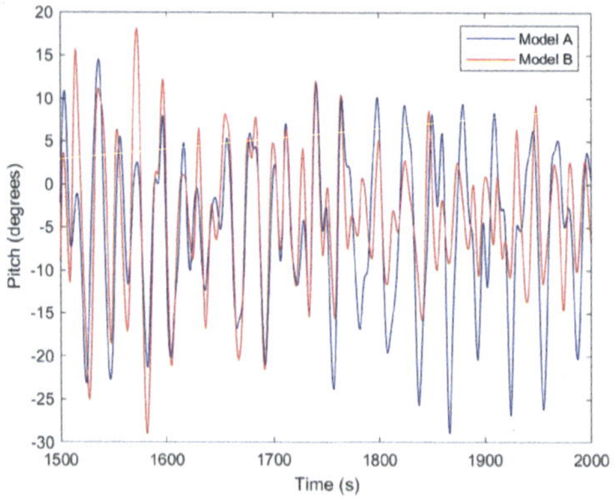

Fig. 8. Time history comparison curve of pitch motion under LC2

In the comparative analysis of pitch time histories for two semi-submersible wind turbine models under extreme operational conditions, as presented in LC2 and depicted in Fig. 8, significant insights into the design effectiveness and operational stability of these models are garnered. Model A, with a standard deviation of 9.19° in pitch motion, exhibits a pronounced response to environmental stresses, indicative of a design potentially less optimized for harsh marine conditions. This increased standard deviation indicates a significant variation in the pitch angle, suggesting a higher sensitivity to pitching movements. This could affect the turbine's alignment with the dominant wind direction, potentially reducing its operational efficiency and increasing structural stress. On the other hand, Model B shows a stronger resilience, managing to lower the standard deviation to 8.02° even under similar harsh conditions. The incorporation of the active pump in Model B, increases structure's ability to withstand loading moments, ensuring greater overall stability in extreme condition.

4.2 Impact of Wave Load on Mooring Tension Response

The assessment of power generation and load performance of a semi-submersible floating wind turbine, in accordance with the IEC 61400-3 offshore wind turbine standard at a depth of 67 m, reveals significant variations between Model A and Model B, as shown in Fig. 9 and 10. In the 50-year extreme wave simulation (Wave Height: 15 m; Period: 19.2 s), the comparative analysis of Model A and Model B underlines significant variations in the structural responses. For Model B, the maximum forces observed between from 21,500 kN to 42,900 kN. The mean forces were recorded between 11,800 kN and 36,900 kN, with standard deviations ranging from 2,620 kN to 4,240 kN. In contrast, Model A exhibited a broader range of maximum forces, from 31,800 kN to 60,600 kN, mean forces ranged from 12,100 kN to 36,500 kN, and the standard deviations varied between 5,310 kN to 6,690 kN. The varying response of the models indicates diverse resiliency in dealing with severe wave conditions, demonstrating the potential impact of active pump systems. The safety factor, computed at a maximum allowable tension of 220,000 kN, demonstrates Model B's better performance. Model B demonstrates a safety factor of roughly 5.13, whereas Model A shows a safety factor of 3.63 in response to the 50-year extreme wave test.

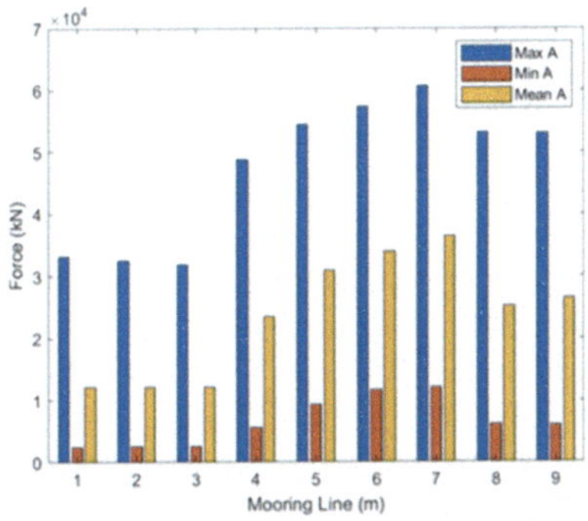

Fig. 9. 50-year wave test of Model A

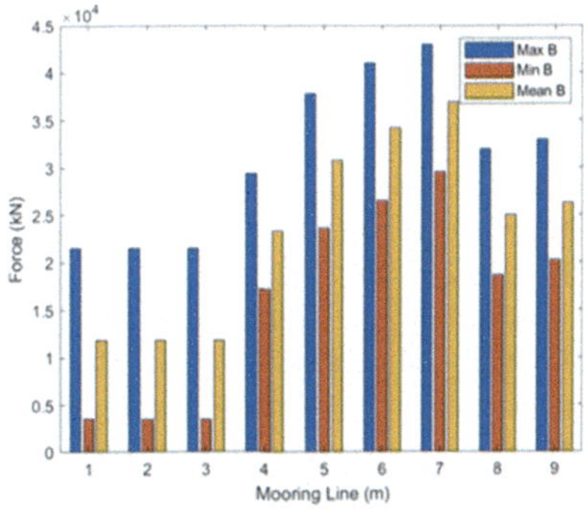

Fig. 10. 50-year wave test of Model B

5 Conclusion

The hydrodynamic performance of two semi-submersible models, Model A and Model B, under possible South China Sea conditions have been examined in this research. Considering intermediate depth marine area at 67 m. The mooring system of floating wind turbines of both model is analyzed for the 50-year extreme test. The load assessments showed that Model B is more stable under the severe test conditions indicated in the IEC 61400-3 standard. The active pumping system in Model B controlled the platform's displacement and angular rotations, lowering high-energy wave-induced structural stresses. These simulations revealed that Model B, with a safety factor of 5.13, was less fragile than Model A, with a safety factor of 3.63, under 50-year extreme wave conditions. This study also compares the motion response of two semi-submersible floating wind turbine platforms under identical design load conditions, thereby demonstrating the relative advantages of the active pump configuration. The results show that Model B performs better than Model A in terms of surge and pitch stabilization, as indicated by the standard deviation of the time-domain surge and pitch under severe operating conditions. Specifically, under LC2 conditions, Model A displays a standard deviation of 9.19°, indicating increased pitch motion due to severe environmental conditions. In contrast, Model B reduces this variability to 8.02°, demonstrating improved damping capabilities. The following conclusion can be taken from the two models research, active pump system reduced dynamic wave impact risks by systematically reducing motion response under simulated harsh conditions. Due to current limitations, this research on the model is confined to the simulation stage.

Acknowledgment. This work was supported by National Key Research and Development Program of China (No.2022YFE0207000); National Engineering Research Center for Offshore Windpower (No.HSFD22001); and Shandong Province science and technology SMES innovation ability improvement project and Rizhao Key Research and Development Project (No. 2022TSGC2504). The authors would like to acknowledge DNV Group for providing license file of Sesam.

References

1. Papi, F., Bianchini, A.: Technical challenges in floating offshore wind turbine upscaling: a critical analysis based on the NREL 5 MW and IEA 15 MW reference turbines. Renew. Sustain. Energy Rev. **162**, 112489 (2022)
2. An, E.: Strategy to harness the potential of offshore renewable energy for a climate neutral future. In: European Commission, Brussels, Belgium (2020)
3. Huo, F., Zhao, Y., Zhang, J., et al.: Study on wave slamming characteristics of a typical floating wind turbine under freak waves. Ocean Eng. **269**, 113464 (2023)
4. Li, Y., Li, H., Wang, Z., et al.: The dynamic response of a Spar-type floating wind turbine under freak waves with different properties. Mar. Struct. **91**, 103471 (2023)
5. Bai, H., Zhang, M., Xu, K.: Conceptual design, parameter optimization and performance investigation of a 10MW semi-submersible floating wind turbine in shallow water. Ocean Eng. **281**, 114895 (2023)
6. Quallen, S., Xing, T.: CFD simulation of a floating offshore wind turbine system using a variable-speed generator-torque controller. Renew. Energy **97**, 230–242 (2016)
7. Liu, L., Guo, Y., Zhao, H., et al.: Motions of a 5 MW floating VAWT evaluated by numerical simulations and model tests. Ocean Eng. **144**, 21–34 (2017)
8. Zhou, Y., Xiao, Q., Liu, Y., et al.: Exploring inflow wind condition on floating offshore wind turbine aerodynamic characterization and platform motion prediction using blade resolved CFD simulation. Renew. Energy **182**, 1060–1079 (2022)

9. Wang, Y., Chen, H., Koop, A., et al.: Verification and validation of CFD simulations for semi-submersible floating offshore wind turbine under pitch free-decay motion. Ocean Eng. **242**, 109993 (2021)
10. Barrera, C., Losada, I., Guanche, R., et al.: The influence of wave parameter definition over floating wind platform mooring systems under severe sea states. Ocean Eng. **172**, 105–126 (2019)
11. Yan, X., Chen, C., Yin, G., et al.: Numerical investigations on nonlinear effects of catenary mooring systems for a 10-MW FOWT in shallow water. Ocean Eng. **276**, 114207 (2023)
12. Alkarem, Y.R., Ozbahceci, B.O.: A complemental analysis of wave irregularity effect on the hydrodynamic responses of offshore wind turbines with the semisubmersible platform. Appl. Ocean Res. **113**, 102757 (2021)
13. Zhang, L., Shi, W., Karimirad, M., et al.: Second-order hydrodynamic effects on the response of three semisubmersible floating offshore wind turbines. Ocean Eng. **207**, 107371 (2020)
14. Chuang, T.C., Yang, W.H., Yang, R.Y.: Experimental and numerical study of a barge-type FOWT platform under wind and wave load. Ocean Eng. **230**, 109015 (2021)
15. DNV: Sesam user manual-Genie. DNV GL Software (2016)
16. DNV: Sesam user manual-Hydrod. DNV GL, Høvik, Norway (2013)
17. DNV: Sesam user manual-Wadam. DNV GL Software (2017)
18. Barooni, M., Ali, N.A., Ashuri, T.: An open-source comprehensive numerical model for dynamic response and loads analysis of floating offshore wind turbines. Energy **154**, 442–454 (2018)
19. DNV GL: DNVGL-OS-E301 Position Mooring (2018)
20. Zou, Q., Lu, Z., Shen, Y.: Short-term prediction of hydrodynamic response of a novel semi-submersible FOWT platform under wind, current and wave loads. Ocean Eng. **278**, 114471 (2023)

Key Technologies for Deep-Sea Mining and Oil and Gas Development

Analysis of Ore Particle Phase Content in Deep Sea Mining Pipeline Riser Using ERT Technology

Wenbin Ma[1,2](✉), Bo Zhou[1,2,3], Yuwen Chen[1,2,4], Yongqi Ma[1,2], and Yijun Shen[1,2]

[1] Faculty of Marine Science and Engineering, Hainan University, Haikou 570228, Hainan, China
wenbinma_edu@163.com

[2] State Key Laboratory of Marine Resources Utilization in the South China Sea, Haikou 570228, Hainan, China

[3] Mechanical and Electrical Engineering, Hainan University, Haikou 570228, Hainan, China

[4] Faculty of Marine Biology and Fisheries, Hainan University, Haikou 570228, Hainan, China

Abstract. With the over-exploitation of mineral resources on land, mineral resources are scarce, so people have set their sights on the deep sea, which is rich in mineral resources. This paper establishes a two-dimensional model for the transportation process of granular ore in the pipeline during the lifting process of deep-sea mining pipelines. Using the method of electrical resistance tomography, the empirical formula of the solid phase content of granular ore in solid-liquid two-phase flow is obtained by analyzing the characteristic voltage. Different models are constructed under this working condition, and their results are compared with those obtained by the empirical formula, which proves the feasibility of this method and provides a technical reference for pipeline lifting in deep-sea mining.

Keywords: Deep-sea mining · Electrical resistance tomography · solid phase content · Solid-liquid two-phase flow · Characteristic voltage

1 Introduction

The deep-sea area is rich in mineral resources, including various metals (such as copper, nickel, cobalt, zinc, etc.), rare metals (such as rare earth elements), minerals and energy resources (such as methane hydrate). Exploring and developing these resources is expected to meet future human needs and reduce the pressure of over-exploitation of terrestrial resources [1]. In recent years, many countries have made the development of deep-sea mineral resources a new strategic goal. However, the long-distance transportation of ore particles in thousands of meters of pipelines is prone to blockage (the schematic diagram of the deep-sea mining system is shown in Fig. 1 [2]). The main reason is related to the solid phase content of the ore particles. For hydraulic lifting systems, the deep-sea mining lifting pipeline is mainly solid-liquid two-phase flow. Measuring the phase content of the two-phase flow is one of the important parameters for studying and controlling its behavior [3]. The following are some commonly used phase content

D. Jeng and B. Cai (Eds.): ICCOE 2024, LNCE 423, pp. 237–252, 2025.
https://doi.org/10.1007/978-981-96-0181-3_22

measurement methods: (1) Gravity separation method [4] (2) Optical attenuation method [5] (3) Radioactive tracer method [6]. The accurate measurement of its parameters is directly related to the stability and economic operation of the deep-sea mining system.

Electrical resistance tomography (ERT) is a type of process tomography (PT) technology, and its origin can be traced back to the 1920s [7]. ERT technology has been widely used in the field of two-phase flow, and some researchers have achieved remarkable results in this regard. Zhang et al. used the ERT system to monitor the transportation process of ice-water two-phase flow online [8]; Tao et al. used ERT technology to study the characteristics of microchannel gas-liquid two-phase flow [9]; Dong et al. combined ERT technology and related technologies to successfully realize the measurement of two-phase flow parameters related to pipeline cross-sectional flow information [10, 11].

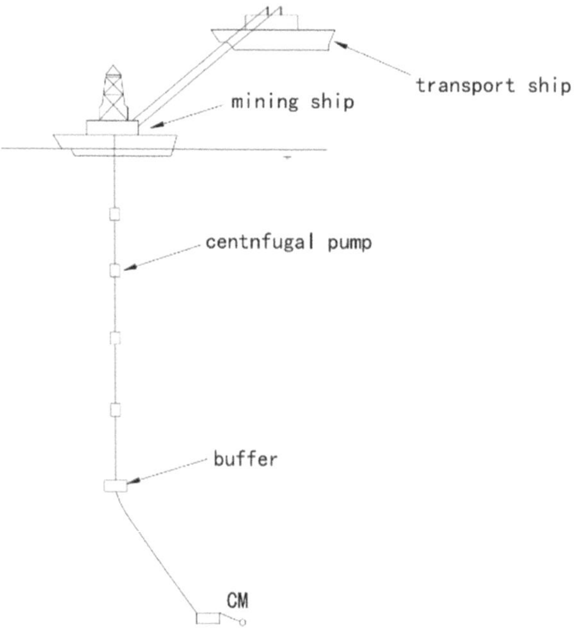

Fig. 1. Schematic diagram of deep-sea mining system

2 The Electrical Resistance Tomography System

2.1 Components of the Electrical Resistance Tomography System

The electrical resistance tomography system mainly consists of three parts, namely the electrode array part, the measurement and data acquisition part, and the computer reconstruction image part. Figure 2 is the schematic diagram of the electrical resistance tomography system. By applying current or voltage excitation to generate a sensitive field, the voltage or current signal on the boundary is measured, and the conductivity distribution of the medium in the sensitive field is inferred. With the help of a specific imaging algorithm, the visualization imaging of the two-phase flow is finally achieved.

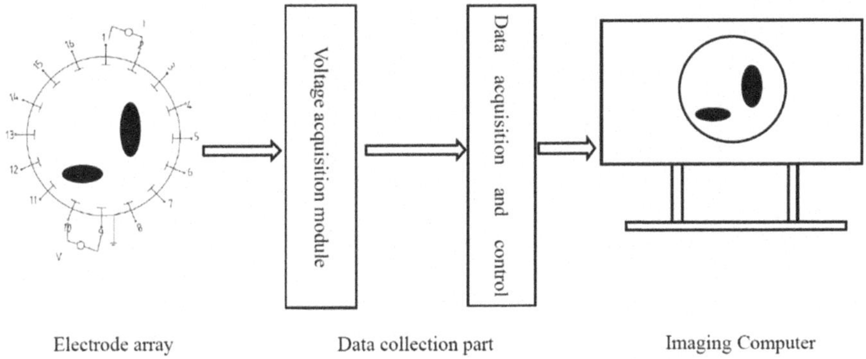

Fig. 2. Schematic diagram of the electrical resistance tomography system

2.2 Mathematical Principles of ERT Technology

In the ERT system, the mining pipeline being tested is regarded as a constant current system and can be treated as a steady-state field. At this time, the distribution of current in the pipeline is mainly affected by the conductivity of the medium. The distribution of current and conductivity can be expressed by Maxwell's equations [12]:

$$\begin{cases} \nabla \cdot D = \rho \\ \nabla \times E = -\frac{\partial B}{\partial t} \\ \nabla \cdot B = 0 \\ \nabla \times H = J + \frac{\partial D}{\partial t} \end{cases} \tag{1}$$

Where D is the electric displacement vector; ρ is the charge density; E is the electric field intensity; B is the magnetic induction intensity; H is the magnetic field intensity; and J is the current density. Equation (1) is Maxwell's first equation, and Eq. (2) is Maxwell's second equation. The relationship between the parameters is as follows:

$$\begin{cases} B = \mu H \\ D = \varepsilon E \\ J = \sigma E \end{cases} \tag{2}$$

In the formula, μ is the dielectric constant; ε is the magnetic permeability; σ is the electrical conductivity.

2.3 Direct and Inverse Problems of Electrical Resistance Tomography

The direct problem mainly refers to the process of inputting the excitation current to solve the potential distribution of any point in the sensitive field under the known medium distribution in the pipeline and the initial conditions of the sensitive field. The mathematical formula (3) of the ERT sensitive field is described as follows:

$$\begin{cases} \nabla \cdot (\sigma \cdot \nabla \phi) = 0 & In\ the\ sensitive\ field\ \Omega \\ \int_{S1} \sigma \frac{\partial \phi}{\partial n} ds = +I & On\ the\ current\ input\ electrode \\ \int_{S2} \sigma \frac{\partial \phi}{\partial n} ds = -I & On\ the\ current\ outflow\ electrode \\ \frac{\partial \phi}{\partial n} |_{S3} = 0 & Other\ electrodes \end{cases} \tag{3}$$

Where n is the external normal vector of any point on the boundary, σ is the distribution of conductivity, ϕ is the potential distribution, $S1$ and $S2$ are the surfaces of the current flowing into and out of the electrode respectively, and $S3$ is the insulating surface of the sensor.

The calculation of the direct problem is divided into analytical method and numerical calculation. The analytical method can obtain accurate results, but the derivation process is very complicated and the scope of application is limited. The numerical analysis methods include: finite difference method, finite element method [13] and boundary element method [14–16].

The inverse problem refers to the problem of reconstructing the conductivity distribution in the measured field based on the measurement data obtained by the sensor and the sensitivity matrix obtained in the direct problem. The difficulty of the inverse problem lies in its non-linear [17], under- determined [18]and ill-conditioned [19].

3 Experiment and Result Analysis

3.1 Experimental Principle

This experiment uses COMSOL software to build models of mining pipelines and media in pipelines. The specific steps are as follows: Construct pipeline and ore particle models in COMSOL, as shown in Fig. 3(a). The large circle represents the mining pipeline, the small circle represents the ore particles, the rest is water, and the blue part represents the 16 copper electrodes attached to the pipeline. According to the electrode arrangement, set appropriate boundary conditions and electrode parameters, define material properties and resistivity. Then the field is meshed as shown in Fig. 3(b), and then the current excitation is applied to the electrode as shown in Fig. 3(c), and the voltage on the adjacent electrode pairs is collected. The radius of the constructed pipeline model R = 200 mm, the electrode is a copper electrode, the central angle of each electrode is 9°, the 16 electrodes are evenly distributed on the pipeline, and the excitation current used is 0.001 A.

The solid phase used in the experiment is non-conductive sand and gravel, which is replaced by its main component silicon dioxide, and the liquid phase is water.

In this experiment, the phase content of solid ore particles is defined as formula (4):

$$\alpha = \frac{S_{\text{solid}}}{S_0} \tag{4}$$

Where S_0 represents the cross-sectional area of the pipe, S_{solid} represents the area occupied by the solid ore particles, and α represents the phase content of the solid ore particles.

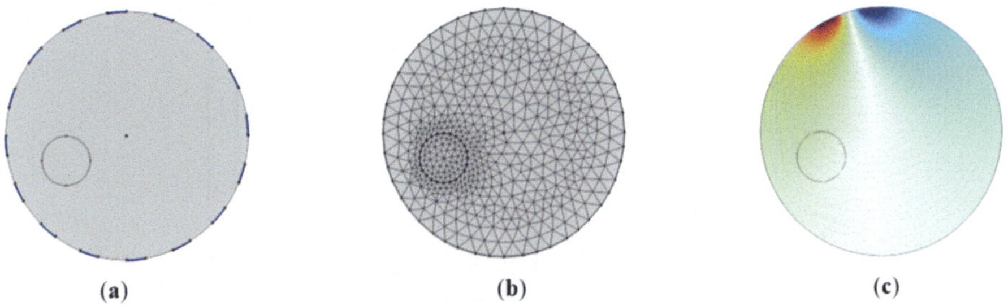

Fig. 3. (**a**) Pipeline model built in COMSOL (**b**) Field meshing (**c**) Electric field line distribution after excitation

Due to the nonlinear characteristics of the ERT sensitivity field, the relationship between the solid phase content and the measured data cannot be simply described by a linear formula [20, 21]. Even at the same solid phase content, different flow forms and solid phase positions will lead to significant differences in the measured data. In order to extract the eigenvalues containing the cross-sectional distribution information, the relative change summation method can be used. This method is closely related to the solid phase content and can effectively reduce the dimensionality of the original measured data. The relative change summation method is defined as summarizing the relative changes of all ERT measurement data [10, 11].

$$V_R = \sum_{i,j} \frac{U_{ij} - U_0}{U_0} \tag{5}$$

In the formula, V_R is the characteristic value. U_{ij} is the voltage data sampled by the j-th pair of electrodes under the excitation current of the i-th pair of electrodes when there are solid ore particles in the measurement pipeline. U_0 is the boundary voltage data when there is pure water.

3.2 Simulation Results

Three cases are constructed for a single particle, so that it is distributed at different positions in the pipeline. The centers of the ore particles are (80, 80), (100, 50), and (0, 0), respectively. The radius of the ore particles in each case is gradually increased, and its solid volume fraction increases by 0.02 from 0.02 to 0.12 each time. The boundary voltage U_0 in the empty field (without solid ore particles) is shown in Fig. 4, and the boundary voltage when there are solid ore particles in the pipeline is shown in Fig. 5, where the X-axis represents 208 points on the boundary, the Y-axis represents the measured boundary voltage data U_1, and the Z-axis 1–6 distribution represents the six cases where the radius of the solid ore particles increases from 28.28 mm to 69.20 mm.

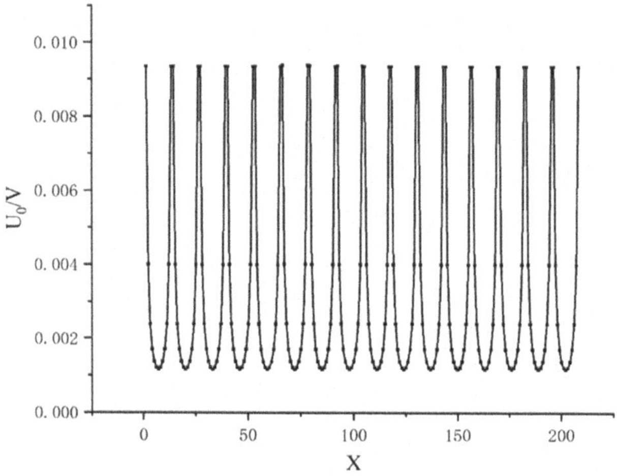

Fig. 4. Boundary voltage U_0 when empty field

According to the obtained U_0 and U_1 data, the characteristic voltage V_R can be obtained using formula (5). The α and V_R of different situations are shown in Table 1. The relationship between α and V_R obtained based on this data is shown in Fig. 6.

By comparing Fig. 6 with the corresponding relationship between bubble void fraction α and characteristic voltage V_R made by Dong Feng [10], it can be seen that when α is between 0 and 0.2, V_R and α are approximately in a linear function relationship, which is basically consistent with the relationship between V_R and α in this experiment when α is between 0 and 0.12, proving the feasibility of this method.

This method is applied to actual working conditions. Three cases are constructed for multiple particles, and they are distributed at different positions in the pipeline as shown in Fig. 7. The initial centers of the ore particles are A (−165, 0), B (−108, 0), and C (−60, 0), respectively. The solid phase content is increased by 0.02 to 0.12 every time it is rotated 60° counterclockwise. The boundary voltage when there are solid ore particles in the pipeline is shown in Fig. 8. The X-axis represents 208 points on the boundary, the Y-axis represents the measured boundary voltage data U_1, and the Z-axis 1–6 distribution represents 6 cases where the number of solid ore particles increases from 1 to 6.

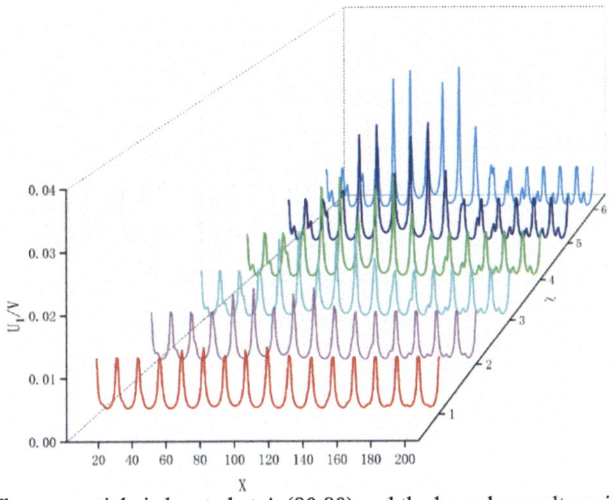

(a)The ore particle is located at A (80,80) and the boundary voltage is U_1

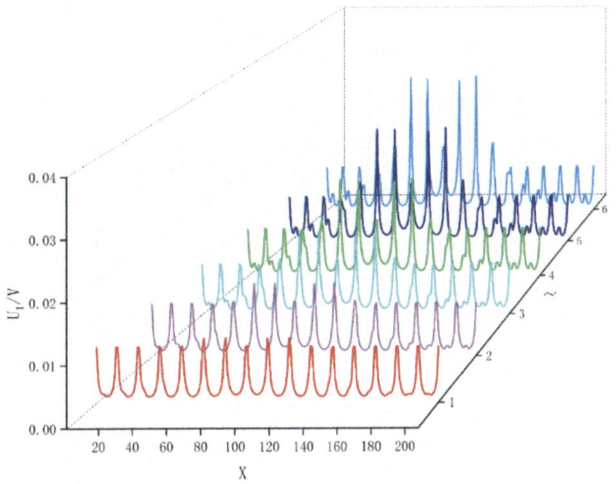

(b) The ore particle is located at B (100,100) and the boundary voltage is U_1

Fig. 5. Boundary voltage U_1 for cases A, B, and C

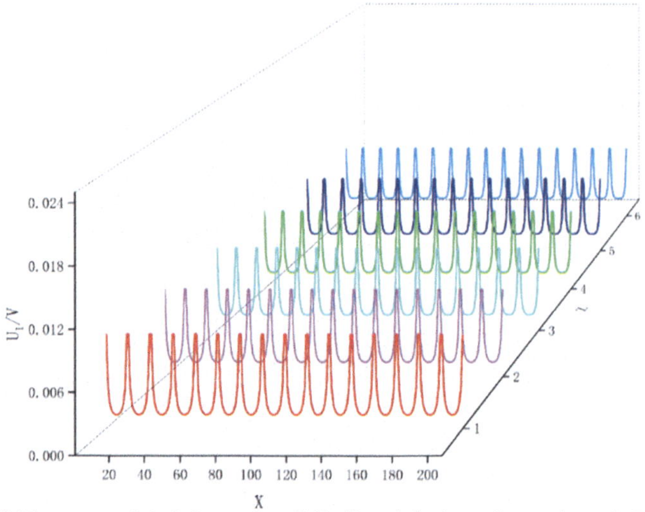

(C) The ore particle is located at C (0, 0) and the boundary voltage is U_1

Fig. 5. (*continued*)

Table 1. The three cases of single particles α with V_R

α	0.02	0.04	0.06	0.08	0.10	0.12
$V_{R(A)}$	10.315	21.389	33.273	45.903	59.435	74.460
$V_{R(B)}$	10.291	21.326	33.151	45.682	59.269	73.900
$V_{R(C)}$	10.289	21.203	32.532	44.172	56.319	68.988

According to the obtained U_0 and U_1 data, the characteristic voltage V_R can be obtained using formula (5). The α and V_R of different situations are shown in Table 2, where s represents the number of solid ore particles. The relationship between α and V_R obtained from this data is shown in Fig. 9.

It can be seen from Fig. 9 that the corresponding relationship between α and V_R for the three cases is basically the same as that obtained for a single particle. The average value of V_R in the three cases is taken as the value of the fitting function V_R, as shown in Table 2. According to the above analysis, the α − V_R relationship function based on polynomial regression can be obtained as formula (6):

$$\alpha = 6.832 \times 10^{-8} V_R^3 - 8.346 \times 10^{-6} V_R^2 + 0.002066 V_R + 0.0002366$$

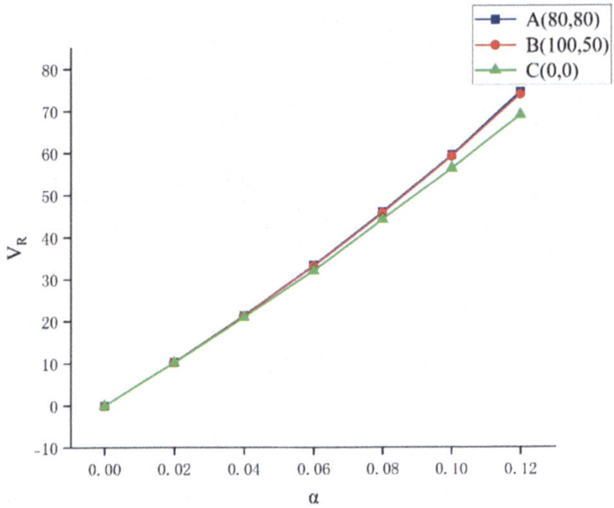

Fig. 6. Corresponding relationship between solid phase content α and V_R

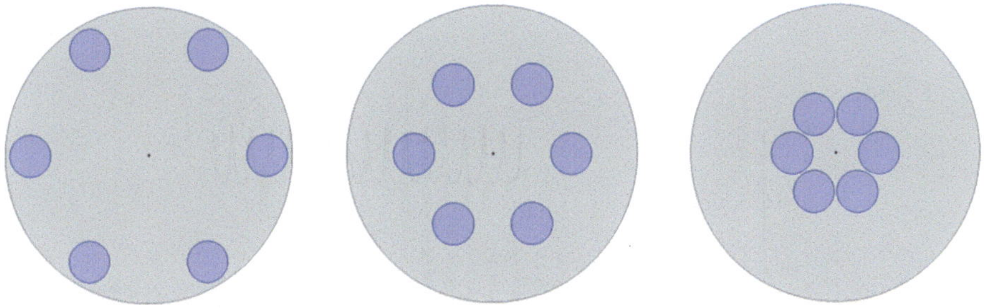

Fig. 7. Pipeline model and ore particle distribution for cases A, B, and C

3.3 Empirical Formula Verification

The empirical formula is verified by constructing a randomly distributed ore particle model, which includes two cases: single particle and multiple particles:

Case 1: Single particles of different sizes are randomly distributed in different locations of the pipe. The phase content is obtained by using the empirical formula (6) obtained through its characteristic voltage V_R, and is recorded as α_m. The actual value of the phase content of particles of different sizes is recorded as α_v. The results are shown in Table 3. Case 2: Multi-particle model as shown in Fig. 11, particles of different sizes and numbers are randomly distributed at different positions in the pipeline. The phase content is obtained by using the empirical formula (6) through their characteristic voltage V_R, and is recorded as α_m. The actual value of the phase content of particles of different sizes is recorded as α_v. The results are shown in Table 4.

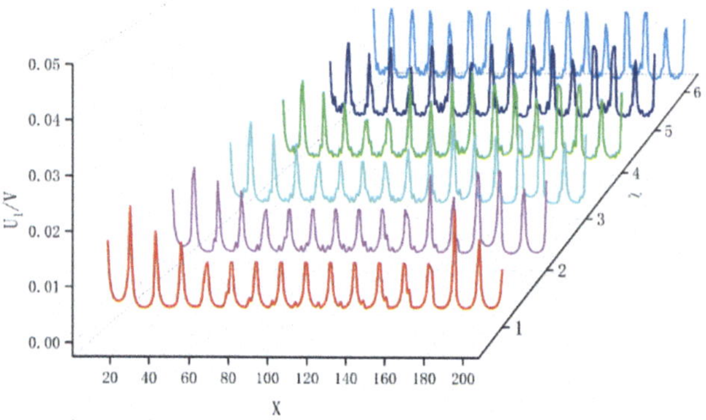

(A) Ore particles are located at the pipe boundary (-165,0) boundary voltage U_1

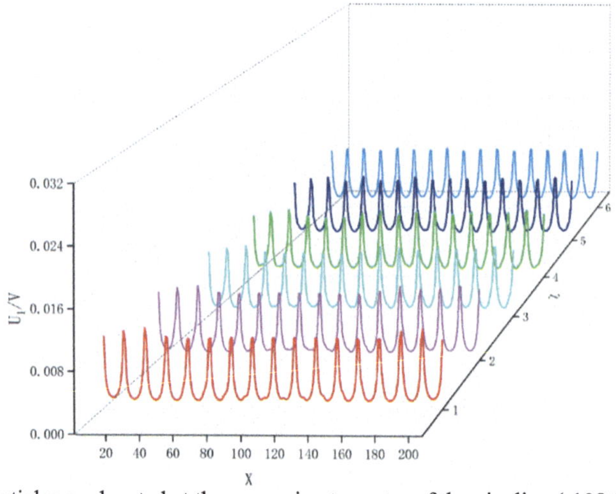

(B) Ore particles are located at the approximate center of the pipeline (-108,0) boundary voltage U_1

Fig. 8. Boundary voltages for cases A, B, and C

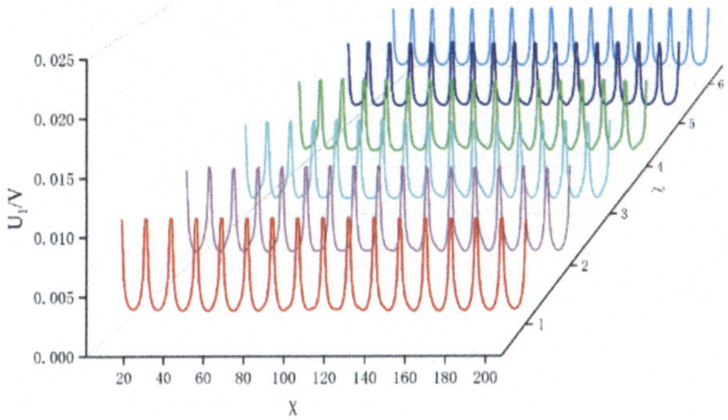

(C) The boundary voltage U_1 of the ore particle at the center of the pipe (-60,0)

Fig. 8. (*continued*)

Table 2. The three cases of multi-particle α with V_R

S	1	2	3	4	5	6
α	0.02	0.04	0.06	0.08	0.10	0.12
$V_{R(A)}$	8.343	17.339	25.842	34.443	43.917	52.609
$V_{R(B)}$	10.314	21.215	32.429	43.842	55.479	67.665
$V_{R(C)}$	10.392	23.634	38.514	52.781	65.331	78.897
$V_{R(Average)}$	9.683	20.729	32.262	43.689	54.909	66.390

It can be seen from Figs. 10 and 12 that for a single particle, the predicted value of the phase content obtained by the empirical formula is highly consistent with the actual value. For multiple particles, the increase in the number of particles and the random distribution of their positions will affect the characteristic voltage V_R, and the predicted value obtained by the empirical formula has a slight deviation from the actual value, but is within 5%.

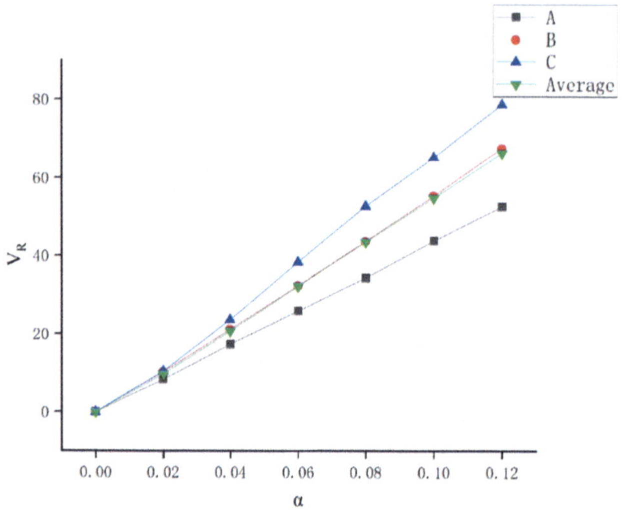

Fig. 9. Correspondence between α and V_R

Table 3. Predicted and actual values of different single particle phase contents

r	30 mm	45 mm	55 mm	60 mm	65 mm	70 mm
α_m	0.0261	0.0521	0.0778	0.0921	0.1089	0.1267
α_v	0.0255	0.0506	0.0756	0.0900	0.1056	0.1225

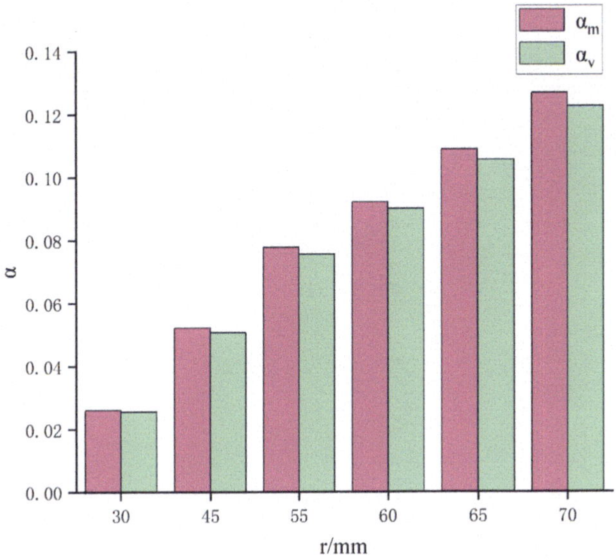

Fig. 10. Predicted and actual values of phase content of single particles of different sizes

Table 4. Predicted and actual values of multi-particle phase fraction

x	1	2	3	4	5	6
α_m	0.0739	0.0736	0.0825	0.0926	0.0997	0.1208
α_v	0.0758	0.0773	0.0849	0.0949	0.1017	0.1187

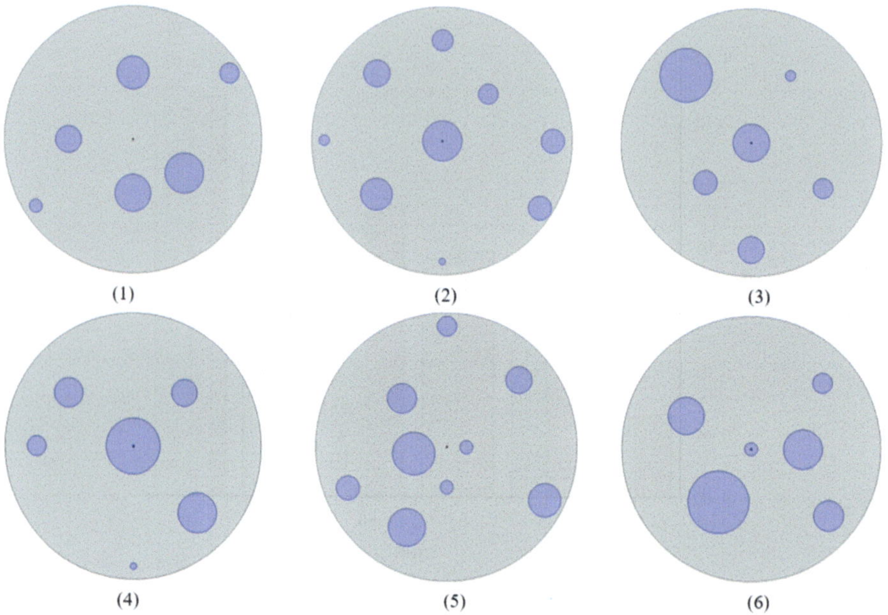

Fig. 11. Multi-particle random distribution model

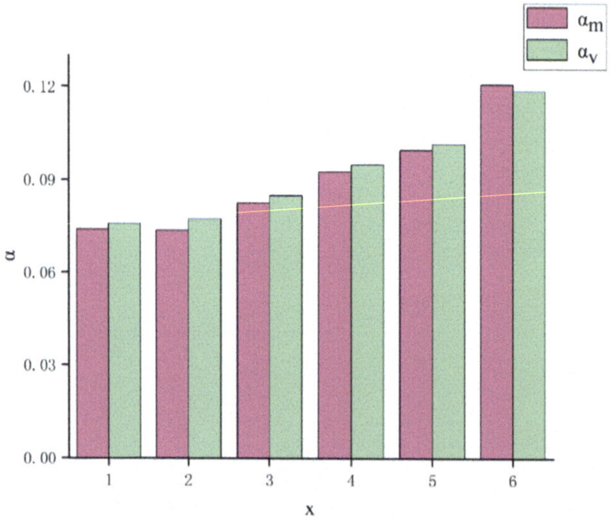

Fig. 12. Predicted and actual values of multi-particle phase content

4 Conclusions

This paper establishes a two-dimensional model of deep-sea mining lifting pipeline and solid ore particles, extracts the ERT boundary voltage signal, uses the relative change summation method to fit the empirical formula, constructs particle models for different situations, compares its predicted value with the accurate value, and obtains better results. It is feasible to use this method to solve the solid phase content of the solid-liquid two-phase flow in the deep-sea mining lifting pipeline, which provides a technical reference for the pipeline lifting of deep-sea mining, thus proving that the electrical resistance tomography method has a good application in deep-sea mining.

Acknowledgement. This work was supported by National Foundation Committee of China - Key Project [research project number 52231012], Hainan University High-level Talents Research Start-up Funding [grant number KYQD(ZR)-22060], International Association of Maritime Universities [research project number 20220304] and Hainan Provincial Natural Science Foundation High-level Talent Project [funding number 423RC436], and the Tianjin University-Hainan University Joint Innovation Fund [funding number HDTD202302].

References

1. Shirayama, Y., Itoh, H., Fukushima, T.: Recent developments in environmental impact assessment with regard to mining of deep-sea mineral resources. In: Deep-Sea Mining: Resource Potential, Technical and Environmental Considerations, pp. 445–463 (2017)
2. Ma, W., Lodewijks, G., Schott, D.: Analysis of a green transport plant for deep sea mining systems. J. Min. Sci. **54**(2), 254–269 (2018)
3. Yan, Y.: Flow rate measurement of bulk solids in pneumatic pipelines problems and solutions. Bulk Solids Handl. **15**(3), 447–457 (1995)
4. Rabinovich, A., Bedrikovetsky, P., Tartakovsky, D.M.: Analytical model for gravity segregation of horizontal multiphase flow in porous media. Phys. Fluids **32**(4) (2020)
5. Pant, H.J., Goswami, S., Sharma, V.K., et al.: Flow investigation in an industrial-scale soaker using radiotracer technique. Appl. Radiat. Isot. **124**, 119–123 (2017)
6. Mokhtari, M., Chaouki, J.: New technique for simultaneous measurement of the local solid and gas holdup by using optical fiber probes in the slurry bubble column. Chem. Eng. J. **358**, 831–841 (2019)
7. Dickin, F., Wang, M.: Electrical resistance tomography for process applications. Meas. Sci. Technol. **7**(3), 247 (1996)
8. Zhang, J.G., Ma, F.C.: Application of electrical resistance tomography to ice-water two-phase flow parameters measurement. KEM **562–565**, 686–690 (2013)
9. Tao, F.F., Jin, H., He, G., et al.: Two-phase flow characteristics of gas-liquids in microchannels using electrical resistance tomography. Heat Mass Transf. **58**(1), 99–114 (2022)
10. Dong, F., Jiang, Z.X., Qiao, X.T., et al.: Application of electrical resistance tomography to two-phase pipe flow parameters measurement. Flow Meas. Instrum. **14**(4–5), 183–192 (2003)
11. Dong, F., Xu, Y.B., Xu, L.J., et al.: Application of dual-plane ERT system and cross-correlation technique to measure gas–liquid flows in vertical upward pipe. Flow Meas. Instrum. **16**(2–3), 191–197 (2005)
12. Udosen, N.I., George, N.J.: A finite integration forward solver and a domain search reconstruction solver for electrical resistivity tomography (ERT). Model Earth Syst. Environ. **4**, 1–12 (2018)

13. Trlep, M., Hamler, A., Jesenik, M.: The FEM-BEM analysis of complex grounding systems. IEEE Trans. Magn. **39**(3), 1155–1158 (2003)
14. Kagami, S., Fukai, I.: Application of boundary-element method to electromagnetic field problems. IEEE Trans. Microw. Theory Tech. **32**(4), 455–461 (1984)
15. de Munck, J.C., Faes, T.J.C., Heethaar, R.M.: The boundary element method in the forward and inverse problem of electrical impedance tomography. IEEE Trans. Biomed. Eng. **47**(6), 792–800 (2000)
16. Xu, Y.Y., Dong, F.: Galerkin boundary element method for the forward problem of ERT. Flow Meas. Instrum. **21**(3), 172–177 (2010)
17. Gehre, M., Jin, B.: Expectation propagation for nonlinear inverse problems–with an application to electrical impedance tomography. J. Comput. Phys. **259**, 513–535 (2014)
18. Cui, Z., Wang, Q., Xue, Q., et al.: A review on image reconstruction algorithms for electrical capacitance/resistance tomography. Sens. Rev. **36**(4), 429–445 (2016)
19. Guo, Q., Wang, C., Pang, R., et al.: A new truncation strategy for regularized D-bar method used for ERT image reconstruction. IEEE Sens. J. **22**(8), 8164–8174 (2022)
20. Hewitt, G.F.: Measurement of two phase flow parameters. NASA STI/Recon Techn. Rep. A **79**, 47262 (1978)
21. Heikkinen, L.M., Vauhkonen, M., Savolainen, T., et al.: Utilizing internal electrodes in volume fraction estimation by electrical process tomography. In: Proceedings of 2nd World Congress on Industrial Process Tomography (2001)

Experimental Study on Jet Pump Lifting Procedures in Marine Natural Gas Hydrate

Xia Jia[1], Hualin Liao[1](✉), Jun Wei[2], and Yuhang He[1]

[1] School of Petroleum Engineering, China University of Petroleum (East China), Qingdao, China
liaohualin2003@126.com

[2] Oil and Gas Engineering Research Institute, CNPC Tarim Oilfield Company, Korla, China

Abstract. In response to the problems of liquid accumulation and sand production during offshore natural gas hydrate extraction, this paper proposes a technical solution that combines concentric coiled tubing closed-loop circulation, jet pump lifting, sand flushing and plugging removal, etc. It can achieve pressure control and lifting of natural gas hydrates, control the recovery rate of hydrates, and remove the sand bed at the bottom of the well at any time according to on-site requirements, with high operational efficiency. Through experimental research on the lifting and suction volume of jet pumps, it has been shown that the distance between the throat and the power nozzle has a significant influence on the suction volume of jet pumps. For distances between 6 mm and 12 mm, the suction volume of the jet pump significantly increases, while the lifting pressure slightly decreases. When the distance is between 12 mm and 18 mm, the suction volume of the jet pump first increases and then decreases, while the lifting pressure decreases. The fluid pressure at the inlet significantly affects the suction volume of the jet pump. Under the action of power fluid pressure, the suction volume of the jet pump significantly increases, while the lifting pressure of the jet pump remains relatively constant. Compared with other lifting methods, it can achieve stable production of natural gas hydrates by reducing bottomhole pressure, solving the problems of severe sand blockage and extending the production cycle of natural gas hydrates.

Keywords: natural gas hydrate · artificial lifting · sand removal · jet pump · experimental research

1 Introduction

Natural gas hydrates are cage shaped crystalline compounds formed by hydrocarbon gases and water under specific temperature and pressure conditions. They are widely distributed in seabed sediments and permafrost regions on land [1]. More than 95% of the resources are found within the marine continental shelf, and are considered the most promising clean energy source to replace conventional oil and gas resources in the 21st century [2, 3].

If the formation water in the wellbore is not discharged in a timely manner during the process of hydrate extraction, the bottomhole pressure will increase, which will inhibit

D. Jeng and B. Cai (Eds.): ICCOE 2024, LNCE 423, pp. 253–262, 2025.
https://doi.org/10.1007/978-981-96-0181-3_23

the continuous decomposition of hydrates. At the same time, after the decomposition of hydrates, the formation near the wellbore loses support and becomes unstable. During the extraction of natural gas hydrates, the reservoir framework falls off, causing a large amount of formation sand to enter the wellbore, blocking the flow channels at the bottom of the well, and affecting the extraction of hydrates. Drainage and sand removal both require manual lifting technology, therefore, manual lifting technology is a key link in the process of natural gas hydrate development [4].

Artificial lifting methods such as jet pump and electric submersible pump have been applied in hydrate trial production experiments both domestically and internationally. Electric submersible pumps are currently the most commonly used artificial lifting method. By continuously discharging liquid through an electric submersible pump, the pressure at the bottom of the well is reduced to facilitate the decomposition of hydrates, and the depressurization method has been verified to be an effective method for extracting marine hydrates. At the same time, due to the easy sand production during the trial production process, the production capacity of the trial production wells is reduced or stopped, the wear and tear of the well and surface equipment, and the collapse of the wellbore wall damage the casing. The early natural gas hydrate trial production wells in Canada, the United States, and Japan were all completed due to sand production from the formation. Therefore, how to solve the problem of sand production during artificial lifting is one of the key issues for promoting the application of Natural gas hydrate extraction [5–7].

2 Methodology

2.1 Principle of Jet Pump Drainage and Extraction

On the basis of traditional jet pump lifting technology, this paper proposes a depressurization and drainage technology scheme by utilizing the characteristics of jet pump lifting and vacuum sand removal. Using concentric coiled tubing [8–10], the jet pump is lowered into the vicinity of the hydrate reservoir, bringing hydrates, sediments, etc. back to the surface. Monitor the pressure at the wellhead of the casing, regulate the bottomhole pressure by adjusting the pump displacement, control the rate of hydrate decomposition at the bottom of the well, and improve the stability and production capacity of gas hydrate extraction [11]. At the same time, based on the sand production situation of the formation, the forward nozzle of the jet pump is used to clean the sand bed at the bottom of the well, and it is sucked into the concentric coiled pipe annulus through the jet pump before finally returning to the surface. The specific implementation steps are as follows:

Hydrate Pressure Control and Extraction. Use concentric coiled tubing to provide pumping and return channels. The power fluid enters the jet pump through the inner tube and utilizes the negative pressure suction effect of the jet pump to suck the bottomhole liquid into the tool and mix with the power fluid. Then, it is discharged back to the surface through the annular space between the inner and outer pipes. By adjusting the position of the power fluid and the pressure of the pump, the underground negative pressure suction capacity can be adjusted. Combined with casing head pressure monitoring, the bottom hole pressure value can be adjusted in real time to ensure the decomposition rate of formation hydrates.

Bottomhole Sand Removal. The lower end of the jet pump is equipped with a nozzle, which can stir the sand bed at the bottom of the well, mix with liquid and enter the interior of the jet pump, and then discharge back to the ground through a concentric coiled tubing. Based on the situation of sand blockage at the bottom of the well, lifting and lowering the concentric coiled tubing to remove the sand bed at the bottom of the well, in order to prevent blockage in the production channel from affecting the discharge of hydrates (Fig. 1).

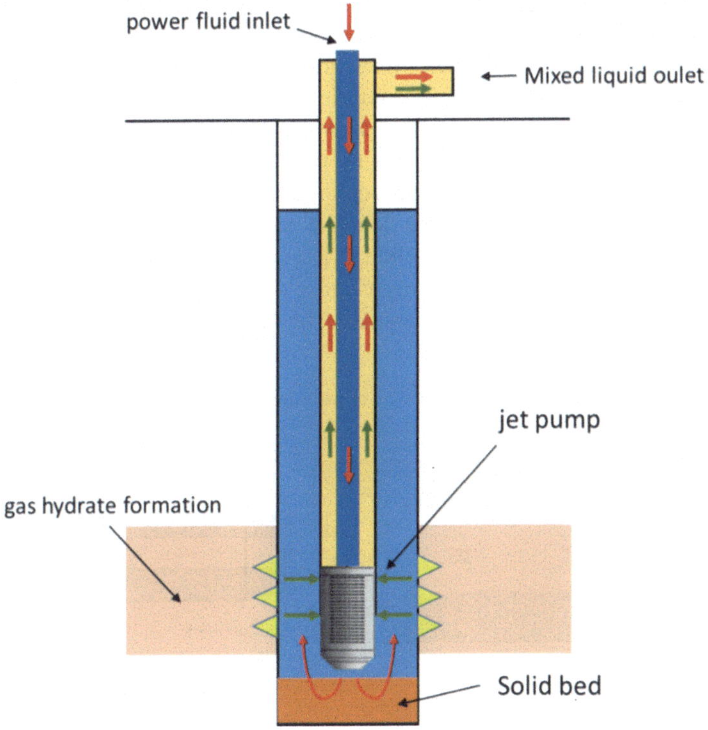

Fig. 1. Schematic diagram of jet pump extraction principle

The following process explains the work procedure.

- Place the concentric coiled tubing equipment and pump truck, and install the jet pump at the end of the concentric coiled tubing.
- Use concentric coiled tubing to lower the jet pump to the depth of the wellbore liquid level.
- Pumping power fluid, lifting the accumulated fluid in the wellbore that return to the surface through the annular between concentric coiled tubing.
- Estimate the bottomhole pressure through the wellhead pressure sensor, adjust the depth and pump parameters of the jet pump, control the wellbore liquid level height, and achieve the bottomhole pressure required for hydrate decomposition;

- According to the backflow of hydrates at the wellhead, regularly lower a jet pump to the level of sand accumulation surface at the bottom of the well, to remove the sand.
- Observe the amount of sand flowing back from the wellhead, confirm that it has been removed, and then lift the jet pump near the reservoir to continue the lifting operation.

2.2 Key Structural Parameters of Jet Pump

The schematic diagram of structural parameters of jet pump is as follows. The power fluid passes through the jet pump and forward nozzle, which agitates the suspended sand bed at the bottom of the well. The jet pump sucks and lifts the suspended sand bed in the wellbore to the ground [12, 13]. Under the condition that the hydraulic pressure of the power fluid flowing through the forward nozzle is p_1 and the flow rate is q_4, the power fluid is pumped through the nozzle with a flow area of A_f. Under the condition that the hydraulic pressure of the power fluid flowing through the jet pump is p_1 and the flow rate is q_1, the power fluid is pumped through a nozzle with a flow area of A_j. In the well, the pressure of the liquid is P_3, the flow rate is q_3, and the liquid is accelerated and sucked into the pipeline. After the pipeline, the accelerating fluid and the suction fluid continuously mix, resulting in a pressure change of discharge pressure P_2 and a flow rate change of q_2, ultimately lifting it to the surface, as shown in Fig. 2.

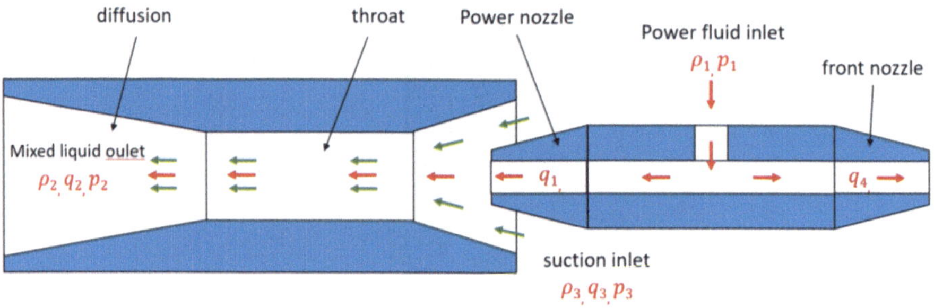

Fig. 2. Schematic diagram of structural parameters of jet pump

2.3 Experimental Setup and Procedures

Due to limitations in experimental conditions, it is not possible to consider the influence of frictional resistance along concentric coiled tubing in simulation experiments. The lifting and suction capabilities of the jet pump were tested through a clear water experiment. The feasibility of sand removal by downhole jet pump was verified through simulated wellbore sand flushing experiments.

As shown in Fig. 3, this is the experimental setup for wellbore simulation testing. This paper obtains the tool structure parameters based on preliminary theoretical calculations [14–17] of relevant operating conditions. In this experiment, a jet pump nozzle with a diameter of 2.5 mm was used, and the pipeline diameters were set to 3.5 mm and 4 mm, respectively. The nozzle distances are set to 12 mm, 15 mm, and 18 mm. During the experiment, the area ratio of the jet pump was switched by changing the inner diameter size of the throat. Considering the presence of throttling pressure conditions, different area ratio configurations, throat distance, and the suction and lifting capabilities of the jet pump at different pump pressures were tested.

Determine the parameters of the power nozzle diameter and throat through theoretical calculations [17], and conduct experiments by adjusting the throttle valve to simulate the lifting head of different jet pumps. Analyze the working performance of the jet pump based on different tool structure parameters, working displacement, throttling pressure, and parameters such as the pre - and post pressurization capacity and suction volume of the jet pump. The performance experiment of the jet pump tool does not consider the flow rate and pressure of the front nozzle.

(a) Physical image

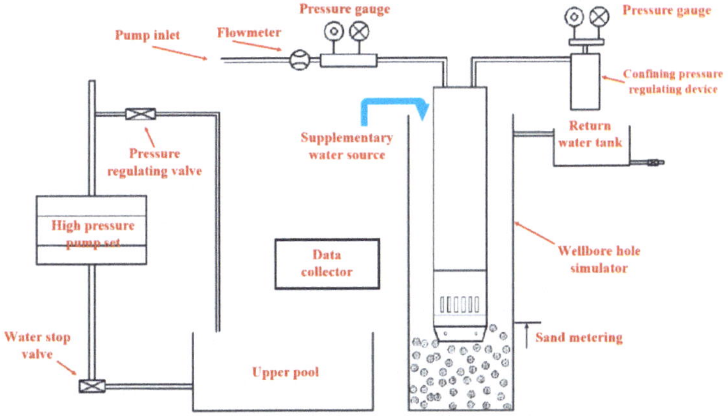

(b) Structural schematic diagram

Fig. 3. Experimental setup for wellbore simulation test

3 Results and Analysis

3.1 The Influence of Pump Pressure on the Performance of Jet Pump

As shown in Fig. 4, under the experimental conditions of a nozzle diameter of 2.2 mm, a throat diameter of 3.5 mm, and a nozzle to throat distance of 6 mm, as the lifting pressure increases, the suction volume of the jet pump remains relatively stable and then significantly decreases. As the pump pressure increases, the lifting pressure P_2 of the jet increases significantly, while the maximum suction volume of the jet pump remains basically unchanged. This indicates that under these specific structural parameters, increasing pump pressure can significantly improve lift pressure, allowing bottom hole fluids and sand beds to be discharged to higher ground levels, thereby increasing the backflow capacity of the jet pump. However, the suction volume of the jet pump does not increase with the increase of pump pressure, indicating that it is impossible to increase the suction volume of the jet pump. At the pump pressure of 30 MPa, the maximum suction volume of the jet pump is approximately 4.9 L/min, and the lifting pressure is approximately 16 MPa.

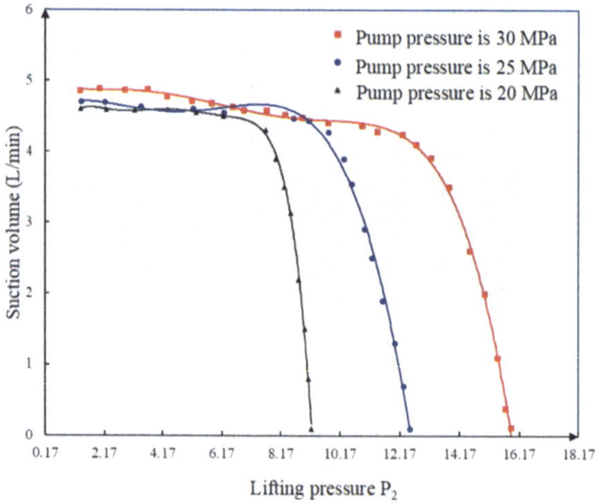

Fig. 4. The Curve of jet pump suction volume under different lifting pressures

3.2 The Influence of the Nozzle to Throat Distance on the Performance of Jet Pump

As shown in Fig. 5, under the experimental conditions of a nozzle diameter of 2.2 mm, a throat diameter of 4 mm, and a pump pressure of 30 MPa, as the nozzle to throat distance increases from 12 mm to 18 mm, the maximum suction volume of the jet pump increases and the maximum lifting pressure slightly decreases. This indicates that increasing the distance between the nozzle and the throat appropriately under the same lifting pressure conditions can accelerate the efficiency of bottom hole fluid and sand removal. When the throat distance is 18 mm, the maximum suction volume of the jet pump is about 9.8 L/min, corresponding to a lifting pressure of about 7.8 MPa.

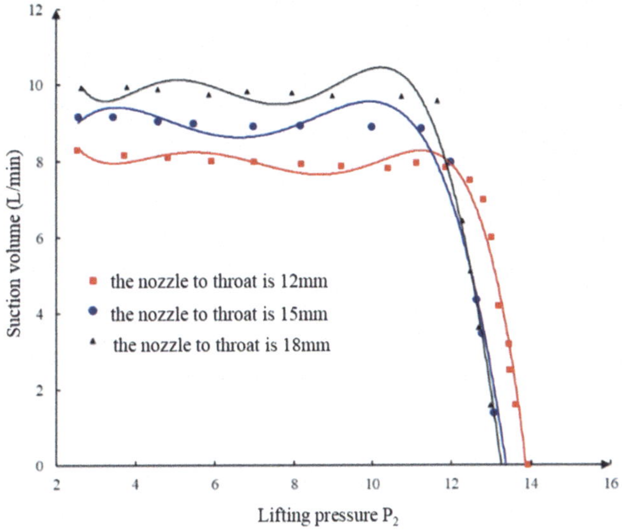

Fig. 5. The curves of jet pump suction volume at various nozzle throat distances

3.3 The Influence of Suction Inlet Pressure on the Performance of Jet Pump

In Fig. 6, the experiment was conducted under conditions of no suction inlet pressure and lower suction inlet pressure. At lower inlet pressures, the sand suction capacity and efficiency of the jet pump will be greatly improved. The jet pump achieved the suction volume of up to 14 L/min and the maximum lift pressure P2 of 9 MPa at the pump pressure of 30 MPa, with the pump efficiency of 18%. Under conditions of equal pump pressure, the throat diameter, and the nozzle-to-throat distance, increasing the suction inlet pressure resulted in a greater suction volume of the jet pump, while the maximum lift pressure P_2 remained relatively constant. The efficiency of the jet pump also improved when operating with powered suction inlet pressure. According to the actual working conditions, it can be known that the suction inlet pressure is equal to the static liquid column pressure in the casing annulus. This indicates that under the condition of constant surface pumping pressure, increasing the relatively low pressure of the casing annulus can significantly accelerate the efficiency of downhole fluid and sand removal.

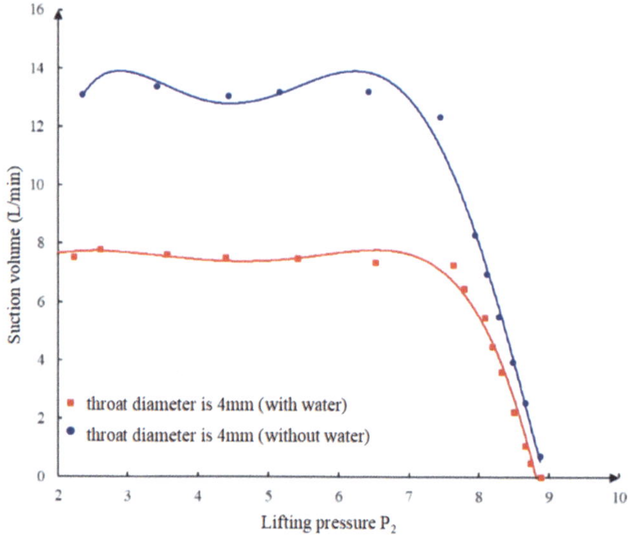

Fig. 6. The curves of jet pump suction volume at various suction inlet power pressures

3.4 The Influence of Area Ratio on the Performance of Jet Pump

In the preliminary theoretical analysis of the jet pump conducted in this project, it was found that the area ratio of the jet pump is an important factor influencing its efficiency and lifting performance. Jet pumps with different area ratios represent different lifting pressure and suction volume. As shown in Figs. 7, the area ratio of the jet pump was manipulated by changing the diameters of the throat and the nozzle. Specifically, the jet pump configuration include throats with inner diameters of 3.5 mm and 4 mm, as well as power nozzles with inner diameters of 2.5 mm, with corresponding area ratios of 0.3 and 0.4, respectively. The influence of area ratio on the lifting capacity and suction efficiency of jet pumps under different pump pressure conditions was studied with a nozzle throat distance of 12 mm.

The influence of area ratio variations on the suction volume and lifting pressure of jet pump with a nozzle diameter of 2.2 mm and nozzle throat distance of 12 mm, under no inlet pressure conditions, is relatively small. However, it has a significant influence on the lift pressure ratio and suction efficiency of the jet pump. A higher area ratio of 0.3 in the jet pump results in increased lifting pressure and suction volume.

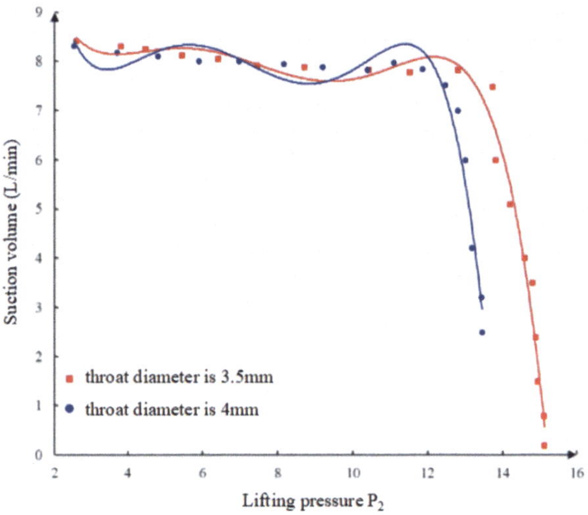

Fig. 7. The curves of jet pump suction volume at different area ratios

4 Conclusion

1. The use of concentric coiled tubing and jet pump to establish a closed-loop lifting method can achieve pressure control lifting of natural gas hydrates, control the rate of hydrate extraction, and remove the sand bed at the bottom of the well at any time according to on-site requirements, resulting in high operational efficiency. The jet pump has a simple structure and high erosion resistance, allowing debris of a certain diameter to pass through. Compared with other lifting methods, it is more suitable for wells with severe sand production.
2. The distance between the throat and the power nozzle has a significant influence on the suction volume of the jet pump. For the distances of 6mm to 12 mm, the suction volume of the jet pump significantly increases, while the lifting capacity slightly decreases. The distance is between 12 mm and 18 mm, the suction volume of the jet pump first increases and then decreases, while the lifting capacity decreases.
3. The fluid pressure at the suction inlet significantly affects the suction volume of the jet pump. Subjected to the power fluid pressure, the suction volume of the jet pump increases significantly, while the lifting capacity of the jet pump remains relatively constant.

References

1. Ye, Y., Liu, C.: Natural Gas Hydrate: Experiential Techniques and Their Applications. Springer, New York (2013). https://doi.org/10.1007/978-3-642-31101-7
2. Acosta, M.J.: Advances in Energy Research, vol. 33. Nova Science Pub Inc., New York (2021)
3. Tang, X., Chen, J., Guo, Y., et al.: Development status of combustible ice mining and test production technologies. Nat. Gas Oil **38**(1), 7–15 (2020)

4. Gao, G., Li, L., Song, X.: A new technique for the real-time monitoring of the sand producing in heavy oil wells. J. Xi'an Petrol. Univ. (Nat. Sci. Ed.) (2020)
5. Zhang, X., Lu, X., Li, P.: A comprehensive review in natural gas hydrate recovery methods. Scientia Sinica (Physica Mechanica Astronomica) **49**(3), 38–59 (2019)
6. Wu, J., Dan-Qing, Y., Guan-Jun, W., et al.: Current situation and prospect of jet pumps. J. Drain. Irrig. Mach. Eng. **121**(02), 65–68 (2007)
7. Walker, C., Bodkin, G.C.I.: Empirical wear relationships for centrifugal slurry pumps-Part1: side-liners. Wear **242**(1), 140–146 (2000)
8. Yong, W., Huaiwen, L., Yingbin, Z., et al.: Sand control technology for low pressure lost-circulation wells and application in South Sudan. J. Phys. Conf. Ser. **2152**(1) (2022)
9. Rafferty, P., Ennis, J., Skufca, J., et al.: Enhanced solids removal techniques from ultra low pressure wells using concentric coiled tubing vacuum technology. SPE 107010 (2007)
10. Citrini, D.: Study of the theory of the jet pump. Energia Elettriea **25**(12) (1948)
11. Falk, K., Fraser, B.: Concentric coiled tubing application for sand cleanouts in horizontal wells. J. Can. Petrol. Technol. **37**(10), 46–50 (1998)
12. Gosline, J.E., Orrien, M.P.: The water jet pump. Univ. Califonia Pub. Eng. **3**(3), 160–200 (1934)
13. Yan-Jun, G.E., Qiang, G.E., Jun, Y.: Numerical simulation and confirming of optimal range on the nozzle-to-throat clearance of liquid-gas jet pump. Fluid Mach. **40**(04), 21–24 (2012)
14. Shao-Feng, L., Pei-Lin, F., Qiang, R., et al.: The research and application of sand washing and temporary blocking technology in Bohai oilfield horizontal wells. Fine Special. Chem. **25**(03), 336 (2017)
15. Wibisono, K., Sitaresmi, R.: Evaluation of hydraulic jet pump application in Sembakung field. J. Earth Energy Sci. Eng. Technol. **2020**(3) (2020)
16. Xianghan, H.: Research of the design theory for liquid state jet pumps. Sci. China Ser. A **1983**(02), 105–115 (1983)
17. Zhu, R., Su, B., Yang, A.: Numerical investigation of unsteady pressure fluctuation in centrifugal pump. Trans. Chin. Soc. Agric. Mach. **41**, 447 (2010)
18. Xu, S., Wang, J., Chen, W., et al.: Removal of field-collected Microcystis aeruginosa in pilot-scale by a jet pump cavitation reactor. Ultrason. Sonochem. **83**, 105924 (2022)

Hydrodynamic Coefficients Analysis of Deepwater BOP Stack Based on CFD

Jinlong Wang[1](✉), Jinghao Chen[2], Leixiang Sheng[1], and Mingchun Wang[1]

[1] Drilling and Production Research Department, CNOOC Research Institute Co. Ltd., Beijing 100028, China
wangjinlong132@126.com

[2] School of Mechanical Engineering, Beijing Institute of Petrochemical Technology, Beijing 102600, China

Abstract. BOP stack is important device of deepwater drilling riser system. Due to the complex structure of BOP stack, the hydrodynamic coefficients are not accurate by experience or standard recommended as a range, not a determined value. The hydrodynamic coefficients of deepwater BOP stack needs to be analyzed to get more accurate coefficients. The analysis method for hydrodynamic coefficients is introduced in detail based on computational fluid dynamics (CFD) theory. The analysis models are established by large eddy turbulence model (LES). The moving mesh technology is used to simulate the dynamic motion performance of deepwater BOP stack under different current velocity. The hydrodynamic coefficients: drag force coefficient and inertia force coefficient of deepwater BOP stack are obtained. The hydrodynamic coefficients analyses of deepwater BOP stack can provide valuable reference for deepwater drilling engineering applications.

Keywords: BOP stack · Hydrodynamic Coefficients · CFD · Moving Mesh Technology

1 Introduction

BOP stack is important device of deepwater drilling riser system, which is widely used in the offshore oil and gas field development. The BOP stack is affected by fluid force during the drilling operation, which has certain influence on its normal operation. The BOP stack will produce large fluid force induced by ocean current. Therefore, it is of great significance to evaluate the safety of the BOP stack through studying the hydrodynamic performance of BOP stack and mastering the calculation method of hydrodynamic coefficients of BOP stack. Some scholars have used the model test method and the approximate theory evaluation method to calculate the hydrodynamic coefficients of the structure. The model test method has high cost and poor universality. The approximate theory evaluation method simplifies the geometric model and is not very accurate for the calculation of complex structures [1, 2]. At present, the Computational Fluid Dynamics (CFD) method is becoming more and more mature, and it is a common method to calculate the hydrodynamic coefficients of a structure, which can save a lot of costs.

D. Jeng and B. Cai (Eds.): ICCOE 2024, LNCE 423, pp. 263–270, 2025.
https://doi.org/10.1007/978-981-96-0181-3_24

Wang [3] studied the hydrodynamic coefficients of a certain type of propeller at different propulsion speeds and the difference met the application requirements of engineering by comparing it with the experiment. Wang and Tang [4, 5] used FLUENT software to study the resistance coefficient of the manifold and made a comparative analysis with the empirical value. Li [6] studied the unsteady aerodynamic characteristics of the rotating wing, and adopted multi-layer embedded grid technology to improve the overall calculation speed. Tian and Qi [7, 8] studied the hydrodynamic coefficients in the process of lowering the manifold and established a standard model to compare with DNV-RP-H103, and the difference was less than 3%. Based on the theory of computational fluid mechanics, Feng [9] adopted fluid software CFX and its moving mesh technology to give a calculation method for the additional mass of an object under different boundary conditions, and summarized the variation law of the additional mass coefficient of an object in a circular tube. Yu [10] proposed a simple and effective test design method to determine the resistance coefficient of the anchor chain. Huang [11] studied the hydrodynamic coefficients and its variation rules of the horizontal and vertical double riser structure. Wang [12] used LES model to analyze the drag force coefficient of anchor chain and its variation with Reynolds number, and gave engineering suggestions through analysis.

In summary, CFD is an effective method to calculate the hydrodynamic coefficients of underwater structures, but no relevant research has been carried out on the complex structure such as BOP stack. CFD method and the CFD analysis software FLUENT is used to build the flow field simulation of the BOP stack and calculate the hydrodynamic coefficient of BOP stack, which provides a basis for the hydrodynamic performance calculation of BOP stack.

2 Analysis Method

CFD is used to calculate the hydrodynamic coefficients of the BOP stack. The commercial CFD software FLUENT is used to calculate the drag force of BOP stack, and then dimensionless, and finally the hydrodynamic coefficients of BOP stack is obtained. The Reynolds number is used to distinguish whether the fluid flow is laminar or turbulent, and can also be used to determine the resistance of the object flowing in the fluid, so as to determine the computational model of FLUENT. The Reynolds number is calculated by Eq. (1).

$$R_e = \rho v L/\mu \tag{1}$$

where, ρ is the fluid density, μ is the dynamic viscosity coefficient, v is the characteristic velocity of the flow field, and L is the characteristic length of the flow field. The Reynolds number represents the ratio of inertial and viscous force. For the outflow problem, v and L generally take the flow speed and the main size of the object.

Then non-dimensional treatment is done based on the resistance of the fluid passing through BOP stack. In the steady flow zone, the normal drag coefficients is shown in Eq. (2).

$$C_{DS} = \frac{2F_e}{\rho A_c \mu_0^2} \tag{2}$$

where, μ_0 is the relative current velocity of BOP stack, ρ is the density of the fluid, A_c is the projected area in the direction of the current, and C_{DS} is the total resistance coefficient under the action of the steady current. FLUENT can be used to simulate the relative motion of the object in three directions in water, and the total resistance F_e is obtained and then C_{DS} is calculated by Eq. (2).

As an object accelerates through water, the water adheres to the surface of the object to form additional mass. The movement of the fluid consumes energy, causes drag and increases the inertia of the object. When solving the additional mass, it is necessary to simulate the unsteady motion of the object. The fluid domain is stationary, and the object moves in unsteady motion. The additional mass coefficient C^* is calculated by Eq. (3).

$$C^* = \frac{m^*}{m_P} = \frac{F_z - F_D}{m_P a} \tag{3}$$

where, F_z is the total resistance of BOP stack at a certain time in unsteady motion, F_D is the total resistance of BOP stack in steady motion at the same transient, m_P is the mass of BOP stack after entering the water, a is the instantaneous acceleration, and m^* is the additional mass. The inertia force coefficient C_M can be calculated by Eq. (4).

$$C_M = C^* + 1 \tag{4}$$

3 CFD Model

BOP stack of Fenjinhao is taken for the analysis. The whole model is 1:1, the model is 17.5 m high, 5.5 m wide and 4.8 m long. The model of BOP stack should be simplified, and the established model was shown in Fig. 1.

As shown in Fig. 1, the BOP stack model simplifies the pipe and connecting device of BOP, but ensures that the projected area of BOP stack is consistent in all directions to ensure the accuracy of the calculation of the hydrodynamic coefficients. The blue axis is the x direction, the green axis is the y direction, and the red axis is the z direction. After the model is established and the coordinate system is determined, the fluid field is established. The boundary conditions of velocity inlet and pressure outlet are used, and the analyzed boundary conditions are shown in Fig. 2.

Structured mesh is used for analysis, and the number of mesh are 1.29 million. Three flow speeds of 0.5 m/s, 1 m/s and 1.5 m/s are applied at the inlet for the calculation of drag force. An acceleration of 0.1 m/s^2 is set for BOP stack during additional mass analysis. The Reynolds number of BOP stack model is calculated with the characteristic diameter of 4.8m, and the calculation results are shown in Table 1.

It can be seen from Table 1 that calculation model corresponding to several velocity are turbulence model, and large eddy turbulence model (LES) can be used for analysis. The LES model is a variation of the N-S equation. By adding a stress term to the LES model, the more accurate calculation results can be obtained, as shown in Eq. (5) and Eq. (6).

$$\frac{\partial \rho}{\partial t} + \frac{\partial(\rho U_j)}{\partial x_j} = 0 \tag{5}$$

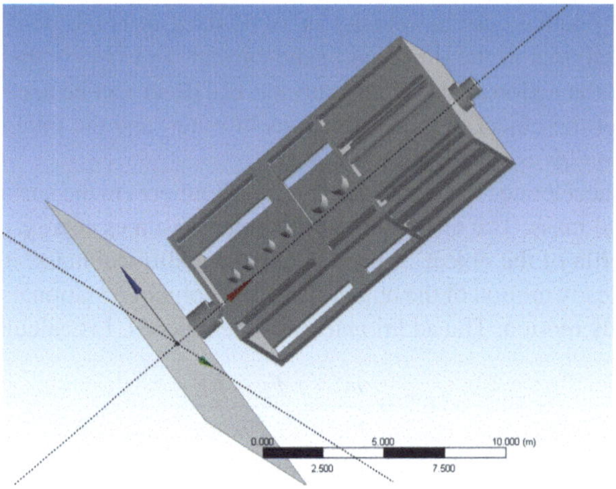

Fig. 1. BOP stack model

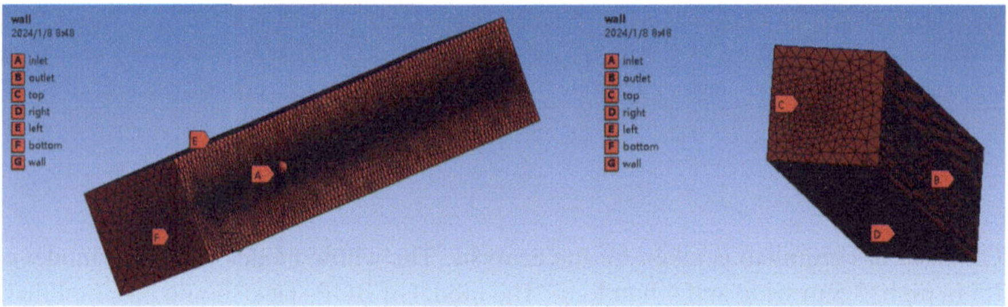

Fig. 2. Boundary condition for hydrodynamic analysis of BOP stack

Table 1. Reynolds number of BOP stack model

Number	Flow velocity(m/s)	Reynolds number	Calculation model
1	0.5	2306250	Turbulence
2	1	4612500	Turbulence
3	1.5	6918750	Turbulence

$$\frac{\partial(\rho U_i U_j)}{\partial x_j} + \frac{\partial(\rho U_i)}{\partial t} = -\frac{\partial P}{\partial x_i} + \frac{\partial(\tau_{\mathrm{ij}} + \tau_{sgs})}{\partial x_j} \tag{6}$$

where, $\frac{\partial(\tau_{\mathrm{ij}}+\tau_{sgs})}{\partial x_j}$ is viscous sub-grid, U_i, U_j is velocity component, P is fluid pressure.

4 Results Discussion

4.1 Results of Steady Motion

The inlet flow rate is set to 1m/s, and the LES analysis model is used to carry out the analysis of the steady flow. The time step is defined as 0.01 s, and contains 3000 analysis steps in total. The flow field distribution of the structure is shown in Fig. 3.

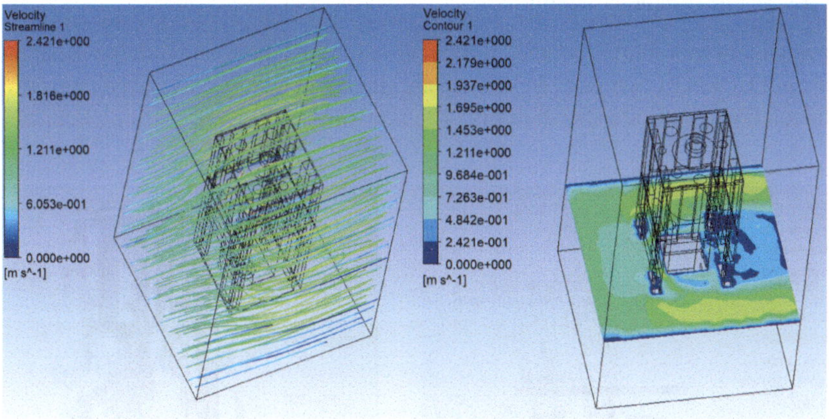

Fig. 3. Velocity nephogram of BOP stack

According to the calculation results in Table 2, tangential drag force coefficient and normal drag force coefficient are not stable at different Reynolds numbers. The reason is that there is a certain gap in the analysis model, and the flow near has a certain influence on the drag force. It can be seen that there is a certain relationship between drag force coefficient and Reynolds number. On the whole, the drag force coefficient of BOP stack varies within a reasonable range, and the variation range is not large. An average value can be used as the drag force coefficient of BOP stack in this direction.

Table 2. Drag force coefficient of BOP stack

Number	Flow speed (m/s)	Normal drag force coefficient	Tangential drag force coefficient
1	0.5	3.34	0.108
2	1	2.78	0.193
3	1.5	2.66	0.268

It can be seen from Table 3 that there is a certain gap between the calculated results hear and the recommended value of DNV. Compared with DNV standard, the difference of drag force coefficient is 25%. Compared with reference [13], the difference of drag force coefficient is 33%. Mainly the Reynolds number and computational model are both different, and there are certain differences in the ratio of height to length.

Table 3. Comparison on drag force coefficient of BOP stack

Number	Normal drag force coefficient	Reference [13]	Difference	DNV-RP-H103 [14]	Difference
1	2.92	1.96	33%	2.2	25%

4.2 Results of Unsteady Motion

Unsteady motion of fluid is adopted to flow around BOP stack, and moving mesh technology of FLUENT is adopted. BOP stack are simulated with 0.1 m/s^2 acceleration, and the calculation results are shown in Fig. 4.

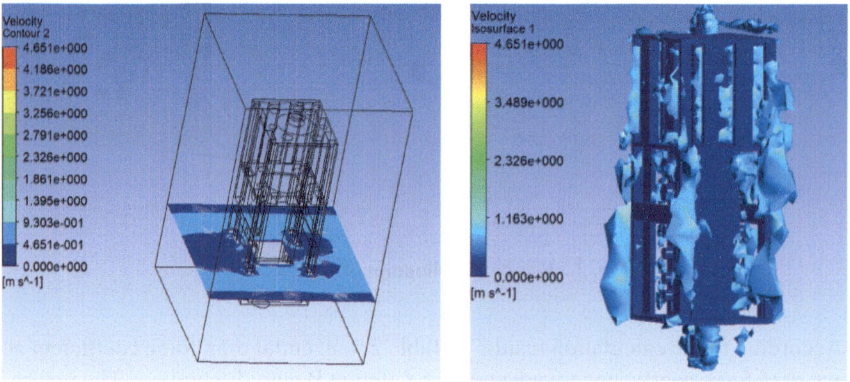

(a) Velocity contour of BOP stack **(b) Surface velocity of BOP stack**

Fig. 4. Calculation results of unsteady motion analysis

The inertia force coefficient of BOP stack are calculated according to Eq. (3) and Eq. (4), as shown in Table 4. Hydrodynamic coefficients of BOP stack at different flow speed are presented.

Table 4. Inertia force coefficient of BOP stack

Number	Flow speed (m/s)	Normal inertia force coefficient	Tangential inertia force coefficient
1	0.5	1.39	1.07
2	1	1.43	1.08
3	1.5	1.48	1.15

The calculation results show that the inertia force coefficient of BOP stack fluctuates within a certain range, and the inertia force coefficient of BOP stack increases with the increase of flow velocity. The fluid resistance to BOP stack gradually increases and exceeds the additional mass force on BOP stack. On the whole, the inertia force coefficient does not change much, and an average value can be used as the inertia force coefficient in this direction.

Table 5. Comparison on inertia force coefficient of BOP stack

Number	Normal inertia force coefficient	Reference [13]	Difference	DNV-RP-H103 [14]	Difference
1	1.43	1.25	13%	1.19	16%

As shown in Table 5, when the aspect ratio of object referred to DNV-RP-H103 is 4, the additional mass coefficient is 1.19 and the difference is about 16%. The main reason is that the aspect ratio is different, and there is a certain gap between the geometric model given in the specification and the model of BOP stack used in this paper. Compared with reference [13], the difference of inertia force coefficient is 13%. The Reynolds number and computational model is different, and there are certain differences in the ratio of height to length.

5 Conclusion

The outflow field computational model of BOP stack is established by using commercial CFD software FLUENT. The steady flow and unsteady flow of BOP stack are analyzed by using LES model. Thus, the drag force coefficient and inertia force coefficient of BOP stack are obtained, and the hydrodynamic coefficients of BOP stack at different flow rates is preliminary obtained.

The hydrodynamic coefficients are obtained by analyzing the flow field of BOP stack, which lay a foundation for the whole performance analysis of the drilling system and provides reference data. It is suggested that the study of hydrodynamic coefficients should be carried out in combination with experiments in the later stage, and the CFD calculation model should be further optimized and designed to improve the reliability of CFD calculation and analysis. It is of great significance to save the test cost and reduce the test period of BOP stack, and can provide better reference data for engineering applications.

Acknowledgement. This research is financially supported by the project of National Key Research and Development Program of China (2022YFC2806501-4).

References

1. Wang, D.S.: Key technology advances and development suggestion for deepwater drilling Riser system. China Petrol. Mach. **46**(7), 39–44 (2018)

2. Chen, G.M., Liu, X.Q., Chang, Y.J.: Advances in technology of deepwater drilling riser and wellhead. J. China Univ. Petrol. **37**(5), 129–139 (2013)
3. Wang, C., Huang, S., Xie, X.S.: Hydrodynamic performance prediction of some propeller based on CFD. J. Naval Univ. Eng. **20**(4), 107–112 (2008)
4. Wang, Y.H., Liu, Z., Zhao, R.Y., et al.: Experimental design of hydrodynamic coefficient measurement for underwater manifold. Exp. Technol. Manag. **38**(10), 29–32 (2021)
5. Tang, K., Duan, M.L., Zhang, X.H., et al.: Numeral simulation of hydrodynamic drag coefficients using CFD. Oil Field Equip. **43**(3), 34–37 (2014)
6. Li, P., Zhao, Q.J., Zhu, Q.X.: CFD calculations on the unsteady aerodynamic characteristics of a tilt-rotor in a conversion mode. Chin. J. Aeronaut. **28**(6), 1593–1605 (2015)
7. Tina, Y., Luo, X.L., Qi, Y., et al.: CFD based drag coefficient and additional mass coefficient calculation of subsea manifold. Oil Field Equip. **46**(6), 16–21 (2017)
8. Qi, Y., Luo, X.L., Zhang, D.H., et al.: Calculation of hydrodynamic coefficients in the process of manifold deploying. China Petrol. Mach. **46**(8), 43–47 (2018)
9. Feng, S.S., Sha, Y.X., Wei, X.J., et al.: Numerical method of added mass based on CFD theory. In: Proceedings of 2014 China Congress of Computational Mechanics, Guiyang (2014)
10. Yu, Y.: Experimental design for the determination of anchor chain drag coefficients in current. J. Exp. Mech. **25**(3), 246–250 (2010)
11. Huang, W.P., Chen, H.M.: Study on lift coefficient and drag coefficient of elastic multi-riser based on CFD. Ship Ocean Eng. **40**(2), 94–97 (2011)
12. Wang, G.X., Zhang, M., Wang, T.G.: Numerical analysis and experimental study on drag coefficient of mooring anchor chain of floating offshore platform. Eng. Mech. **37**(S1), 347–351 (2020)
13. Qiao, C.: Dynamic response analysis of riser considering hydrodynamic characteristics of subsea BOP stack. China Univ. Petrol. (2021)
14. DNV-RP-H103. Modelling and Analysis of Marine Operations (2011)

Penetration Mechanism and Limiting Cavity Depth of Caisson Foundation for Jack-Up in Clay

Xi Wei[1,2], Ning Jia[1], and Teng Wang[1(✉)]

[1] China University of Petroleum, (East China) No. 66, Changjiang West Road, Huangdao District, Qingdao, China
wteng73@upc.edu.cn

[2] Drilling Technology Research Institute, Sinopec Oilfield Service Shengli Corporation, No. 827, Beiyi Road, Dongying District, Dongying, China

Abstract. The adoption of caisson foundations for offshore jack-up rigs is attributed to their capability to prevent punch-through failure and enhance stability during drilling operations. The presence of the caisson wall alters the penetration mechanism, which differs from that of saucer-shaped spudcans, thus affecting the method for calculating the limiting cavity depth and bearing capacity. In this study, the Coupled Eulerian-Lagrangian (CEL) finite element method was employed to conduct large deformation simulations of the entire preload penetration process. The influence of caisson wall length on the penetration mechanism and the limiting cavity depth of caissons was investigated. The analysis of penetration resistance and soil deformation mechanisms surrounding the advancing caisson has revealed a three-stage caisson penetration process: the caisson wall penetration stage, shallow penetration stage, and deep penetration stage. In the wall penetration stage, the penetration resistance originates from internal and external friction resulting from the interaction between the wall and soil. During the shallow penetration stage, soil gradually flows back into the cavity until reaching a stable cavity depth. In the deep penetration stage, a fully localized flow occurs around the embedded caisson, while the cavity remains unchanged. This paper provides a comparative analysis of the cavity depth between caisson and spudcan foundations. Considering the normalized strength formula and the length of the caisson wall, an equation is presented to estimate the limiting cavity depth for caisson foundations in homogeneous clay.

Keywords: Caisson Foundation · Penetration Mechanism · Limiting Cavity Depth

1 Introduction

Jack-up rigs are typically utilized in marine with water depths between 0 and 120 m, which composed by a hull and pile legs. Spudcans connect to the bottom of the pile legs (see Fig. 1(a)). During offshore operations, spudcans penetrated into the seabed to provide support for the rig [1–3].

D. Jeng and B. Cai (Eds.): ICCOE 2024, LNCE 423, pp. 271–280, 2025.
https://doi.org/10.1007/978-981-96-0181-3_25

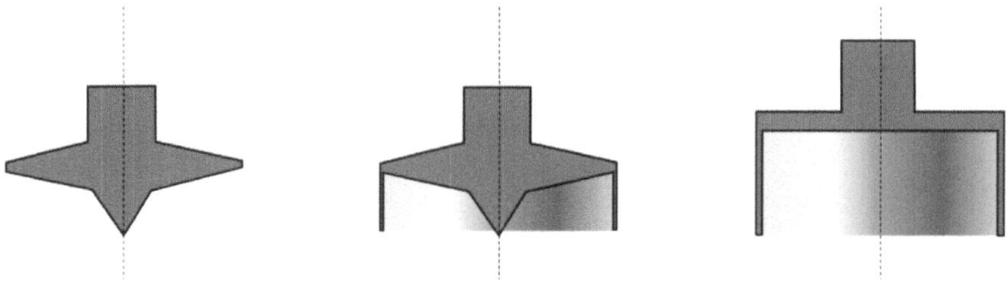

Fig. 1. Type of different spudcans

The limiting cavity depth of the spudcan is an essential parameter for analyzing the spudcan penetration mechanism, determining the spudcan penetration stage, and calculating the penetration resistance. Hossain et al. [4] introduced a calculation method for the limiting cavity depth of the saucer spudcan, as shown in Eq. 1.

$$\frac{H_c}{D} = (S)^{0.55} - \frac{1}{4}S \tag{1}$$

$$\mathrm{S} = \left(\frac{s_{um}}{\gamma' D}\right)^{\left(1-\frac{k}{\gamma'}\right)} \tag{2}$$

where D is the diameter of the spudcan, and γ' is the effective unit weight of the clay. The undrained shear strength of non-homogeneous clay seabed varies with depth. s_{um} is the intact undrained shear strength of top layer clay, k is dimensionless strength increasing parameter for non-homogeneous clay. Hossain and Randolph [5, 6] explained the penetration mechanism of saucer spudcans in clay. For a spudcan embedded shallower than the limiting cavity depth H_c, Failure is in shear mechanism with an open cavity above. As the spudcan penetrates past the limiting cavity depth, soil backflow commences and the footing gradually becomes buried. With the onset of backflow, there is also a gradual transition from the general shear mechanism to the continuous flow-around mechanism. This theory has been successfully utilized in studies by Zhang et al. [7] and Zheng et al. [8]. The researchers propose that the limiting cavity depth H_c is a crucial factor in determining the calculation method for spudcan penetration resistance.

In recent years, punch-through failure phenomenon during spudcans penetrate into stiff-over soft soil becomes the factor of accidents involving jack-up rigs [9]. Researchers have improved the shape and penetration technique of traditional spudcans to reduce the risk of punch-through failure and bolster the stability of offshore platforms. Zheng et al. [10] designed a spudcan with a square main body and four movable plates, allowing for flexible adjustments to the stress distribution across various operational stages by controlling the rotation of the plates. Wang et al. [11] introduced a large rectangular spudcan along with a graded-load method to prevent punch-through failures.

However, these methods are too complicated to be practically applied. A simpler and more effective solution is to add a skirt around the saucer spudcan (see Fig. 1(b)) [12–14]. By centrifuge test and numerical simulations, researchers [15, 16] confirmed that the height of the soil plug and the length of the slip planes were lower for the skirted spudcans, leading to less severe or easing punch-through. Combining the advantages of skirt in preventing the punch-through failure with the basic types of other marine structures, researchers [17, 18] proposed the design of caisson foundation (see Fig. 1(c)). The length of caisson wall alters the soil movement pattern during penetration. Therefore, it is crucial to refine Eq. 1 by the penetration mechanism of caisson foundations. In addition, saucer spudcans ignore the impact of interface frictional resistance during penetration, whereas it is imperative to consider the contact surface between the caisson wall and soil in caisson foundations. The length of caisson wall alters the soil movement pattern during penetration. Therefore, it is crucial to refine Eq. 1 by the penetration mechanism of caisson foundations. In addition, saucer spudcans ignore the impact of interface frictional resistance during penetration, whereas it is imperative to consider the contact surface between the caisson wall and soil in caisson foundations.

In this study, the commercial finite element software Abaqus is utilized to employ the explicit large deformation numerical simulation method CEL (Coupled Eulerian-Lagrangian). The penetration mechanism and the limiting cavity depth of caisson foundation with varying caisson wall lengths in homogeneous clay are analyzed. A calculation equation for the limiting cavity depth in homogeneous clay is developed.

2 Finite Element Model

The CEL method was used to simulate the large deformation process for caisson foundations. In this study, the caisson foundation is discretized by Lagrangian mesh, while the soil is discretized by Eulerian mesh. The numerical simulation cases are shown in Table 1.

Table 1. Numerical simulation cases and material parameters

Analysis		Undrained shear strength of the clay, s_u (kPa)	Bucket wall length, T (m)	Material parameters
1	1–1	10	0	Effective unit weight: $\gamma' = 8.5$ kN/m^3 Young's modulus: $E = 100s_u$ Poisson's ratio: $v = 0.49$ Frictional coefficient: $a = 0.5$
	1–2	10	3.75 (0.25D)	
	1–3	10	7.50 (0.5D)	
2	2–1	15	0	
	2–2	15	3.75 (0.25D)	
3	3–1	20	0	
	3–2	20	3.75 (0.25D)	

A 8° slice of the axisymmetric caisson foundation penetration problem was considered (see Fig. 2). The caisson foundation was modelled as a rigid body to simplify the calculation considering its small deformation. Values of the key model parameters were shown in Fig. 2. The model for caisson with wall length T = 0.5D is shown in Fig. 2.

The inner diameter of the caisson foundation was D = 15 m, the bucket wall thickness was 0.12 m, the height of the caisson foundation was t = 1.25 m, and the diameter of the pile leg was $D_L = 4.3$ m. For the simulation, a constant penetration speed v = 0.5 m/s was maintained until reaching the total depth of L = 2D (30 m).

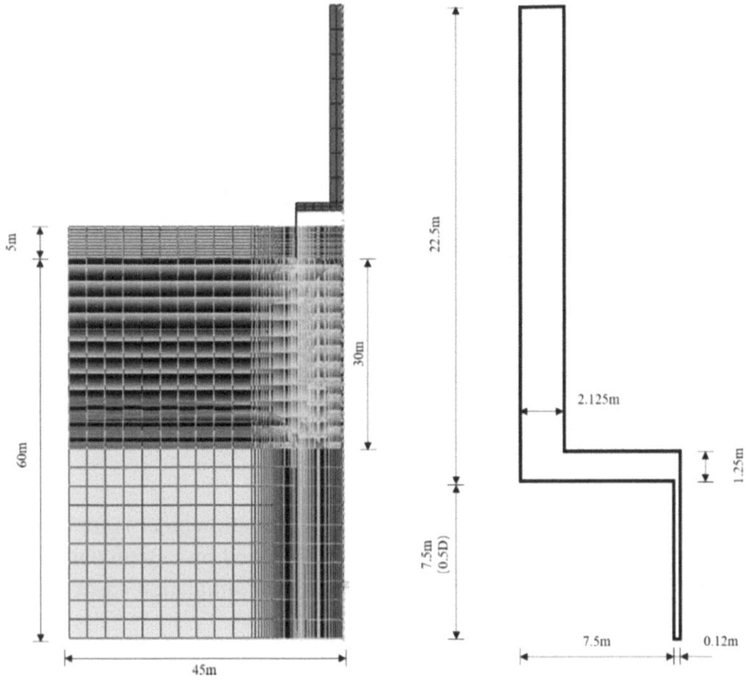

Fig. 2. Numerical simulation process dimensions

The soil domain was with a depth of 2L and a radial length of 3D to reduce the boundary effect after a domain study. The soil was modelled as an elastic-perfectly plastic undrained clay which met the Von-Mises yield criterion. The undrained clay was assumed to be saturated using a Poisson ratio υ of 0.49, with a Young's modulus of $E = 300s_u$. The unit weight of the soil is $\gamma' = 8.5$ kN/m^3. A void layer was set at the top of soil, which allowed the soil squeezed by caisson foundation to heave and move into it. Corresponding constraints are enforced at the boundaries.

The interaction between the bucket and the soil is taken into consideration during the calculations. The contact between the bucket spudcan, as well as the pile leg, and the soil are modeled using a general contact approach, with rigid contact in the forward direction and Coulomb friction law in the tangential direction, described as $\tau_{max} = \alpha s_u$, where α denotes the interface friction coefficient, which is the reciprocal of the sensitivity S_t. For marine clays, the standard sensitivity ranges from 2 to 5. In the simulation the sensitivity value is 2, and an interface friction coefficient of $\alpha = 0.5$ is applied.

3 Numerical Simulation Results

3.1 Penetration Mechanism of Caisson

The penetration resistance varies with depth for caissons with three different wall lengths: T = 0, T = 0.25D, and T = 0.5D as illustrated in Fig. 3. Homogeneous clay with an undrained shear strength of $s_u = 10$ kPa was considered. From Fig. 3, it is evident that the penetration process can be classified into a three-stage penetration process: the caisson wall penetration stage, shallow penetration stage, and deep penetration stage.

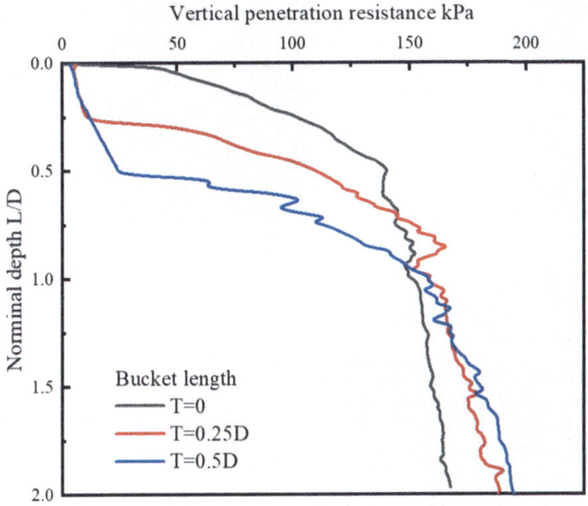

Fig. 3. Vertical penetration reisistance of spudcans with different caisson wall lengths

During the wall penetration stage, as depicted in Fig. 4(a), the penetration resistance primarily stems from internal and external friction resulting from the interaction between the caisson wall and the soil. This interaction leads to a gradual increase in penetration resistance with depth. Moreover, as the length of the caisson wall increases, frictional resistances also exhibit a linear increase. This observation emphasizes the importance of caisson wall length in influencing frictional forces and, consequently, the penetration process.

During the shallow penetration stage, soil gradually flows back into the cavity until a complete seal cover is formed above the caisson, as depicted in Fig. 4(b). In this stage, when the soil comes into contact with the top plate of the caisson, the penetration resistance increases abruptly, as indicated by the inflection point shown in Fig. 3. As the penetration depth increases further, an open cavity forms, and at a certain depth, soil backflow is initiated, accompanied by a reduction in the rate of increase resistance.

In the deep penetration stage, a fully localized flow occurs around the embedded caisson, while the cavity remains unchanged, as shown in Fig. 4(c). In this stage, the penetration resistance increases slowly and does not approach a stable value with increasing penetration depth. This is because the frictional resistance on the leg sides increases

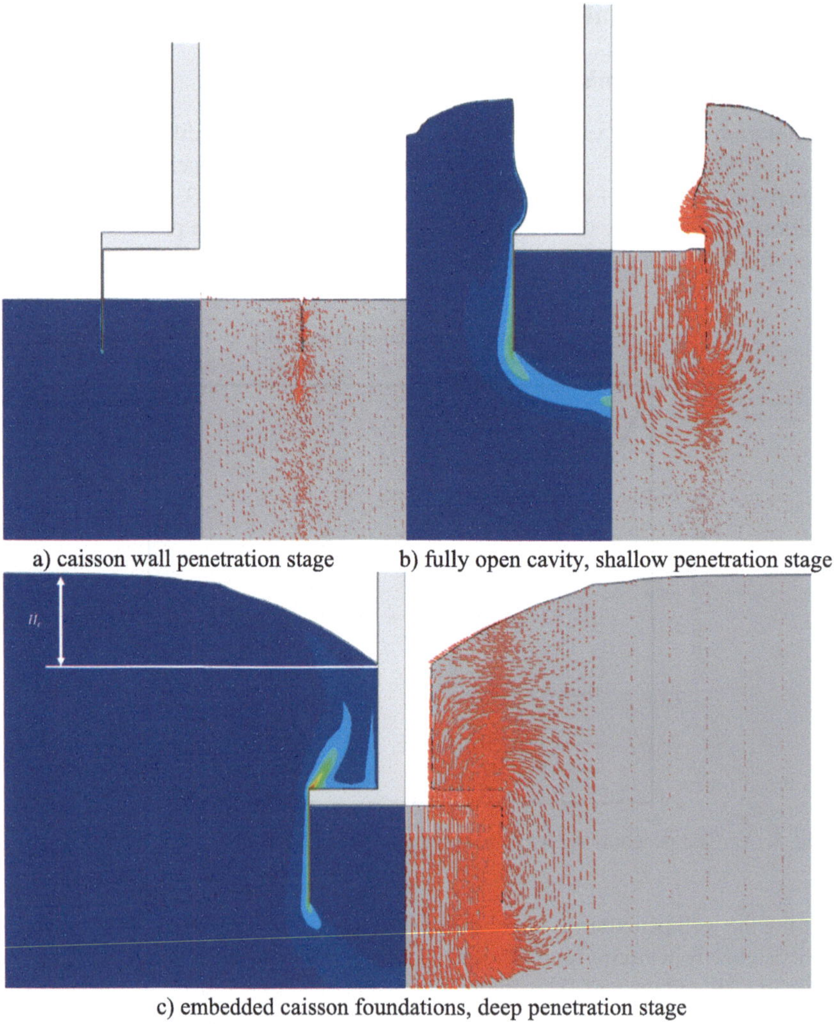

Fig. 4. Equivalent plastic strain and displacement cloud map of soil at each stage in T = 0.5D

with penetration depth. Additionally, Fig. 4 illustrates that the deep penetration resistance also increases with the wall length, attributed to the larger frictional resistance.

In conclusion, as shown in Fig. 4, for the caisson penetration stage the caisson foundation embeds shallow than the caisson wall depth, $L \leq T$. As the caisson foundation penetrates past the caisson wall depth T but less than $T+H_c$, $T \leq L \leq T+H_c$, the caisson foundation is in the shallow penetration stage. As the caisson penetrates deeper than $T + H_c + 0.5D$, the caisson foundation is in the deep penetration stage. The penetration stage is affected by the limiting cavity depth H_c.

3.2 Limiting Cavity Depth

Figure 5 presents the plastic strain and the limiting cavity depth of caisson foundations for three different wall lengths: T = 0, T = 0.25D, and T = 0.5D. This visualization offers valuable insights into how the influence of wall length affects the corresponding limiting cavity depths.

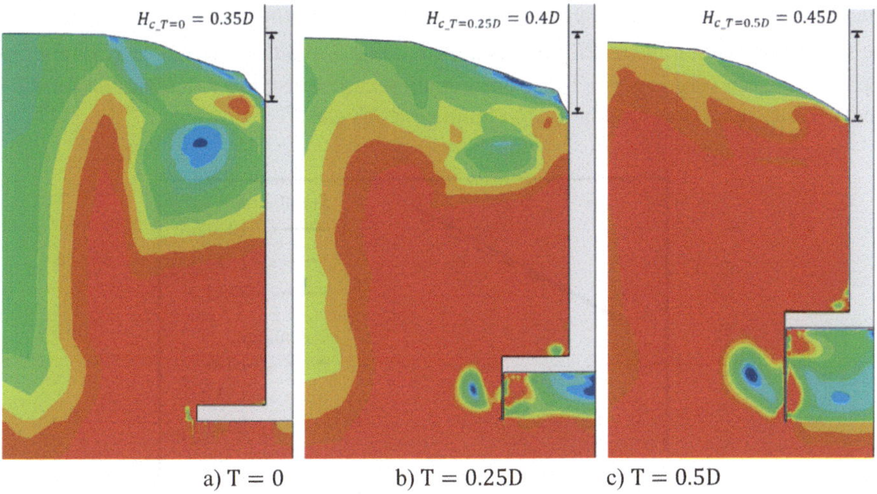

Fig. 5. The limiting cavity depth for different bucket wall lengths in $s_u = 10$ kPa.

As shown in Fig. 5, the limiting cavity depth of the caisson increased by the caisson wall length increase longer. The flow path of the soil in longer caisson wall length is further than T = 0. The simulation result of the limiting cavity depth of T = 0 is $H_c = 0.35D$ in $s_u = 10$ kPa, while result calculated by Eq. 1 is $H_c = 0.227D$. Thus, the original calculation result of saucer spudcan cannot be utilized for the caisson foundation. Based on the simulation results, the calculation equation for caisson foundation need to be optimized considering the length of the caisson wall length T and the undrained shear strength s_u of the clay.

Table 2. Summary of the limiting cavity depth H_c

The limiting cavity depth H_c	T = 0	T = 0.25D	T = 0.5D
$s_u = 10$ kPa	0.35D	0.40D	0.45D
$s_u = 15$ kPa	0.46D	0.54D	——
$s_u = 20$ kPa	0.59D	0.70D	——

Table 2 shows the limiting cavity depth H_c results under the variation of the undrained shear strength s_u and caisson wall length T. The increase in the limiting cavity depth is linearly related to the increase in bucket wall length. The limiting cavity depth increases when the undrained shear strength s_u increased. Based on Eq. 1, the limiting cavity depth of the caisson foundation is derived in Eq. 3.

$$\frac{H_c}{D} = (S)^a - \frac{1}{4}S + 2.55\frac{T}{D}S \tag{3}$$

where $S = \frac{s_u}{\gamma' D}$ in homogeneous clay, nondimensional parameter $b = 0.55 - 1.9125S$ is the soil strength coefficient. Plot the data from Table 2 and the calculation results from Eq. 3 as a curve (see Fig. 6). Thus, this equation can be used to calculate the limiting cavity depth of caisson foundations in homogeneous clay.

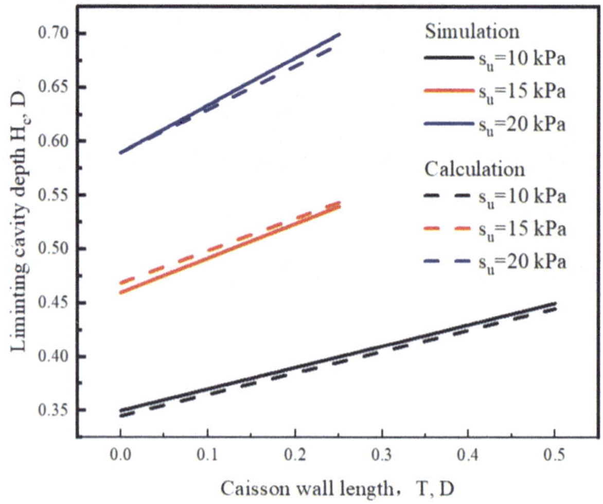

Fig. 6. The comparison between the calculated and numerical simulation results

4 Conclusion

This study employed large deformation finite element method to investigate the penetration process of caisson foundation of jack-up rigs in homogeneous clay. The results unveiled the penetration mechanism of caisson foundation and presented the calculation equation for the limiting cavity depth of caisson in homogeneous clay. The specific research conclusions are as follows:

(1) Based on the penetration resistance curve, the penetration process of caisson foundation can be categorized into three stages: the caisson wall penetration stage, the shallow penetration stage and the deep penetration stage. The growth rates of penetration resistance in the three stages are different. The penetration resistance undergoes a significant increase during the shallow penetration stage, while it increases slowly in the caisson wall penetration stage and deep penetration stage.

(2) The limiting cavity depth increases linearly with the increase in caisson wall length. For example, the limiting cavity depth of the saucer spudcan is $H_c = 0.227D$ with $s_u = 10$ kPa, which is lower than the $H_c = 0.35D$ of the caisson foundation when T = 0. This indicates that the original calculation equation for spudcan is not applicable for caisson foundations.

(3) The limiting cavity depth increases with the normalized undrained shear strength increased based on the simulation results, an equation is presented to estimate the limiting cavity depth for caisson foundations in homogeneous clay.

References

1. Yi, J.T., Lee, F.H., Goh, S.H., et al.: Eulerian finite element analysis of excess pore pressure generated by Spudcan installation into soft clay. Comput. Geotech. **2012**(42), 157–170 (2012)
2. Zhang, Y., Wang, D., Cassidy, M.J., et al.: Effect of installation on the bearing capacity of a spudcan under combined loading in soft clay. J. Geotech. Geoenviron. Eng. **140**(7), 04014029 (2014)
3. Hossain, M.S., Hu, Y., Ekaputra, D.: Skirted foundation to mitigate spudcan punch-through on sand-over-clay. Géotechnique. **64**(4), 333–340 (2014)
4. Hossain, M.S., Randloph, M.F.: New mechanism based design approach for Spudcan foundations on single layer clay. J. Geotech. Geoenviron. Eng. **135**(9), 1264–1274 (2009)
5. Hossain, M.S., Hu, Y., Randolph, M.F., et al.: Limiting cavity depth for Spudcan foundations penetrating clay. Géotechnique **55**(9), 679–690 (2005)
6. Hossain, M.S., Randolph, M.F., Hu, Y., et al.: Cavity stability and bearing capacity of Spudcan foundations on clay. In: Offshore Technology Conference. Houston (2006)
7. Zhang, T., Yi, J., Wang, Z., Liu, F. Numerical study on Spudcan penetration-consolidation-uplift in soft soil using large deformation simulation. In: Casini, M. Proceedings of the 2nd International Civil Engineering and Architecture Conference. CEAC 2022. Lecture Notes in Civil Engineering, vol. 279. Springer, Singapore (2023). https://doi.org/10.1007/978-981-19-4293-8_23
8. Zheng, J.B.: Numerical modelling of Spudcan and cone penetration in multilayer soils, UWA (2015)
9. Li, Y.P., et al. Centrifuge study of the effects of lattice leg on penetration resistance and bearing behavior of spudcan foundations in NC clay. In: International Conference on Offshore Mechanics and Arctic Engineering, vol. 44915. American Society of Mechanical Engineers (2012)
10. Zheng, J., Chen, Y., Chen, X., et al.: Improved prediction of peak resistance for Spudcan penetration in sand layer overlying clay. J. Geotech. Geoenviron. Eng. **148**(1), 04021174 (2022)
11. Wang, D.S., Fan, L.D., Wang, F., et al.: Deep penetration of rectangular Spudcan into single-layer clay. Int. J. Offshore Polar Eng. **32**, 236–243 (2022)
12. Hu, P., Wang, D., Stanier, S.A., et al.: Assessing the punch-through hazard of a spudcan on sand overlying clay. Géotechnique **65**(11), 883–896 (2015)
13. Lee, FH., Fu, Y., Li, Y., Yi, J., Zhang, X.Y.: Lattice leg–soil interaction effects in deeply embedded Spudcan foundations. In: Haldar, S., Patra, S., Ghanekar, R. (eds.) Advances in Offshore Geotechnics . LNCV, vol 92. Springer, Singapore (2020). https://doi.org/10.1007/978-981-15-6832-9_4
14. Lee, K.K., Randolph, M.F., Cassidy, M.J.: Bearing capacity on sand overlying clay soils: a simplified conceptual model. Géotechnique **63**(15), 1285–1297 (2013)

15. Ullah, S.N., Stanier, S., Hu, Y., et al.: Foundation punch-through in clay with sand: centrifuge modelling. Géotechnique **67**(10), 870–889 (2017)
16. Qiu, G., Henke, S.: Controlled installation of spudcan foundations on loose sand overlying weak clay. Mar. Struct. **24**(4), 528–550 (2011)
17. Lee, J.: Development of a novel spudcan shape for easing punch-through (2019)
18. Yang, N.X., et al.: Bearingcapacities of buried bucket foundations in marine tidal flat subjected to combined loading. Int. J. Geomech. **22**(8), 04022110 (2022)

Analysis of the Connection Operation Window for Riserless Drilling Mud Return Line

Zhaowei Liu, Di Wu, and Xiuquan Liu(✉)

Centre for Offshore Engineering and Safety Technology, China University of Petroleum, Qingdao 266580, China
lxqmcae@163.com

Abstract. The connection operation window of the mud return line can determine the environmental limits of safe operation, which is very important to ensure the safety of the riserless drilling system. To analyze the safe operation window of the mud return line, a mechanical model of the mud return line is established based on the infinitesimal method. The safe operation criteria of the mud return line under connected condition is proposed. The operation window analysis process of the mud return line based on the dichotomous method is established. According to the analysis of the operation window for the riserless mud return line, it is found that the safety of the mud return line decreases with the increase of current velocity and wave amplitude. The operation window of the mud return line is sensitive to the change in wave amplitude. The riserless drilling system is recommended when the current velocity is less than 0.8 m/s and the wave amplitude is less than 5 m.

Keywords: Riserless drilling · Mud return line · Mechanical model · Operation window

1 Introduction

The riserless mud return drilling system overcomes the shortcomings of traditional drilling riser systems, such as kick and loss. The riserless mud return drilling system is getting more and more applications. However, the mud return line, which is the key equipment of the riserless mud return drilling system, is very fragile. The fragility of the mud return line threatens the operation safety of the riserless mud return drilling system [1]. To improve the operation safety of the riserless mud return drilling system, the operation window of the mud return line needs to be analyzed.

In recent years, scholars have carried out a series of studies on the mud return line. Qin et al. analyzed and summarized the impact of factors such as winds, currents, drilling ship motion, and mud pump vibration on the stability of the mud return line [2]. Zhang et al. established a finite element model for the deepwater mud return line and analyzed the vortex-induced response of the mud return line [3]. Xia et al. established a dynamic model to analyze the dynamic response of the mud return line under drilling ship motion [4]. Wang discretized the mud return line, which consisted of multiple seabed pumping modules in series, using the concentrated mass method, and analyzed

D. Jeng and B. Cai (Eds.): ICCOE 2024, LNCE 423, pp. 281–289, 2025.
https://doi.org/10.1007/978-981-96-0181-3_26

its vibration response [5]. Wang analyzed the variation in natural frequencies of the mud return line under different influencing factors [6]. Currently, the operation window for connecting the mud return line has not been analyzed.

In this paper, a mechanical model of the mud return line is established to analyze the operation window. The operation window analysis method is based on dichotomy. The analysis factors include current, wave amplitude, and operation depth. The research results can provide support for the safe operation of the riserless mud return drilling system.

2 Mechanical Model of the Mud Return Line

The riserless mud return drilling system mainly consists of a drillship, drill pipe, suction module, lift pump group, mud return line, and submarine anchoring device, as shown in Fig. 1 [7].

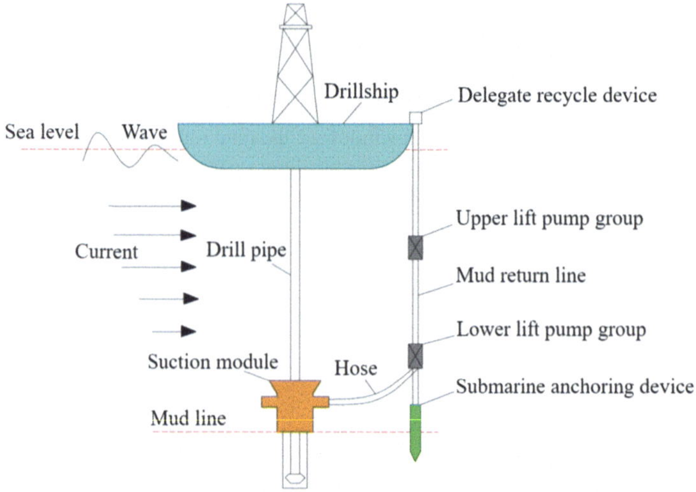

Fig. 1. Riserless mud return drilling system.

An equivalent physical model of the mud return line is established to facilitate analysis. The bottom of the mud return line is anchored to the sea floor, and the top is connected to a hydraulic platform. The hydraulic platform can be equivalent to a spring [8]. The bottom of the mud return line is equivalent to a fixed constraint. The mud return line is equivalent to a beam. The equivalent physical model of the mud return line is shown in Fig. 2.

Under the influence of current, the mud return line bends, and it vibrates due to the wave and the motion of the drilling ship. Force analysis of the mud return line element is conducted, as shown in Fig. 3.

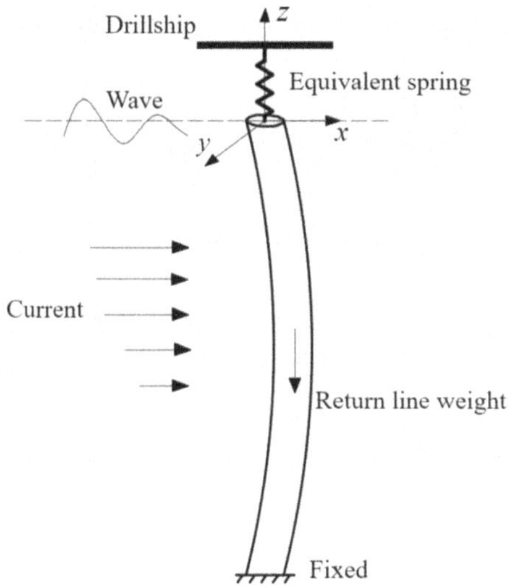

Fig. 2. Equivalent physical model of the mud return line.

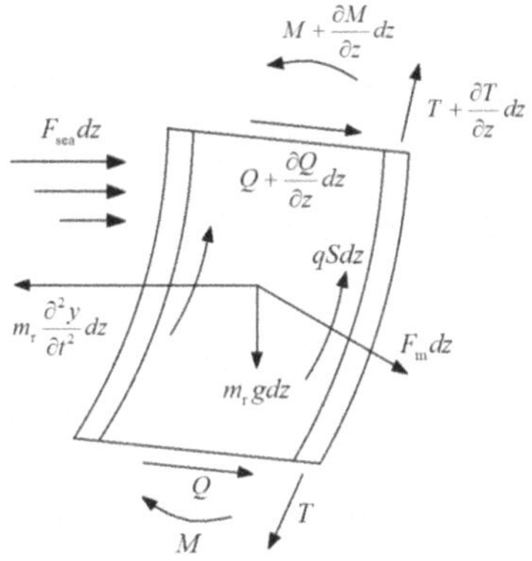

Fig. 3. Force analysis of return line element.

Assuming the mud return line undergoes extremely small deformations, the negligible impact of higher-order quantities allows us to derive the force balance equation in the y-direction for the mud return line element segment [9].

$$-m_r \frac{\partial^2 y}{\partial t^2} + \frac{\partial Q}{\partial z} + \frac{\partial}{\partial z}(T \frac{\partial y}{\partial z}) + F_{sea} = 0 \tag{1}$$

where m_r is the mass per unit length of the line; Q is the shear forces at both ends; T is the axial tension; F_{sea} is the external force due to wave and current.

Based on the beam theory [10]

$$Q = -EI \frac{\partial^3 y}{\partial z^3} \tag{2}$$

where E is the elastic modulus of the return line; I is the moment of inertia of the return line.

The y-direction force equilibrium equation of the mud return line is derived as

$$EI\frac{\partial^4 y}{\partial z^4} - T\frac{\partial^2 y}{\partial z^2} + m_{\mathrm{r}}\frac{\partial^2 y}{\partial t^2} = F_{\mathrm{sea}} \tag{3}$$

The expression for the effective axial force in the mud return line is

$$T = T_{\mathrm{t}} - \int_0^z m_r g \mathrm{d}z \tag{4}$$

where T_{t} is the tension at the top of the mud return line.

During the operation of the mud return line, wave, and current act together on the return line. The modified Morison equation is used to represent the wave and current forces [11].

$$\begin{aligned} F_{\mathrm{sea}} = &\frac{\pi}{4}\rho_{\mathrm{sea}}C_{\mathrm{m}}D_e^2\frac{\partial u_{\mathrm{w}}}{\partial t} - \frac{\pi}{4}\rho(C_{\mathrm{m}} - 1)D_{\mathrm{e}}^2\frac{\partial^2 y}{\partial t^2} \\ &+\frac{1}{2}\rho C_{\mathrm{D}}D_e(u_{\mathrm{w}} + u_{\mathrm{c}} - \frac{\partial y}{\partial t})\left|u_{\mathrm{w}} + u_{\mathrm{c}} - \frac{\partial y}{\partial t}\right| \end{aligned} \tag{5}$$

where ρ_{sea} is the density of seawater, kg/m^3; D_{e} is the outer diameter of the return line, m; C_{m} is the inertia force coefficient; C_{D} is the drag force coefficient; u_{w} is the velocity of the water particle, m/s; u_{c} is the velocity of the steady-state ocean current m/s.

The motion of the drillship is described by the response amplitude operator [12].

$$S_{\mathrm{p}}(t) = A_0 R(\omega)\cos(\omega t + \varphi(\omega)) \tag{6}$$

where A_0 is the wave amplitude; R is the response amplitude operator; φ is the phase difference between wave and drillship; ω is the wave frequency.

3 Operation Window Analysis Method of the Mud Return Line

The main factors to be considered when analyzing the return line connection operation window are current, wave amplitude, and operation depth. These factors are combined two by two for operation window analysis. Before conducting operation window analysis, safety criteria should be established first. The main limiting factors for the safe operation of the mud return line under connection conditions include the equivalent stress, horizontal displacement, and axial deformation. The safety criteria for the mud return line are presented in Table 1 [13, 14].

The analysis of the operation window for the mud return line requires extensive calculations. To enhance the efficiency of the calculations, a dichotomous method is used to search for parameters. The process of the operation window analysis based on the dichotomous method is shown in Fig. 4 [15].

Based on the analysis process of the safe operation window, conduct studies on the operation window of the mud return line under connection conditions, including research on the current-operation depth operation window, wave amplitude-operation depth operation window, and current-wave amplitude operation window.

Table 1. Safety criteria for the mud return line.

Item	Safety criteria
Maximum equivalent stress of line/MPa	67% of yield stress
Maximum horizontal displacement of line/m	10% of operation water depth
Axial deformation of line/m	> 0

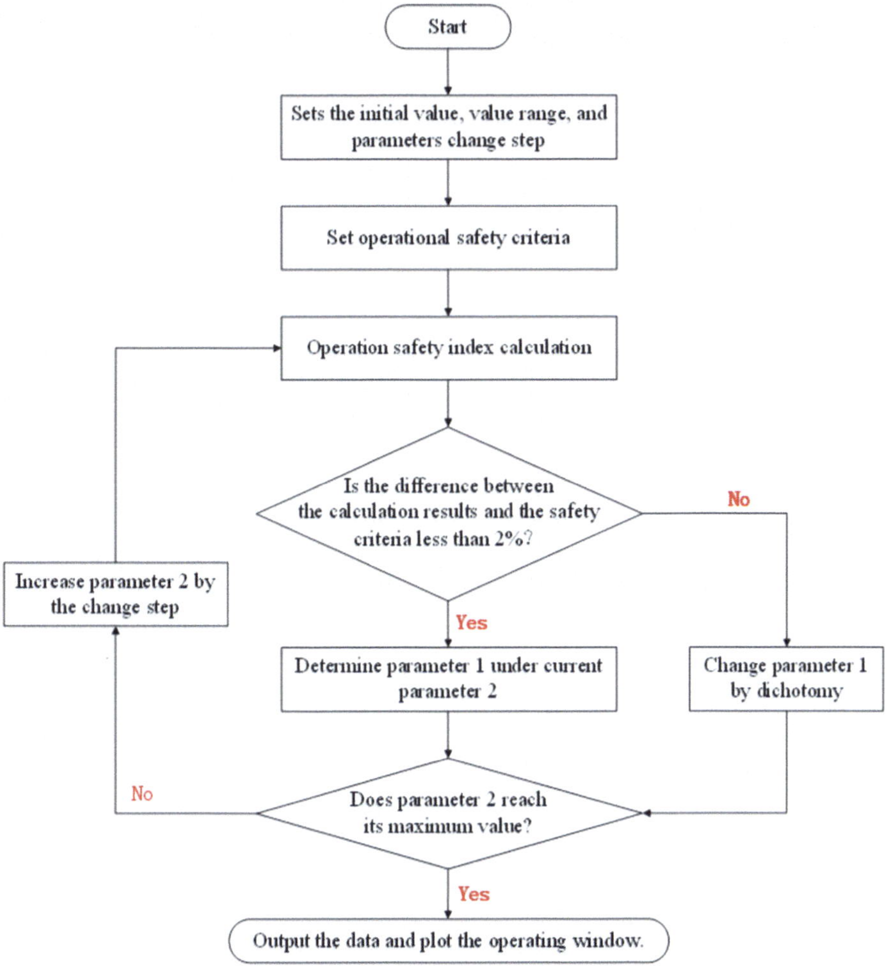

Fig. 4. Operation window analysis process.

4 Analysis of the Operation Window for the Mud Return Line

4.1 Basic Parameters

The basic parameters for analyzing the operation window of the mud return line are presented in Table 2. The response amplitude operator of the drillship motion is shown in Fig. 5.

Table 2. Basic parameters

Name	Value
Outside diameter of return line	168.3 mm
Inner diameter of return line	152.4 mm
Density of return line	7850 kg/m^3
Line weight of return line	568 N/m
Elastic modulus of return line	206 GPa
Surface current velocity	0.6 m/s
Wave amplitude	1 m
Wave period	10 s

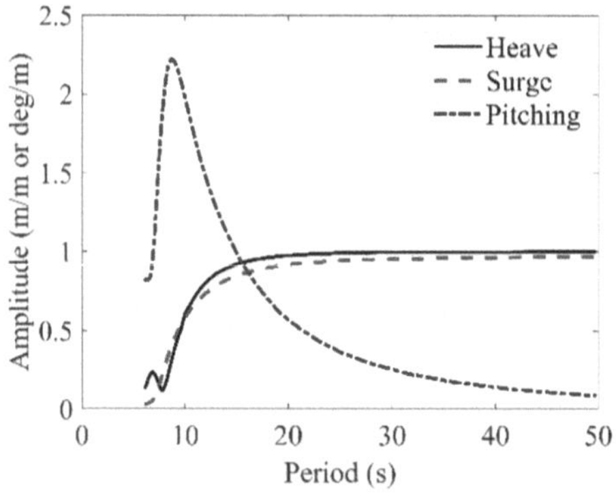

Fig. 5. The response amplitude operator of the drillship.

4.2 Analysis of the Operation Window

(1) The operation window of current-operation depth

The analysis is conducted on the safety operation window for the mud return line concerning current and operation depth. Parameter 1 is assigned to the current, with a value ranging from 0.1 m/s to 1m/s. Parameter 2 is assigned to operation depth, with a

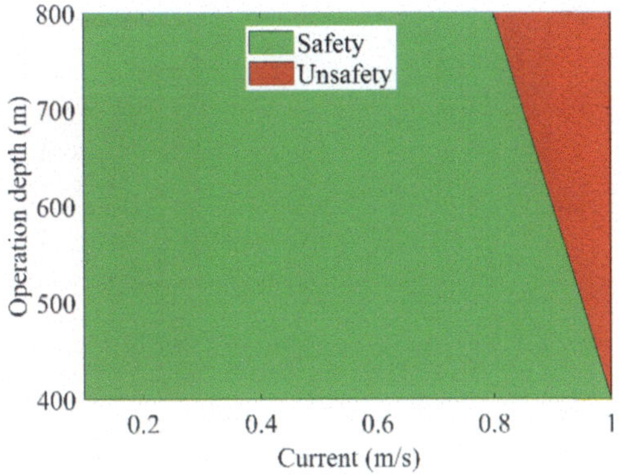

Fig. 6. The operation window of current-operation depth.

value ranging from 400 m to 800 m. The operation window of current-operation depth is shown in Fig. 6. The green area indicates that the mud return line can operate safely. Red areas indicate that the mud return line is not safe to operate.

According to Fig. 6, it is observed that when the surface current velocity does not exceed 0.8m/s, safe operations can be conducted within a water depth range of 800 m. However, as the current velocity exceeds 0.8 m/s, the permissible water depth range for the safe operation of the mud return line rapidly decreases with increasing current velocity. Moreover, when the current velocity surpasses 1.0 m/s, severe sea conditions arise, and the significant force of the current causes the maximum equivalent stress on the return pipeline to exceed the established limits. As a result, drilling operations involving mud recovery through an open-ended pipe are not permitted in any water depth under such adverse sea conditions.

(2) The operation window of wave amplitude-operation depth

The analysis is conducted on the safety operation window for the mud return line concerning wave amplitude and operation depth. Parameter 1 is assigned to wave amplitude, with a value ranging from 0.5 m to 8 m. Parameter 2 is assigned to operation depth, with a value ranging from 400 m to 800 m. The operation window of wave amplitude-operation depth is shown in Fig. 7.

According to Fig. 7, it is evident that when the wave amplitude does not exceed 5.5 m, safe operations can be carried out within a water depth range of 800 m. However, if the wave amplitude exceeds 5.5 m, operations are not permitted at any water depth. The mud return line is sensitive to wave amplitude during connection operations. The operational water depth has minimal impact on the safety operating window of the mud return line.

(3) The operation window of current-wave amplitude

The analysis is conducted on the safety operation window for the mud return line concerning current and wave amplitude. Parameter 1 is assigned to the current, with a value ranging from 0.1 m/s to 1.0 m/s. Parameter 2 is assigned to wave amplitude, with

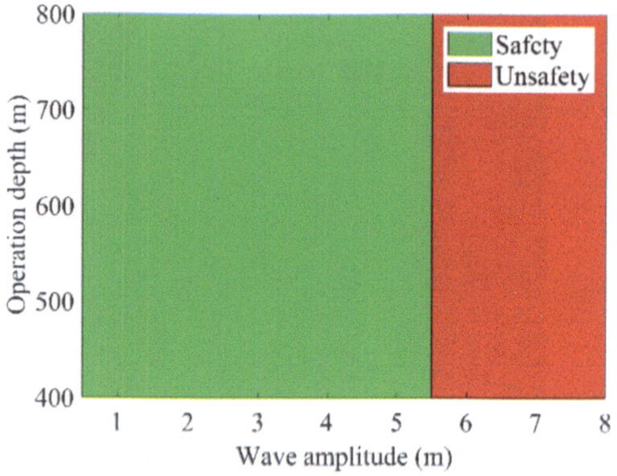

Fig. 7. The operation window of wave amplitude-operation depth.

a value ranging from 0.5 m to 8 m. The operation window of current-wave amplitude is shown in Fig. 8.

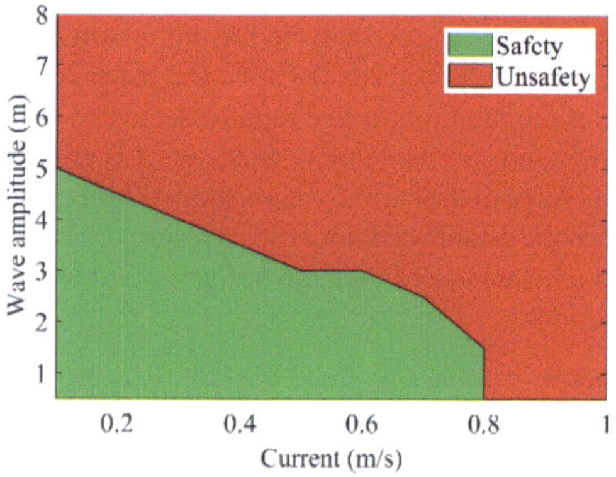

Fig. 8. The operation window of current-wave amplitude.

According to Fig. 9, it can be observed that as the current velocity increases, the permissible wave amplitude for the safe operation of the mud return line gradually decreases. When the current speed ranges from 0.5 m/s to 0.6 m/s, the permissible wave amplitude for safe operation of the mud return line remains at 3 m. If the wave amplitude exceeds 5 m or the current speed is greater than 0.8 m/s, operations of the mud return line are not allowed.

5 Conclusion

(1) A mechanical model of the mud return line is established based on the infinitesimal method. The axial and lateral responses of the mud return line can be calculated by the mechanical model. The marine environmental load is calculated using the

modified Morison equation. The motion of the drillship is described by the response amplitude operator.

(2) The safe operation criteria of mud return line under connected condition is proposed. The indexes that describe the safe operation criteria of the mud return line include the maximum equivalent stress, the maximum horizontal displacement, and the axial deformation. The operation window analysis process of the mud return line based on the dichotomous method is established.
(3) The factors that influence the safe operation of the mud return line, including current, wave amplitude, and operation depth, are combined two by two for operation window analysis. The result shows that the safety of the mud return line decreases with the increase of current velocity and wave amplitude. The operation window of the mud return line is sensitive to the change in wave amplitude. The riserless drilling system is recommended when the current velocity is less than 0.8 m/s and the wave amplitude is less than 5 m.

References

1. Jonggeun, C.: Analysis of riserless drilling and well-control hydraulics. SPE Drill. Complet **14**(1), 71–81 (1999)
2. Qin, R.L., Lu, Q.P., He, G.L.: Quantitative analysis of the stability of a mud-return circulation system in a riserless mud-recovery drilling system. Appl. Sci. **13**(16), 9320 (2023)
3. Zhang, P.F., Chen, G.M., Xie, L.W.: Simulation-based subsea drilling fluids circulation system. China Petrol. Mach. **39**(5), 23–26
4. Xia, K.L., Chen, G.M., Wang, Z.W.: Dynamic analysis of mud return line in deep water drilling ship. China Petrol. Mach. **48**(6), 57–62 (2020)
5. Wang, G.D.: Research on the System Design and Monitoring Strategy for the Device Of Subsea Mudlift Drilling in Deepwater. China University of Petroleum (Huadong), Qingdao (2015)
6. Wang, Z.W.: Design and Study on the Riser-Less Mud Recovery Drilling System for the Ocean Exploration Ship. China University of Petroleum (Huadong), Qingdao (2019)
7. Wang, L., Zhang, H., Ke, K.: Optimization of anchor positioning scheme for deep water riserless drilling fluid lifting system based on genetic algorithm. China Offshore Oil Gas **34**(6), 142–148 (2022)
8. Liu, X.Q., et al.: Recoil control of deepwater drilling riser system based on optimal control theory. Ocean Eng. **220**, 108473 (2021)
9. Liu, Z.W., Wu, D., Wang, X., Chang, Y.J., Chen, G.M., Liu, X.Q.: Mechanical analysis of mud return on deepwater drilling riser system. Ocean Eng. **302**, 117657 (2024)
10. Qiu, N., et al.: Dynamic stability evaluation of subsea wellhead based on fully coupled model. Ocean Eng. **257**, 111720 (2022)
11. Qiu, N., et al.: Dynamic catastrophe analysis of deepwater mooring platform/riser/wellhead coupled system under ISW. Reliab. Eng. Syst. Saf. **246**, 110084 (2024)
12. Qiu, N., et al.: Mechanical model and characteristics of deep-water drilling riser-wellhead system under internal solitary waves. Mar. Struct. **90**, 103439 (2023)
13. Wang, Y.B., Gao, D.L., Fang, J.: Mechanical behavior analysis for the determination of riser installation window in offshore drilling. J. Nat. Gas Sci. Eng. **24**, 317–323 (2015)
14. Ju, S.D., Chang, Y.J., Chen, G.M., Chen, L.M., Xu, L.B.: Determination of the hang-off window for deepwater drilling riser. Acta Petrolei Sin. **33**(1), 133–136 (2012)
15. Ju, S.D., Chang, Y.J., Chen, G.M., Liu, X.Q., Xu, L.B., Wang, R.Y.: Envelopes for connected operation of the deepwater drilling riser. Pet. Explor. Dev.Explor. Dev. **39**(1), 105–110 (2012)

Deep-Sea Mining Baseline Surveys and Monitoring Research

Wenbin Ma[1,2](✉), Yuwen Chen[1,2,4], Bo Zhou[1,2,3], Yongqi Ma[1,2], and Yijun Shen[1,2]

[1] Faculty of Marine Science and Engineering, Hainan University, Haikou 570228, Hainan, China
wenbinma_edu@163.com

[2] State Key Laboratory of Marine Resources Utilization, South China Sea, Haikou 570228, Hainan, China

[3] Mechanical and Electrical Engineering, Hainan University, Haikou 570228, Hainan, China

[4] Faculty of Marine Biology and Fisheries, Hainan University, Haikou 570228, Hainan, China

Abstract. The concept of deep-sea mining has been proposed for over half a century, yet the answer to whether it can solve the global mineral resource crisis has not been determined. The main issue remains environmental concerns. One ongoing research topic that runs through this issue is environmental baseline surveys and monitoring. This article uses a literature review method to compare and summarize previous related research. Firstly, it explains the importance of baseline survey and monitoring in the deep-sea environment. Then, it selects key content such as water quality and sediment analysis for detailed analysis, involving the technologies, methods, and frameworks used in the research. The article also focuses on analyzing several world-renowned deep-sea mining environmental pollution sea trials, summarizing their limitations. The article concludes by analyzing the indispensable importance of baseline survey and monitoring research topics in Environmental Impact Assessment (EIA) work for deep-sea mining. The significance of this research is to promote stakeholders' investment in baseline survey and monitoring work for deep-sea mining, rather than focusing solely on technological research and development. By drawing on past experiences and lessons, one can find a suitable framework for deep-sea environmental research. This effort is crucial for sustainable development in deep-sea mineral extraction.

Keywords: Deep-sea Mining · Baseline Survey and Monitoring · Water Quality · Pollution · Sustainable Development

1 Introduction

Deep-sea mining refers to the extraction of valuable minerals and resources from the ocean floor at depths greater than 200 m. It typically involves the use of remotely operated vehicles (ROVs), dredging equipment, and other specialized machinery (e.g., production support vessel, multistage pump) to access and collect mineral deposits found on the

W. Ma—Associate Professor of Hainan University.

D. Jeng and B. Cai (Eds.): ICCOE 2024, LNCE 423, pp. 290–300, 2025.
https://doi.org/10.1007/978-981-96-0181-3_27

seabed [1]. The significance of deep-sea mining lies in its potential to provide access to valuable resources that are becoming increasingly scarce on land. These resources include polymetallic nodules, polymetallic sulphides, and cobalt-rich ferromanganese crusts, which contain high concentrations of metals such as copper, nickel, cobalt, and rare earth elements [2]. These metals are crucial for the production of modern technologies, including smartphones, electric vehicle batteries, renewable energy infrastructure, and other high-tech applications. However, deep-sea mining also raises significant environmental concerns. It could disrupt fragile marine ecosystems, destroy deep-sea habitats, and have unknown long-term consequences on marine biodiversity. Critics argue that the potential environmental risks and uncertainties surrounding deep-sea mining warrant caution and careful regulation to ensure that any extraction activities are conducted in a sustainable and responsible manner [3]

Baseline surveys serve as a reference point against which future changes can be compared and evaluated during and after mining activities [4]. Baseline surveys and monitoring are critical components of the Environmental Impact Assessment (EIA) process for deep-sea mining projects [5]. These activities are conducted to assess the existing environmental conditions of the proposed mining area and to monitor any changes that may occur over time due to mining activities. Monitoring programs are designed to track environmental changes and impacts resulting from deep-sea mining operations. This ongoing assessment is crucial for detecting any adverse effects early, implementing mitigation measures, and ensuring compliance with environmental regulations.

Fig. 1. NORI and Allseas Deep-sea Mining Sea Trials on Nov 14th 2022. Figure 1(a) represents Seabed mineral ores collectors on Hidden Gem waiting for deployment. Figure 1(b) represents 3000 tonnes of manganese nodules collected in the trials. Figure 1(c) represents the pilot nodule collection system with a production rate of 86.4 tonnes per hour [6].

Baseline surveys and monitoring play crucial roles in ensuring that deep-sea mining activities are conducted in an environmentally responsible manner, with a focus on sustainable resource extraction and protection of marine ecosystems. The purpose of this literature review paper is to critically evaluate existing research, methodologies, and findings related to the establishment of baseline data and ongoing monitoring activities in the context of deep-sea mining operations and finally derives its challenges and future research directions.

2 Importance of Baseline Surveys and Monitoring

2.1 Importance of Establishing Baseline Data Before Initiating Deep-Sea Mining Operations

Figure 2 explains the importance of baseline surveying during the exploration and exploitation stages of deep-sea mining projects. It serves as the basis for the before-after-control-impact design, determines the monitoring sites, selects the suitable assessment criteria and start the duplicated measurements of pollution and impacts caused by deep-sea mining activities [7]. Understanding Environmental Conditions: Baseline data provide a snapshot of the existing environmental conditions in the deep-sea mining area. This includes parameters such as water quality, sediment composition, seabed topography, biodiversity, and ecosystem health [8]. Understanding these conditions is essential for assessing potential impacts and establishing benchmarks for comparison during and after mining activities [9]. This assessment informs decision-making regarding the approval, modification, or cessation of mining activities based on their environmental consequences. Baseline data lay the foundation for designing effective environmental monitoring programs.

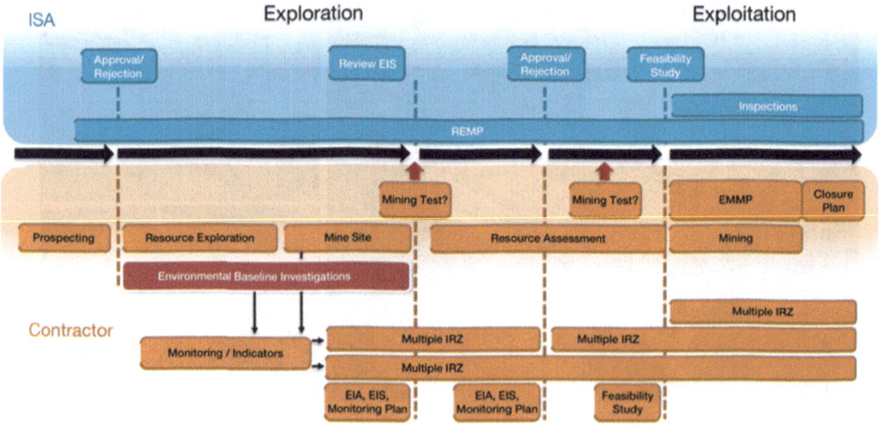

Fig. 2. ISA Baseline Survey for Environmental Management [7].

2.2 Importance of Establishing Baseline Data Before Initiating Deep-Sea Mining Operations

Baseline surveys provide a comprehensive assessment of the environmental conditions in the deep-sea mining area before any mining activities commence. This includes gathering data on factors such as water depth, seabed topography, geology, currents, and physical properties of sediments. These surveys help establish a baseline understanding of the natural state of the marine environment, including any existing natural variations or gradients [10] . Baseline surveys are crucial for assessing the biodiversity of marine ecosystems in the mining area. This involves identifying and documenting species diversity, abundance, distribution, and ecological roles of different organisms, including fish, invertebrates, and microbial communities. Surveys may use methods such as visual surveys, trawling, net sampling, and DNA analysis to characterize the biological diversity and understand species interactions.

2.3 Long Term Monitoring Project

Long-term monitoring is essential in the context of deep-sea mining and other industrial activities to track changes, assess impacts, and implement adaptive management strategies. Long-term monitoring allows scientists and stakeholders to track environmental changes and trends over extended periods. This includes changes in water quality, sediment composition, biodiversity, habitat health, and ecosystem functioning. By collecting data at regular intervals over years or decades, trends and patterns can be identified, providing insights into gradual shifts or disturbances in the marine environment. Long-term monitoring helps in early detection of emerging environmental risks or unexpected changes that may result from mining operations. This early warning system allows stakeholders to respond promptly, implement corrective measures, and mitigate potential harm to marine ecosystems [11]. Detecting trends or thresholds that may indicate ecosystem stress or degradation enables proactive management strategies. Long-term monitoring generates scientific evidence and data-driven insights that inform policy development, regulatory frameworks, and decision-making processes related to deep-sea mining and environmental management [12].

3 Methods and Techniques in Baseline Surveys

Baseline surveys for deep-sea mining projects require a range of methods and techniques to comprehensively assess the environmental conditions, biodiversity, habitat characteristics, and water quality parameters. These surveys are typically conducted before mining operations commence to establish a baseline understanding of the marine ecosystem.

3.1 Multi-beam Bathymetry and Seafloor Mapping

Multibeam sonar systems are used to map the seabed topography and create high-resolution bathymetric maps. Side-scan sonar and acoustic imaging technologies can also be employed. These techniques help identify features such as underwater mountains, valleys, ridges, seamounts, and potential mining sites. They provide a baseline for understanding seafloor geomorphology and geological features. Wölfl et al. (2019) conducted a literature review on the techniques of seafloor mapping of the global ocean bathymetry [13]. The research showed the practices of bathymetry data collection, compilation and open data sharing with Nippon Foundation-GEBCO (general bathymetry chart of the oceans) and emphasized on the non-traditional new and innovative techniques with a high efficiency of bathymetric data collection. Pillay et al. (2020) gave an innovated seafloor mapping technique, which integrated multibeam bathymetry, backscatter, side-scan sonar data [14]. This new technique serves as prototype of a holistic predictive tool and could also be used to model the biological community distribution and produce the benthic habitat mapping.

3.2 Sediment Sampling and Analysis

Sediment samples are collected using grab samplers, corers (e.g., box corer, gravity corer), or sediment traps deployed from research vessels or remotely operated vehicles (ROVs). Sediment analysis helps assess sediment composition, grain size distribution, organic content, mineralogy, and potential presence of contaminants. It provides insights into sediment stability, benthic habitats, and potential impacts of mining activities on sediment quality. During 2018–2019, two small scale deep-sea mining field tests were conducted using Apollo II mining vehicle of Royal IHC focusing on the sediment plume transport simulation. The experiments were carried out in a shallow water area of 300 m deep on the northwestern Alboran Sea continental slope ($<1°$). The monitoring equipment consists of ROV Zonnebloem (for video survey), underwater GAPS positioning equipment (for sediment coring positioning) and CTD Rosette (for the data measurement of sediment and water property). In this experiment, it shows that the sediment plume mass concentration caused by the mineral collector has a linear relationship with the surrounding turbidity [15]. However, the measurement data did not show the single logarithmic relationship between SPM and acoustic backscatter data as planned.

3.3 Water Column Sampling

Water column samples are collected using Niskin bottles, CTD (conductivity, temperature, depth) rosettes, or plankton nets deployed at different depths. Water column sampling measures parameters such as temperature, salinity, dissolved oxygen, nutrient concentrations (nitrate, phosphate), pH, turbidity, and metal content. It helps characterize water quality, nutrient dynamics, and potential sources of contamination [16, 17].

3.4 Biodiversity Surveys

Biodiversity surveys involve visual observations, video recordings, and species sampling using trawls, dredges, ROVs, or manned submersibles. These surveys identify and quantify marine organisms, including fish, invertebrates, corals, sponges, and microbial communities. They assess species diversity, abundance, distribution, and habitat associations to establish baseline biodiversity data [18].

3.5 Habitat Mapping and Characterization

Habitat mapping combines data from bathymetry, sediment samples, and biological surveys to map and classify different habitats (e.g., rocky reefs, soft sediments, hydrothermal vents). Habitat mapping helps identify critical habitats, ecological zones, and areas of high biodiversity or sensitivity. It provides information on habitat structure, complexity, and connectivity, informing conservation and management strategies [19].

3.6 Chemical and Biological Analyses

Laboratory analyses of water, sediment, and biological samples are conducted to measure concentrations of nutrients, metals, pollutants, biomarkers, and biological indicators. Chemical and biological analyses assess baseline concentrations, pollution levels, ecological health indicators, and potential impacts of contaminants on marine organisms and ecosystems [18].

3.7 Remote Sensing and Satellite Imagery

Remote sensing technologies and satellite imagery are used to gather data on sea surface conditions, water turbidity, chlorophyll levels, and surface currents [20]. Remote sensing provides synoptic views of large areas, detects spatial patterns, and helps monitor environmental parameters on a regional scale, complementing in-situ surveys and data collection efforts. These above methods and techniques are often integrated into a comprehensive survey plan tailored to the specific objectives, geographical characteristics, and regulatory requirements of deep-sea mining projects. Each survey technique has its unique strengths and applications, and they are often combined and integrated into a comprehensive baseline survey strategy to provide a holistic understanding of the pre-mining environmental conditions in deep-sea mining projects. Collaboration among experts from different disciplines is key to optimizing the use of these techniques and interpreting data effectively for environmental assessment and management purposes.

4 Habitat Mapping and Biodiversity Assessment

Deep-sea mining is a relatively new industry with limited operational experience, but there have been several case studies and examples of baseline surveys and monitoring associated with deep-sea mining projects.

4.1 Deep-Sea Mining Environmental Studies (DOMES) (1970s)

The objective of DOMES is to try to figure out the answer to the question of potential environmental impact of commercial scale manganese nodule mining on the abyssal regions of the ocean and the fauna and flora of the oceanic surface water. It consists of 2 phases: Phase I – Study prior to disturbance by mining utilizing the towed hydraulic system (1976), and Phase II – Monitoring of precommercial test of mining systems (1978). The biggest feature of this project is the beginning of an attempt to understand the characterizing the immediate and cumulative ecological effects results from the sediment plume discharge [21]. Its limitations consist: I – Short term scale studies (confined within 5 years). Their research report also clearly acknowledged that the data from this project can only be more real and valid in the context of a longer-term observation. II – Limited measurement and data availability. The type and number of parameters measured in the project are very limited (e.g., near-bottom current, dissolved oxygen concentration, etc.). Additionally, many measurements have been made, but the data are not publicly available even now. The measurement stations in the project are also highly biased selected. There is no proof whether repeated measurement data at limited stations can represent the entire mining area. III – Limited technology utilized. In the 1970s, there were no advanced equipment capable of long-term, multi-dimensional submarine environmental data monitoring. Therefore, each data measurement time is very short (10–20 days). Although this project has many limitations, it has extraordinary scientific research value as an early human study on deep-sea mining baseline surveying and monitoring. The research methods, framework and research content involved in this study are still being referenced today.

4.2 Disturbance and Recolonization Experiment (DISCOL) in a Deep-Sea Benthic System (1980s-1990s)

The Disturbance and Recolonization Experiment (DISCOL) was a real-life scientific experiment conducted in the late 1980s and early 1990s in the deep-sea environment of the eastern Pacific Ocean. Before the disturbance, baseline data about the benthic communities, sediment characteristics, and environmental conditions were collected. During and after the disturbance, intensive sampling and monitoring were conducted to track changes in biodiversity, species abundance, sediment composition, water quality, and other parameters. One of the key aspects of DISCOL was to observe how benthic organisms recolonized the disturbed area over time. Researchers monitored the process of recolonization to understand the resilience of deep-sea ecosystems and the factors influencing recovery rates. As a pioneering study in deep-sea ecology, specifically focused on simulating the effects of seabed disturbance akin to deep-sea mining activities, DISCOL experiment also have its limitations as follows: I - Scale of Disturbance: While DISCOL simulated seabed disturbance on a large scale, it was still a controlled experiment within a relatively confined area compared to actual deep-sea mining operations. The scale of disturbance in real mining scenarios can be much larger, potentially leading to different ecological responses and recovery rates. II - Short-Term Study: The duration of the DISCOL experiment and subsequent monitoring period were limited compared to the long-term impacts that deep-sea mining activities can have on marine ecosystems. Long-term ecological changes, including species dynamics, habitat recovery, and ecosystem stability, may unfold over decades or even centuries, which the DISCOL experiment could not fully capture. III - Single Disturbance Event: DISCOL involved a single disturbance event mimicking deep-sea mining, but in reality, mining operations can involve repeated disturbances over time due to ongoing extraction activities, maintenance, and equipment deployment. The cumulative effects of multiple disturbances on benthic communities and sediment ecosystems may differ from a single disturbance event. IV - Species-Specific Responses: The experiment focused on broad community-level responses and recolonization patterns, but it may not have captured species-specific responses or the nuances of ecological interactions within benthic communities. Different species may exhibit varying tolerance levels to disturbance and recovery rates, which could influence ecosystem dynamics. V - Limited Spatial Representation: The DISCOL experiment was conducted in a specific region of the eastern Pacific Ocean, and while it provided valuable insights, extrapolating its findings to other deep-sea environments globally requires caution. Deep-sea ecosystems exhibit considerable variability in terms of biodiversity, geological features, hydrothermal activity, and environmental conditions, which can influence ecological responses to disturbance. VI - Technological and Methodological Constraints: The technology and methodologies available during the DISCOL experiment era (1980s-1990s) had limitations compared to modern advancements in deep-sea research. Advances in remote sensing, genetic analysis, underwater robotics, and environmental monitoring have since provided researchers with more sophisticated tools for studying deep-sea ecosystems.

4.3 Benthic Impact Experiments (BIE) (1991–1993)

The BIE experiment was conducted by Ocean Minerals and Energy Division (OME) of National Academy of Science to analyze the small-scale fine particle suspension phenomenon impact on the seabed biology system as a consequence of burial and food dilution. The towing disturb area was set to be 150 m in width and 3000 m in length, towing 49 times within 19 days. In BIE experiment, 4888 metric ton of sediment was pumped. Five UNSEL box, roughly 0.25 m^2, were selected as the baseline surveying areas. Based on experimental data, the project estimates that a total of about 5000 tons of particles are suspended in the region approximately 13–17 feet above the seabed surface in this project, and most of the sediment will settle on the seabed within a short period of time. However, there is still a portion of nanoparticles that will remain suspended for a longer period and may potentially disperse to further locations with ocean currents [22].

4.4 Japan Deep-Sea Impact Experiment (JET) (1994)

JET experiment, belonging to Environmental Research for Manganese Nodule Development Project managed by Ministry of International Trade and Industry of Japan, was conducted to evaluate the effects of sediment suspension and deposition caused by deep-sea mining activities. The results figured out the negative impact of seafloor disturbance on the meio-benthos abundance, undersea organisms showed signs of returning after a few years, but the composition of the organisms is different. In this project, conducting surveys to establish baseline data on the environmental conditions of the mining area before any operations begin. This helps in monitoring changes over time. Implementing monitoring programs to track environmental changes during and after mining operations. Ensuring compliance with environmental regulations and standards [18, 22]. This experiment also has many shortcomings and does not involve in-depth cumulative environmental impact analysis. There is a difference in magnitude between the scale of disturbance on the seabed and the scale of commercial exploitation.

4.5 Indian Deep-Sea Environment Experiment (INDEX, 1995–2001)

INDEX experiment was conducted by National Institute of Oceanography of India in the Central Indian Basin, which consists of 3 phases: I – Evaluation of baseline data in the selected seabed area; II – Disturbance using a hydraulic device to make the plume suspension and re-sedimentation phenomenon; III – Environmental data monitoring and comparison with the baseline data. INDEX selected a disturbance area of 3000 m in length and 200 m in width with the depth 5300 – 5350 m. Through the box core sampling, macrofauna (classified in Polychaeta, Harpacticoida, Tanaidacea, Nematoda, Sipunculida, Oligochaeta, Unidentified) at the sediment depth ranging from 0 – 20 cm were analyzed. The macrofauna's average density, biomass, community structure and distribution data were also figured out. The results proved that the community of benthic macrofauna are extremely sensitive to even a slight environment disturbance [23, 24].

4.6 Solwara 1 Project (Papua New Guinea, Early 2000s to 2019)

The Solwara 1 Project refers to a deep-sea mining venture in Papua New Guinea (PNG) operated by Nautilus Minerals (In 2019, Nautilus declared bankruptcy). The project faced numerous challenges, including financial difficulties and environmental concerns. Baseline survey and monitoring runs throughout this project. The baseline surveys covered the project area in the Bismarck Sea where mineral deposits were targeted for mining. This included specific regions near hydrothermal vents where high concentrations of valuable metals like copper, gold, silver, and zinc were found. Baseline surveys establish the environmental conditions in an area before any major activity begins. They provide a reference point for assessing changes caused by deep-sea mining. Monitoring continues after baseline surveys to track changes over time during and after project implementation. It helps assess whether predicted impacts are occurring and informs adaptive management strategies [25].

In this section, we only focus on analyzing a few selected typical deep-sea environment baseline surveying and monitoring cases. However, individual countries and regions are also advancing this topic at different levels. For instance, Okinawa Experiment Trough Seafloor Massive Sulfide (SMS) Deposits (Japan, 2017): This project utilized the deep-towed marine electrical resistivity tomography to get the detailed environment information of near-seafloor hydrothermal field images, which gage a deep understanding of a new mechanism of SMS evolution.

5 Conclusion and Recommendation

Baseline surveys and monitoring play crucial roles in the deep-sea mining industry, providing essential data and insights for environmental assessment, impact mitigation, and sustainable resource management. An in-depth analysis of the topic has the following benefits: Understanding Environmental Baselines, Identifying Sensitive Areas and Species, Predicting and Mitigating Impacts, Supporting Regulatory Compliance, Long-term Environmental Monitoring, Advancing Scientific Knowledge, Engaging Stakeholders and Communities, Enhancing Sustainability and Best Practices.

While baseline surveys and monitoring are essential components of deep-sea mining environmental impact assessments, they also face several limitations and challenges that need to be addressed: Data Availability and Quality, Technological Limitations, Spatial and Temporal Scale, Complexity of Ecosystems, Regulatory and Governance Challenges, Cumulative Impacts.

Addressing these limitations requires collaborative efforts among scientists, policymakers, industry stakeholders, and environmental organizations. Investing in research and technology development, improving data sharing and transparency, strengthening regulatory frameworks, and adopting adaptive management approaches can help overcome some of these challenges and promote more sustainable practices in deep-sea mining.

Acknowledgement. This work was supported by National Foundation Committee of China - Key Project [research project number 52231012], Hainan University High-level Talents Research Start-up Funding [grant number KYQD(ZR)-22060], International Association of Maritime Universities [research project number 20220304] and Hainan Provincial Natural Science Foundation High-level Talent Project [funding number 423RC436], and the Tianjin University-Hainan University Joint Innovation Fund [funding number HDTD202302].

References

1. Childs, J.: Extraction in four dimensions: time, space and the emerging geo (-) politics of deep-sea mining. Geopolitics **25**(1), 189–213 (2020)
2. Sakellariadou, F., Gonzalez, F.J., Hein, J.R., Rincón-Tomás, B., Arvanitidis, N., Kuhn, T.: Seabed mining and blue growth: exploring the potential of marine mineral deposits as a sustainable source of rare earth elements (MaREEs)(IUPAC Technical Report). Pure Appl. Chem. **94**(3), 329–351 (2022)
3. Heinrich, L.: Interdisciplinary assessment of deep-sea mining impacts: from science to policy and implications for sustainable development (Doctoral dissertation, Jacobs University Bremen) (2021)
4. Christiansen, B., Denda, A., Christiansen, S.: Potential effects of deep seabed mining on pelagic and benthopelagic biota. Mar. Policy **114**, 103442 (2020)
5. Cordes, E.E., et al.: Environmental impacts of the deep-water oil and gas industry: a review to guide management strategies. Front. Environ. Sci. **4**, 58 (2016)
6. TMC: The Metals Company, NORI and Allseas Lift Over 3,000 Tonnes of Polymetallic Nodules to Surface from Planet's Largest Deposit of Battery Metals, as Leading Scientists and Marine Experts Continue Gathering Environmental Data, reviewed on April 29th, 2024 (2024). https://investors.metals.co/news-releases/news-release-details/nori-and-allseas-lift-over-3000-tonnes-polymetallic-nodules
7. Christiansen, S., Bräger, S., Jaeckel, A.: Evaluating the quality of environmental baselines for deep seabed mining. Front. Mar. Sci. **9**, 898711 (2022)
8. Clark, M.R., Durden, J.M., Christiansen, S.: Environmental Impact Assessments for deep-sea mining: Can we improve their future effectiveness? Marine Policy, **114** (2020). 10.1016/j.marpol.2018.11.026
9. Gann, G.D., et al.: International principles and standards for the practice of ecological restoration. Restor. Ecol. **27**(S1), S1–S46 (2019)
10. Mostafa, M.T., et al.: Geochemical baseline and pre-mining environmental assessment of heavy metals at iron exploration area, Northeastern Aswan. Egypt. Water, Air Soil Pollut. **234**(7), 456 (2023)
11. Cormier, R.: Ecosystem approach for the management of deep-sea mining activities. In: Environmental issues of deep-sea mining: impacts, consequences and policy perspectives, pp.381–402 (2019)
12. Shih, Y.C.: Ocean governance in practice: a study of the application of marine science and technology research techniques to maritime law enforcement in Taiwan. Mar. Policy **163**, 106081 (2024)

13. Wölfl, A.C., et al.: Seafloor mapping–the challenge of a truly global ocean bathymetry. Front. Mar. Sci. **6**, 434383 (2019)
14. Pillay, T., Cawthra, H.C., Lombard, A.T.: Characterisation of seafloor substrate using advanced processing of multibeam bathymetry, backscatter, and sidescan sonar in Table Bay. South Africa. Marine Geol. **429**, 106332 (2020)
15. Haalboom, S., et al.: Monitoring of a sediment plume produced by a deep-sea mining test in shallow water, Málaga Bight, Alboran Sea (southwestern Mediterranean Sea). Mar. Geol. **456**, 106971 (2023)
16. Washburn, T.W., Turner, P.J., Durden, J.M., Jones, D.O., Weaver, P., Van Dover, C.L.: Ecological risk assessment for deep-sea mining. Ocean Coast. Manag. **176**, 24–39 (2019)
17. Tekman, M.B., et al.: Tying up loose ends of microplastic pollution in the Arctic: distribution from the sea surface through the water column to deep-sea sediments at the HAUSGARTEN observatory. Environ. Sci. Technol. **54**(7), 4079–4090 (2020)
18. Jones, D.O., et al.: Biological responses to disturbance from simulated deep-sea polymetallic nodule mining. PLoS ONE **12**(2), e0171750 (2017)
19. Ware, S., Downie, A.L.: Challenges of habitat mapping to inform marine protected area (MPA) designation and monitoring: an operational perspective. Mar. Policy **111**, 103717 (2020)
20. Li, X., et al.: Deep-learning-based information mining from ocean remote-sensing imagery. Natl. Sci. Rev. **7**(10), 1584–1605 (2020)
21. Sharma, R.: Development of environmental management plan for deep-sea mining. In: Deep-Sea Mining: Resource Potential, Technical and Environmental Considerations, pp.483–506 (2017)
22. Fukushima, T., Tsune, A.: Long-term monitoring of environmental conditions of benthic impact experiment. In: Environmental Issues of Deep-Sea Mining: Impacts, Consequences and Policy Perspectives, pp.191–211 (2019)
23. Sharma, R.: Indian deep-sea environment experiment (INDEX): an appraisal. Deep Sea Res. Part II **48**(16), 3295–3307 (2001)
24. Sharma, R., Nath, B.N.: Selection of test and reference areas for the Indian Deepsea environment experiment (INDEX). Mar. Georesour. Geotechnol. **18**(3), 177–187 (2000)
25. Phillips, H.: A sea of voices: deep sea mining and the Solwara 1 Project in Papua New Guinea (Doctoral dissertation, The University of Waikato) (2019)

Optimization Analysis of Structural Stiffness of Flexible Riser for Deep-Sea Mining Based on SBP-BNSGA-II Model

Haibo Sun[1], Yingying Wang[1](✉), Wujun Tong[1,2], Yufang Li[1], Zeqing Lin[1], and Zhen Wang[1]

[1] College of Safety and Ocean Engineering, China University of Petroleum (Beijing), Beijing 102249, China
wyy@cup.edu.cn

[2] CNOOC Ener Tech Drilling & Production Co., Ltd., Haikou 518067, China

Abstract. Flexible riser is a key equipment for ensuring the continuous transportation of deep-sea mineral resources. Composite materials are more suitable for flexible risers due to their advantages such as high strength and weight reduction, and their structural design parameters directly affect their load-bearing performance. This paper takes the structural stiffness of composite flexible risers as the research object, and explores the influence of structural design parameters on the structural stiffness. A structural stiffness prediction model for flexible risers is established based on multi-layer artificial neural networks. A SBP-BNSGA-II model is proposed to obtain the Pareto frontier solution set for optimization objectives. Provide important reference for the structural design of composite flexible risers.

Keywords: Deep-sea Mining · Structural Stiffness · Artificial Neural Networks · Multi-objective Optimization

1 Introduction

There are abundant mineral resources in the deep-sea, which have great development value. Pipeline is a key equipment for continuous transportation of deep-sea minerals. Traditional steel risers and metal reinforced flexible risers are limited in their wear resistance and empty weight, making them difficult to apply in the development of deep-sea mineral resources. In recent years, composite materials have been widely used in marine equipment due to their high strength and weight reduction advantages. The load-bearing performance of composite materials is affected by the angle and thickness of fiber winding. Therefore, it is crucial to optimize the structural design parameters of composite flexible risers.

Many scholars have conducted relevant research on the optimization of the load-bearing performance of flexible risers. Roy et al. [1] established an optimization model for fiber wound pipelines using both micromechanics and macromechanics methods, and conducted optimization analysis on design parameters and material utilization. Tabakov

D. Jeng and B. Cai (Eds.): ICCOE 2024, LNCE 423, pp. 301–312, 2025.
https://doi.org/10.1007/978-981-96-0181-3_28

et al. [2] established a control algebraic equation system for multi-layer pressure vessels using the stress function method, determined the radial stress, axial stress, and shear stress under plane stress state, and optimized the relevant parameters using genetic algorithm. Farshi et al. [3] focused on the natural frequency of laminated plates, considered all types of boundary conditions, and optimized the minimum thickness of laminated plates. Apalak et al. [4] optimized the stacking sequence by taking the first-order natural frequency of laminated plates as the objective function. In addition, to shorten the search time, the optimal stacking sequence was predicted based on an artificial neural network model. Naik et al. [5] proposed a new failure criterion based on maximum stress and Tsai-Wu failure criterion, and used genetic algorithm to optimize the minimum weight of laminated panels. Xing et al. [6] optimized the fiber winding angle using an improved optimization algorithm with the strength ratio coefficient as the optimization objective. Tabakov and Summers [2] optimized the fiber winding angle based on Tsai-Wu failure criterion and genetic algorithm.

The optimization research on the load-bearing performance of the flexible riser mentioned above mainly focuses on its strength, natural frequency, material utilization, and other aspects. The prediction of the structural stiffness of flexible risers is a crucial step in the early stages of structural design. This paper takes the structural stiffness of composite flexible risers as the optimization objective. Firstly, based on the finite element method (FEM), the influence of thickness and fiber winding angle on the structural stiffness of flexible risers is explored. Secondly, a predictive model for the structural stiffness of flexible risers is established. Finally, multi-objective optimization is performed on the structural design parameters to obtain the optimal solution set under constraint conditions.

2 Analysis of the Load-Bearing Performance for Flexible Risers

2.1 Establishment of Numerical Model

This paper takes glass fiber reinforced flexible risers as the research object. Flexible risers are divided into inner liner, reinforcement layer, and outer sheath. The structural and Material parameters of the flexible riser are shown in Table 1.

Table 1. Structure and Material Parameters of Flexible Risers.

Layers name	Internal Radius (mm)	External radius (mm)	Thickness (mm)	Material	Elastic modulus (MPa)	Poisson's ratio
Inner liner	26.0	29.5	3.5	UHMW-PE	536	0.41
Reinforcement layer	29.5	35.5	6.0	glass fibre	49600	0.27
Outer sheath	35.5	37.5	2.0	Polyurethane	41.4	0.49

A numerical model of composite flexible riser was established based on the FEM, as shown in Fig. 1. The effective calculation length of the model is set to 1300mm. The length is greater than twice the pitch of the spiral strip of the reinforcement layer. The mesh types of the inner liner and outer sheath are set as C3D8R solid elements. The mesh type of the enhancement layer is set to SC8R continuous shell elements. The number of grid elements in the inner liner, reinforcement layer, and outer sheath is 13600, 15200, and 15200 respectively. The equivalent stacking method of composite materials was adopted, which can consider the angle and thickness of fiber winding. Set the elastic parameters of the material based on the attribute "Lamina". The two adjacent layers are unbonded. Therefore, considering the complexity of contact, this model adopts the "All with itself" algorithm in General Contact. Set the normal contact attribute to "Hard contact" and the friction coefficient to 0.2. Establish reference points RP-1 and RP-2 at the centers of the two end faces of the flexible riser. Two end faces are coupled to the reference point separately. Apply internal pressure to the surface of the inner liner; Axial tension is applied to the RP-1.

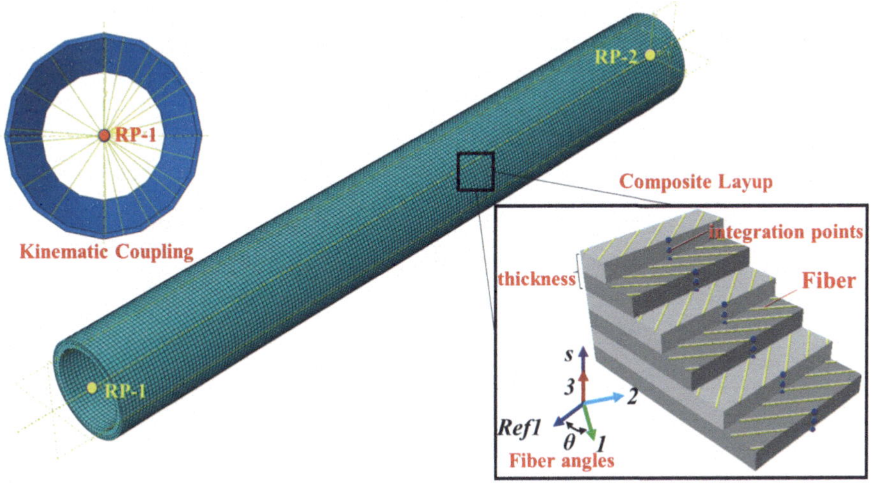

Fig. 1. Numerical model of flexible riser.

2.2 Analysis of Structural Design Parameters

The reinforcement layer plays a major load-bearing role in flexible risers, therefore, it is necessary to consider both axial and radial stiffness comprehensively. The structural design parameters of the reinforcement layer mainly include material, thickness, fiber winding angle, etc. The radial displacement and axial strain of the reinforcement layer vary with the fiber winding angle as shown in Fig. 2. As the angle of fiber winding increases, the radial displacement of the reinforcement layer decreases and the axial strain of the reinforcement layer increases.

The radial and axial stiffness of the reinforcement layer varies with the fiber winding angle as shown in Fig. 3. As the angle of fiber winding increases, the radial stiffness of the reinforcement layer increases and the axial stiffness of the reinforcement layer decreases.

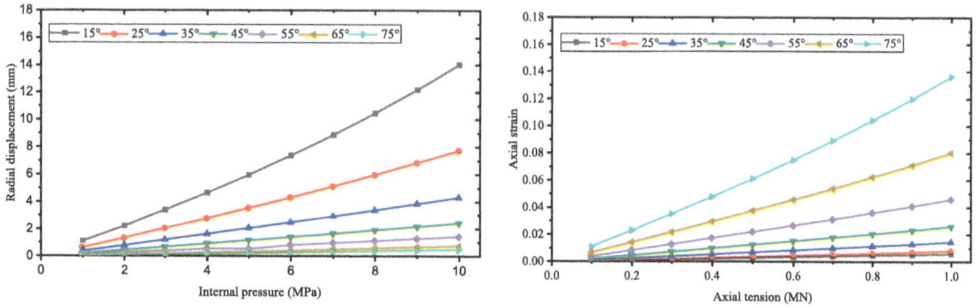

Fig. 2. The radial displacement and axial strain of the reinforcement layer.

There is a contradiction between radial stiffness and axial stiffness. Therefore, it is necessary to comprehensively consider the fiber winding angle to ensure both radial and axial stiffness. When the radial and axial stiffness do not meet the design requirements, it is necessary to further adjust the thickness of the reinforcement layer. As shown in Fig. 3, as the thickness of the reinforcement layer increases, the radial stiffness and axial stiffness increase. But an increase in thickness will suppress its bending performance.

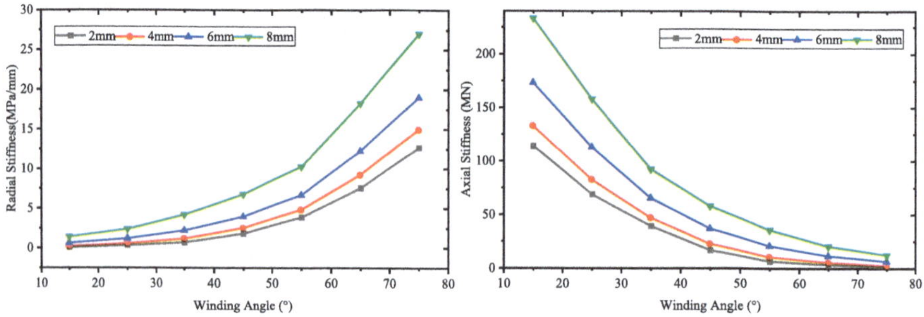

Fig. 3. The radial and axial stiffness of the reinforcement layer varies with thickness.

3 Construction of a Structural Stiffness Prediction Model

Using the winding angle and thickness as input parameters, the radial and axial stiffness of the flexible riser were initially obtained through FEM. Obtaining sample data using the FEM requires a significant amount of time, therefore, in order to obtain more sample data, a structural stiffness prediction model for flexible risers is established. As shown in Fig. 4, an artificial neural network is used to map the nonlinear relationship between input and output. The dataset is divided using random criteria. The number of Hidden Layers is determined using an automatic optimization algorithm. In order to improve the accuracy and generalization ability of the model prediction, an SBP model using the Scaled Conjugate Gradient algorithm to optimize the BP neural network is established. Mean Squared Error (MSE) is used as an evaluation metric for the regression model. The minimum MSE of the training objective is set to 0.0001. The maximum number of training sessions is set to 1000. In addition, dynamic learning rates were adopted.

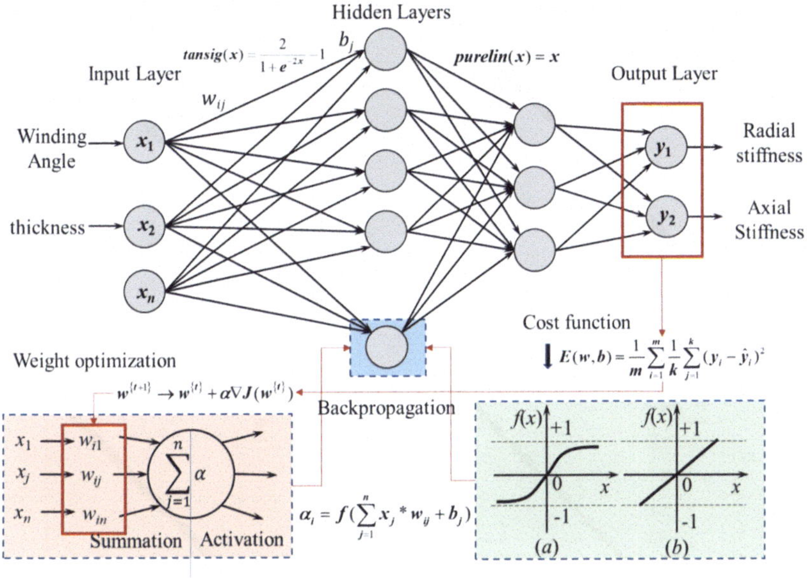

Fig. 4. Structural stiffness prediction model for flexible risers.

When the number of hidden layer nodes is 9, the MSE of the model is the smallest. As shown in Fig. 5, the training, validation, and testing errors of the model all steadily decrease. The SBP model was trained 127 times and the MSE reached 1.7719e-5.

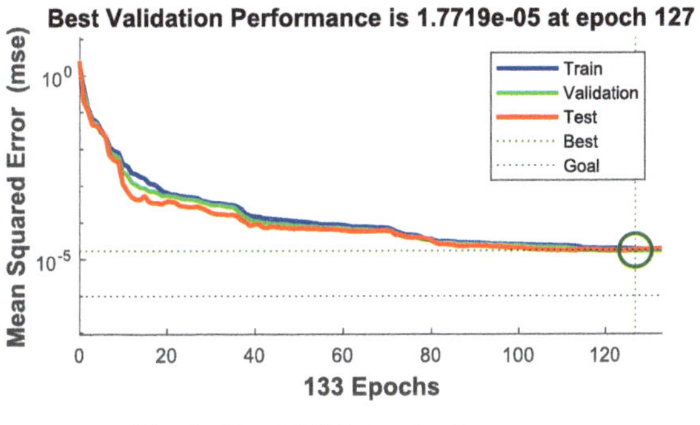

Fig. 5. Best Validation Performance.

The performance of the SBP model is shown in Fig. 6. The SBP model performs well in the training set, validation set, and test set. The correlation coefficient R between the predicted output and the target output in all datasets reached 0.99998, indicating excellent predictive performance of the SBP model.

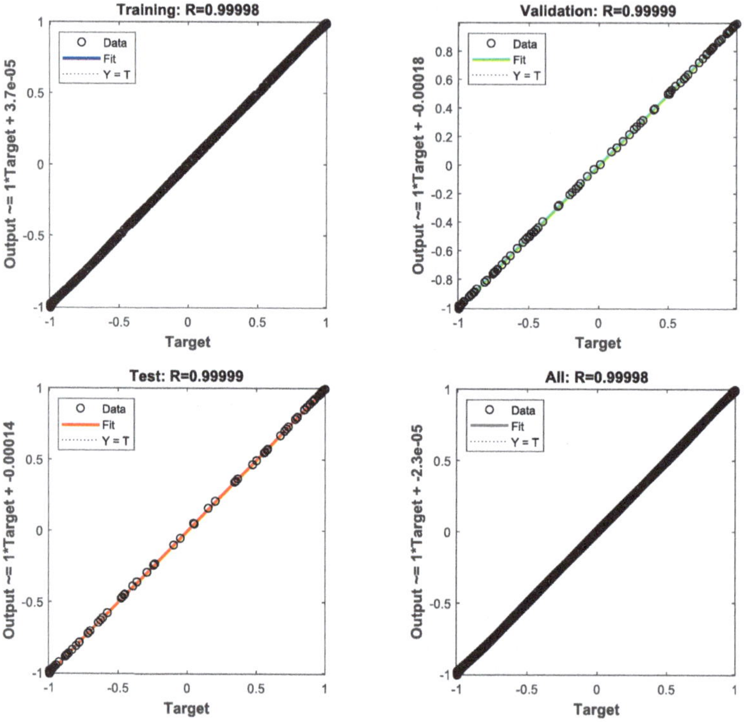

Fig. 6. Regression performance of the SBP model.

4 Construction of BP-BNSGA-II Model

This paper selects radial stiffness and axial stiffness as optimization objectives. Find the winding angle and thickness corresponding to the optimal structural stiffness. P_{design} is the design internal pressure. T_{design} is the design tension. t_1 and t_2 are the upper and lower limits of the thickness of the enhancement layer, respectively. α_1 and α_2 are the upper and lower limits of the winding angle of the reinforcement layer, respectively. The mathematical expression of the optimization model is as follows.

$$
\begin{gathered}
max(P/\triangle r, T/\triangle \varepsilon) \\
s.t.\begin{cases} P \leq P_{design} \\ T \leq T_{design} \\ t_1 \leq t \leq t_2 \\ \alpha_1 \leq \alpha \leq \alpha_2 \end{cases}
\end{gathered}
\tag{1}
$$

In order to solve the problem of poor quality and stability of the initial population in the traditional NSGA-II algorithm, a BNSGA-II algorithm based on the best point set is proposed. n represents the spatial dimension of the population. m represents the population size. $P_n(i)$ represents the sample set. The steps for initializing the population using the best point set method are as follows.

Step 1: Calculate the value of r. mi represents the i-th individual.

$$\begin{cases} r = (r_1, r_2, \ldots r_n) \\ r_j = mod(2cos(2\pi j/7)m_i, 1), \ 1 \leq j \leq n \end{cases} \quad (2)$$

Step 2: Construct the best point set.

$$P_n(i) = (r_1 i_1, r_2 i_2, r_3 i_3 .., r_n i_n), i = 1, 2, 3, \ldots n \quad (3)$$

Step 3: Finally, map $P_n(i)$ to the domain where the population is located. a_j and b_j are the lower and upper limits of the current dimension, respectively.

$$x_i^j = a_j + P_n(i)(b_j - a_j) \quad (4)$$

In addition, the population size is set to 200, the optimal individual coefficient is set to 0.3, the maximum number of iterations is set to 500, and the fitness function deviation is set to 1e-10. Using the best point set method to improve the uniformity and diversity of the initial population. Uniformly distributed points map the search space, making individuals more evenly distributed within the search space. The comparison of two methods for initializing populations is shown in Fig. 7. The random method makes the population initialization chaotic and disorderly, while the population distribution based on the best point set method is more uniform.

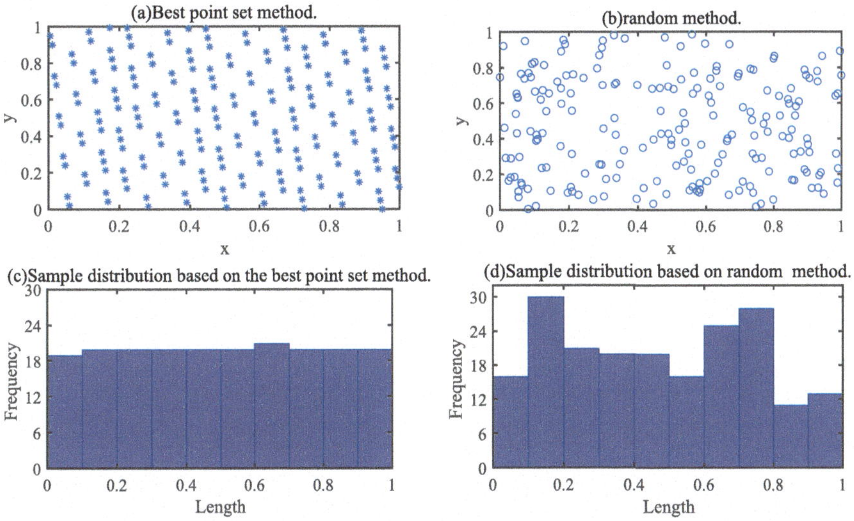

Fig. 7. Population distribution characteristics.

The multi-objective optimization process based on the SBP model is shown in Fig. 8.

(a) Determine the structural design parameter variables of the reinforcement layer. Construct an appropriate number of sample points based on the design parameter range to ensure their representativeness and uniformity. Obtain the radial and axial stiffness of the reinforcement layer based on the FEM.

(b) Based on the sample dataset obtained by the finite element method, the SBP model was constructed to predict the structural stiffness of the reinforcement layer. The reliability of the SBP model is evaluated by evaluation indicators.
(c) The SBP-BNSGA-II model was constructed to search for Pareto frontiers that minimize multiple targets simultaneously.

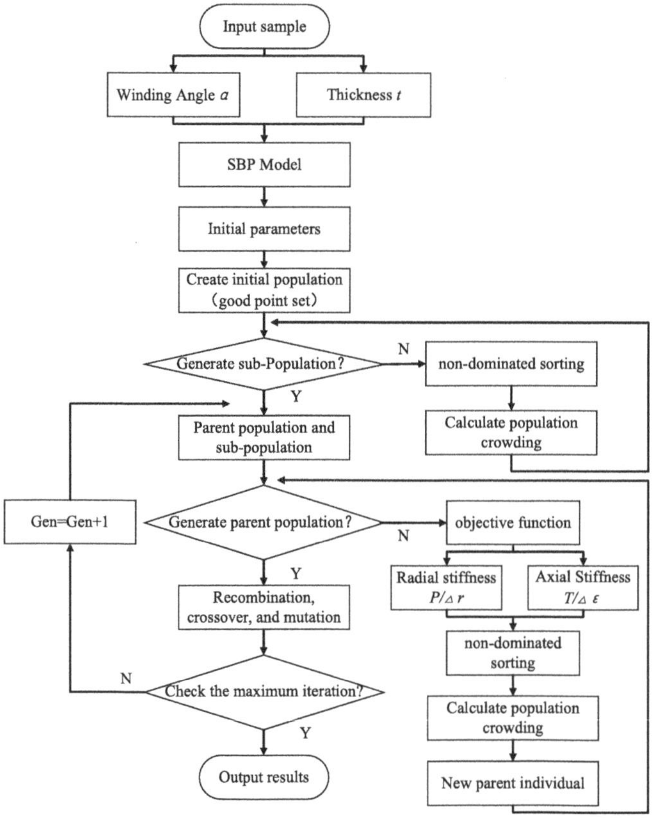

Fig. 8. The multi-objective optimization process of SBP-BNSGA-II model.

5 Results

5.1 SBP Model Validation

The radial stiffness predicted by the SBP model is shown in Fig. 9. The prediction results of the SBP model are in good agreement with the FEM calculation results, and this model can replace FEM to predict the radial stiffness of flexible risers.

To verify the superiority of the SBP model, several typical regression prediction models were compared. Firstly, a regression model combining wavelet packet decomposition and Support Vector Machine (SVM) was studied [7]. Secondly, traditional BP

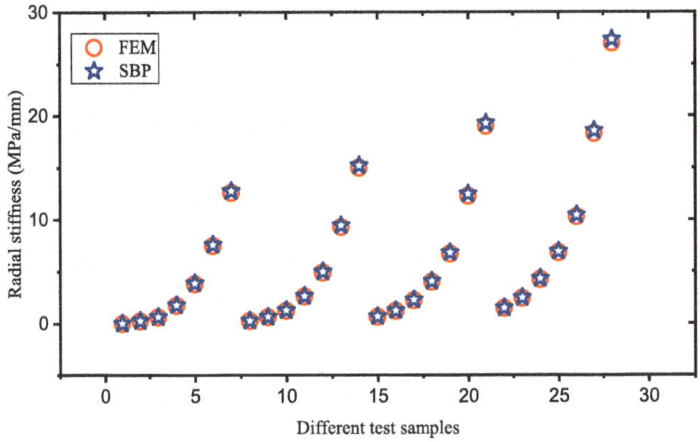

Fig. 9. Prediction results of radial stiffness of SBP model.

models based on the training function "trained" were compared. Thirdly, the M-BP regression model based on Markov chain extraction of data features was studied [8]. Fourthly, a K-Nearest Neighbor (KNN) regression model was compared by selecting K known samples that are closest to the unknown samples [9]. Finally, a decision tree (DT) model was studied by partitioning the feature space [10]. All test models use the same test data. To prevent overfitting in the SBP model, a new dataset was generated using FEM as the test data.

In order to evaluate the predictive performance of different regression models, four evaluation indicators were introduced to measure the degree of deviation between the predicted and true results, namely the sum of squared error (SSE), MSE, mean absolute error (MAE), and mean absolute percentage error (MAPE).

$$SSE = \sum_{i=1}^{m} (y_i - \hat{y}_i)^2 \tag{5}$$

$$MSE = \frac{SSE}{n} = \frac{1}{n}\sum_{i=1}^{m} (y_i - \hat{y}_i)^2 \tag{6}$$

$$MAE = \frac{1}{n}\sum_{i=1}^{n} |\hat{y}_i - y_i| \tag{7}$$

$$MAPE = \frac{1}{n}\sum_{i=1}^{n} \left|\frac{\hat{y}_i - y_i}{y_i}\right| \tag{8}$$

The prediction performance of different regression models is shown in Table 2. The traditional BP neural network has the worst prediction performance, while the improved BP neural network has significantly improved prediction accuracy. Among all evaluation indicators, the SBP model has the highest prediction accuracy and can be used as a proxy model for predicting the structural stiffness of flexible risers. In addition, the testing error of the new test data also indicates that the SBP model has extremely strong robustness.

Table 2. Regression performance of different models.

evaluation indicators	SBP	SVM	BP	M-BP	KNN	DT
SSE	0.4050	68.3850	151.7697	103.1041	136.2636	60.8953
MSE	0.0227	0.2953	0.4400	0.3626	0.4170	0.2787
MAE	0.0836	1.0814	1.5989	1.3158	1.5057	1.0125
MAPE	0.0132	0.1691	0.2505	0.2069	0.2345	0.1599

5.2 Global Optimization Based on the SBP-BNSGA-II Model

The nonlinear mapping relationship between structural stiffness and independent variables was obtained based on the SBP model. And use the SBP model as the objective function for multi-objective optimization. Perform global parameter optimization based on the SBP-BNSGA-II model within the given range of structural design parameter variables. A two-dimensional non-dominated Pareto front considering radial stiffness and axial tensile stiffness was obtained, as shown in Fig. 10. The radial stiffness of the reinforcement layer is inversely proportional to the axial stiffness.

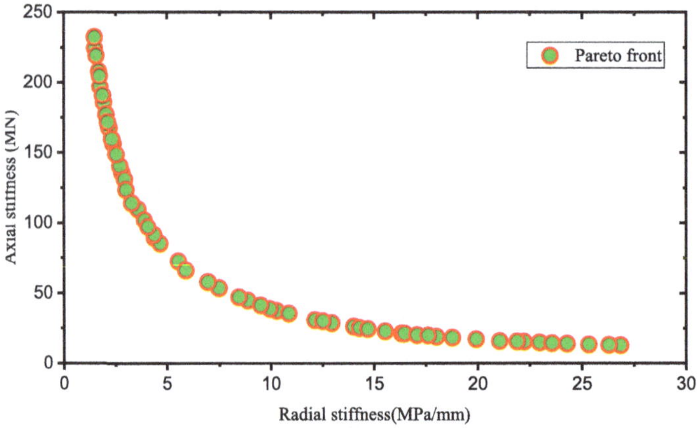

Fig. 10. Two-dimensional non-dominated Pareto front.

Based on the SBP-BNSGA-II model, 67 sets of global optimal solutions were obtained, providing a reasonable selection space for the structural design of flexible risers. When selecting structural design parameters, it is necessary to comprehensively consider the structural stiffness. Further optimal solutions can be obtained from the solution set based on the design load requirements.

6 Conclusions

This paper takes radial stiffness and axial stiffness as the optimization objectives for flexible risers. Find the winding angle and thickness corresponding to the optimal structural stiffness. A SBP-BNSGA-II model is proposed to optimize the structural design parameters of flexible risers. This paper can provide specific inputs for the design and dynamic analysis of flexible risers.

(1) The structural design parameters of the reinforcement layer are the fiber winding angle and thickness, which will significantly affect the load-bearing performance of the reinforcement layer. The radial stiffness and axial stiffness under different winding conditions are contradictory to each other.
(2) Based on the sample data obtained from FEM, an SBP model was established to predict the radial and axial stiffness of the reinforcement layer. The SBP model can effectively describe the nonlinear relationship between design parameters and optimization objectives, with a MSE of 1.7719e-5. In addition, the SBP model has the best predictive performance among different prediction models and exhibits excellent robustness.
(3) Based on the SBP-BNSGA-II model, multi-objective optimization is carried out on the structural design parameters of the reinforcement layer, and 67 sets of global optimal solutions are obtained. The optimal solution can be further obtained from the solution set based on design bias, design load requirements, etc.

Acknowledgement. The authors thank the support provided by: The National Key Research and Development Program of China: Design and Key Technology Research of Non-metallic Flexible Risers for Deep Sea Mining (2022YFC2803701) and The General Program of National Natural Science Foundation of China (52071336, 52374022).

References

1. Roy, A.K., Tsai, S.W.: Design of thick composite cylinders. J. Pressure Vessel Technol. **110**(3), 255–262 (1988)
2. Tabakov, P.Y., Summers, E.B.: Lay-up optimization of multilayered anisotropic cylinders based on a 3-D elasticity solution. Comput. Struct. **84**(5), 374–384 (2006)
3. Farshi, B., Rabiei, R.: Optimum design of composite laminates for frequency constraints. Compos. Struct. **81**(4), 587–597 (2007)
4. Apalak, M.K., Yildirim, M., Ekiei, R.: Layer optimization for maximum fundamental frequency of laminated composite plates. Compos. Sci. Technol. **68**(2), 537–550 (2008)
5. Naik, G.N., Gopalakrishnan, S., Ganguli, R.: Design optimization of composites using genetic algorithms and failure mechanism based failure criterion. Steel Construct. **83**(4), 354–367 (2008)
6. Xing, J.Z., Geng, P., Yang, T.: Stress and deformation of hybrid filament-wound thick cylinder vessel under internal and external pressure. Compos. Struct. **131**(1), 868–877 (2015)
7. Qu, Z., Feng, H., Zeng, Z.: A SVM-based pipeline leakage detection and pre-warning system. Measurement **43**(4), 513–519 (2010)
8. Liu, J., Zang, D., Liu, C., Ma, Y., Fu, M.: A leak detection method for oil pipeline based on markov feature and two-stage decision scheme. Measurement **138**, 433–445 (2019)

9. Arian, R., Hariri, A., Mehridehnavi, A., Fassihi, A., Ghasemi, F.: Protein kinase inhibitors' classification using K-Nearest neighbor algorithm. Comput. Biol. Chem. **86**, 1–7 (2020)
10. Sabah, M., Talebkeikhah, M., Agin, F., Agin, F., Talebkeikhah, F., Hasheminasab, E.: Application of decision tree, artificial neural networks, and adaptive neuro-fuzzy inference system on predicting lost circulation: A case study from Marun oil field. J. Petrol. Sci. Eng. **177**, 236–249 (2019)

Study on Mechanical Properties of Reinforced Thermoplastic Pipes Under Pure Bending

Chengwei Cui[1], Min Lou[1](✉), Yangyang Wang[1], Xiaoliang Qi[2], and Xiaolong Guo[1]

[1] China University of Petroleum (East China), Qingdao Shandong 266580, China
B23020048@s.upc.edi.cn, shidaloumin@163.com

[2] CNOOC Energy Technology and Services Limited, Binhai New Area, Tianjin 300451, China

Abstract. This Study focuses on the bending performance and cross-sectional deformation of reinforced thermoplastic pipes (RTPs) used in offshore oil and gas transmission. Based on the nonlinear cylindrical shell theory, the elliptical deformation of the RTPs' cross-section is parametrically analyzed, the geometric equations for the deformation of the flexible pipe are derived, and the principal structure equations of the flexible pipe are established considering the material nonlinearity and failure criterion of the composite materials. Further, the nonlinear equations of RTPs in equilibrium at any loading moment are established by combining the geometric equations, intrinsic constitutive equations and applying the principle of virtual work, and the equations are solved numerically by Newton's iterative method to complete the establishment of the three-dimensional theoretical analysis model of RTPs under pure bending. The model is validated through mechanical property tests. Buckling stress analysis, failure mode analysis, and stiffness degradation analysis are conducted to investigate the stress distribution, bending failure mechanism. The study identifies five failure modes under pure bending conditions and presents a four-stage degradation process of RTPs' stiffness. It contributes to understanding the behavior of RTPs under pure bending loads and provides insights for improving the design and reliability in offshore applications.

Keywords: Reinforced Thermoplastic Pipes · Pure Bending · Composites
Failure Behavior

1 Introduction

Reinforced Thermoplastic Pipe (RTP) is a type of bonded composite flexible pipe. It offers a promising application prospect due to their lighter weight and economic convenience compared to traditional metal flexible pipes [1–3]. RTP frequently experiences bending loads during production, installation, and service, primarily caused by factors like coiling and complex marine environments. The bending process of RTP can easily cause local buckling, potentially resulting in localized structural failure or complete pipeline replacement and leading to substantial economic losses. Therefore, it is necessary to study the mechanical properties under pure bending to guarantee the safe and economic operation of these RTPs.

D. Jeng and B. Cai (Eds.): ICCOE 2024, LNCE 423, pp. 313–325, 2025.
https://doi.org/10.1007/978-981-96-0181-3_29

The stress-strain state of reinforced thermoplastic pipes (RTPs) under bending loads is highly complex due to their unique structural characteristics and material properties. The cross-sectional mechanical behavior of RTPs differs significantly from that of common metal pipelines. Li et al. [4] introduced a four-layer theoretical model grounded in Donnel's theory to examine the mechanical behavior of steel wire-reinforced plastic composite pipes when subjected to pure bending. Bai et al. [5] extended Kyriakides' model [6, 7] by proposing a nonlinear ring theory model that incorporated variations in average radius, wall thickness, transverse shear deformation, and pre-buckling deformation. Wang [8] and Yu [9] developed a two-dimensional plane theoretical model utilizing the nonlinear ring theory, together with the J2 plastic flow theory and the principle of virtual work, for analyzing severe buckling challenges encountered by RTPs. Nevertheless, the study overlooked the impact of fiber strength on the bending resistance of the pipe, simplifying the reinforcement layer material, thus causing certain limitations. Yu et al. [10] investigated the mechanical response of RTPs to combine external pressure and bending through the utilization of ABAQUS software. Zhang et al. [11] investigated the influencing factors on the bending capacity of steel wire-reinforced plastic composite pipes under combined loading conditions of internal pressure and bending through experiments and finite element analysis. Szabó et al. [12] studied the overall buckling of RTPs under axial compression using a nonlinear finite element model, and analyzed the cross-sectional instability caused by bending. Sharma et al. [13] extended the hyperbolic shear deformation theory to theoretically model cylindrical laminated plates made of elastic orthotropic materials.

For RTPs reinforced glass fiber tape, it has a high anisotropy, while for the reinforced layer bonded by multi-layer glass fiber tape, the impact of damage is limited to a certain extent due to load sharing and restraint between adjacent glass fiber tape. As a result, the reinforced layer has suffered complex damage before reaching the ultimate strength and resulting in final failure [14]. In the process of damage failure prediction, the instantaneous damage theory is adopted to reduce the material properties, which leads to inaccurate calculation results and increases the difficulty of numerical computation convergence [15]. Therefore, Nagesh [16] first optimized the instantaneous damage degradation model and proposed a progressive damage model, which was better matched with the experiment. Geng [17] respectively used different failure criteria combined with progressive damage failure model to simulate the damage propagation of fiber reinforced composite laminates, which had a better agreement with the experiment and finite element simulation, fully demonstrating the rationality of the progressive failure model. Han [18] built a damage prediction model for fiber-reinforced resin laminates based on continuous damage mechanics, and used acoustic emission method to detect the damage, further verifying the progressive damage failure process of composite materials.

Currently, further research on the bending characteristics of RTPs remains to be explored. There are limited theoretical studies that simplify the model to a two-dimensional plane, which cannot accurately calculate the mechanical response in three-dimensional directions of the structure, let alone predict the damage of each layer of materials. This study proposes a three-dimensional theoretical analysis model based on nonlinear cylindrical shell theory. It accounts for the elliptical deformation of the cross-section, initial pipe defects during bending, material nonlinearity, and the failure criterion of the composite materials. The model will facilitate deeper research into the buckling stress, failure modes, and stiffness degradation of RTPs during the pure bending process.

2 Theoretical Model

Assumption Reinforced thermoplastic pipes (RTPs) consist of a tightly bonded multi-layer composite structure. Hence, this study assumes that RTPs as a whole exhibit deformation continuity. To simplify the analysis and calculation of the pipe section's structural characteristics, material properties, and elliptical deformation process, the following basic assumptions are made:

(1) The inner lining layer, outer protective layer, and matrix material HDPE in the reinforcing layer do not crack, while the glass fiber reinforcing material remains unbroken. Both materials possess uniform and continuous physical properties.
(2) The glass fiber reinforcing material is firmly bonded to the HDPE matrix material, with no separation at the interface. The layers are also tightly bonded to each other, preventing relative sliding at the interlayer interface. The RTPs, therefore, sustain displacement continuity.
(3) During the bending process, the cross-section of the pipe maintains flatness, undergoing only elliptical deformation and overall angular deviation. It consistently intersects with the longitudinal axis of the pipe and extends towards the curvature center.
(4) The bending-induced section deformation adheres to the standard elliptical deformation process. The transverse axis (x-axis) and longitudinal axis (y-axis) of the section remain stationary without displacement. Throughout the deformation process, the section preserves symmetry around both the x-axis and y-axis. Consequently, points *a* and *n* experience radial displacement exclusively, without any circumferential displacement.

2.1 Geometrical Equation

Geometric equations for the deformation of RTPs during the bending process are analyzed to derive the geometric strain equations at any position inside the pipeline. A circular arc element is selected for analysis of a single-layer bent section of RTPs. Let *r* be the radius of the cylindrical surface at any point on the single-layer pipe section of RTPs, and *R* be the radius of the neutral surface of the circular arc.

Based on the Von-Karman geometric nonlinear theory and the first-order shear deformation theory [19], the strain components ε_i at any point on a single-layered cylindrical section can be determined from the strain components ε_i^0 at the neutral surface. The specific expression is given as follows:

$$\varepsilon_i = \frac{1}{1+\frac{l}{R_i}}\varepsilon_i^0(\overline{u}_\theta, \overline{u}_r), (i = \theta, z) \tag{1}$$

where, l is the distance from any point to the neutral cylindrical surface; R_i is the principal curvature radius of each surface; $\overline{u}_\theta$ is the circumferential displacement of any point; $\overline{u}_r$ is the radial displacement of any point.

According to Fig. 1, the circumferential principal curvature radius $R_\theta = R$, and the axial principal curvature radius $R_z = R_w + Rsin\theta$. The values of the strain components on the neutral surface of the circular arc are substituted into Eq. (1) to obtain the circumferential strain, axial strain, and axial shear strain components. In addition, there is a circumferential strain component that is solely caused by bending deformation and is not influenced by cross-sectional ellipticity deformation, which can be calculated based on the Poisson's ratio of each layer material and axial strain; As the magnitude of the radial strain component is extremely small, this study does not perform geometric analysis or derive relevant expressions for it. Similarly. Considering the aforementioned analysis, the expressions for the directional strain components at any point along the pipeline are given below:

$$\begin{cases} \varepsilon_\theta = \left[\frac{\partial \overline{u}_\theta}{R\partial\theta} - \frac{\overline{u}_{\mathrm{r}}}{R} + \frac{1}{2R^2}\left(\frac{\partial \overline{u}_\theta}{\partial\theta} - \overline{u}_{\mathrm{r}}\right)^2 + \frac{1}{2R^2}\left(\frac{\partial \overline{u}_{\mathrm{r}}}{\partial\theta}\right)^2\right]\frac{R}{R+(r-R)} + \varepsilon_{\theta 0} \\ \varepsilon_{\mathrm{z}} = \left[\varepsilon_{\mathrm{z}0} + \frac{(R-\overline{u}_{\mathrm{r}})\sin\theta + \overline{u}_\theta\cos\theta}{R_{\mathrm{w}}}\right]\frac{R_{\mathrm{w}} + R\sin\theta}{R_{\mathrm{w}} + R\sin\theta + (r-R)} \\ \varepsilon_{\mathrm{r}} = \varepsilon_{\mathrm{r}0} \\ \gamma_{\mathrm{r}\theta} = \frac{\partial \overline{u}_{\mathrm{r}}/\partial\theta}{R + \partial \overline{u}_\theta/\partial\theta - \overline{u}_{\mathrm{r}}}\frac{R}{R+(r-R)} \end{cases} \quad (2)$$

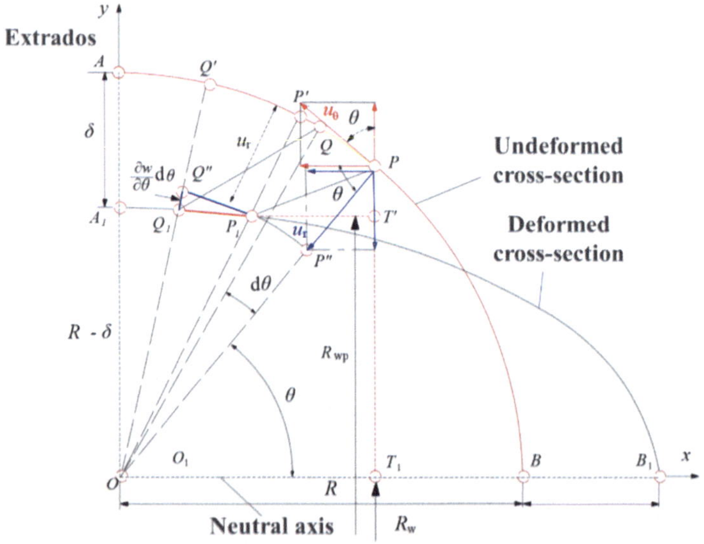

Fig. 1. Geometrical analysis of ovalization of RTPs cross-section

The displacement ($\overline{u}_\theta$ and $\overline{u}_r$) at any point along a pipeline can be derived based on the displacement at the neutral axis. The expressions are as follows:

$$\begin{cases} \overline{u}_\theta = u_\theta + (r - R)\left(\frac{u_\theta}{R} + \frac{\partial u_r}{R\partial\theta}\right) \\ \overline{u}_r = u_r \end{cases} \tag{3}$$

To satisfy the displacement boundary conditions, at $\theta = 0$, π/2, π, and 3π/2, $u_\theta = \partial u/\partial\theta = 0$. Additionally, rigid body motion of RTPs during plastic bending is excluded. The radial displacement u_r and tangential displacement u_θ are discretized using a trigonometric series, resulting in the Eq. 4 for the displacement function:

$$\begin{cases} u_r = R\sum_{n=1}^{N} a_n \cos(2n\theta) \\ u_\theta = R\sum_{n=1}^{N} b_n \sin(2n\theta) \end{cases} \tag{4}$$

In the actual production process, the pipeline cross-section may exhibit a certain level of initial ovality denoted as d. This initial ovality leads to the presence of certain initial displacements u_{r0}^0 and $u_{\theta 0}^0$ in the pipeline before it even bears any load. The expressions are as follows:

$$\begin{array}{c} u_{r0}^0 = -rd\cos 2\theta \; u_{\theta 0}^0 = r\frac{d}{2}\sin 2\theta \\ d = \frac{D_{max} - D_{min}}{D_{max} + D_{min}} \end{array} \tag{5}$$

The initial displacement represents the initial state of the pipeline and does not induce stress and strain within the pipeline. Hence, in calculations, the strain ε_θ^0 and $\gamma_{r\theta}^0$ caused by the initial displacement should be subtracted. The expressions are as follows:

$$\begin{cases} \varepsilon_\theta^0 = \dfrac{\left(5rd\cos 2\theta\frac{r}{R} - 3rd\cos 2\theta\right)^2}{2rR} \\ +\dfrac{2rd\cos 2\theta}{r} + \dfrac{2r^2d^2\sin^2 2\theta}{rR} + \dfrac{5rd\cos 2\theta(r - R)}{rR} \\ \gamma_{r\theta}^0 = \dfrac{2rd\sin 2\theta}{\frac{r}{R}\left[R + 2rd\cos 2\theta - 5rd\cos 2\theta\left(1 - \frac{r}{R}\right)\right]} \end{cases} \tag{6}$$

2.2 Constitutive Equation

The materials used for the inner liner and outer protective layer are HDPE, which is an isotropic material. The reinforcing layer is made of glass fiber-reinforced composite, which is an orthotropic material. The constitutive relationships, namely the stress-strain relationships, of these materials can be described using Hooke's law.

Due to the presence of a certain winding angle in the reinforced thermoplastic pipes (RTPs), there is a certain angular deviation between the material's orthogonal direction and the pipeline's axial direction. Therefore, it is necessary to perform a certain conversion between the material's orthogonal coordinate system and the pipeline's cylindrical coordinate system. For more details, please refer to the work by Wang et al. [20].

2.3 Failure Criterion

Based on the Hashin-Yeh failure criterion, considering the three-dimensional stress state and incorporating a laminated failure factor to predict the failure modes of the reinforced glass fiber strip. Furthermore, considering that the Von Mises stress criterion is widely used for failure prediction of isotropic materials, it is applied to evaluate the yield failure of the inner liner and outer protective layer. For more details, please refer to the work by Wang et al. [20].

2.4 Equation of Virtual Work and Solution

The principle of virtual work is adopted to construct the equilibrium equation. The virtual strain energy of the unit length flexible pipe is calculated, Due to the pre-set curvature during the loading process, the virtual work done by external forces $\delta W = 0$, thus the virtual strain energy of the RTPs structure $\delta U_m = 0$. By employing the Rayleigh-Ritz method and taking partial derivatives with respect to the unknown variables, the following system of equations is obtained.

$$
\begin{aligned}
\frac{\partial U_{\mathrm{m}}}{\partial a_n} &= \sum_f \int_0^{2\pi} \int_{r_1^k}^{r_2^k} \left(\hat{\sigma}_{\mathrm{z}} \frac{\partial \dot{\varepsilon}_{\mathrm{z}}}{\partial a_n} + \hat{\sigma}_\theta \frac{\partial \dot{\varepsilon}_\theta}{\partial a_n} + \hat{\sigma}_{\mathrm{r}} \frac{\partial \dot{\varepsilon}_{\mathrm{r}}}{\partial a_n} + 2\hat{\tau}_{\mathrm{r}\,\theta} \frac{\partial \dot{\gamma}_{\mathrm{r}\theta}}{\partial a_n} \right) r dr d\theta = 0 \\
\frac{\partial U_{\mathrm{m}}}{\partial b_n} &= \sum_f \int_0^{2\pi} \int_{r_1^f}^{r_2^k} \left(\hat{\sigma}_{\mathrm{z}} \frac{\partial \dot{\varepsilon}_{\mathrm{z}}}{\partial b_n} + \hat{\sigma}_\theta \frac{\partial \dot{\varepsilon}_\theta}{\partial b_n} + \hat{\sigma}_{\mathrm{r}} \frac{\partial \dot{\varepsilon}_{\mathrm{r}}}{\partial b_n} + 2\hat{\tau}_{\mathrm{r}\theta} \frac{\partial \dot{\gamma}_{\mathrm{r}\theta}}{\partial b_n} \right) r dr d\theta = 0 \\
\frac{\partial U_{\mathrm{m}}}{\partial \varepsilon_{\mathrm{z}0}} &= \sum_f \int_0^{2\pi} \int_{r_1^k}^{r_2^k} \left(\hat{\sigma}_{\mathrm{z}} \frac{\partial \dot{\varepsilon}_{\mathrm{z}}}{\partial \varepsilon_{\mathrm{z}0}} + \hat{\sigma}_\theta \frac{\partial \dot{\varepsilon}_\theta}{\partial \varepsilon_{\mathrm{z}0}} + \hat{\sigma}_{\mathrm{r}} \frac{\partial \dot{\varepsilon}_{\mathrm{r}}}{\partial \varepsilon_{\mathrm{z}0}} + 2\hat{\tau}_{\mathrm{r}\theta} \frac{\partial \dot{\gamma}_{\mathrm{r}\theta}}{\partial \varepsilon_{\mathrm{z}0}} \right) r dr d\theta = 0
\end{aligned}
\tag{7}
$$

where, $\hat{\sigma}_i (i = z, \theta, r, r\theta)$ represents the stress components in various directions in the current analysis step. $\dot{\varepsilon}_i (i = z, \theta, r, r\theta)$ denotes the incremental strains in different directions between the current and previous analysis steps. k indicates the number of layers in the flexible pipe.

The theoretical model is solved using the aforementioned methods and implemented through MATLAB mathematical software. The computational flow of the model is illustrated in Fig. 2.

2.5 Validation of Theoretical Model

A four-point bending test on RTPs was conducted based on the ASTM 6272–17 [21] standard. The experimental device is illustrated in Fig. 3 The total length of each specimen is 1800 mm, and the winding angle of the reinforcement layer is ± 55°. The specimens are divided into three sizes based on the number of layers in the reinforcement layer: 4 layers, 8 layers, and 14 layers. Among them, the 4-layer specimen has an external protective layer, while the rest of the specimens do not. The specific structural parameters are presented in Table 1. The material parameters are listed in Table 2.

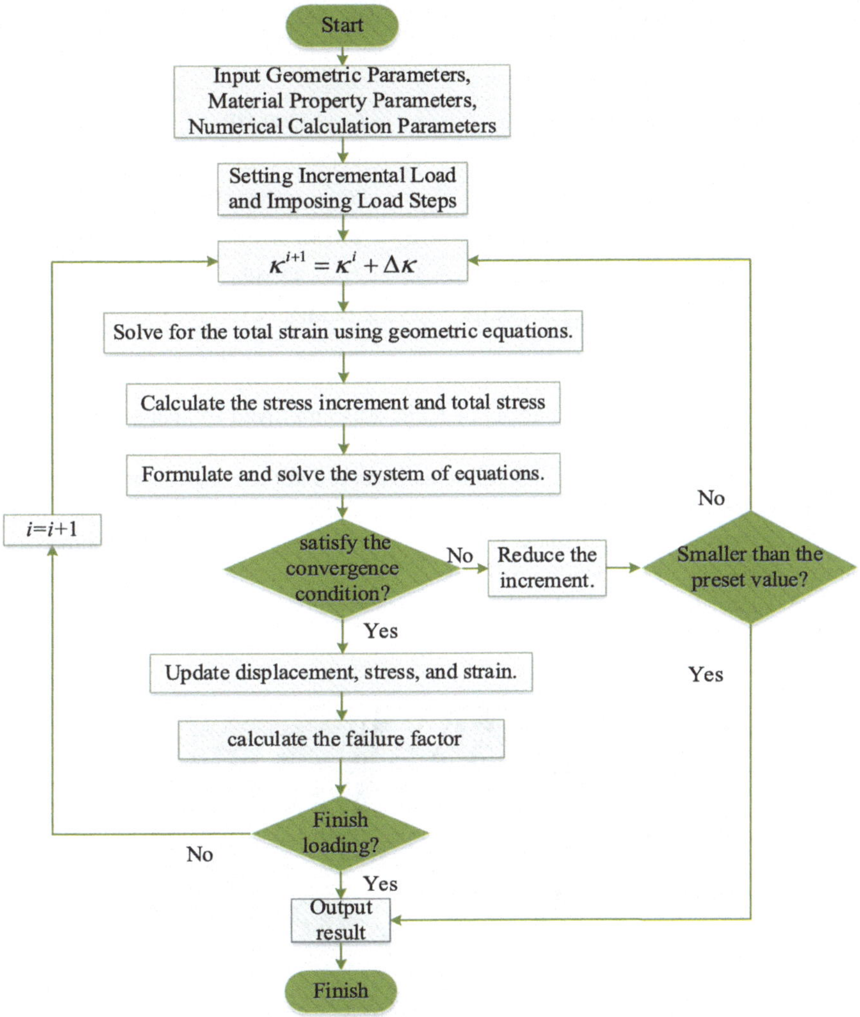

Fig. 2. Computational Flowchart of the Theoretical Model.

Table 1. The parameters of specimen pipe structure

Number of Layers	Pipe Length	Outer Protective Layer Thickness	Inner Lining Layer Thickness	Inner Diameter	Outer Diameter
4	1800 mm	2 mm	5 mm	48 mm	64 mm
8	1800 mm	0 mm	5 mm	50 mm	64 mm
14	1800 mm	0 mm	5 mm	50 mm	67 mm

Table 2. The parameters of sample pipe materials

	Reinforced Pipe		HDPE
Elastic Constants	$E_1 = 28000$ MPa	$E_2 = E_3 = 3200$ MPa	
	$\nu_{23} = 0.3$	$\nu_{12} = \nu_{13} = 0.0343$	$E = 1200$ MPa
	$G_{23} = 1230$ MPa	$G_{12} = G_{13} = 2700$ MPa	$\nu = 0.38$
Intensity Parameter	$X_T = 829$ MPa	$X_C = 146$ MPa	$\sigma_s = 19$ MPa
	$Y_T = Z_T = 25$ MPa	$Y_C = Z_C = 49$ MPa	
	$X_{12} = X_{13} = 18$ MPa	$X_{23} = 18$ MPa	

The bending performance analysis of RTPs was employed to analyze three types of pipelines listed in Table 1. The theoretical model calculated the bending stiffness of these pipes by applying the same curvature as observed in the experimental process. A comparison between the calculated results from the theoretical model and the experimentally measured bending stiffness is presented in Table 3. Based on this comparison, it can be observed that the bending stiffness calculated by the theoretical model has an error within 5% when compared to the experimental values. The average error is found to be 3.77%, with a maximum error of 4.22%.

Fig. 3. Schematic diagram of experimental device and sample pipe deformation.

Table 3. The bending stiffness of three types of RTPs.

Number of Layers	Theoretical Model	Experiment	Error
4	1072.25 N·m^2	1115.33 N·m^2	3.06%
8	1452.58 N·m^2	1513.52 N·m^2	4.03%
14	2351.70 N·m^2	2455.32 N·m^2	4.22%

3 Results and Discussion

3.1 Buckling Stress Analysis

Stress analysis is conducted on the working condition of the pipe under the ultimate bending moment, using a four-layer reinforced structure as an example (with parameters shown in Tables 1 and 2). The distribution of various stresses along the circumferential direction of the RTPs is calculated separately. As depicted in Fig. 4, the reinforced layer bears higher axial, radial, and shear stresses, indicating its primary load-bearing role. Due to the phenomenon of cross-sectional ovalization during bending, the stress distribution at 90° and 270° sections of the pipe exhibits varying degrees of convexity or concavity. The circumferential stress shows the highest degree of concavity or convexity, suggesting that it is primarily controlled by cross-sectional ovalization. Additionally, the curves depicting the distribution of axial, radial, and shear stresses show relatively small degrees of convexity or concavity, indicating that these three types of stresses are mainly governed by the upward and downward deformations induced by bending.

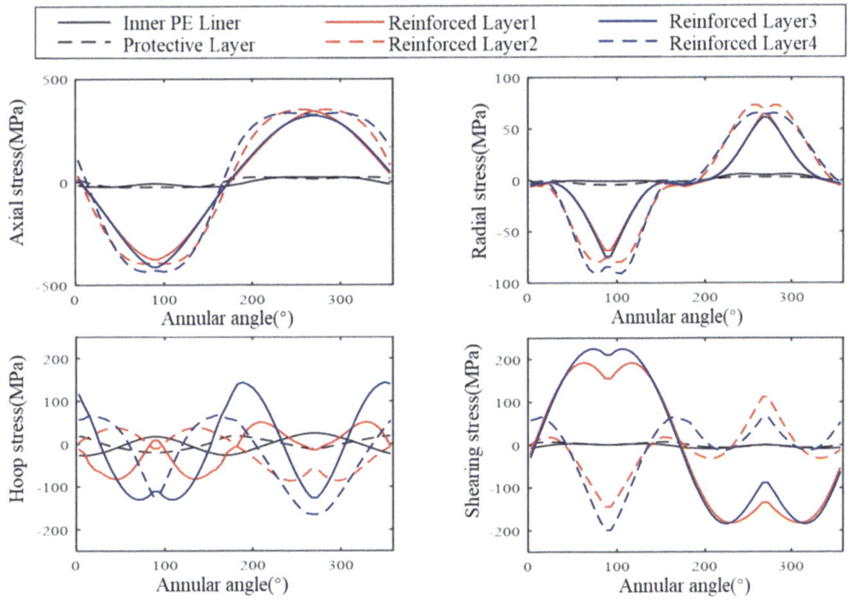

Fig. 4. Stress distributions when buckling occurs

3.2 Failure Mode Analysis

The stress distribution in each direction of the strengthening layers in the local coordinate system is analyzed when the bending moment is 500 Nm. The local coordinate system is established with the *x*-axis along the fiber winding direction. The specific calculation results are shown in Fig. 5.

According to Fig. 5, it can be observed that the stress levels in the fiber direction, perpendicular to the fiber direction, and in the fiber-matrix shear direction are significantly higher than the radial stress. This indicates that the radial stress contributes minimally to the damage and failure of the strengthening layers. Different failure modes are mainly controlled by the stress in the fiber direction, perpendicular to the fiber direction, and in the fiber-matrix shear direction. Although the stress perpendicular to the fiber direction is lower than the stress in the fiber direction, the difference between them is relatively small. However, the strength of the matrix material in the strengthening layers is considerably lower than that of the fiber-reinforced material, and the difference is significant. Therefore, the first failure mode to occur is the tensile failure of the matrix material in the direction perpendicular to the fibers, followed by the tensile failure of the fiber-reinforced material in the direction of fiber winding. The shear strength of the fiber-matrix interface is relatively close to the tensile strength of the fibers, and the stress level in the fiber-matrix shear direction is higher than that in the fiber direction. As a result, the matrix shear failure occurs before the fiber tensile failure.

Furthermore, from Fig. 5, it can be observed that when subjected to bending loads, the stress levels on the tension side and compression side are similar. According to the failure criterion, when the stress values are comparable, the calculated compression failure factor of the material is smaller than the tensile failure factor. Therefore, the material experiences compression failure in the end, following a similar pattern to tensile failure. Due to the significant difference in strength between the matrix material and the fiber-reinforced material, the matrix material undergoes compression failure before the fiber-reinforced material.

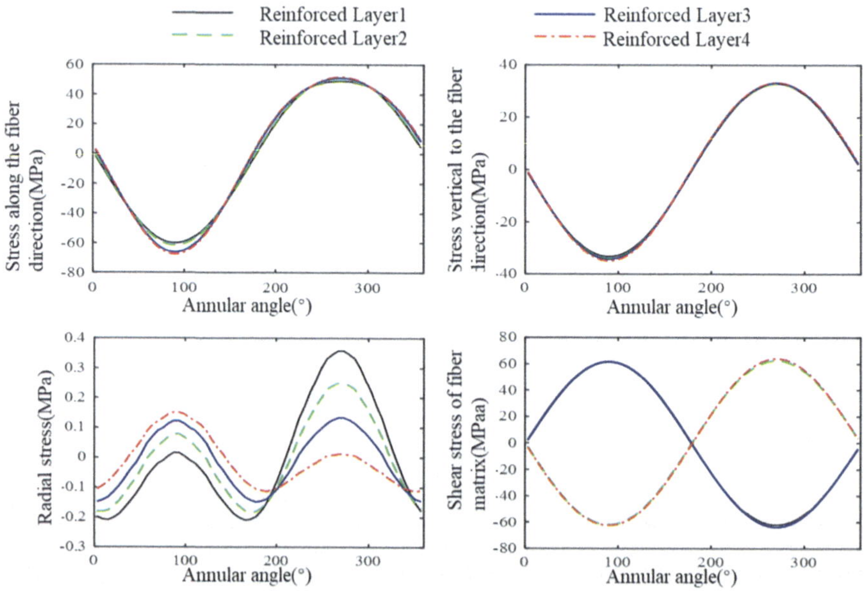

Fig. 5. Stress distribution in the local coordinate system of the reinforcement layer when the bending moment is 500 N·m

3.3 Stiffness Degradation Analysis

Taking a 4-layer reinforced RTP as an example (with parameters shown in Tables 1 and 2), the changes in bending moment, bending stiffness of the pipe were calculated. According to the calculation results from Figs. 6 and 7, it can be observed that when the curvature of the pipe is less than 0.5, it is the first stage. The decrease in the bending stiffness of the pipeline is very small. The reinforcement layer material has not yet failed during this stage, and its stiffness has not deteriorated. Additionally, the elliptical deformation of the cross-section is minimal, and the decrease in the moment of inertia is also very small. Therefore, the bending stiffness remains almost unchanged during this stage. When the curvature of the pipe ranges from 0.5 to 1.5, it enters the second stage. The bending stiffness of the pipe decreases rapidly. The reinforcement layer material gradually starts to fail as the pipe experiences increasing moments. In addition, during this stage, the elliptical deformation of the pipe cross-section starts to increase, and there is a slight decrease in the moment of inertia. Hence, the rapid decline in bending stiffness during this stage is mainly influenced by the degradation of the reinforcement layer material.

When the curvature of the pipe exceeds 1.5 until buckling occurs (the third stage), the bending stiffness of the pipe continues to decrease at a slightly slower rate compared to the previous stage. During this stage, the degradation of the reinforcing layer material stiffness is relatively small, while the cross-sectional ellipticity of the pipe rapidly increases and the sectional moment of inertia decreases. Therefore, the main factor affecting the decrease in bending stiffness during this stage is the deformation caused by cross-sectional ellipticity. After the pipe undergoes buckling (the fourth stage), the cross-sectional ellipticity increases rapidly, indicating that the pipe section collapses, resulting in complete structural instability.

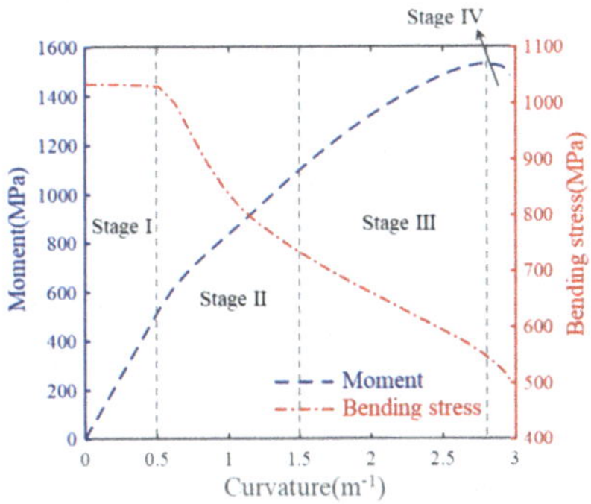

Fig. 6. Variation of bending moment and bending stiffness with curvature

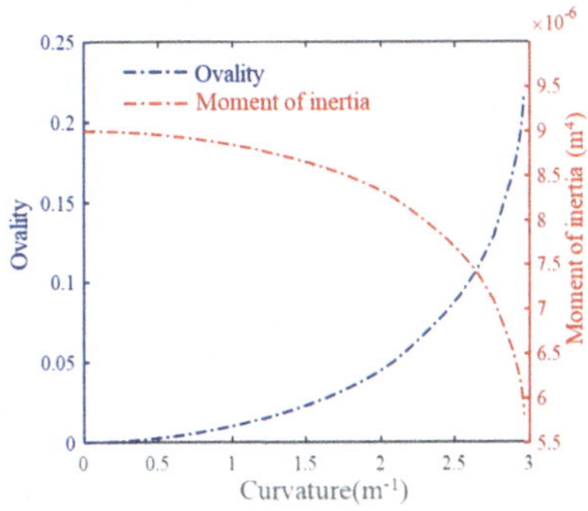

Fig. 7. Variation of ellipticity and moment of inertia with curvature

4 Conclusion

Accounted for the elliptical deformation of the cross-section, initial defects, material nonlinearity, and the failure criterion of the reinforcement layer, a three-dimensional theoretical analysis model for RTPs under pure bending load was established based on geometric equations, constitutive relationships, and the principle of virtual work. The accuracy of the theoretical model was verified by comparing it with four-point bending tests. Based on this, the buckling characteristics of RTPs were investigated, and the following conclusions were drawn:

(1) During the buckling failure process, the reinforcing layer is the main load-bearing structure of RTPs. Axial stress, radial stress, and shear stress are mainly controlled by the tensile-compressive deformation induced by bending, while circumferential stress is mainly controlled by section ovalization deformation. The internal lining layer and external protective layer also bear a portion of this deformation load.
(2) There are five failure modes under bending conditions, which occur in the following sequence: matrix tensile failure, fiber-matrix shear failure, fiber tensile failure, matrix compression failure, and fiber compression failure. The initial damage and failure of different failure modes occur at the outermost layer of the reinforcing layer.
(3) The overall stiffness degradation process of RTPs can be divided into four stages: the initial stage with minimal section ovalization deformation and almost unchanged bending stiffness when the material is intact; the rapid decrease in bending stiffness dominated by material failure; the continued decrease in bending stiffness dominated by section ovalization deformation after material failure is complete; and the stage of complete structural instability after pipe buckling occurs.

References

1. Bai, Y., Xu, W., Cheng, P., et al.: Behavior of reinforced thermoplastic pipe (RTP) under combined external pressure and tension. Ships Offshore Struct. **9**(4), 464–474 (2014)
2. Wang, Y.Y., Lou, M., Yang, L.Z., et al.: Study on the tensile properties of reinforced thermoplastic pipes under different internal pressures and temperatures. Int. J. Press. Vessels Pip. **200**, 104820 (2022)
3. Wang, Y.Y., Lou, M., Wang, Y.: Stochastic analysis of failure pressure in reinforced thermoplastic pipes under axial loading and external pressure. Ocean Eng. **240**, 109963 (2021)
4. Li, X., Zheng, J.Y., Shi, J.F., et al.: Experimental investigation on buckling of plastic pipe reinforced by winding steel wires under external pressure. J. Thermoplast. Compos. Mater. **23**(6), 827–843 (2010)
5. Bai, Y., Wang, N.S., Cheng, P.: Collapse of RTP (reinforced thermoplastic pipe) subjected to external pressure. In: International Conference on Pipelines and Trenchless Technology, Location of Conference, China, October 19–22 (2012)
6. Kyriakides, S., Ju, G.T.: Bifucration and localization instabilities in cylindrical shells under bending Part l: experiments. Int. J. Solids Struct. **29**(9), 1117–1142 (1992)
7. Ju, G.T., Kyriakides, S.: Bifucration and localization instabilities in cylindrical shells under bending Part 2: Predictions. Int. J. Solids Struct. **29**(9), 1143–1171 (1992)
8. Wang, N.S.: Investigation on Buckling and Collapse Mechanisms of Filament-wound Fiber-reinforced Composite Pipe under External Pressure and Combined Load. Master's Thesis, ZJU, Zhejiang, China (2013)

9. Yu, B.B.: Investigation on Nonlinear Buckling Mechanisms of Filament-wound Fiber-reinforced Composite Pipe during Installation. Master's Thesis, ZJU, Zhejiang, China (2013)
10. Yu, K., Morozov, E.V., Ashraf, M.A., et al.: Numerical analysis of the mechanical behaviour of reinforced thermoplastic pipes under combined external pressure and bending. Compos. Struct. **131**, 453–461 (2015)
11. Zhang, J.: Study on Mechanical Properties of Plastic pipe Reinforced by Cross Helically Wound Steel Wires under Combined Loads of Internal Pressure and Bending Load. Master's Thesis, ZJU, Zhejiang, China (2017)
12. Szabó G., Váradi K.: Instability analysis of a filament-wound composite tube subjected to compression/bending. Adv. Compos. Lett., 28 (2019)
13. Sharma, L.K., Grover, N., Purohit, A., et al.: A non-polynomial axiomatic framework for modelling and bending analysis of doubly curved spherical and cylindrical shells: an analytical solution. Proc. Institut. Mech. Eng., Part L: J. Mater.: Design Appl. **235**(9), 2083–2099 (2021)
14. Talreja, R., Waas, A.: Concepts and definitions related to mechanical behavior of fiber reinforced composite materials. Compos. Sci. Technol. **217**, 109081 (2022)
15. Wang, Y.Y.: Study on Mechanical Properties and Optimization Design of Deepwater Thermoplastic Composite Flexible Pipe. Doctor's Thesis, UPC, Qingdao, China (2022)
16. Nagesh: Finite-element Analysis of Composite Pressure Vessels with Progressive Degradation. Defence Sci. J. **53**, 75–86 (2003)
17. Geng, P.: The bearing capacity analysis and optimization design of variable winding angle fiber spiral winding structure. Master's Thesis, TJPU, Tianjin, China (2015)
18. Han, W.Q.: Experimental and Numerical Analysis on Damage Evolution of Composite Laminates. Doctor's Thesis, JSU, Zhenjiang, China (2017)
19. Qatu, M.S., Boston, A., London, H., et al.: Shallow Shells. Vibration of Laminated Shells and Plates **6**, 183–255 (2004)
20. Wang, Y.Y., Lou, M., Dong, W., Wang, Y.: Predicting failure pressure of reinforced thermoplastic pipes based on theoretical analysis and experiment. Compos. Struct. **270**, 114039 (2021)
21. ASTM D6272–17, Standard Test Method for Flexural Properties of Unreinforced and Reinforced Plastics and Electrical Insulating Materials by Four-Point Bending. Available online: astm.org (Accessed 14 Jul 2020)

Underwater Wireless Communication and Deep-Sea Information Detection Technology

Detection of Underwater Acoustic Information Based on FMCW Millimeter-Wave Radar Cross the Sea-Air Medium

Tengyuan Cui[1], Xiaolong Cao[1(✉)], Yiguang Yang[2], Yuchen Du[1], Tongchang Zhang[1], Jiaqi Yuan[1], Zhenyuan Zhu[1], and Jianquan Yao[2,3(✉)]

[1] College of Ocean Science and Engineering, Shandong University of Science and Technology, Qingdao 266590, China
caoxiaolong63@126.com

[2] Laoshan Laboratory, Qingdao 266237, China
jqyao@tju.edu.cn

[3] School of Precision Instrument and Opto-Electronics Engineering, Tianjin University, Tianjin 300072, China

Abstract. Frequency Modulated Continuous Wave (FMCW) millimeter wave radar directly detects underwater sound field information from the air, providing an innovative solution for cross medium wireless communication from underwater to air. At present, there is no systematic theoretical model for the physical process of acoustic water surface micromotion, and it is difficult for radar detection to fully align with the center of water surface waves, which greatly affects communication quality. In this article, an acoustic water surface micromotion model and a diffusion model are established to conduct theoretical model simulation and experimental verification of the process of detecting the water surface by millimeter-wave radar. The results show that as the sound source level increases, the amplitude of the water surface wave increases. At the same sound source level, the amplitude and wave diffusion range of the water surface wave decrease with the increase of the sound source frequency. The signal characteristics of the experimental results of radar for water surface detection match the simulation results of the theoretical model, and an amplitude of 2.13 μm is detected, and the signal frequency is consistent with the frequency of the acoustic source, which enables the detection of acoustic water surface micromotion signals of −70 dB/Hz. The experiment has verified that the theoretical model can accurately simulate the shape of acoustic water surface micromotion, proved the feasibility of radar to detect underwater sound field information from the air, and provided a theoretical and experimental basis for building cross-medium communication links.

Keywords: Underwater acoustic information · FMCW millimeter-wave radar · Acoustic water surface micromotion · Cross sea air medium communication

D. Jeng and B. Cai (Eds.): ICCOE 2024, LNCE 423, pp. 329–340, 2025.
https://doi.org/10.1007/978-981-96-0181-3_30

1 Introduction

Directly penetrating the air-sea interface to achieve wireless communication is significant for deep-sea exploration and underwater communication [1–3]. Current cross-medium communication solutions include direct laser communication [4, 5], intermediary-based communication [6], and acoustic wave-millimeter wave communication [7]. However, laser signals attenuate more in water compared to sound signals, limiting propagation distance. Intermediary-based communication requires pre-arranged relay devices, which can hinder real-time communication. Acoustic wave-millimeter wave communication enables automatic, pure physical signal conversion through surface wave vibrations at the air-sea interface, known as Translational Acoustic-RF Communication (TARF). This approach offers a reliable solution to overcome significant underwater attenuation of electromagnetic waves in cross-medium communication. Frequency Modulated Continuous Wave (FMCW) millimeter wave radar, with its low collimation and good anti-drift capabilities, it can improve the problem that radar is difficult to align the center of water surface wave in complex sea conditions. TARF has great application potential in the field of cross medium communication in deep sea environment [8].

In 1972 [9], Tremain D.E. explored the feasibility of detecting underwater sound source frequencies by illuminating the water surface with microwave radiation under turbulent surface conditions. In 1979 [10], Weisbuch, G., and Garbay, F. experimentally demonstrated the interaction between light waves and surface tension waves, providing a method to measure surface tension. In 1998 [11], Lee M.S. from Tulane University proposed a technique to detect underwater acoustic signals through water surface wave disturbances using amplitude modulation effects of laser beams. In 2018 [7], Tonolini F. and F. Adib introduced an air-sea communication scheme, TARF, based on microwave vibration measurement. In 2022 [12], Qu Fengzhong from Zhejiang University conducted acoustic-millimeter wave communication experiments, establishing a closed-form end-to-end channel, but the communication distance was only 15 cm. And the highest wave is only 5 cm. In 2023 [13], Shi Qiang developed a finite element model to simulate small surface oscillations caused by underwater low-frequency sound sources. The study analyzed sound source parameters and their effects on surface perturbations, but experimental verification focused on the center frequency, lacking an investigation into the relationship between frequency and surface wave perturbations. Acoustic-millimeter wave communication faces challenges from depth, environmental factors, complex sea conditions, multipath effects, and data rates. To apply this technology in deep-sea environments, it is necessary to model the entire process from underwater to air, study the signal conversion at the water surface, and validate the theoretical model through experiments.

This study established a sound-induced water surface micro-motion model and a diffusion model to examine the relationship between amplitude, sound source frequency, and sound source level. Simulations demonstrated the correlation between water surface wave diffusion and sound source frequency, offering theoretical guidance for selecting parameters in diverse underwater environments. A millimeter wave radar detection model was simulated, and an experiment validated the theoretical results. The feasibility of detecting underwater sound signals from the air using millimeter wave radar was confirmed, providing an experimental basis for constructing cross-medium communication links from underwater to air.

2 Theoretical Modeling and Simulation of Millimeter-Wave Radar Detection of Acoustic Water Surface Micromotion

When underwater acoustic signals propagate to the air-sea interface, the sound waves will excite water surface ripples with the same frequency at the water surface [14]. By analyzing and processing the water surface ripples, the underwater sound field information can be obtained to achieve cross-sea-air medium communication.

2.1 Acoustic Water Surface Micromotion Model

When the underwater sound wave reaches the water surface, part of the sound wave is transmitted into the air, and the other part is reflected back on the water surface. Let the incident sound pressure be Pi, the transmitted sound pressure be P_t, and the reflected sound pressure be P_r. $P_i = P_t + P_r$.

Sound propagates along the Z-axis direction and is a one-dimensional longitudinal wave. The expression is as shown in Eq. 1:

$$\begin{cases} \frac{\partial^2 P_i}{\partial t^2} = c_1 \frac{\partial^2 P_i}{\partial z^2} \\ \frac{\partial^2 P_t}{\partial t^2} = c_2 \frac{\partial^2 P_t}{\partial z^2} \end{cases} \tag{1}$$

Among them, t is time in s; c_1 is the speed of sound propagating in water, and c_2 is the speed of sound propagating in air in m/s. Incident sound waves, reflected sound waves, and projected sound waves are shown in Eq. 2:

$$\begin{cases} P_i = p_i e^{j(\omega t - k_1 z)} \\ P_r = p_r e^{j(\omega t - k_1 z)} \\ P_t = p_t e^{j(\omega t - k_1 z)} \end{cases} \tag{2}$$

Among them, p is the sound pressure amplitude; k is the wave number, and the expression is $k = \omega/c$.

The vibration speed of the particle excited by different sound pressures at the center of the water surface is shown in Eq. 3:

$$\begin{cases} u_i = \frac{p_i}{\rho_1 c_1} \\ u_r = -\frac{p_r}{\rho_1 c_1} \\ u_t = \frac{p_t}{\rho_2 c_2} \end{cases} \tag{3}$$

Among them, ρ_1 and c_1 are the density of water and the speed of sound in water respectively. ρ_2 and c_2 are the density of air and the speed of sound in air respectively. The acoustic impedance $Z = \rho c$. Because the acoustic impedance of water and air is very different, the transmitted sound pressure $P_t = 0$, $P_i = -P_r$.

By combining Eq. 2 and Eq. 3, the vibration velocity of the particle at the center of the sea-air interface is obtained, as shown in Eq. 4:

$$u_i = \frac{2p_i}{\rho_1 c_1} \tag{4}$$

At the sea-air interface, the amplitude of the particle is shown in Eq. 5:

$$A = \frac{2p_i}{\omega \rho_1 c_1} \tag{5}$$

It can be seen from Eq. 5 that the amplitude of water surface waves excited by sound waves is directly proportional to the incident sound pressure and inversely proportional to the frequency. When the underwater sound source is 1 m away from the water surface, the sound source level is set to 180 dB, and the change of the water surface wave amplitude with the sound source frequency is obtained, as shown in Fig. 1.

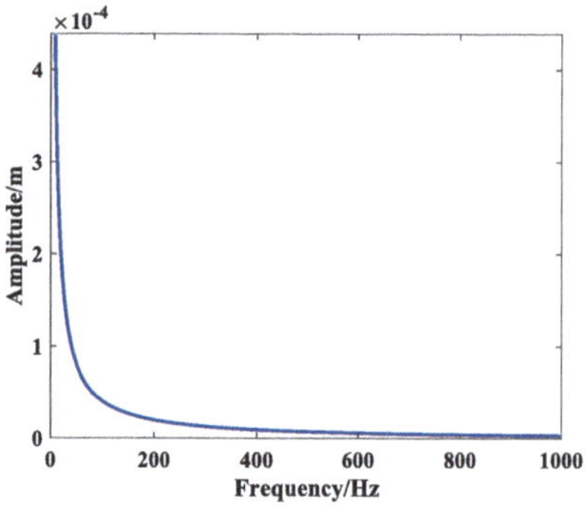

Fig. 1. Variation of water surface wave amplitude with sound source frequency

As can be seen from Fig. 1, when the frequency of the sound source is lower than 100 Hz, the amplitude of the water surface wave decreases rapidly with the increase of the frequency of the sound source; when the frequency of the sound source is higher than 100 Hz, the amplitude of the water surface wave decreases relatively slowly. The two major factors affecting the amplitude are sound source level and frequency. The simulation results of the change of the maximum amplitude with the sound source level at different frequencies are shown in Fig. 2.

In Fig. 2, the horizontal coordinate is the size of the sound source level, in dB, and the vertical coordinate is the acoustic water surface freting amplitude, in m. The curves in different colors represent the change of the maximum amplitude with the sound source level at the frequency of 10 Hz ~ 500 Hz. As can be seen from the figure, the amplitude of the water surface wave is in the order of μm, which is positively correlated with the size of the sound source level and inversely correlated with the frequency of the sound source. The amplitude directly affects the signal-to-noise ratio of the cross-medium communication. Selecting the appropriate size of the sound source level and the frequency of the sound source is the basis and prerequisite for realizing the cross-medium communication from underwater to air.

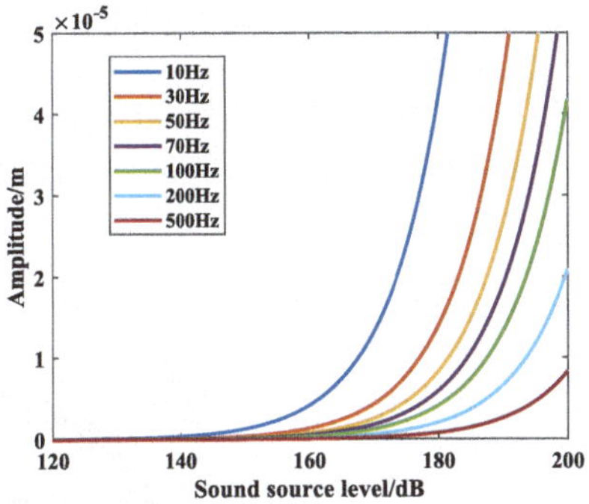

Fig. 2. The maximum amplitude varies with the sound source level at different frequencies

2.2 Acoustic Water Surface Diffusion Model

When the underwater sound wave reaches the water surface, it will generate vertical water surface vibration at the center. Due to the existence of water surface tension, a transverse wave will be generated which fluctuates and diffuses outwards, and its fluctuation rule is in accordance with the law of fluid motion [15]. The linear partial differential equation of the velocity potential of an ideal fluid under the condition of surface tension is shown in Eq. 6:

$$\begin{cases} -\frac{\sigma\prime}{\rho}\frac{\partial^3\varphi}{\partial x^2\partial y} + \frac{\partial^2\varphi}{\partial y} = 0 & y = 0 \\ \frac{\partial^2\varphi}{\partial x^2} + \frac{\partial^2\varphi}{\partial y^2} & -h_0 < y < 0 \\ \frac{\partial\varphi}{\partial y} = 0 & y = -h_0 \end{cases} \quad (6)$$

The solution of equations obtained by solving equations in Eq. 6 is shown in Eq. 7:

$$\begin{cases} \eta_k = Acos(kx - \omega t) \\ \varphi_k = \frac{A(Tk^2+\rho g)}{\rho\omega\cosh kh_0}\cosh k(h_0 + y)\sin(kx - \omega t) \end{cases} \quad (7)$$

By bringing φ_k into Eq. 6, the dispersion expression of acoustically induced water surface micro-motion is obtained, as shown in Eq. 8:

$$\omega^2 = \frac{\sigma^{\prime k^3} + \rho g k}{\rho} \quad (8)$$

Combined group velocity $v_g = \mathrm{d}\omega/\mathrm{d}t$ and Eq. 8, the wave group velocity on the water surface is obtained as shown in Eq. 9:

$$v_g = \frac{\left(g + \frac{3\sigma\prime k^2}{\rho}\right)}{2\sqrt{kg + \frac{\sigma\prime k^3}{\rho}}} \quad (9)$$

Finally, the attenuation coefficient of the acoustically induced water surface wave diffusion model is shown in Eq. 10:

$$\alpha = \frac{4\mu k^3\sqrt{kg + \omega^2}}{\rho kg + 3\omega^2} \quad (10)$$

As can be seen from Eq. 10, the attenuation degree of acoustic-induced water surface diffusion is mainly affected by frequency. When the sound source is located 1 m underwater, the sound source level is set to 180 dB, and the sound source frequencies are set to 10 Hz, 50 Hz and 100 Hz respectively. The simulation results of acoustic-induced water surface wave diffusion are shown in Fig. 3.

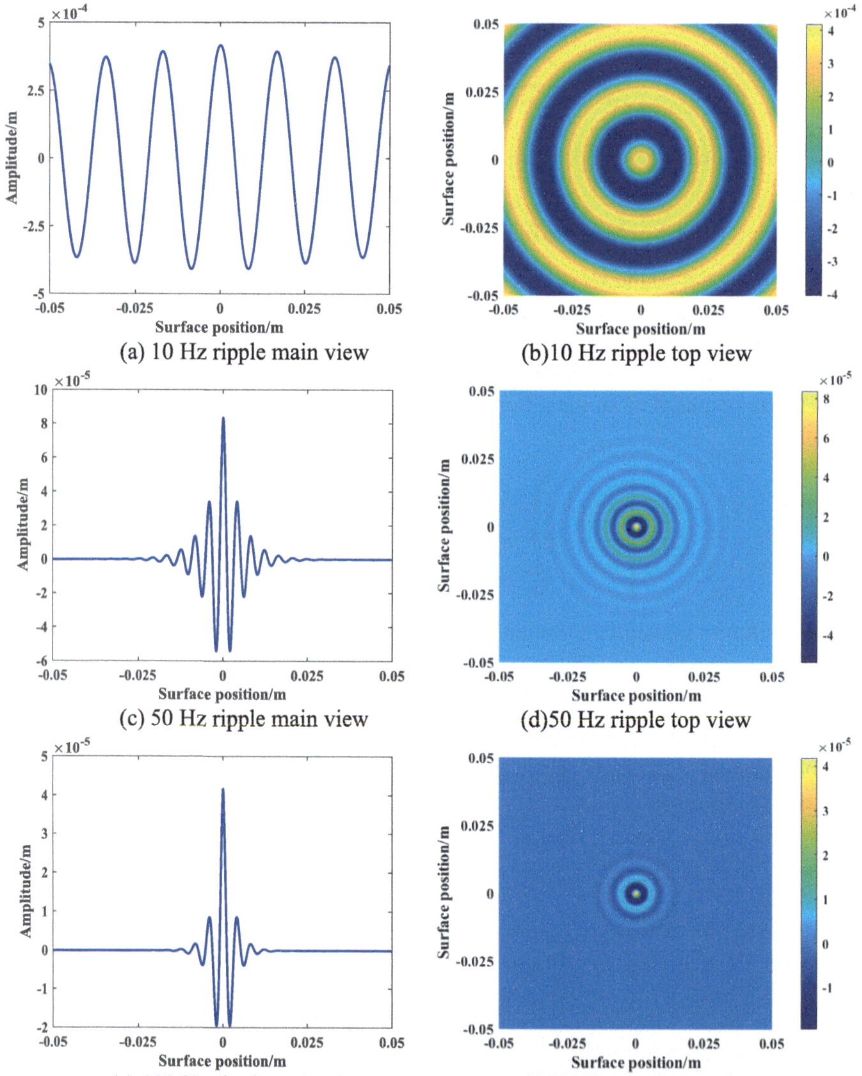

(a) 10 Hz ripple main view (b)10 Hz ripple top view

(c) 50 Hz ripple main view (d)50 Hz ripple top view

(e) 100 Hz ripple main view (f)100 Hz ripple top view

Fig. 3. Simulation results of acoustic wave diffusion on water surface

As can be seen from the Fig. 3, under the same sound source level, when the frequency is 10 Hz, the maximum amplitude of the water surface wave is 416.75 μm, and the wave diffusion range is 1600 mm. When the frequency is 50 Hz, the maximum amplitude is 83.637 μm and the diffusion range is 73 mm. When the frequency is 100 Hz, the maximum amplitude is 41.819 μm and the diffusion range is 20.6 mm. This shows that under the premise of constant sound source level, the higher the frequency, the smaller the amplitude of water surface wave and the smaller the diffusion range. The selection of appropriate communication frequency is the basis for ensuring the low bit error rate of cross medium communication. The simulation results of diffusion range changing with frequency are shown in Fig. 4.

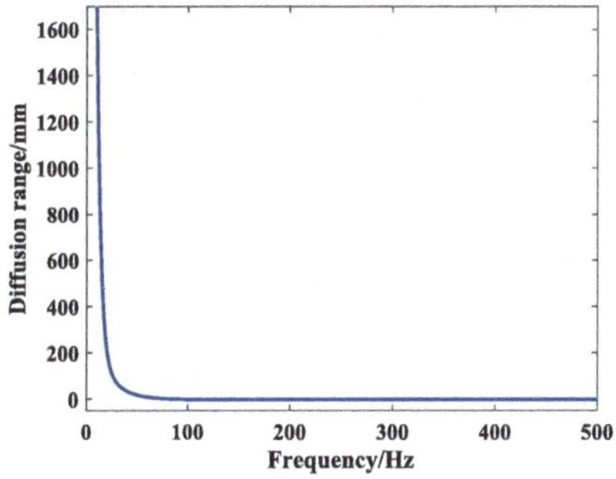

Fig. 4. The diffusion range varies with frequency

In Fig. 4, at 6 Hz ~ 50 Hz, the diffusion range drops sharply, which poses a challenge to the ability of millimeter wave radar to detect micro-motion signals. Further research on cross medium detection of millimeter wave radar is required.

2.3 Millimeter Wave Radar Detection Model

The amplitude of acoustic-induced water surface microwaves is in the order of μm, which is much smaller than the natural water surface fluctuation in the order of cm. In order to study the detection performance of FMCW millimeter wave radar, the process first needs to be simulated and analyzed.

Figures 5, 6 and 7 shows the simulation results of water surface detection by mmwave radar. The underwater speaker is located 1m below the water surface, the sound source level is set to 180 dB, and the frequency is set to 600 Hz. Under first-order sea state, the natural water surface fluctuations below 0.1 m approximately follow a sinusoidal pattern. The amplitude of sound-induced water surface micromotion is much smaller than that of natural water surface fluctuations, making it undetectable in Fig. 5. Figure 6 shows the time-domain simulation results after filtering out natural water surface fluctuations. In Fig. 6 (a), due to the characteristics of the Butterworth filter, a substantial amount of low-frequency signals remain unfiltered in the first cycle. Figure 6 (b) illustrates that the subsequent filtering aligns with expectations, concentrating amplitude mainly between 40 μm and 100 μm, resembling the simulation results of sound-induced water surface micromotion. Figure 7 is a time-frequency analysis of the millimeter wave radar simulation results. The color scale represents the power frequency ratio of the signal,

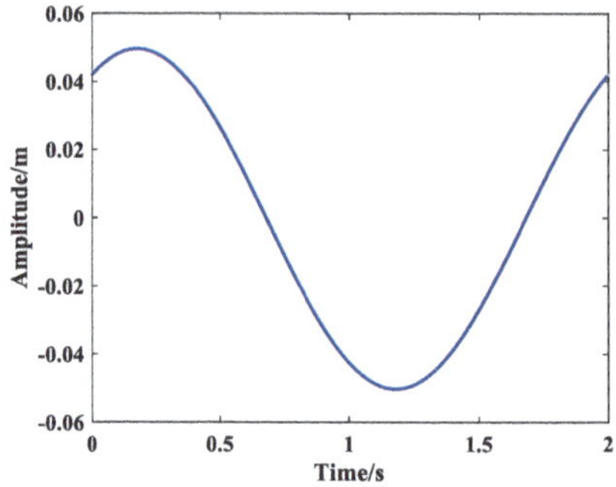

Fig. 5. Time domain simulation results of water surface detection by millimeter wave radar

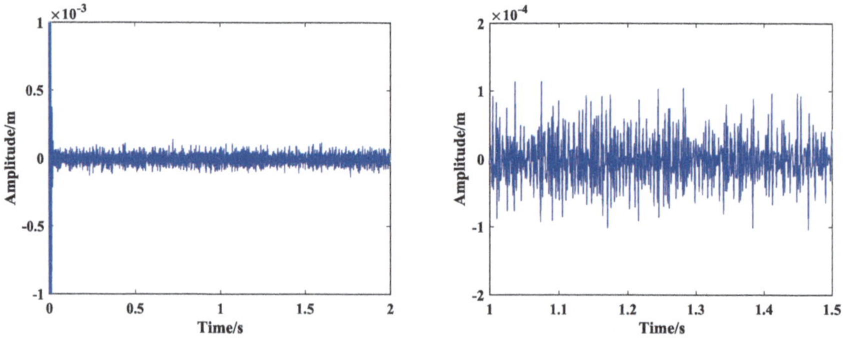

(a) Time domain simulation results (b) Local amplification of simulation results

Fig. 6. Time domain simulation results after filtering out natural fluctuation of water surface

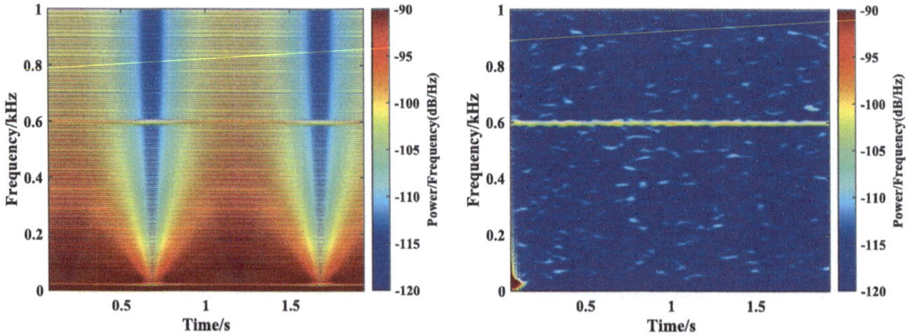

(a) The original acquired signal (b) The filtered signal

Fig. 7. Time-frequency analysis of simulation results of water surface detection by millimeter wave radar

with units in dB/Hz. Figure 7 (a) indicates that the power of natural water surface fluctuations and other low-frequency noise interference is significantly greater than that of the sound-induced water surface micromotion signals. Figure 7 (b) demonstrates the effectiveness of filtering, reducing the power frequency ratio of noise interference to below −120 dB/Hz, while maintaining the sound-induced water surface micromotion signals at −95 dB/Hz. The effective filtering confirms the feasibility of using FMCW millimeter wave radar to detect underwater sound signals from the air from a simulation perspective.

3 Acoustic Wave and Millimeter Wave Cross Sea-Air Medium Detection Experiment

3.1 Introduction to Experimental Environment

The experimental system diagram is shown in Fig. 8.

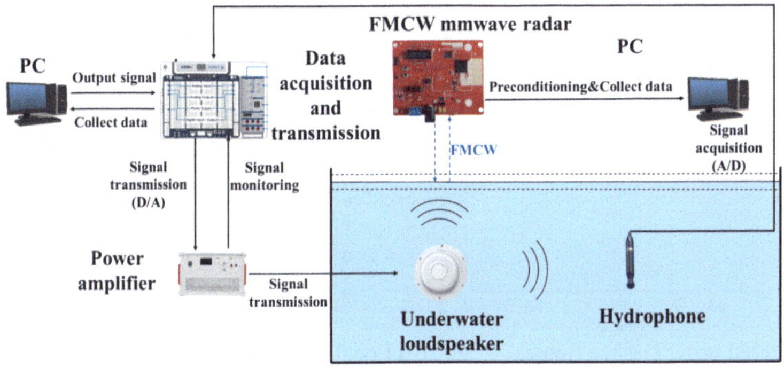

Fig. 8. Experimental system diagram

3.2 Experimental Process and Results of Radar Cross Medium Detection

An omnidirectional underwater speaker is placed 1 m underwater, with a set frequency of 600 Hz. The hydrophone is located 1 m away from the center of the underwater speaker. An FMCW millimeter-wave radar is positioned 1 m above the water surface, conducting vertical detection of the water surface. The time-domain plot of the detection results is shown in Fig. 9.

The meaning of horizontal and vertical coordinates in Fig. 9 is consistent with that in Fig. 5. It can be seen from the figure that the fluctuation of water surface detected by FMCW millimeter wave radar is roughly consistent with the law of sine wave. The maximum amplitude of water surface wave is 60 μm, and the frequency of natural fluctuation of water surface is 0.3 Hz.

Figure 10 is the time-domain diagram of water surface ripples obtained after the filtering of Fig. 9. The meaning of horizontal and vertical coordinates is consistent with that of Fig. 6. It can be seen from the figure that Butterworth filter also has the problem of incomplete filtering of low-frequency interference in the first cycle in the experiment, and the signal waveform is similar to that of the simulation diagram in Fig. 6 (a). It

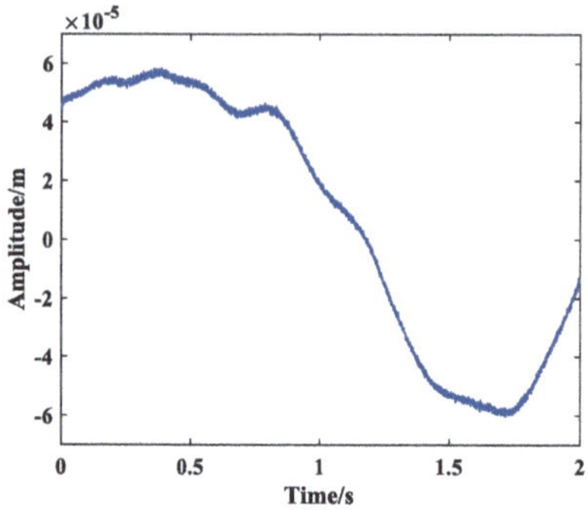

Fig. 9. Time domain diagram of water surface ripple detection by millimeter wave radar

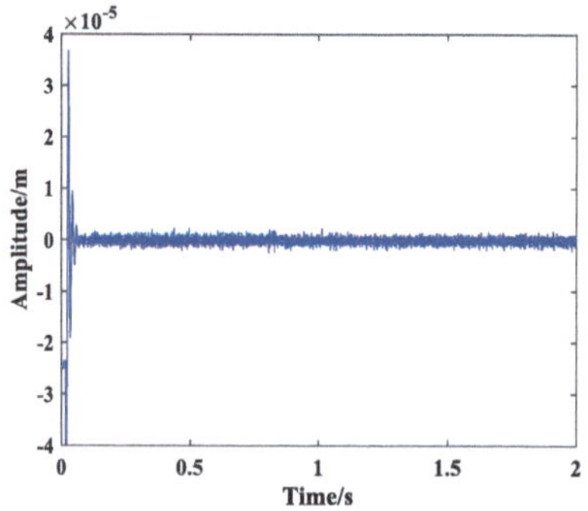

(a) Filtered water surface ripples

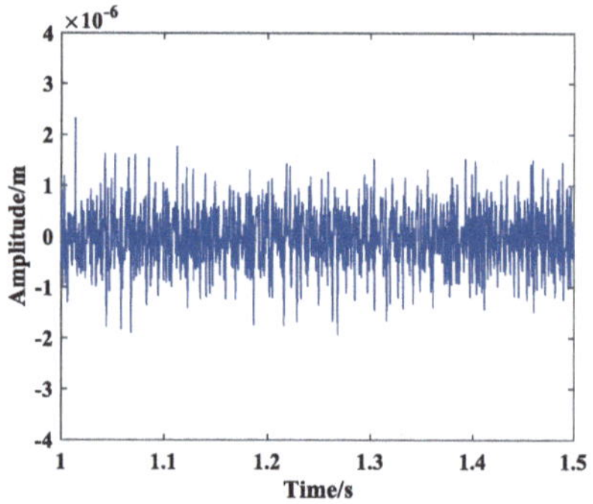

(b) Local amplification of water surface ripples

Fig. 10. Time domain map of water surface ripples after filtering

can be seen from Fig. 10 (b) that the maximum amplitude of the acoustic water surface micromotion signal is 2.13μm, and the waveform is similar to that of Fig. 6 (b), which verifies the correctness of the mmwave radar detection model.

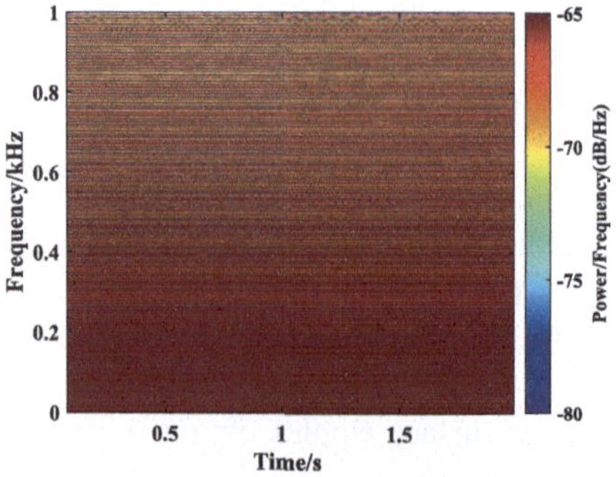

(a) The original acquired signal

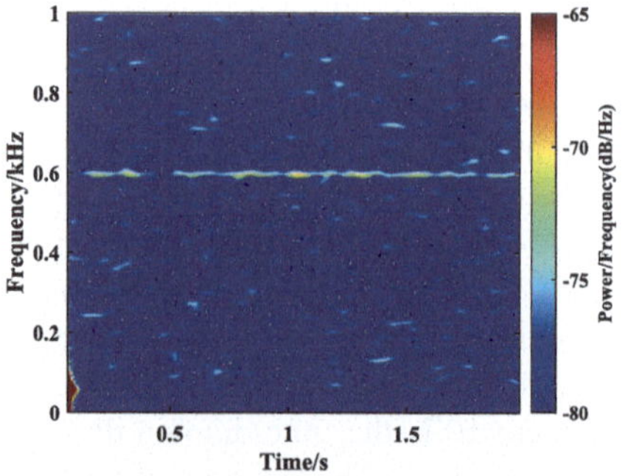

(b) The filtered signal

Fig. 11. Time-frequency analysis of millimeter wave radar detection experiment results

Figure 11 is the time-frequency analysis of the detection results of millimeter wave radar, and Fig. 11 (a) corresponds to Fig. 9. It can be seen from the figure that there is noise interference of up to −10 dB/Hz in the water surface signal detected by millimeter wave radar, and the frequency distribution is consistent with the characteristics of 1/f noise. All acoustic water surface micromotion signals of 600 Hz are annihilated in the noise and cannot be observed. The power frequency ratio of noise interference such as natural fluctuation of water surface is greater than −60 dB/Hz. Figure 11 (b) corresponds to Fig. 10 (a). It can be seen from the figure that the 600Hz acoustic-induced water surface fidget signal is clear, its power-frequency ratio is −70 dB/Hz, most of the interference signals such as natural fluctuation of water surface are filtered out, and its power-frequency ratio is reduced to below −90 dB/Hz, and the acoustic-induced water surface fidget signal is effectively separated.

Underwater noise is mainly a combination of 1/f noise and white noise, due to the severe attenuation of high-frequency signals in water and the lower diffusion range of water surface fluctuations after reaching the water surface. In fact, FMCW millimeter wave radar is difficult to face the center of water surface waves, and increasing frequency unilaterally will increase the difficulty of radar detection and cannot guarantee communication quality. Through simulation analysis of radar detection, the theoretical feasibility of cross medium communication based on acoustic millimeter waves has been proven. In the experiment, underwater sound field information was successfully detected directly from the water surface, providing theoretical and experimental basis for building cross medium communication links and transmitting information.

4 Conclusion

In this paper, the acoustic water surface micromotion model and wave diffusion model are established, and the process of water surface detection by millimeter wave radar is simulated and verified by experiments. At the same frequency, the amplitude of water surface wave increases with the increase of sound source level. Under the same sound source level, the lower the sound source frequency, the larger the amplitude of the water surface wave and the wider the wave diffusion range, which provides a theoretical basis for selecting suitable parameters in different underwater environments. In the experiment, the signal waveform detected by the millimeter wave radar is consistent with the simulation results of the theoretical model, and the acoustic water surface micro-signal with an amplitude of 2.13 μm is detected. The filtering effect is in line with the expectation, and the signal frequency is consistent with the sound source frequency. The acoustic water surface micro-signal of −70 dB/Hz can be detected in the air. It is verified that the theoretical model can accurately simulate the morphology of acoustic microwaves on water surface, and the feasibility of FMCW millimeter wave radar to detect underwater sound field information from the air is proved, which provides an important technical support for the bi-directional communication between air and sea medium.

Acknowledgements. This work was supported by the Opening Funding of National Key Laboratory of Electromagnetic Space Security, and Laoshan Laboratory Science and Technology Innovation Project (No. LSKJ202200801).

References

1. Luo, H., Wang, J., Bu, F., et al.: Recent progress of air/water cross-boundary communications for underwater sensor networks: a review. IEEE Sens. J. **22**(9), 8360–8382 (2022)
2. Kong, M., Kang, C.H., Alkhazragi, O., et al.: Survey of energy-autonomous solar cell receivers for satellite–air–ground–ocean optical wireless communication. Progress Quantum Electron. **74**, 100300 (2020)
3. Khan, M.A., Alzahrani, B.A., Barnawi, A., et al.: A resource friendly authentication scheme for space–air–ground–sea integrated maritime communication network. Ocean Eng. **250**, 110894 (2022)
4. Carver, C.J., Tian, Z., Zhang, H., et al.: Amphilight: Direct air-water communication with laser light. GetMobile: Mobile Comput. Commun. **24**(3), 26–29 (2021)
5. Sun, X.B., Kong, M.W., Alkhazragi, O.A., et al.: Field demonstrations of wide-beam optical communications through water-air interface. IEEE Access **8**, 160480–160489 (2020)
6. Chen, H., Chen, H.F., Zhang, Y., et al.: Decentralized estimation of ocean current field using underwater acoustic sensor networks. J. Acoust. Soc. Am. **149**(5), 3106–3121 (2021)
7. Tonolini, F., Adib, F.: Networking across boundaries: enabling wireless communication through the water-air interface. In: Proceedings of the 2018 Conference of the ACM Special Interest Group on Data Communication, pp. 117–131 (2018)
8. Saeed, N., Celik, A., Al-Naffouri, T.Y., et al.: Underwater optical wireless communications, networking, and localization: a survey. Ad Hoc Netw. **94**, 101935 (2019)
9. Tremain, D.E., Angelakos, D.J.: Detection of underwater sound sources by microwave radiation reflected from the water surface. Proc. IEEE **60**(6), 741–742 (1972)
10. Weisbuch, G., Garbay, F.: Light scattering by surface tension waves. Am. J. Phys. **47**(4), 355–356 (1979)
11. Lee, M., Bourgeois, B., Hsieh, S., et al.: A laser sensing scheme for detection of underwater acoustic signals. In: Proceedings of the Conference Proceedings' 88, pp. 253–257. IEEE Southeastcon (1988)
12. Qu, F.Z., Qian, J.Y., Wang, J., et al.: Cross-medium communication combining acoustic wave and millimeter wave: theoretical channel model and experiments. IEEE J. Oceanic Eng. **47**(2), 483–492 (2022)
13. Shi, Q., He, Z., Xu, H., et al.: Research on information transmission technology through the water-air interface combining a sonar and a radar. IEEE Sens. J. **23**(7), 7616–7625 (2023)
14. Dai, Z., Sun, J., Sui, P.: Theoretical study of transverse microwaves on water surface caused by underwater sound source. J. Natl. Univ. Defense Technol. **26**(01), 95–98 (2004)
15. Ma, M., Ye, Z.: A numerical method for calculating unsteady flow using Euler equation. Chin. J. Aerodyn. **21**(04), 459–463 (2003)

A Leakage Localization Method for Subsea Manifolds Based on Adaptive Median Filtering Algorithm

Xuelin Liu, Baoping Cai(✉), Guowei Ji, Yi Jiang, and Kaizheng Wu

College of Mechanical and Electronic Engineering, China University of Petroleum, Qingdao 266580, Shandong, China
caibaoping@upc.edu.cn

Abstract. Pipeline leakage is a common problem in industrial production, and it is a key factor restricting the safety and reliability of pipeline system, which may lead to safety accidents, environmental polluted and resource waste. Therefore, accurate and rapid detection and location of pipeline leakage points is very important to ensure production safety and environmental protection. A pipeline leak location method based on adaptive median filter algorithm is proposed, and an adaptive median filter model is established, which realizes noise filtering and data adaptive picking, and accurately extracts signal features. The generalized weighted quadratic cross correlation algorithm is used to improve the signal-to-noise ratio, obtain the time delay estimation and locate the leakage point. Subsea Manifolds are used as a case to demonstrate the proposed method. The experimental results show that the method can detect and locate the leakage point of pipeline quickly and accurately, and has high reliability and practicability. This study provides an effective solution for pipeline leakage monitoring and location, and has important engineering application prospects.

Keywords: Leakage Localization · Adaptive Median Filtering · Subsea Manifolds

1 Introduction

Pipeline systems play a vital role in modern industry and are widely used to transport liquids and gases. However, pipeline leakage has always been one of the key factors restricting the safety and reliability of pipeline systems [1]. Leakage will not only lead to environmental pollution and waste of resources, but also may cause serious accidents, resulting in casualties and property losses [2, 3]. Therefore, timely and accurate detection and location of pipeline leakage points is very important to ensure the safety of pipeline system operation.

Pipeline leak detection methods can be roughly divided into three categories: method based on signal processing, method based on state estimation and method based on knowledge [4]. The signal processing-based approach involves analyzing different types of measurements collected from different sensors. Among different sensing schemes,

D. Jeng and B. Cai (Eds.): ICCOE 2024, LNCE 423, pp. 341–349, 2025.
https://doi.org/10.1007/978-981-96-0181-3_31

detecting transient pressure waves in pressure data is considered to be one of the most mature methods in laboratory and field Settings [5].

The leak location method based on negative pressure wave (NPW) signal has become one of the research hotspots. NPW method has the advantages of simple deployment, economy, high sensitivity, low false alarm rate and high stability accuracy. Therefore, NPW method is widely used in pipeline leak detection and location [2]. When a leak occurs, the NPW propagates to upstream and downstream sites, where they are captured by pressure sensors. Since the arrival time of upstream and downstream sensors is different, the leakage is detected and located according to the wave speed and time difference [6]. However, in general, the pressure signal will undergo many changes during transmission, which usually mask the information carried by the signal [7]. The signal will add a lot of noise in the process of transmission, and the level of noise is related to the pipeline environment. In signal processing, it is not only necessary to avoid signal distortion and obtain complete information from the signal, but also to enhance the robustness of signal processing [8].

A pipeline leak location method based on adaptive median filter algorithm is proposed. The adaptive median filter model is established to realize noise filtering and data adaptive picking. The generalized weighted quadratic cross correlation algorithm is used to improve the signal-to-noise ratio, accurately estimate the time delay and locate the leak point. The subsea production manifold is used as a case to demonstrate the proposed method. The rest of the paper is organized as follows. The proposed leak detection and location methods are described in detail in Sect. 2. Section 3 demonstrates the method using subsea production manifold as a case. Section 4 summarizes this paper.

2 The Proposed Method

As shown in Fig. 1, this paper proposed a pipeline leak location method based on adaptive median filtering algorithm. Firstly, an adaptive median filter model is established to filter the noise data collected from upstream and downstream pipelines. Secondly, the data set after noise reduction is analyzed to find the signal drop point and select the signal data point before and after the drop. Finally, the quadratic correlation algorithm is used to analyze the autocorrelation and cross-correlation of the adaptive time series, and the time delay is calculated to locate the leak point.

The flow chart of pipeline leak location method combining adaptive median filtering and generalized weighted quadratic cross-correlation algorithm is shown in Fig. 2. The median filtering algorithm of fusion adaptive data extraction is proposed to process the data collected by upstream and downstream sensors, and to eliminate the influence of extreme noise by calculating the median of samples in the local window.

First, all data points are traversed in each data set and a series of median filtering Windows are formed centered on each pixel point. Define the current signal value as C_{ij}. The signal value in the window frame centered on C_{ij} is defined as F_{xy}, and x and y are the coordinate values of the corresponding pixels in the frame. The average p of the difference between the current data value and the data value in the search frame is defined as follows:

$$p_{ij} = \frac{1}{N}\sum_{a=0}^{N}\left|C_{ij} - F_{xy}\right| \tag{1}$$

N is the number of data points in the window frame, and a is the neighborhood of the 2D filter window.

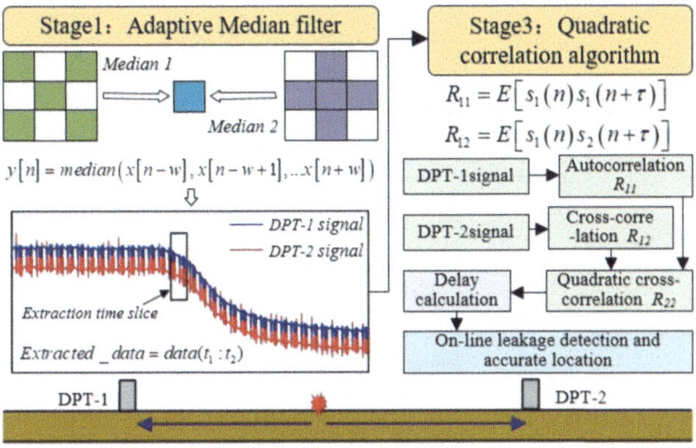

Fig. 1. Pipeline leak location method based on adaptive median filter algorithm

Secondly, read out each data value in the filtering window, and then the data group *P[ij]* after data filtering is obtained:

$$P[ij] = \{p_1, p_2, p_3, ...p_n\} \tag{2}$$

Analyze the filtered set, calculate the pressure difference *PD* of adjacent data points, and obtain the pressure drop array *PD[ij]*. Found the maximum pressure drop point $PD[ij]_{max}$, and the signals before and after sudden drop are selected as the symmetric point, and the data set AP after adaptive selection is output. Then, the generalized weighted quadratic cross correlation is calculated. In order to enhance the anti-noise performance of the basic cross-correlation algorithm, a quadratic correlation algorithm is proposed to effectively reduce the influence of noise on the performance of delay estimation when the signal-to-noise ratio decreases.

The signals received by the upstream and downstream sensors M1 and M2 are $x_1(t)$ and $x_2(t)$, respectively. It is assumed that the noise $n_1(t)$ and $n_2(t)$ are uncorrelated white Gaussian noise, and there is no correlation between the noise and the original signal.

$$\begin{aligned} x_1(t) &= a_1 s(t) + n_1(t) \\ x_2(t) &= a_2 s(t-\tau) + n_2(t) \end{aligned} \tag{3}$$

τ is the delay between the signal reaching the sensor M1 and M2.

The $R_{x1x2(\tau)}$ composed of two signals $x_1(t)$ and $x_2(t)$ is:

$$R_{x1x2}(\tau) = \int_0^{2\pi} X_1(w)X_2^*(w)e^{-iw\tau}dw \tag{4}$$

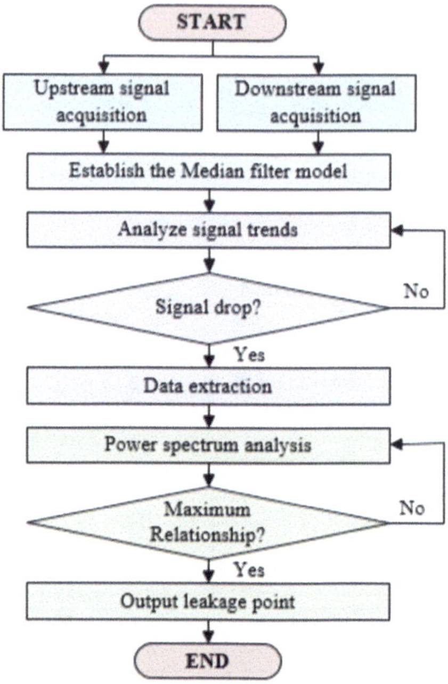

Fig. 2. Flow chart of pipeline leak location method based on adaptive median filter algorithm

$X_1(w)$ is the frequency domain signal of the time domain signal $x_1(t)$, $X^* 2(w)$ is the signal obtained by the conjugation of the frequency domain signal $x_2(t)$ of the time domain signal. The function $R_{x1x2}(\tau)$ is transformed from frequency domain to time domain by Fourier inversion, and the maximum peak value is detected and the delay is estimated.

The autocorrelation function of $x_1(t)$ is:

$$R_{11}(\tau) = E[x_1(t)x_1(t - \tau)] \tag{5}$$

The cross-correlation function of c $x_1(t)$ and $x_2(t)$ is:

$$R_{12}(\tau) = E[x_1(t)x_2(t - \tau)] \tag{6}$$

$R_{11}(\tau)$ and $R_{12}(\tau)$ are still functions of time τ as the dependent variable, and carry out the second cross-correlation operation on these two functions. The quadratic correlation function $R_{RR}(\tau)$ as follows:

$$R_{RR}(\tau) = E[R_{11}(t)R_{12}(t - \tau)] \tag{7}$$

Assuming that the target signal is not correlated with the noise, it can be expressed as:

$$R_{RR}(\tau) = R_{RS}(\tau - D) + R_{RN}(\tau) \tag{8}$$

R_{RS} and R_{RN} are quadratic correlation functions of target signal and noise respectively. Assuming that the noise is uncorrelated, $R_{RN} = 0$. It can be seen from the formula that when τ = D, $R_{RS}(\tau\text{-}D)$ gets the maximum value, and the quadratic correlation function $R_{RR}(\tau)$ gets the maximum value, τ is the desired delay value.

Through the above generalized quadratic correlation calculation, the interference of noise to the target signal is relatively weakened, and the signal-to-noise ratio is improved to a certain extent. Compared with the first correlation method, the time delay can be obtained more accurately in a smaller SNR environment.

3 Case Study

Leakage phenomenon in liquid environment exists widely in natural and industrial fields, especially in Marine environment, which has great research value in the field of Marine development. With the development of exploration and development of oil and gas resources gradually from shallow water to deep water, the subsea oil and gas production system has changed from the early single satellite well development model to the cluster satellite well development model with underwater manifold as the core. Subsea production manifold is the key facility of subsea production system. The leakage of subsea manifold will not only cause a huge waste of resources, but also have a bad impact on Marine ecological environment, Marine climate and Marine economic development. Only when the leak source is found in time and the leak point is accurately located, effective measures can be taken to deal with and deal with it, and the harm of the leak to the environment and human beings can be minimized.

As shown in Fig. 3, a prototype of the principle of subsea production manifold is built in the laboratory.

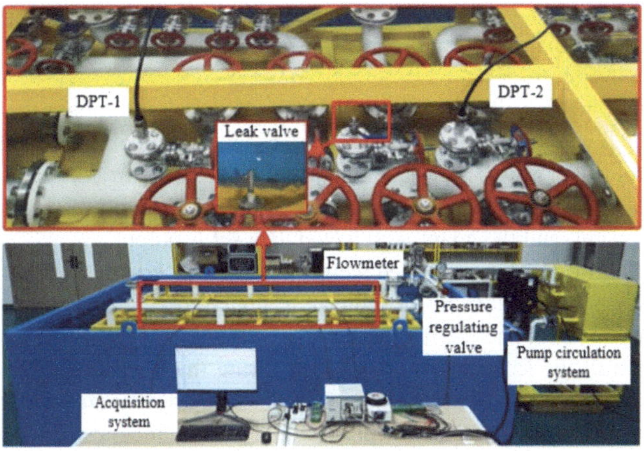

Fig. 3. Principle prototype of subsea manifolds

Eight branch pipes are connected to the main pipe to realize liquid circulation, and manual leakage valves are set up to simulate underwater leakage conditions. The pump circulation system is a complete manifold conveying medium, simulating the oil recovery process. The upstream dynamic pressure transmitter (DPT-1) and downstream dynamic pressure transmitter (DPT-2) collect the pressure signal inside the manifold with a sampling frequency of 50 kHz. The leakage valve simulates underwater leakage and adjusts the pressure inside the pipe through the pressure regulator.

The pressure data collected by DPT-1 and DPT-2 are shown in Fig. 4. There is a pressure difference in the upstream and downstream pipelines. Open the leak valve, the pressure in the pipeline continues to drop. The pressure data signal fluctuates greatly and the noise is noisy, so it is difficult to obtain the accurate information of the leakage signal.

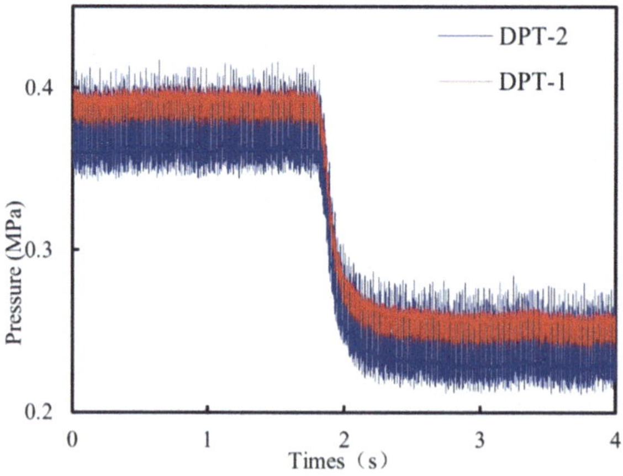

Fig. 4. Original pressure data of manifolds

The adaptive median filtering algorithm was used to filter and extract the pressure data, as shown in Fig. 5.

The adaptive median filtering algorithm can effectively eliminate the noise of the fixed frequency band and analyze the filtered data set. By calculating the pressure drop value of adjacent data points, the data adaptive selection is carried out, and the signal before and after the sudden drop is selected with the maximum pressure drop point as the symmetric point, and the adaptive selected data set is output.

As shown in Fig. 6, the generalized weighted quadratic cross correlation analysis of data sets can reduce the interference of noise to the target leakage signal and improve the signal-to-noise ratio, and the time delay can be obtained more accurately.

The location of the leak is estimated by capturing the time difference between the leak point and the pressure wave of the upstream and downstream dynamic pressure transmitter. The inner diameter of the pipe is 0.08 m, the thickness of the pipe wall is 0.01 m, the material is 304 stainless steel, and the experimental medium is water. The propagation speed of the pressure wave can be obtained as 1420 m/s [2].

The overall circulating pressure is 0.4 Mpa (P4), 0.6 MPa(P6) and 0.8 MPa(P8). Ten groups of repeated tests are carried out respectively, and 30 groups of experimental data were analyzed. The positioning calculation is carried out according to the obtained time

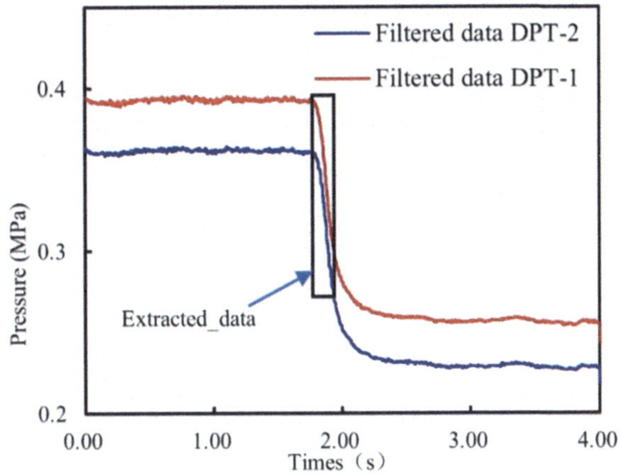

Fig. 5. Result of adaptive median filtering algorithm

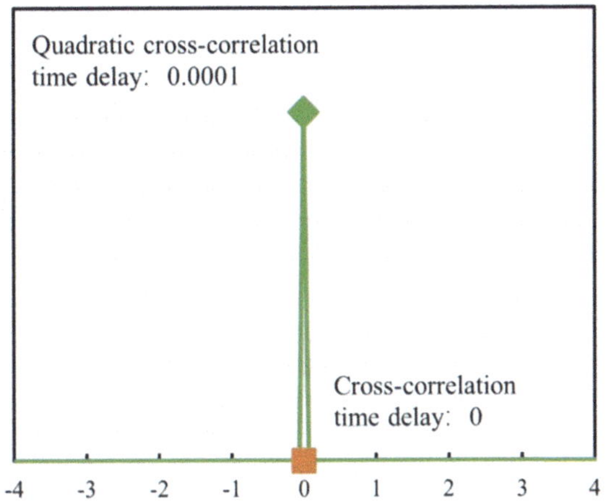

Fig. 6. Quadratic cross-correlation algorithm to find the time delay

delay. The positioning accuracy of the proposed adaptive median filter combined with the generalized weighted quadratic cross-correlation algorithm is shown in Fig. 7.

The average positioning error of P4 dataset is 0.077 m, the average positioning error of P6 dataset is 0.105 m, and the average positioning error of P8 dataset is 0.0966 m. The average positioning error of the ten experiments is 0.093 m. The results show that the proposed algorithm has strong adaptability and robustness to underwater pipeline leakage location under different working pressures. The method can filter the non-leakage noise, improve the anti-interference ability of the system and improve the positioning accuracy.

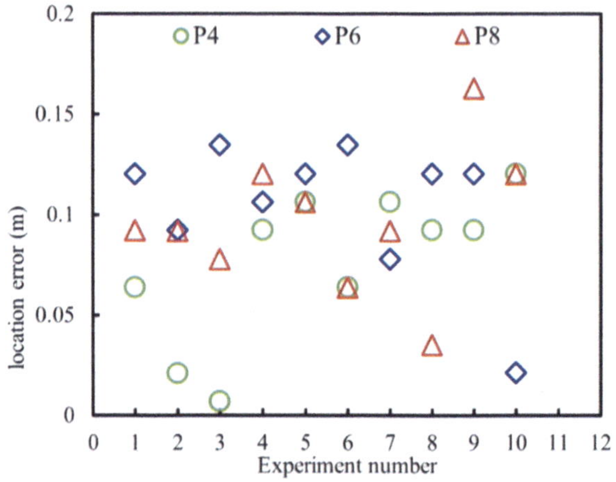

Fig. 7. The error of leakage locating

4 Conclusion

In this work, an adaptive median filter and generalized weighted quadratic cross correlation method for pipeline leak location is proposed. An adaptive median filter model is established to reduce data noise, find the signal drop point, and select the signal data point before and after the drop point. The quadratic correlation algorithm is used to analyze the autocorrelation and cross-correlation of the adaptive time series, calculate the time delay and locate the leak. The results show that the method has strong adaptability to the leakage location of the subsea production manifold under different pressures, and the positioning error is only 0.0966 m. The method has good robustness and anti-interference ability, and the accuracy of leakage location is higher.

Acknowledgements. This work was supported by the National Key Research and Development Program of China (No. 2022YFC2806102), the National Natural Science Foundation of China (No. 52171287, 52325107), High tech Ship Research Project of Ministry of Industry and Information Technology (No. 2023GXB01–05-004–03, No. GXBZH2022–293), the Science Foundation for Distinguished Young Scholars of Shandong Province (No. ZR2022JQ25), the Taishan Scholars Project (No. Tsqn201909063).

References

1. Liao, Z., Yan, H., Tang, Z., Chu, X., Tao, T.: Deep learning identifies leak in water pipeline system using transient frequency response. Process Saf. Environ. Prot., 155 (2021)
2. Li, J., Zheng, Q., Qian, Z., Yang, X.: A novel location algorithm for pipeline leakage based on the attenuation of negative pressure wave. Process. Saf. Environ. Prot. **123**, 309–316 (2019)
3. Yuan, F., Zeng, Y., Luo, R., Khoo, B.C.: Numerical and experimental study on the generation and propagation of negative wave in high-pressure gas pipeline leakage. J. Loss Prev. Process Ind. **65**, 104129 (2020)

4. Kim, S.H.: Multiple leak detection algorithm for pipe network. Mech. Syst. Signal Process. 139 (2020)
5. Chu, J., et al.: Experimental platform for blockage detection and investigation using propagation of pressure pulse waves in a pipeline. Measurement **160**, 107877 (2020)
6. Yan, X.F., Duan, H.F., Wang, X.K., Wang, M.L., Lee, P.J.: Investigation of transient wave behavior in water pipelines with blockages. J. Hydraul. Eng. **147**, 04020095 (2021)
7. Yang, B., Deng, J., Yuan, W., Wang, Z.: Investigation on continuous pressure wave in a periodic transient flow using a three-dimensional CFD model. J. Hydraul. Res.Hydraul. Res. **58**, 172–181 (2020)
8. Wang, X., et al.: Pipeline leak localization using matched-field processing incorporating prior information of modeling error. Mech. Syst. Signal Process., 143 (2020)

Research on the Strength of Ocean Detection Communication Signals Based on Laser Induced Sound

Tongchang Zhang[1], Xiaolong Cao[1](✉), Yiguang Yang[2], Yuchen Du[1], and Jianquan Yao[2,3](✉)

[1] College of Ocean Science and Engineering, Shandong University of Science and Technology, Qingdao 266590, China
Caoxiaolong63@126.com
[2] Laoshan Laboratory, Qingdao 266237, China
jqyao@tju.edu.cn
[3] School of Precision Instrument Opto-Electronics Engineering, Tianjin University, Tianjin 300072, China

Abstract. Laser acoustic signal used in detection and communication has been widely concerned by people, and it is particularly important to analyze and calculate its intensity. The experimental platform of laser induced sound is built, and the intensity of laser induced sound signal is simulated and compared with the experimental results. The results show that the initial velocity of the signal generated by 1J laser energy under the influence of 32° convergence Angle is 5839m/s and the sound intensity is 1.468×10^{10}pa. The attenuation of shock wave pressure and transmission velocity is studied, and it decays to constant within 1ms. The relationship between sound intensity and energy is calculated theoretically. The velocity and pressure of sound signal are positively correlated with laser energy. It is proposed and proved that the laser acoustic signal intensity can be improved by shortening the laser focal length through the simultaneous action of multiple lasers.

Keywords: Plasma shock wave · Laser Array · Laser Induced sound

1 Introduction

Ocean research has gradually become a hot research issue in recent years [1], and ocean detection and communication are one of the hot issues, and laser acoustic signal has the advantages of wide spectrum [2], long transmission distance, and high signal-to-noise ratio [3], so the research on ocean detection and communication based on laser acoustic has gradually become the focus of future development. In 1980, Hansson I. et al. experimentally demonstrated that the principle of laser sound is the formation and collapse of cavitation and can produce sound waves in solid liquids [4–6]. In 1985, J.G. Fujimoto et al. used Nd: YAG lasers to achieve optical breakdown in salt water and generate ultrasonic waves [7]. In 1991, Sacchi C. A. discovered that the mechanism of

D. Jeng and B. Cai (Eds.): ICCOE 2024, LNCE 423, pp. 350–357, 2025.
https://doi.org/10.1007/978-981-96-0181-3_32

laser light breakdown in water is due to avalanche ionization in the liquid [8]. Ni Xiaowu et al. first proposed in 1998 that the weak limit of laser plasma in solids is sound wave, and defined laser plasma acoustic wave (LPAW) [9]. In addition, laser-induced sound can also be used for temperature measurement [10], pipeline damage measurement [11] and other directions. In 2023, Heesuk Jang [12] used 1064nm Nd: YAG laser and 2940nm Er: YAG laser respectively to conduct laser acoustic experiments, and the results showed that the shock wave generated by 1064nm laser only propagated in the horizontal direction, while the wave generated by 2940nm laser had better longitudinal penetration. Hou [13] proved through research in 2021 that the intensity of sound wave stimulated by array laser is about 4 times stronger than that stimulated by single light source laser, and high-power laser [14] can also effectively improve the pulse width of laser acoustic signal. Lu [15] obtained the transmission characteristics of acoustic signals through simulation and mathematical calculation in 2015. The current research on the principle and intensity of laser sound signal has not considered the absorption of laser signal in water, so the authenticity of the results is weak.

In this paper, the laser energy is modified by using the parameter ζ, which is related to the laser convergence Angle. The experimental results show that the model is effective.

2 Experiment

To study the relationship between the intensity of laser-induced ocean detection communication signals and laser energy, we built an experimental platform as shown in Fig. 1:

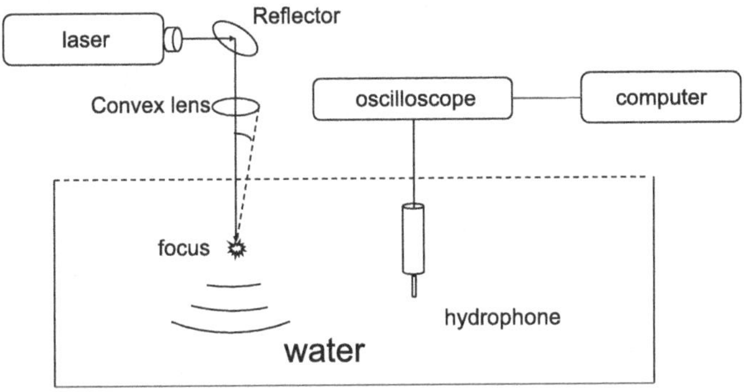

Fig. 1. Experimental diagram

This system used Nd: YAG pulse laser (energy 100mJ-400mJ continuously adjustable, pulse width 6-8ns, repetition frequency 1Hz) as a laser source, laser source emits laser through the mirror and convex lens to focus the laser on water to generate plasma, and stimulate sound waves, which are received by RHSA-5 hydrophone. The oscilloscope with a sampling rate of 2 GHz collects the received signal of the hydrophone, stores it, transmits it to the computer and further processes it.

The laser is emitted vertically into the water at a repetition frequency of 1Hz, the focus is located at 0.01m underwater, and the distance between the hydrophone and the focus is 0.4m. The laser energy is adjusted from 0.1J to 0.4J, measured every 0.02J, and measured 50 times repeatedly. After filtering and noise reduction in the computer, the average value is taken as the actual value of the energy generated.

3 Theoretical Model and Simulation Result a Subsection Sample

3.1 The Distance and Speed of the Shock Wave

When the laser energy is increased to the breakdown threshold of water, light breakdown occurs in water. In general, the breakdown threshold of water is $10^7 w/cm^3$ for pure water. The breakdown threshold of pure water is: $10^{10} w/cm^3$. In the process of optical breakdown, water is ionized by laser, forming plasma, and its own properties are changed, so it is a nonlinear process with poor reproducibility, and it is difficult to establish a reasonable overall model. Therefore, we use the transmission distance formula of shock wave generated by point source explosion in water to analyze the sound wave generated by plasma generation. The theoretical formula of transmission distance of shock wave in water is as follows:

$$R(t) = M_0 c_0 t \left[1 - \left(1 - \frac{1}{M_0} \right) e^{-\left[\frac{R_0}{c_0 t} \right]^{2-n}} \right] + R_0 \tag{1}$$

where c_0 is 1497m/s, n is ideal specific heat capacity of the explosion source, n = 1.2, R0 = 0.001m, $M_0$is the maximum Mach when t = 0, and:

$$M_0 = k(\frac{\zeta E_0}{R_0^3 c_0^2 \rho_0})^{1/5} + 1 \tag{2}$$

where k = 1, ζ is absorption coefficient of water to laser, and related to the convergence Angle θ of the laser, E_0 is laser energy, ρ_0 is density of water, Table 1 is Relation between convergence Angle and absorption coefficient:

Table 1. Relation between convergence Angle and absorption coefficient

Angle	coefficient ζ
15°	37%
22°	43%
32°	46%

The strength of the signal is very important in the application of laser sound in the actual ocean detection and communication. To research the relationship between the intensity of shock wave signal generated by laser sound and laser convergence Angle, the shock wave generated when the laser energy is 1J without considering the influence of the convergence Angle (ζ=1) and θ = 32° is calculated respectively. The simulation of transmission distance, speed and time is shown in Figs. 2 and 3:

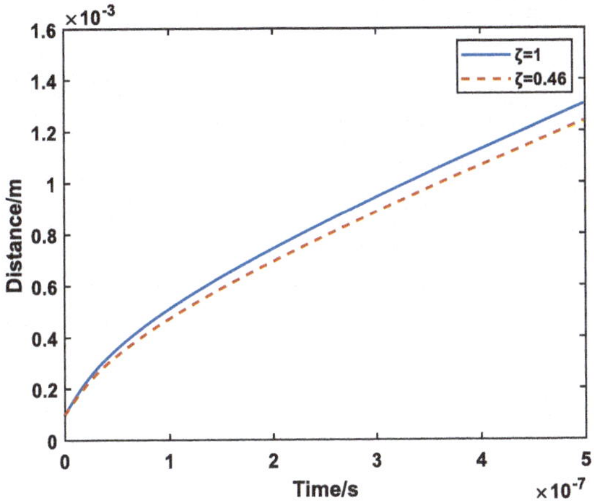

Fig.2. The relationship between the transmission distance of shock wave and time

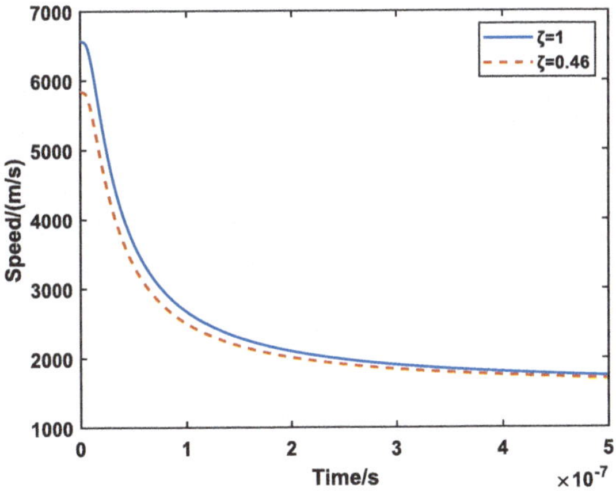

Fig.3. The relation between shock wave propagation velocity and time

As can be seen from Fig. 2, the transmission distance of the signal is positively correlated with time, and is proportional after a period, and at the same time, the transmission distance of the shock wave generated by the 1J laser energy at the same time than the transmission distance of the shock wave generated by the introduction of the convergence Angle laser is farther and farther. The value of the shockwave transmission velocity in Fig. 3 can be obtained by calculating the derivative of the transmission distance with respect to time. The resulting result is shown in the figure as follows: the initial velocity of the shockwave generated by 1J energy is 6568 m/s; when ζ=0.46, the corresponding initial velocity of the shockwave is 5839 m/s; the higher the laser energy is, the higher the velocity of the shockwave is, and it is much higher than the velocity of sound in water. And rapidly decays to the speed of sound within 1 ms.

3.2 The Pressure of Wave

The pressure of the acoustic signal generated by laser sound can be obtained by the R-W model, which can be expressed by Eq. (3):

$$p = v_1 \rho_0 v/t \left[10^{\frac{v - c_0}{v_2}} - 1 \right] + p_0 \tag{3}$$

where v_1=5190 m/s, v_2=25306 m/s, v is wave velocity, p_0 is atmospheric pressure, The change curve of shock wave intensity generated by laser over time is shown in Fig. 4:

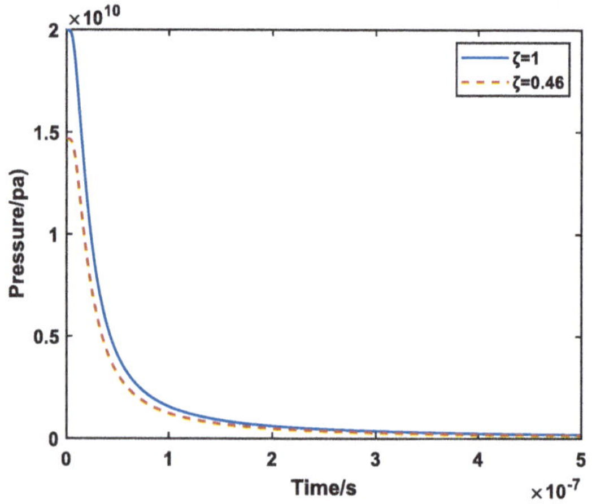

Fig.4. The relationship between the maximum sound intensity and time

As can be seen from Fig. 4, the sound intensity of the shock wave is positively correlated with the velocity of the shock wave, so the pressure is maximum now of generation. As can be seen from Fig. 4, the maximum shock wave pressure generated by the 1J laser energy theory is 1.999×10^{10}pa without considering the influence of the convergence Angle; when the focus is 32°, the maximum shock wave pressure of 1.468×10^{10}pa is generated, which rapidly decayed within 1ms and finally decayed to the standard atmospheric pressure p_0. After the laser absorption coefficient ζ is introduced, the theoretical shock wave sound intensity is modified once to reduce the model error.

3.3 The Relationship Between Sound Intensity and Energy

The relationship between the maximum pressure of the shock wave of the laser acoustic signal and the laser energy can be expressed as:

$$p = \lim_{\Delta t \to 0} v_1 \rho_0 \left(c_0 \Delta t \left[1 - \left(1 - \frac{1}{M_0} \right) e^{-\left[\frac{R_0}{c_0 \Delta t} \right]^{2-n}} \right] \right) / \Delta t \left[10^{\frac{\left(M_0 c_0 t \left[1 - \left(1 - \frac{1}{M_0} \right) e^{-\left[\frac{R_0}{c_0 \Delta t} \right]^{2-n}} \right] \right) / \Delta t - C_0}{v_2}} - 1 \right] + p_0 \tag{4}$$

When ς=0.46, then the laser convergence Angle is 32°, and the theoretical correspondence between the laser energy and the theoretical maximum pressure of the shock wave is shown in Fig. 5:

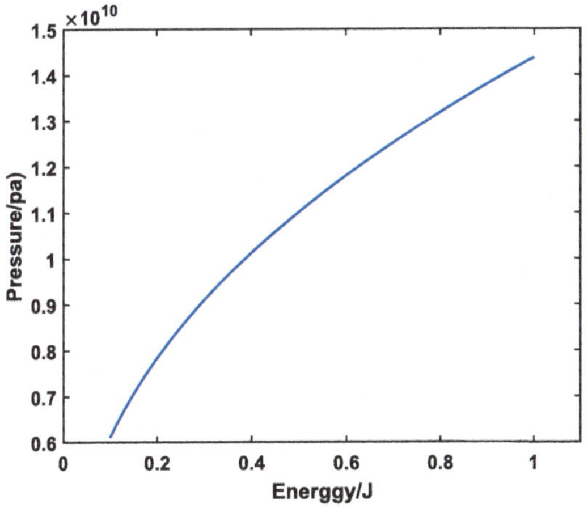

Fig.5. Relationship between laser energy and sound intensity of shock wave

As can be seen from Fig. 5, with the increase of energy, the pressure of shock wave generated by laser also gradually increases, but the growth rate gradually tends to be gentle. When two laser beams with the same total laser energy work at the same time, the minimum sound pressure of laser induced sound signal is single laser operation, and the energy of the other laser is 0. Therefore, under the premise that the energy of each laser is greater than the breakdown threshold and the sum of laser energy is certain, the use of multiple lasers can stimulate louder and stronger shock waves, that is, the problem of low sound intensity of single pulse laser is based on laser induced sound. Ignoring the influence of the laser convergence Angle, the maximum pressure produced by the 400 mJ energy single laser beam is 1.363×10^{10}pa, but the maximum pressure produced by the double laser beam can reach 1.53 times, which significantly improves the work efficiency. Therefore, under the premise that the energy is greater than the breakdown threshold, the number of array lasers can produce higher sound intensity as much as possible. At the same time, ζ is positively correlated with the laser convergence Angle. When the laser convergence Angle is larger, the value of ζ is larger. Therefore, by increasing the laser convergence Angle, the absorption coefficient of liquid to laser is increased, and the sound intensity is increased.

Equations (1), (2) and (3) can calculate the theoretical sound intensity of 400 mJ laser energy after 0.4 m transmission of the signal generated by laser sound, and the experimental results can be obtained from Fig. 6:

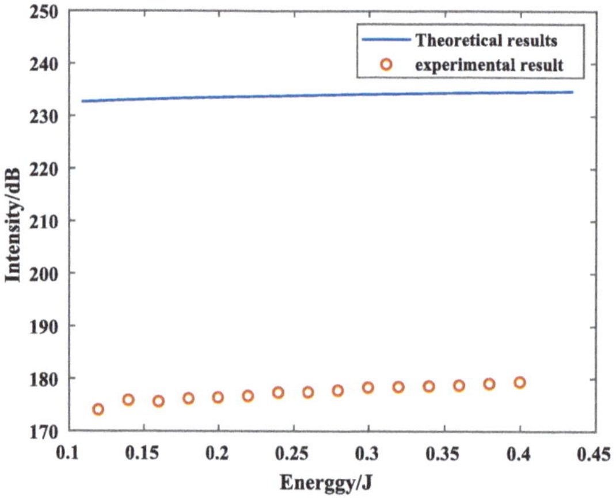

Fig.6. Comparison of theoretical and experimental sound intensity results

As can be seen from Fig. 6 and the above model, the simulated signal of sound intensity generated by higher laser energy has a larger value after 0.4m transmission, but due to the influence of transmission speed, its value gradually tends to atmospheric pressure. Moreover, as can be seen from Eq. (4), the relationship between energy and sound intensity is not linear, so the increase of signal sound intensity tends to be saturated when the energy increases within a certain range. The signal sound intensity measured in the experiment gradually increases with the increase of energy, and the growth trend is like that of the simulation results, but the theoretical sound intensity is about 57dB higher than that measured in the experiment, which is because the problems such as propagation loss and photoacoustic conversion efficiency are not considered in the theoretical calculation process. In the optical breakdown mechanism, the conversion efficiency is generally about 10%–30%. Therefore, to make the study of energy-sound intensity model more accurate, more influencing factors need to be considered.

4 Conclusion

The experimental platform of laser induced sound is built, and the law of laser acoustic signal intensity varying with energy is measured. At the same time, the laser absorption coefficient is induced by zeta, a parameter related to convergence Angle. The intensity and velocity of the shock wave generated by the plasma under the mechanism of optical breakdown are studied. The maximum velocity of the shock wave generated by the 1J pulsed laser at 32° convergence Angle is 5839 m/s and the maximum loud intensity is 1.468×10^{10}pa, and it is pointed out that the velocity will eventually decay to the speed of sound in water. The sound intensity eventually decays to atmospheric pressure p_0. It is concluded that the absorption rate of single laser beam can be improved by increasing the convergence Angle, and the energy limitation of single laser pulse can be solved by using laser array. The theoretical calculation formula between laser energy and sound signal intensity is derived, which provides some theoretical and experimental basis for the follow-up research.

Acknowledgement. This work was supported by the Opening Funding of National Key Laboratory of Electromagnetic Space Security, and Laoshan Laboratory Science and Technology Innovation Project (No. LSKJ202200801).

References

1. Luo, H.J., Wang, J.L., Bu, F.F., et al.: Recent progress of air/water cross-boundary communications for underwater sensor networks: a review. IEEE Sens. J. **22**(9), 8360–8382 (2022)
2. Brelet, Y., Jarnac, A., Carbonnel, J., et al.: Underwater acoustic signals induced by intense ultrashort laser pulse. Acoust. Soc. Am. **137**(4), 288–292 (2015)
3. Li, P., Zhao, Y., Zhou, Z.Q., et al.: Research on laser induced acoustic detection of trans-media aerial-underwater. Infrared Laser Eng. **50**(5), 20200310 (2021)
4. Hansson, I., Morch, K.A.: The dynamics of cavity clusters in ultrasonic (vibratory) cavitation erosion. J. Appl. Phys. **51**(9), 4651–4658 (1980)
5. Lauterborn, W.: Cavitation bubble dynamics - new tools for an intricate problem. Appl. Sci. Res. **38**, 165–178 (1982)
6. Tomita, Y., Shima, A., Takahashi, K.: The collapse of a gas bubble attached to a solid wall by a shock wave and the induced impact pressure. Am. Soc. Mech. Eng. **105**(3), 96 (1983)
7. Fujimoto, J.G., Lin, W.Z., Ippen, E.P., Puliafito, C.A., Steinert, R.F.: Time-resolved studies of Nd: YAG laser-induced breakdown. Plasma formation, acoustic wave generation, and cavitation. Invest. Ophthalmol. Vis. Sci. **26**(12):1771–1777 (1985)
8. Sacchi, C.A.: Laser-induced electric breakdown in water. J. Optical Soc. Am. B **8**(2), 337–345 (1991)
9. Ni, X., Zou, B., Chen, J.: On the generation of laser-induced plasma acoustic waves. Acta Phys. Sinica (002):142–147 (1998)
10. Li, S., Weng, W.B., Kong, C.D., et al.: Dual-Laser-Induced breakdown thermometry via sound speed measurement: a new produced for improved spatiotemporal resolution. Sensors **20**, 2803–2814 (2020)
11. Kajiwara, I., Akita, R., Hosoya, N.: Damage detection in pipes based on acoustic excitations using laser-induced plasma. Mech. Syst. Signal Process. **111**, 570–579 (2018)
12. Jang, H., Song, H.J., Koh, H.S., et al.: Shock wave generation in water by nanosecond pulse irradiation with 1064 and 2940 nm wavelengths. Opt. Laser Technol. **167**, 109670 (2023)
13. Hou, R.J., Xu, B., Xia, Z.Y., Zhang, Y., et al.: Numerical simulation of enhanced photoacoustic generation and wavefront shaping by a distributed laser array. Appl. Sci. **11**, 9497–9513 (2021)
14. He, Q.Y., Zong, S.G.: Analysis and thinking about the development of air and water crossed laser acoustic technology. Laser Infrared **49**(1), 3–8 (2019)
15. Lu, J.Y., Wang, J.A., Zong, S.G., et al.: Research on energy distribution after laser-induced optical breakdown underwater. Chin. J. Laser **40**(10), 1002007 (2015)

Design of Deep-Sea Optical Transmission System for "Jellyfish"

Yingjie Liu[1], Qinghua Sheng[1], Yi Ke[2], Yanbing Jiang[2], and Xiaofang Huang[3](✉)

[1] School of Electronics and Information, Hangzhou Dianzi University, Hangzhou 310018, China
[2] College of Media Engineering, Communication University of Zhejiang, Hangzhou 310018, China
[3] School of Electronic Engineering, Hangzhou Dianzi University Information Engineering College, Hangzhou 311305, China
20221016@hdu.edu.cn

Abstract. The emergence of deep-sea ROVs (Remotely Operated Vehicles) is crucial for marine scientific research and exploration of marine resources. The optical transmission system is an important guarantee for ROVs' reliable and stable operation in the deep-sea environment. For the multi-service demands of the optical transmission system of the "Jellyfish" Mobile Rig Platform, a design solution of the multi-service optical transceiver with a single optical fiber was proposed, realizing the functions of underwater HD 1080P 3G-SDI video transmission, Gigabit Ethernet (GbE), and serial port data communication. Compared with the traditional Wavelength Division Multiplexing (WDM) technology implemented in the multi-module combination optical transceiver, the single-module optical transceiver is more reliable and stable, and its functions are more powerful than those of the traditional ones. The sea trial suggests that the system is designed to meet the demands for underwater communication for deep-water ROVs.

Keywords: Deep-sea ROV · Optical transceiver · 3G-SDI · Ethernet · Serdes

1 Introduction

The ROV is a robot system that can be submerged. It is controlled and observed by an operator on shore or a diver in the water and is applied in operations such as port and marine infrastructure, collection of marine organisms, exploration for marine oil and gas, and underwater search and rescue [1]. The operator can obtain underwater information through pictures and videos, locate underwater targets, learn about the underwater environment with all kinds of sensors, and obtain experimental samples by using mechanical components [2]. Different ROVs have different shapes and unique functions, playing their parts in exploring

D. Jeng and B. Cai (Eds.): ICCOE 2024, LNCE 423, pp. 358–373, 2025.
https://doi.org/10.1007/978-981-96-0181-3_33

oceans, lakes, and rivers [3]. In deep-sea operation, the underwater ROV system and the control equipment on the water are connected by photoelectric composite cables, also called umbilical cables. To ensure the reliability and stability of communication in places with large water depths (up to 10km), copper cables are used only for power transmission, and optical fibers transmit data signals. First, it avoids the attenuation and interference of electrical signals in the copper medium in long-distance transmission and data errors caused at the receiving end. Second, when a single optical fiber is used to transmit multiple channels of data signals, it can effectively decrease the pressure on umbilical cables [4]. As shown in Figs. 1 and 2, the "Jellyfish" Mobile Rig Platform, jointly developed by the China Geological Survey, Zhejiang University, and Guangzhou Marine Laboratory, is designed for geological exploration of carbon dioxide, methane, dissolved oxygen, and conductivity temperature depth the seafloor at greater water depths, as well as sampling of seafloor sediments, pore water, and bottom water. Its maximum submerged depth is up to 10km. To ensure that clear geological features of the seafloor can be obtained in real-time, the optical transmission system of the equipment should be able to transmit multiple channels of 1080P HD videos, one channel of Gigabit Ethernet (GbE), and several channels of RS232/RS485 signals.

Fig. 1. The "Jellyfish" Mobile Rig Platform

Among them, 1080P videos are transmitted by using the 3G-SDI signal. Two signal channels from the underwater HD camera are transmitted to the equipment at the surface receiving end in real time. Gigabit Ethernet and RS232 /RS485 signals are responsible for each channel's data transmission of network cameras, sensors, and valve boxes.

Multiplexing technology refers to the recombination of resources of multiple channels in a certain way and transmission in a single medium to achieve the sharing of medium resources. The corresponding demultiplexing technology is used at the terminal to restore multiple channels of original resources. The multiplexing technology is like a tunnel for the transmitted data resources. There is a degree of delay in the data at both ends of the tunnel, but the content is identical. Currently, the common multiplexing schemes are as follows:

(1) Wavelength division multiplexing (WDM): Each channel's data is transmitted on a different wavelength optical carrier, combined at the transmitting end and separated at the receiving end[5]. This process causes optical attenuation, reducing signal quality, and the independence of each optical transmitter and receiver reduces resource efficiency and reliability.

(2) Time division multiplexing (TDM): Data from each channel is transmitted on the same optical carrier through time slicing, then restored at the receiving end[6]. TDM reduces optical attenuation but complicates data processing, requiring different methods based on data type to meet protocol requirements.

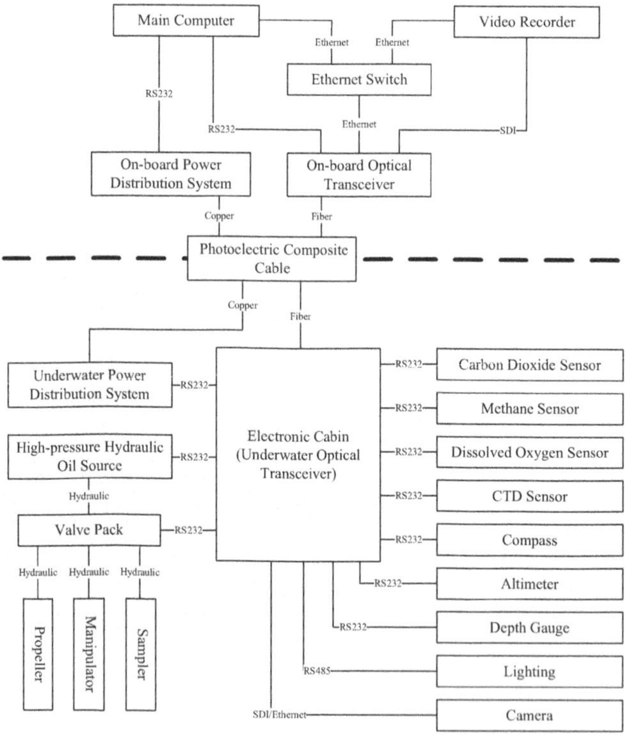

Fig. 2. System block diagram of "Jellyfish" Mobile Rig Platform

Table 1 shows the comparison table of video parameters of underwater optical transceivers in different countries. The video transmission capability provided

Table 1. Comparison of video parameters of underwater optical transceivers in different countries

Equipment name	Multiplexing technology	Video parameter
China Haidou ROV [7]	Unknown	HD
China Sea Horse ROV [8]	WDM	HD, 1080I
France Victor 6000 ROV [9]	WDM	SD, Analog
Japan ASSS11000 ROV [10]	Unknown	SD, Analog
US Jason & Medea ROV [11]	Unknown	4K UHD, local storage
924-HDE	TDM	Full HD, 1080P

by the device is basically limited to 1080P. Even on equipment supporting 4K UHD video, videos can only be stored locally and cannot be transmitted in real-time. Based on the above scheme, a new RGIF bus, and VDMA structure are proposed in this paper to optimize the transmission of UHD video signals from the perspective of data transmission efficiency. The main innovation points and work contents of the scheme proposed in this paper are as follows:

(1) Lossless acquisition, transmission, and restoration of 1080P60 3G-SDI videos on a single fiber dual channel is achieved.

(2) The TDM technology is used for lossless transmission of 2-channel 3G-SDI signals and provides 1-channel Gigabit Ethernet and 16-channel RS232/RS485 signal transmission.

(3) A non-homologous restoration scheme of video data based on the RGIF interface and arbitration algorithm design is proposed, effectively improving the transmission efficiency and stability of the bus in the system and realizing the lower data transmission latency.

(4) With a multi-fiber redundant backup design, it can automatically switch to the backup line when the main optical fiber line is interrupted, which can ensure that the deep-sea operations are reliable and stable.

The remaining parts of this paper are organized as follows. In the second part, the system as a whole and the specific implementation process of each part are presented; in the third part, the core design of the receiving end is introduced; in the fourth part, the process and results of the experiment are shown; and in the fifth part, the conclusion is given.

2 System Scheme

2.1 Introduction to the Overall Scheme

The system's overall structure in this paper is shown in Fig. 3. The system includes the underwater transmitting and receiving end on the water's surface. The transmitting end receives 3G-SDI images, Ethernet data, and serial port data. The receiving end is responsible for data restoration and is controlled by the operation platform on the water's surface. FPGA is the core of this system.

The task of non-homologous restoration of image data is completed through the RGIF VDMA logic inside the FPGA.

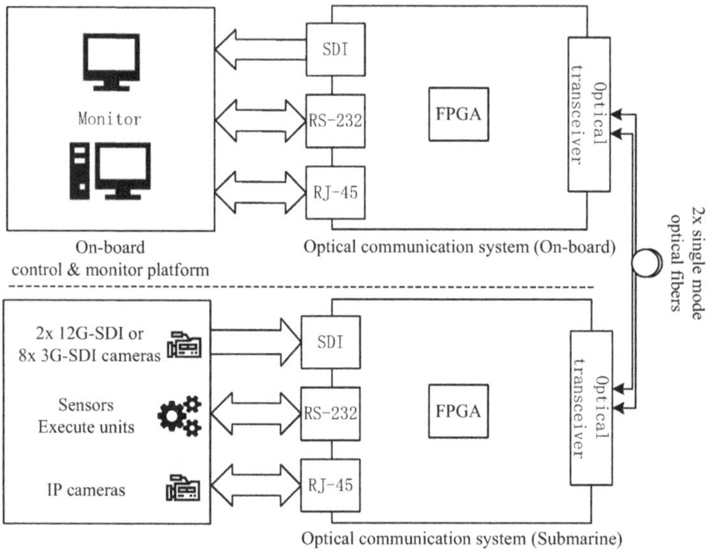

Fig. 3. Overall scheme of the system

2.2 Non-Homologous Data Restoration Scheme

At the receiving end, the 3G-SDI video bitstream should be output without interval. The local crystal oscillator should be used to restore the images and ensure the phase noise is relatively low. This will lead to different rates on both sides and image flicker. The digital Voltage-Controlled Crystal Oscillator (VCXO) or triple buffering technology can be used to adjust the output rate to ensure that the input and output rates are balanced.

In this design, the VDMA-based triple buffering technology is applied. The local clock source is used at the receiving end to restore and output data. In this caching method, the following features of SDI are utilized. Each frame is not correlated and is cached independently. The readout logic ensures that a frame of complete and reliable image is read each time. Due to different transmitting and receiving rates, the distance between the reading and writing pointers accumulates, and the images will not overwrite. The VDMA technology has better time stability compared with external VCXO. Auxiliary information can be inserted into the image for various application scenarios.

This design realizes the triple buffering technology through RGIF VDMA (Request and Grant Interface Video Direct Memory Access). The address controller in VDMA controls the read and write pointers independently and keeps them from crossing each other. After a system reset, the writing pointer returns to its original position and writes an image frame in the first memory area. After writing, the reading pointer is informed of starting to read the data. The writing pointer switches to the second memory area after writing a frame of the image. It is repeated in this way. The core strategy of triple buffering is as follows. The reading pointer always only reads a frame of images written before, and the writing pointer always only writes in a memory area not occupied by the reading pointer. Compared with dual buffering, triple buffering ensures that the read data is not corrupted by the writing pointer. Even if the writing stops, the reading pointer will continuously read a frame of image previously cached. Triple buffering is a widely used image buffering strategy.

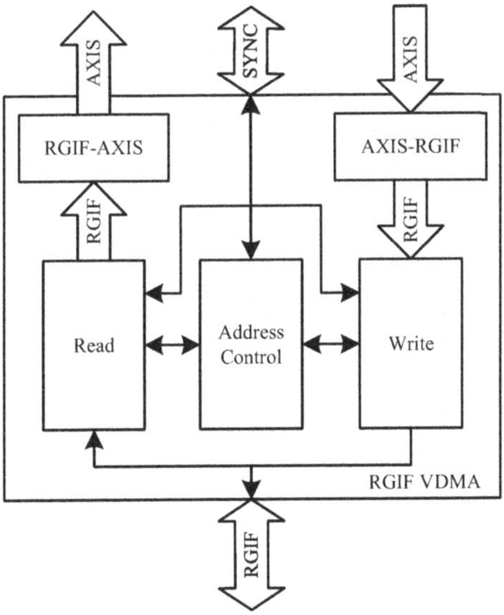

Fig. 4. VDMA internal schematic diagram

Figure 4 is the VDMA internal schematic diagram. It mainly consists of logic at the reading, writing, address controller, RGIF (Request and Grant Interface), and AXI4-S (Advanced eXtensible Interface 4 Stream) conversion interface. The image data enters VDMA as SDI over AXI4-S, and the address controller controls the address to point to the correct location at the reading/writing end. In addition to completing the triple buffering strategy, VDMA also converts the AXI4-Stream code stream to the RGIF code stream or the RGIF code stream to the AXI4-Stream code stream. The RGIF protocol in this design has higher sequential reading and writing efficiency than the AXI4-MemoryMapped protocol, which can lower DRAM's data bit width requirement in practical applications.

In this method, FPGA should have a buffer zone big enough for a multiple-frame image, and the reading/writing bandwidth should be more than twice the video output bandwidth. DDR SDRAM memory is generally used to meet the requirements of a video buffer. When the triple buffering technology is used, the transmission line rate on the optical fiber should be only greater than the sum of video rates. In the actual transmission, TDM technology is used. The types of data transmitted at different times are arbitrated by FPGA.

3 Core Design at the Receiving End

3.1 Highly Efficient RGIF VDMA

RGIF VDMA means a request to allow direct access to the memory of the interface video. It is responsible for the non-homologous restoration of 3G-SDI data and is the core module at the receiving end on the water's surface. The module uses the independent SDI data frames to cache and restore data in the local clock domain to ensure a stable picture. RGIF VDMA is a DMA of video using the RGIF interface. RGIF is a bus interface redesigned according to the features of the video cache. With a dual-channel RGIF arbiter and a length of 344 Burst, the Xilinx MIG DDR4 has a memory bandwidth utilization of 88.23% for reading and writing at full load, which is calculated by the sum of the reading and writing time. The efficiency of the RGIF interface and arbiter is higher than that of the third-party AXI4-MM VDMA under the same conditions. Figure 5 shows the design block diagram of the 3-channel RGIF VDMA and arbiter, and the 2-channel design is actually used. On the left side of the picture is a superposed channel, which overlays OSD characters. The data in the module is exchanged with the RGIF interface, and AXI4-Stream and AXI4-MM are the module's external interfaces. 2-channel of 12G-SDI data enters RGIF VDMA in the format of AXI4-Stream.

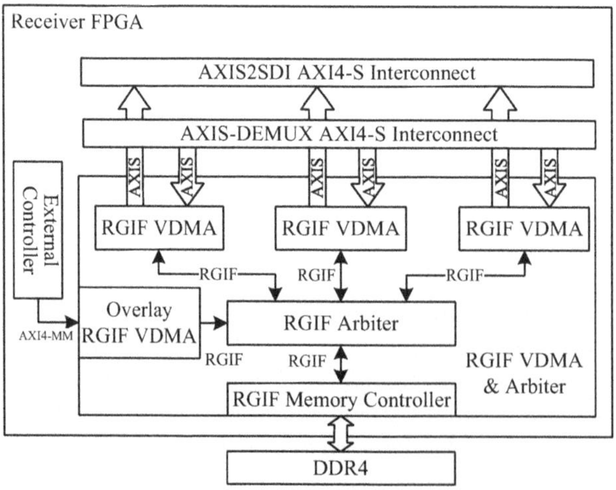

Fig. 5. RGIF VDMA and arbiter

The system's core components are at the receiving end of RGIF VDMA. Because the user-defined RGIF interface is used, an RGIF-based VDMA is redesigned in this design. RGIF VDMA isolates the SDI and local clock domains from each other at the transmitting end. The problem of image tearing due to different clock frequencies is solved. After resetting the system, a triple buffering mechanism is formed in the RGIF VDMA, and three memory areas are created to store the video data of three frames of a single channel, as shown in Fig. 6. RGIF VDMA reads and writes in three memory areas simultaneously. The reading and writing address can point to any of the three memory areas. To make an explanation, the three areas are numbered 1, 2, and 3. First, RGIF VDMA writes a frame of data in Area 1. After a frame is written, the writing address switches to Area 2. At the same time, the reading address points to Area 1, and the reading of the last data frame starts. After writing in Area 2, the writing address switches to Area 3. After reading in Area 1, the reading address switches to Area 2. It is repeated in this way.

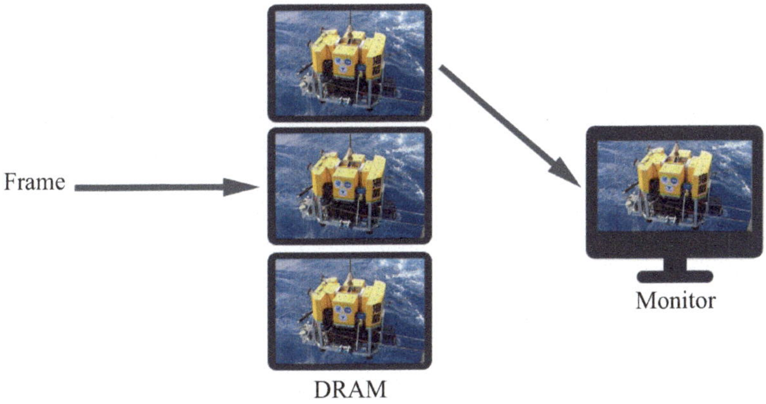

Fig. 6. RGIF VDMA and arbiter

The core of this triple buffering mechanism is as follows. When the address switches, the writing address always switches to the area not pointed to by the reading and writing address, and the reading address always switches to the last written area. This mechanism ensures that the read data is continuous. When the readout rate exceeds the writing rate, the reading address will remain the last written area. RGIF VDMA will alternately write in the two free areas when the writing rate exceeds the readout rate. The above two situations are the forced synchronization strategy of RGIF VDMA. When the difference between remote and local clock frequency is less than 10 ppm, RGIF VDMA requires a synchronization strategy after about 555.56 h, and the synchronization strategy does not affect the normal output of images.

Figure 7 is a complete structure block diagram of RGIF VDMA. The VDMA uses two AXI4-Stream interfaces and an RGIF interface for data exchange with

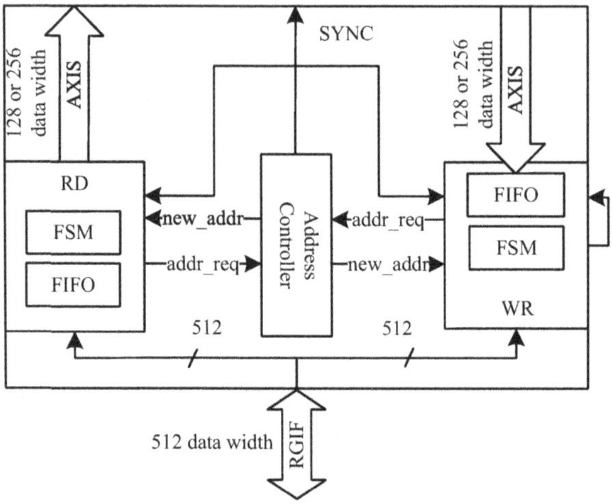

Fig. 7. Detailed structure chart of RGIF VDMA

the outside. The AXI4-Stream interfaces are divided into a Master interface and a Slave interface for output and input image data. The image data of 12G-SDI is transmitted as an AXI4-Stream data package. Because the bit width of the user-side interface data of the native 12G-SDI is 80 bits, in the case of AXI4-Stream packets, the bits need to be rearranged. This step is completed by the SDI2AXIS module. After it is wrapped by the AXI4-Stream protocol, the data enters the logic at the writing end. The logic at the writing end consists of a state machine and FIFO (First Input, First Output) memory. The state machine writes data to the corresponding memory area according to the address output by the address controller. The logic at the reading end is similar. In addition to the AXI4-Stream and RGIF interface, there is a SYNC interface. The signals processed by this interface are actually some secondary signals of AXI4-Stream, which are used to control and debug reading and writing pointers. For example, the "tuser" signal marks the first pixel of a picture and is also the starting signal after the resetting of the state machine. Prempty, prefull, and other signals are used to indicate the empty reading and full writing status of the internal FIFO.

3.2 New RGIF Bus

RGIF (Request and Grant Interface) is a bus interface designed based on AXI4-Stream. Compared with AXI4-Stream, RGIF adds the address signal and improves the reading and writing performance according to the features of the video cache. Video data generally occupies a continuous address space and is rarely random-access memory. We hope to transmit data during most bus operations to improve reading and writing efficiency instead of transmitting some unnecessary address logic. Therefore, RGIF uses a long Burst transmission scheme, with a maximum single transmission length of 1024, and only the first address and the number of data are transmitted in one reading and writing operation. Like most modern buses, each RGIF bus has two independent reading and writing channels. The reading and writing operations can be completed simultaneously. The RGIF interface determines a Burst transmission using the req and fin signals, with req indicating a transmission request and fin indicating the end of the transmission. All signals of the RGIF interface are shown in Table 2.

Table 2. Definition of RGIF bus

Signal name	Position (bit)	Direction	Definition
wr_req	1	Master → Slave	Writing request signal
wr_addr	WA	Master → Slave	Write the first address
wr_num	10	Master → Slave	One less writing data
wr_dat	WD	Master → Slave	Write data
wr_mask	WD/8	Master → Slave	Write data mask
wr_grant	1	Slave → Master	Permit to write
wr_fin	1	Slave → Master	Writing ends
rd_req	1	Master → Slave	Reading request signal
rd_addr	RA	Master → Slave	Read the first address
rd_num	10	Master → Slave	One less reading data
rd_dat	RD	Slave → Master	Read the data
rd_grant	1	Slave → Master	Permit to read
rd_fin	1	Slave → Master	Reading ends

In Table 2, the user defines the bit width WA for writing address, WD for writing data, RA for reading address, and RD for reading data. When MIG IP provided by Xilinx is used as the memory controller of RGIF, WA, and RA, WD and RD have the same bit width, where the address bit width depends on the DDR4 address bit width, and the data bit width is equal to the data bit width of MIG IP.

Master and slave devices are distinguished for the RGIF interface. The master device must make the reading and writing request, and the permission signal must be transmitted from the slave device to the master device. The slave device of the RGIF interface is generally dynamic memory, and the refresh rate limits the continuous reading and writing length of some dynamic memories. In SDRAM, when the full-page reading and writing mode is used, the maximum number of data read and written at a single time is the length of a line of a Bank [8]. Refreshing is required after reading and writing. If the Burst length of RGIF increases, the reading and writing efficiency will not be effectively improved. When Xilinx MIG IP is used as the memory controller of RGIF, the Burst length of RGIF is actually not limited. The reading and writing efficiency of the RGIF interface is related to that of the MIG Native interface. To adapt the

RGIF interface to most memory architectures and reduce the unnecessary bit width cost, the width of the wr_num signal is limited to an acceptable range. In this design, twice the length of the longest consecutive Burst of SDRAM is used as the maximum number of wr_num reads and writes. wr_num of 10 bits is enough for most applications. The Burst length used in this design is the length of one line of a picture. The value of wr_num is 344.

Figure 8 shows the operation time sequence of the RGIF reading and writing channel. A reading and writing operation starts when the req signal is high and ends when the fin signal is high. The host must wait for the slave to set the grant signal high before the reading and writing operation. In the process of reading and writing, if the grant signal is low, the host will be in the waiting state. In the waiting state, the host will keep the current signal state until the grant signal is high again. The data for the RGIF interface is aligned in the Prefetch mode. The host must output the first data to the wr_dat data line before the grant signal is valid.

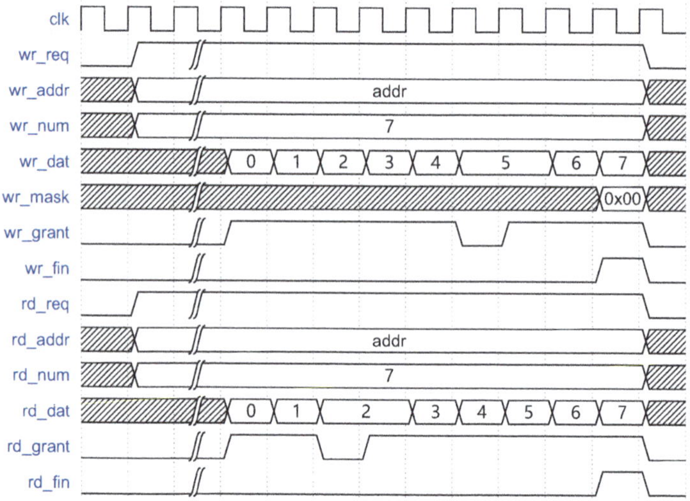

Fig. 8. Sequence diagram of RGIF bus

3.3 RGIF Arbiter

The RGIF arbiter is located between the RGIF VDMA and the RGIF memory controller. It plays a vital role as an essential part of the multi-channel RGIF interface. The RGIF arbiter uses the Round-robin strategy. Round-robin is a circular priority arbiter. This arbiter is characterized by responding to each request in a limited time. Other requests will not fail to respond because one request is too frequent. Requests arriving simultaneously are processed separately using a

Round-robin priority arbiter 1. The RGIF arbiter uses a fixed priority arbiter and a clearer request to achieve the effect of Round-robin. To adapt to the time sequence of the RGIF interface, the arbiter does not output an arbitration result for each clock. It starts the next arbitration after the arrival of the RGIF fin signal. The request signals from other channels must be cached during the arbitration process. Therefore, the RGIF arbiter has an internal request cache FIFO. The depth of the FIFO is equal to the number of channels, and its width is equal to the number of channels.

Figure 9 shows the structure chart of a 4-channel RGIF arbiter. The 4*4 square grid in the chart represents FIFO. The input of FIFO is connected to four request signals. Only when any request signal has a rising edge will the 4 request states be written to the FIFO, and the output will be connected to a standard arbiter with fixed priority. The priority of the standard arbiter becomes lower from the lowest bit to the highest bit. It will output the first position from the lowest bit to the main state machine. The main state machine accepts the output of the standard arbiter and FIFO and interconnects RGIF to the correct channel according to the arbitration result of the standard arbiter. If there are multiple requests in FIFO simultaneously, it suggests that all three requests arrive simultaneously. For example, the three low bits of "0111" in the first line of FIFO on the left in the chart are all 1. At this point, the state machine will first process the lowest bit according to a fixed priority and then clear the lowest bit with the request. The state machine selects the secondary low bit for interconnection according to the latest arbitration result of the standard arbiter and so on.

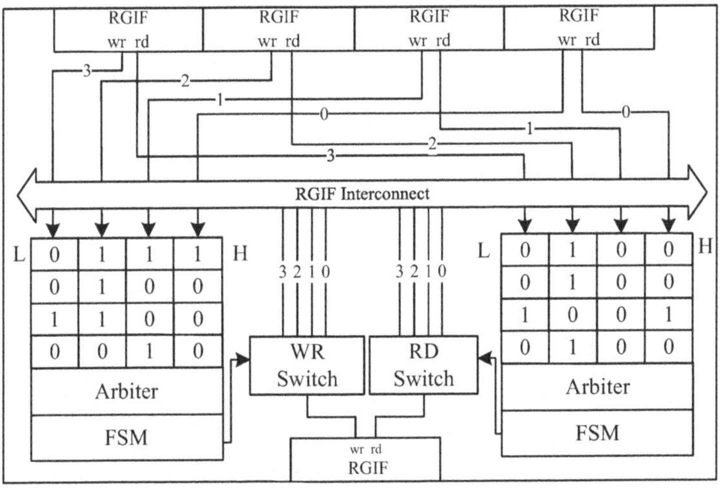

Fig. 9. Structure chart of RGIF arbiter

The RGIF arbiter will consume additional clock cycles while processing the request. Each channel consumes about 4 clock cycles. When the Burst length is 344, the RGIF arbiter will lose about 1.16% of the effective bandwidth.

Figure 10 shows the working waveform of the two-channel RGIF arbiter. rd_req[0] and rd_req[1] in the chart represent the reading request signals of two channels, respectively. wr_req[0] and wr_req[1] represent the writing request signals of two channels, respectively. rd_req and wr_req represent the arbitration results. In the waveform, the rd_req[0] signal sending the request first gets the response first, followed by wr_req[0], rd_req[1], and wr_req[1]. It can be noticed that wr_req[1] is the last to get the response. Its response time is about three times the Burst length, and this time interval is the maximum RGIF response time. The maximum RGIF response time is positively related to the number of channels. The more channels there are, the longer the maximum RGIF response time. If the response time is longer, more temporary data should be cached, so the number of channels and the depth of the FIFO must be increased.

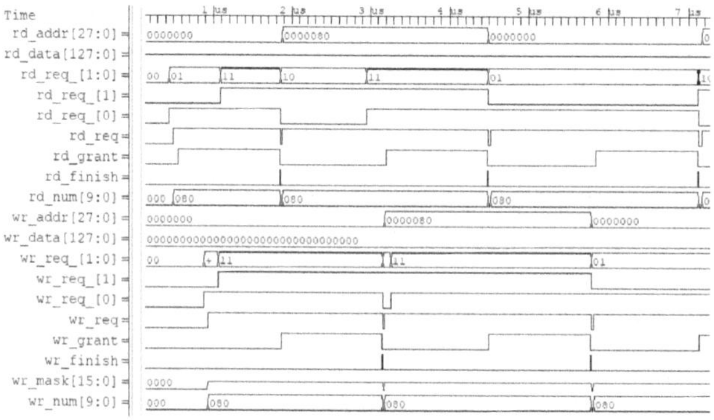

Fig. 10. Simulation waveform chart of two channels of arbitration of RGIF

4 Results

This system uses two Xilinx Kintex-7 XC7K325T FPGAs, which have 16 pairs of 12.5Gbps transceivers and support up to DDR3 memory. The maximum clock frequency of DDR3 is 1066MHz.

4.1 Preparation of Experimental Environment

In the video transmission test, two SDI signal sources are required. In the experiment, one video signal channel is output by a 1080P60 video player, and an SDI signal is obtained with an HDMI-SDI converter. For the other signal channel, a 1080P60 SDI camera directly outputs the SDI signal. The video signal test equipment is a Telestream SDI/IP waveform monitor. The specific model information is shown in Table 3.

Table 3. Video test environment

Name	Version model
SDI video source 1	UGREEN CM132
SDI video source 2	MOKOSE 3G-SDI Camera
SDI/IP waveform monitor	Telestream PRISM MPS-300

4.2 3G-SDI Eye Diagram Measurement

Through the actual test, the eye diagram data at the output end of the SDI signal is shown in Fig. 11[12]. The eye amplitude is 657.8 mV, the rising time is 150 ps, the falling time is 153ps, and the peak-to-peak jitter is 17.0 ps. The test data indicates that the eye amplitude of the SDI output signal and rising and falling time are slightly lower than the standard requirements [13–15], but it works stably in actual operation and can meet the requirements of the system design.

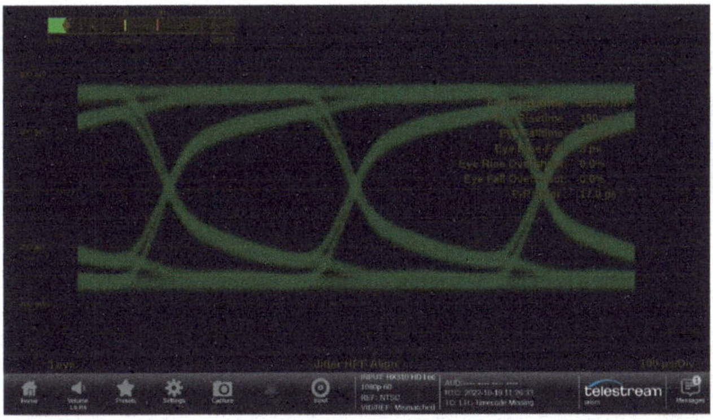

Fig. 11. 3G-SDI eye diagram

4.3 Stability Test

Figure 12 is the test chart of the system performance experiment. The dual-channel 1080P 3G-SDI video transmission experiment and eye diagram test are being done. Figure 13 is the operation interface of the system under the actual working conditions on the sea. The dual-channel 3G-SDI video transmission has a stable rate of 1080P60 through 3 h of long-time operation test and equipment monitoring. No frame loss of the video is observed during the testing process, and the results of the test on the sea are satisfactory. The test results show that the scheme in this paper can be used for a long time in practical applications and meet the requirements of real-time lossless video transmission and system stability.

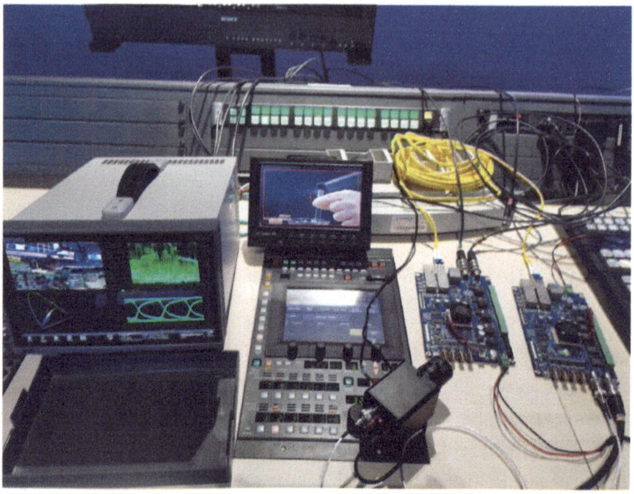

Fig. 12. System performance test

Fig. 13. Work interface of the sea test

5 Conclusions

This paper presents a non-homologous restoration scheme for video data based on the RGIF interface and arbitration algorithm design. The scheme designs a new RGIF bus interface by analyzing the cache features of video data in deep-sea operation optical transmission system to maximize the bus transmission efficiency inside FPGA. The result shows that the design can transmit dual 3G-SDI video over a single fiber and simultaneously transmit Gigabit Ethernet (GbE) and serial data. After joint commissioning and testing, the system works properly, performs stable, and achieves the expected design goals.

References

1. Dalhatu, A.A., de Azevedo, R.C., Udebhulu, O.D., De Tomi, G.: Recent developments of remotely operated vehicle in the oil and gas industry. Holos **3**, 1–18 (2021)
2. Chutia, S., Kakoty, N.M., Deka, D.: A review of underwater robotics, navigation, sensing techniques and applications. Proc. Adv. Robot., 1–6 (2017)
3. Tahir, A.M., Iqbal, J.: Underwater robotic vehicles: latest development trends and potential challenges. Sci. Int. **26**(3) (2014)
4. Liu, P., Wang, L., Hu, F., Wang, Y.: Research and prospect of underwater optical communication technology. Digit. Ocean Underwater War-Fare **5**, 329–334 (2022)
5. Yuli, C., Guangping, Q.: Design of optical fiber communication system under carrier photodiode/compact PCI bus and signal data modulation. J. Nanoelectron. Optoelectron. **17**, 495–504 (2022)
6. Zhuang, Z., Liu, G., Huang, X., Jia, X., Liu, W.-H., Guo, W.: Multi-Stage FPGA Routing for Timing Division Multiplexing Technique, MSFRoute (2020)
7. Tang, Y.G., Wang, J., Lu, Y.: Parametric design method and experimental research on Haidou full-depth ocean autonomous and remotely-operated vehicle. Robot **41**(6), 697–705 (2019)
8. Tao, J., Chen, Z.H.: Development and application of HAIMA (ROV). J. Eng. Stud. **8**(2), 185–191 (2016)
9. Yoshida, H., et al.: 2007 Symposium on Underwater Technology and Workshop on Scientific Use of Submarine Cables and Related Technologies. IEEE (2007)
10. Churiwala, S. (ed.): Designing with Xilinx® FPGAs. Springer, Cham (2017). https://doi.org/10.1007/978-3-319-42438-5
11. Ballard, R.D.: The MEDEA/JASON remotely operated vehicle system. Deep Sea Research Part I: Oceanographic Research Papers **40**(8), 1673–1687 (1993)
12. Park, J., et al.: A novel stochastic model-based eye-diagram estimation method for 8B/10B and TMDS-encoded high-speed channels. IEEE Trans. Electromagnetic Compatibility **60**(5), 1510–1519 (2017)
13. ST 291-1:2011 - SMPTE Standard - Ancillary Data Packet and Space Formatting. pp. 1–17. https://doi.org/10.5594/SMPTE.ST291-1.2011
14. ST 425-5:2019 - SMPTE Standard - Image Format and Ancillary Data Mapping for the Quad Link 3 Gb/s Serial Interface. https://doi.org/10.5594/SMPTE.ST425-5.2019
15. ST 2082-10:2018 - SMPTE Standard - 2160-line and 1080-line Source Image and Ancillary Data Mapping for 12G-SDIe. https://doi.org/10.5594/SMPTE.ST2082-10.2018

Research on Uplink Communication Across Water-Air Medium by a Millimeter-Wave Radar

Yuchen Du[1], Xiaolong Cao[1(✉)], Yiguang Yang[2], Tongchang Zhang[1], and Jianquan Yao[2,3(✉)]

[1] College of Ocean Science and Engineering, Shandong University of Science and Technology, Qingdao 266590, China
Caoxiaolong63@126.com

[2] Laoshan Laboratory, Qingdao 266237, China

[3] School of Precision Instrument and Opto-Electronics Engineering, Tianjin University, Tianjin 300072, China

Abstract. The uplink communication technology across the water-air interface holds significant application value for marine resource surveys, and the establishment of an integrated information network encompassing both sea and air. Currently, an effective wireless communication scheme that utilizes the translational acoustic radio frequency (TARF) system achieves exceptional anti-interference performance by utilizing the detection of acoustically induced water surface microwaves. When an underwater sound wave is transmitted to the water surface, it generates tiny vibrations of the identical frequency. Additionally, the air node can be employed to emit frequency modulated continuous waves (FMCW) towards the water surface for measurement purposes, enabling the retrieval of underwater data. However, the signal transmission process of this model remains incompletely understood, and there is an absence of a cross-medium transmission scheme suitable for complex underwater acoustic signals. Therefore, in this paper, the channel model of communication system is studied, and experiments are designed to verify it. Specifically, an in-depth investigation is conducted on the physical process of acoustic-induced water surface microwaves, focusing on low-frequency sound sources. A cross-medium information transmission system that operates between water and air was established. This system underwent experimental testing to discern and reinstate diverse acoustic signals emitted by underwater nodes through frequency domain transformation. The findings of this investigation offer crucial theoretical frameworks and practical insights for advancing cross-sea and air-medium wireless communication technologies, as well as underwater target exploration methods.

Keywords: Cross-medium communication · Underwater acoustic waves · FMCW · acoustically induced water surface microwaves

1 Introduction

As ocean exploration technology continues to advance, communication across water-air interface plays a crucial role in facilitating marine ecological research, underwater resource exploration, and the development of air-sea Internet infrastructure [1–3]. The

D. Jeng and B. Cai (Eds.): ICCOE 2024, LNCE 423, pp. 374–385, 2025.
https://doi.org/10.1007/978-981-96-0181-3_34

current cross-media communication primarily relies on relay modules such as buoys and ships, however, it suffers from limitations such as limited anti-interference capabilities and substantial costs [4]. Adopting communication technology without a relay node has the potential to significantly address existing challenges. Presently, there are three predominantly researched cross-media wireless communication technologies: visible light communication (VLC) direct communication, electromagnetic wave wireless communication, and underwater information recognition integrated with sound wave and microwave technologies [5].

Signals exhibit varying propagation characteristics across different media, and the methods of communication between air and underwater environments have been relatively well-established. Electromagnetic waves offer significant bandwidth and communication rates in the air. However, underwater, they experience substantial loss and absorption, resulting in limited propagation range and reduced communication rates. Consequently, they are primarily suited for low-frequency band operations [6, 7]. As the light signals traverse the medium, it becomes susceptible to the dynamic environment of the ocean and encounters challenges related to misalignment due to water refraction [8–11]. Sound waves propagate as mechanical waves underwater, offering the benefit of long-distance transmission. Nevertheless, they exhibit limitations such as a narrow bandwidth and a low communication rate. Additionally, sound waves undergo reflection attenuation when transitioning through different media, rendering them ineffective for efficiently transmitting information [12].

Research has demonstrated that when an underwater sound source emits sound waves towards the water-air interface, it generates minute vibrations at the same frequency as the sound source on the water surface, known as surface waves. These waves can be detected by laser sensors or millimeter-wave radar [13]. In 1988, Lee et al. [14] first proposed the theory of laser wave amplitude modulation on the water surface, successfully detecting underwater acoustic signals that elicited vibrations on the surface of the water. Nevertheless, the laser-acoustic detection technology necessitates precise collimation of the channel at both the receiving and sending ends, as the laser signal is highly susceptible to absorption by water.

Fortunately, in 2018, Tonolini et al. [15] introduced an innovative cross-media communication technology that utilizes acoustic wave and millimeter wave to establish a TARF-based communication system, enabling efficient wireless information transmission. The millimeter-wave radar emits frequency-modulated continuous waves onto the water surface, enabling the detection and demodulation of surface wave displacements triggered by underwater sound source signals in the low-frequency range. This technology enables the acquisition of underwater sound source information, boasts robust anti-interference capabilities, and facilitates long-range communication. In 2020, Romero et al. [16] devised an TARF communication system that emulated a water surface. This system utilized a loudspeaker covered with a film to replicate the underwater sound source and water environment. They successfully demonstrated the viability of the TARF channel in the low-frequency range of 100 ~ 300Hz by processing the received echo signal using IQ modulation. In 2022, Qu et al. [17] undertook a mathematical modeling of the TARF system, employing diverse signal carrier modes across both the middle and low frequency bands, to examine the channel performance. In 2023, Gai et al. [18] introduced a novel signal-to-noise ratio model, aiming to assess the performance indicators of TARF systems. Through rigorous experiments, they validated the model's accuracy and conducted a thorough examination of the communication system's underlying physics. In 2024, Fang et al. [19] presented a cross-medium underwater target detection system, leveraging the microwave characteristics of water surfaces and the principles of laser detection. They further conducted simulation analysis to explore cross-medium information transmission. Since then, the ongoing research on acoustic wave-millimeter wave joint cross-sea-air communication has been progressing steadily and is expected to yield further theoretical outcomes.

However, the air-sea transmedia uplink communication technologies mentioned above are primarily investigated with a focus on acoustic wave propagation characteristics, the practicality of surface vibration detection, as well as the final communication performance of the system. The lack of physical processes of acoustic surface microwaves and the cross-media transmission of various acoustic signals are studied. In this study, we present the development of a TARF water-air information transmission system that integrates acoustic and millimeter wave signals. We analyze and simulate the temporal evolution of water surface ripples induced by acoustic signals. Furthermore, we investigate the relationship between sound source frequency, intensity, and the micro-amplitude waves generated on the water surface due to transmitted acoustic signals. To demonstrate the system's capabilities, an experimental platform is established to conduct cross-media information transmission experiments with single-frequency, sweep-frequency, and simple coding signals. This approach enhances the diversity of information transmission in the TARF communication system.

2 Cross-Medium Communication Model

To enable water-to-air information transmission, it is imperative to have a theoretical model that accurately represents the physical process of signal transmission. The parameters of the water environment, including surface tension, viscosity, as well as the frequency and power of the sound source, exert a significant influence on the amplitude of surface vibrations. The physical model offers a dynamic adjustment scheme aimed at enhancing channel performance via system parameters.

Additionally, Fig. 1 illustrates the uplink communication process between water and air, utilizing the TARF system as its foundation. The system is organized into three distinct modules: the RF channel, the water-air interface, and the underwater acoustic channel. The transducer emits the underwater acoustic signal in the form of pressure waves, resulting in slight surface vibrations that propagate to the water surface. Millimeter wave radar emits FM continuous waves towards the water surface, receives the reflected signal, and subsequently processes it to reconstruct the source information.

2.1 Propagation Process of Underwater Acoustic Signals

During the propagation of underwater acoustic signals, attenuation occurs, consisting of both absorption and expansion losses. Absorption loss corresponds to the physical

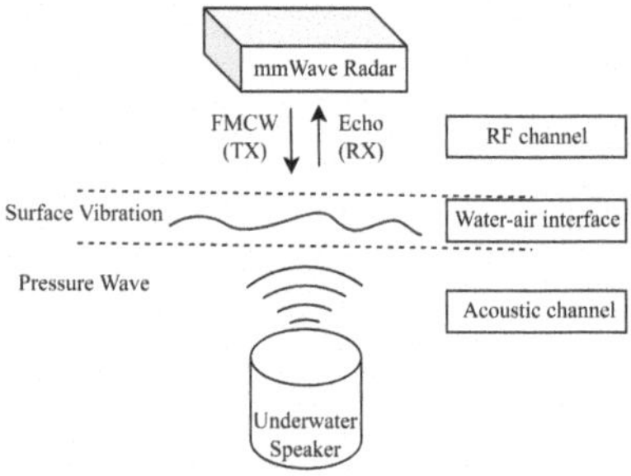

Fig. 1. The process of detecting water surface vibrations by a radar.

attenuation within the dielectric medium, resulting from sound energy dissipation or conversion within the medium's interior or during reflection. Conversely, expansion loss typically refers to the attenuation of sound intensity due to the wave front's expansion during propagation, which is closely tied to the sound signal's frequency and the geometric radiation field of the sound field. Acoustic signals are transmitted underwater through pressure waves designated as $P\ (r,\ t)$, varying both temporally with respect to time t and spatially with respect to distance r, and can be represented as per the formulation provided in[20]:

$$P(\omega, r, t) = A_0(\omega)e^{-\alpha r + \omega(t - r/v_W)}/r \tag{1}$$

The initial amplitude of the signal is denoted as A_0, while the attenuation coefficient of underwater sound is designated as α. The angular frequency is represented by ω, and the propagation speed of the acoustic signal in water is designated as v_W. When considering far-field plane waves, it suffices to consider solely the absorption loss of sound propagation. The absorption loss experienced by sound waves propagating through seawater is primarily attributed to the super-absorption resulting from various relaxation processes of dissolved salts. For low-frequency bands, the acoustic attenuation coefficient in seawater can be derived using the empirical formula outlined in reference[21]:

$$\alpha(f, h) = (\frac{0.109f^2}{1+f^2} + \frac{40.7f^2}{4100+f^2} + 3.01 \times 10^{-4}f^2)(1 - 6.67 \times 10^{-5}h) \tag{2}$$

where α is dB/km, f denotes the frequency of the underwater acoustic signal, and h represents the depth of the sound source. It is evident that the frequency has a negative correlation with the propagation range of sound waves. Shown in Fig. 2 is the alteration of the sound attenuation coefficient of the underwater sound source at a depth of 1m.

2.2 Acoustically Induced Water Surface Microwaves

A sound pressure wave is a longitudinal wave that, when transmitted to the water-air interface, generates lateral micro-vibrations on the surface, resulting in the formation of a water surface wave. This displacement can be derived by solving the boundary conditions of the wave equation. By defining the XOY plane as parallel to the water surface and

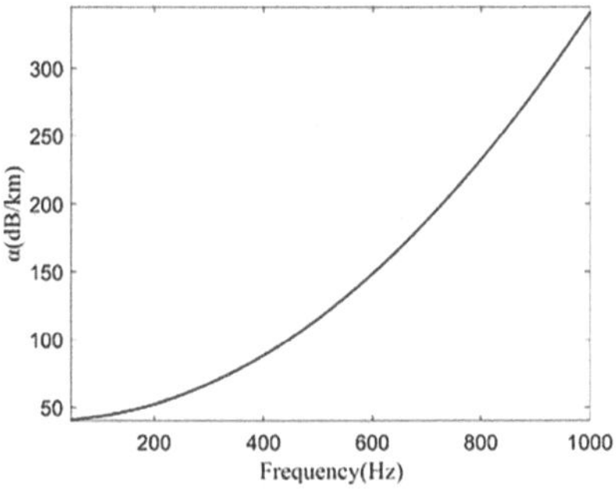

Fig. 2. The attenuation coefficient varies with frequency.

substituting the underwater acoustic pressure P, a three-dimensional mathematical model representing the temporal fluctuations of the water surface δ, can be derived[22]:

$$\delta(x, y, z) = \frac{2P}{\omega \rho v_W} e^{-\alpha_0 \sqrt{x^2+y^2} \cos(k\sqrt{x^2+y^2} - \omega t)} \tag{3}$$

$$\delta_{MAX} = \frac{P}{\pi \rho v_W \cdot f} \tag{4}$$

where k is the wave number of the water surface wave, α_0 is the attenuation coefficient of the surface wave amplitude, μ is the viscosity of the water medium, g is the acceleration of gravity, and σ' is the surface tension coefficient. When the detection point is situated precisely overhead of the sound source, meaning when the sound wave strikes the water-air interface perpendicularly, the maximum ripple amplitude can be achieved. Utilizing the mathematical model of water surface micro-waves, under specific dispersion relationships, modeling and simulation of water surface micro-vibrations were conducted, revealing the amplitude's varying trends in both vertical and horizontal directions. Figure 3 illustrates the mathematical representation of the water surface microwaves when the sound source frequency is set to 50Hz and the sound pressure level (SPL) is maintained at 170dB. It is evident from the figure that the attenuation of acoustic vibrations on the water surface microwaves is pronounced, with a maximum amplitude reaching up to 1.4μm in this particular setting.

Figure 4 illustrates the relationship among amplitude, frequency, and SPL. Notably, as the frequency of the underwater acoustic source rises and the sound source level diminishes, the peak amplitude on the water surface undergoes attenuation. Consequently, the choice of an optimal frequency and intensity for the acoustic signal holds significant influence on the efficient transmission of information across the medium.

2.3 Detection of Surface Vibration by a Millimeter Wave Radar

When the FMCW radar transmits the FM signals to detect a target, the echo signal received exhibits a specific frequency and phase shift in comparison to the initial signal. By mixing the echo signal with the transmitting signal, an intermediate frequency (IF) signal is derived, and its frequency and phase encode the displacement information of

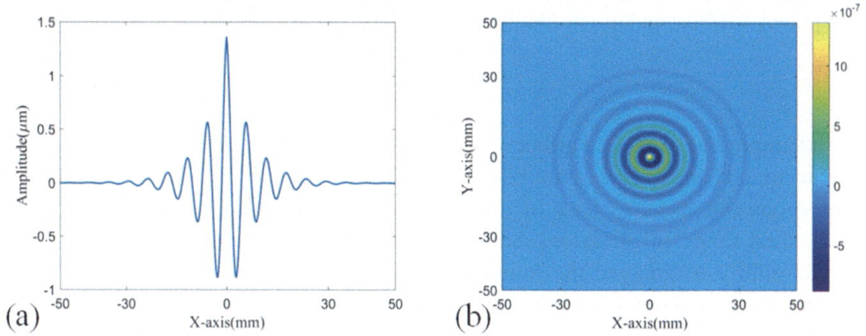

Fig. 3. Theoretical model of water surface microwave. (a)2D profile (b)3D top view.

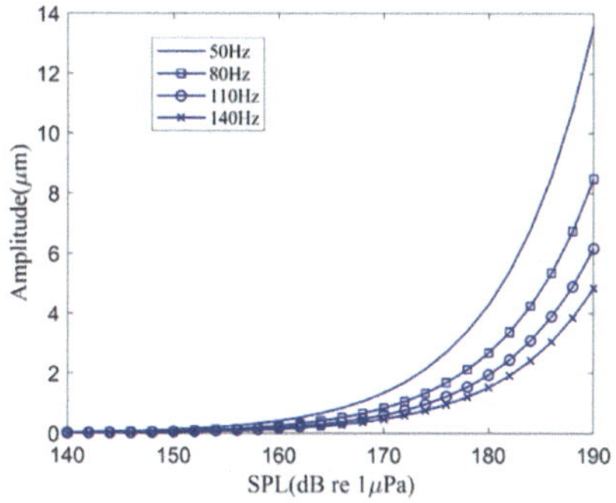

Fig. 4. The maximum amplitude of water surface varies with frequency and SPL.

the object being measured. The FMCW radar exhibits precision ranging capabilities and a superior signal-to-noise ratio, with its transmitted electromagnetic wave shape represented as follows:

$$s(t) = A\cos\left\{2\pi\left[f_0 + \frac{1}{T}Bt\right]t + \varphi_0\right\} \tag{5}$$

where A represents the amplitude of the FMCW signal, B denotes the bandwidth of the sweep signal, φ_0 indicates the initial phase of the signal transmitted by the millimeter wave antenna, T represents the sweep period, and f_0 is the carrier frequency of the signal, typically 24 GHz or 77 GHz for millimeter wave radars. The IF signal is denoted as:

$$X(t) = M\cos(2\pi\Delta ft + \Delta\varphi) \tag{6}$$

where M represents the amplitude, Δf denotes the frequency difference between the transmitted and received signals, and $\Delta\varphi$ signifies the phase difference. Upon transmission of the FMCW radar signal by the system, it is reflected after encountering the object to be measured, and received by the radar system after the delay τ at the transmitting time. The radial distance d between the radar and the measured object is derived from the frequency difference signal, and the relationship between them is given as follows:

$$d = \frac{\lambda \Delta \varphi}{4\pi} \tag{7}$$

when the amplitude of water surface vibrations is on the order of millimeters or less, it can be accurately measured by utilizing frequency differences. To fulfill more precise ranging requirements, the phase of the difference frequency signal can provide even more detailed range information. Additionally, short-time Fourier transform (STFT) offers the ability to analyze the temporal variations in signal frequency, along with instantaneous frequency and amplitude at each moment. This frequency information is obtained by applying a sliding window to the signal under analysis and performing a Fourier transform. The STFT is computed as follows:

$$STFT(t,f) = \int_{-\infty}^{\infty} x(\tau)h(\tau - t)e^{-j2\pi f\tau}d\tau \tag{8}$$

where x represents the initial signal, and $h(\tau - t)$ denotes the analysis window function.

3 Experimental Setup

To assess the viability of the aforementioned scheme in wireless communication between water and air media, an information transmission system was established in a pool environment, utilizing an underwater sound source and airborne millimeter wave radar, as depicted in Fig. 5(a). The system primarily consists of an underwater acoustic transducer, a power amplifier, the AWR1642 millimeter wave radar board from TI, and the DCA1000 acquisition card. The extensive 5-m-deep pool environment is capable of relieving echo vibration interference and enhancing the precision of information transmission.

An underwater sound source produces sound signals, which propagate towards the surface, inducing minute vibrations. Subsequently, the radar board dispatches multiple sets of frequency-modulated pulse frames onto the water surface. These detected signals are then converted by the acquisition card and transmitted to the computer via the network port for digital signal processing. The continuous wave bandwidth B = 4GHz, with a duration of $T_c = 40\mu s$ and a starting frequency of $f_c = 77GHz$. The experimental setup of the pool is depicted in Fig. 5(b), where the underwater sound source is situated 1 m beneath the surface, while the airborne radar is positioned 40cm above the water surface. To investigate the transmission diversity of underwater acoustic signals, it is imperative to calibrate the structural parameters of FMCW pulses, encompassing the start and stop frequencies, thereby enhancing the precision and discrimination of various signal types within a dynamic communication milieu.

4 Cross-Media Uplink Communication Experiment

4.1 Underwater Acoustic Signals Recognition and Processing

Initially, considering underwater acoustic signals as a medium for information transmission, it becomes evident from the propagation characteristics of acoustic waves that the effective transmission distance is frequency-dependent. Elevated frequencies result

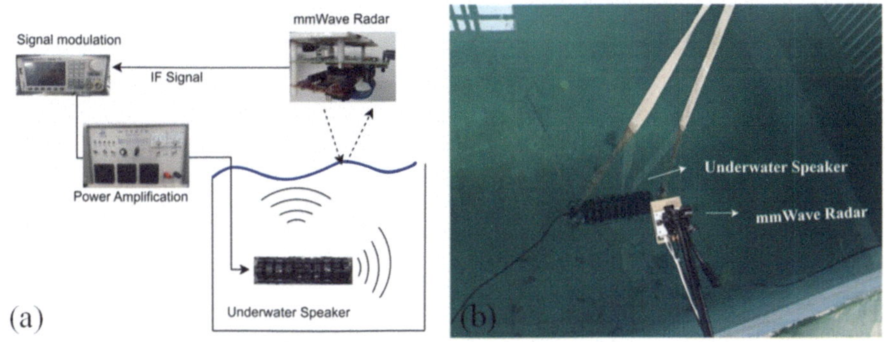

Fig. 5. (a) Cross-media information transmission system (b) Pool environment.

in enhanced attenuation, which hinders the generation of surface microwaves. Furthermore, in both pool and marine environments, the predominant external interference that impacts channel quality is low-frequency noise, including waves and equipment vibrations. Consequently, to enhance communication efficiency, it is imperative to select an optimal frequency band as the signal carrier and devise analog or digital filters to enhance signal transmission identification.

As previously mentioned, for the purpose of uplink communication across the water-air medium in this verification experiment, 166Hz has been chosen as the frequency for the underwater sound source. Subsequently, processing and analysis can be conducted in both the time and frequency domain. Additionally, Fig. 6(a) depicts the initial phase information of the IF signal detected by the radar board, which correlates with the vibration pattern of the water surface. Furthermore, as illustrated in Fig. 6(b), the outcome of the FFT frequency domain transformation of the subtle vibrations clearly identifies the signal emanating from the underwater sound source. Moreover, by designing a digital filter, the low-frequency noise can be effectively eliminated.

To enhance the diversity of information transmitted across interfaces, we design a sound source that emits a linearly frequency-modulated signal onto the water surface. This sinusoidal signal gradually rises from 400Hz to 900Hz, with each data cycle lasting for 5 s. The experimental findings are depicted in Fig. 7, after analyzed in the frequency domain to extract the phase information of the difference frequency signal, measured by the radar board. By applying the STFT, the temporal and frequency variations of water surface vibration information can be observed. To mitigate the interference from environmental noise, visual processing can be implemented by establishing a signal strength threshold, as demonstrated in Fig. 7(b). Furthermore, the findings of this sweeping experiment also confirm the attenuation of underwater acoustic signals during propagation, highlighting the frequency selection characteristic of the TARF system. Additionally, the design of the underwater acoustic transducer plays a crucial role in cross-medium communication.

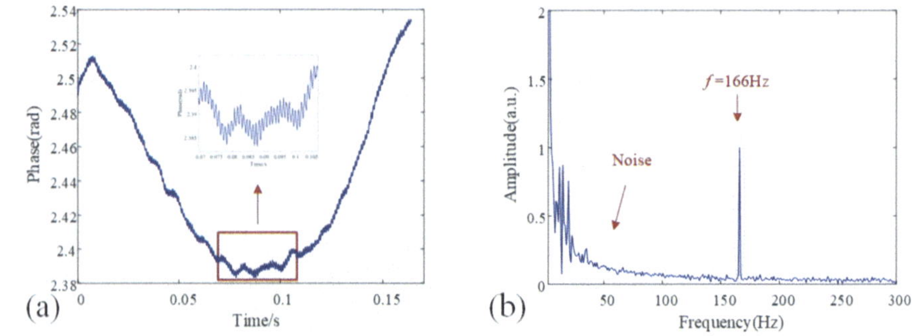

Fig. 6. Single frequency signal transmission, f = 166Hz. (a) Time domain (b) Frequency domain.

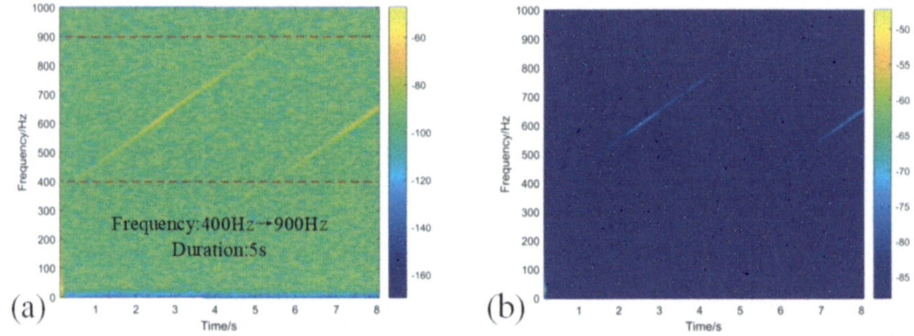

Fig. 7. Linear sweep signal transmission. (a) Initial acquisition (b) Filter processing.

4.2 2FSK Encoded Signals Transmission Across Water-Air Media

The experiments conducted in Sect. 4.1 have successfully demonstrated the feasibility of water-air wireless communication through the TARF system using both single and variable frequency acoustic signals. However, to cater to the practical demands of cross-media communication, it is imperative to transmit complex underwater information, such as terrain and oil resources, to the air node for processing in a precise and timely manner. Additionally, to ensure the confidentiality of the communication process, encoding and modulation of the generated signal are crucial.

Therefore, a straightforward coded signal transmission experiment was devised, utilizing 2FSK modulation. In this experiment, 500Hz represented '0', while 700Hz denoted '1'. The corresponding time-frequency analysis of the detected vibrations on the water surface, with a communication rate of 5bps, is depicted in Fig. 8. It is evident that the modulated acoustic signal can be accurately discerned by the airborne millimeter-wave radar. This enables the restoration of underwater sound source information over time, maintaining signal precision even at specific communication rates.

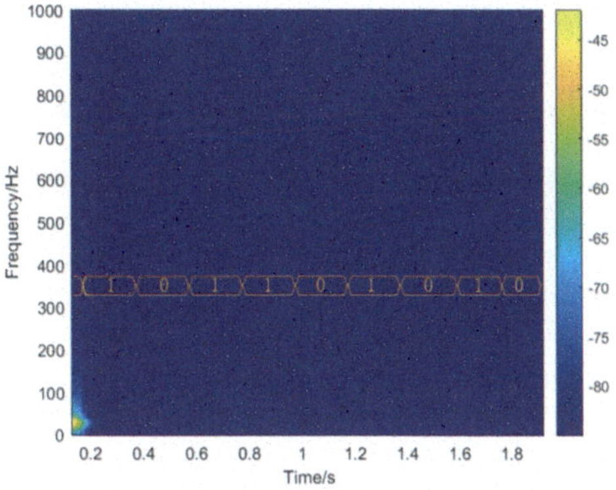

Fig. 8. 2FSK Simple Code Signal Transmission (5bps).

5 Conclusion

In this paper, we investigate the uplink communication technology across water-air medium, simulating and analyzing the characteristics of the TARF channel. Furthermore, we design experiments to assess the viability of cross-media information transmission and explore the complex signal transmission demands of practical applications. Our findings are summarized as follows:

Firstly, during the propagation of underwater acoustic signals, they are not solely restricted to the interior of the water body. Instead, they emit sound waves carrying a certain amount of energy, which propagate towards the water-air interface and generate minute vibrations with the identical frequency as the sound source. The simulation outcomes reveal that the amplitude of these vibrations typically falls within the micron range. Furthermore, this amplitude is influenced by the frequency and intensity of the sound source, subject to specific dispersion relations. This surface vibration can be detected by a laser sensor or millimeter-wave radar, employing micro-ranging technology, thereby enabling the recovery of underwater sound source information. Secondly, due to the optical signal's stringent detection environment requirements and its inherent collimation characteristics, it is not ideal for handling complex aquatic surfaces. Alternatively, the millimeter wave phase ranging principle can be employed to construct a cross-media uplink wireless communication system. This system effectively identifies surface vibrations and recovers sound source frequency information while demonstrating resistance to noise interference. Eventually, the TARF system was developed to conduct experiments, examining the cross-media transmission of various signals, including single-frequency, sweep-frequency, and simple coding signals. In addition to validating the system's viability, the identification of 2FSK coded signals and a frequency domain analysis of a 5bps communication rate were achieved, offering valuable insights for future underwater node signal diversity.

Acknowledgement. This work was supported by the Opening Funding of National Key Laboratory of Electromagnetic Space Security, and Laoshan Laboratory Science and Technology Innovation Project (No. LSKJ202200801).

References

1. Ma, Y., Zhang, Q., Wang, H.: 6G: Ubiquitously Extending to the Vast Underwater World of the Oceans. Engineering **8**(01), 12–17 (2022)
2. Khan, M.A., Alzahrani, B.A., Barnawi, A., et al.: A resource friendly authentication scheme for space-air-ground-sea integrated Maritime Communication Network Ocean engineering 250(Apr.15), 110894.1–110894.8 (2022)
3. Akyildiz, I.F., Pompili, D., Melodia, T.: Underwater acoustic sensor networks: research challenges. Ad Hoc Netw. **3**(3), 257–279 (2005)
4. Chen, H., Chen, H., Zhang, Y., et al.: Decentralized estimation of ocean current field using underwater acoustic sensor networks. Acoustical Society of America **149**(5), 3106–3121 (2021)
5. Luo, H., Wang, J., Bu, F., et al.: Recent Progress of Air/Water Cross-Boundary Communications for Underwater Sensor Networks: A Review. IEEE Sens. J. **22**(9), 8360–8382 (2022)
6. Wang, H., Ren, Y.: EM wave propagation through the seawater-air interface in dynamic marine environment. In: Conference on Global Oceans: Singapore – US Gulf Coast, pp. 809–812. IEEE, Biloxi (2020)
7. Dautta, M., Hasan, M.I.: Underwater Vehicle Communication Using Electromagnetic Fields in Shallow Seas. In: International Conference on Electrical. Computer and Communication Engineering (ECCE), pp. 38–43. Cox's Bazar, Bangladesh (2017)
8. Schirripa Spagnolo, G., Cozzella, L., Leccese, F.: Underwater Optical Wireless Communications: Overview. Sensors (Basel) **20**(8), 2261 (2020)
9. Lv, Z., He, G., Yang, H., et al.: The Investigation of Underwater Wireless Optical Communication Links Using the Total Reflection at the Air-Water Interface in the Presence of Waves. Photonics **9**(8), 525 (2022)
10. Zhu, S., Chen, X., Liu, X., et al.: Recent progress in and perspectives of underwater wireless optical communication. Progress in Quantum electronics 73, 100274.1–100274.28(2020)
11. Junejo, N.U.R., Sattar, M., Adnan, S., et al.: A Survey on Physical Layer Techniques and Challenges in Underwater Communication Systems. Journal of Marine Science and Engineering 11(4), (2023)
12. Song, A., Stojanovic, M., Chitre, M.: Editorial Underwater Acoustic Communications: Where We Stand and What Is Next? IEEE J. Oceanic Eng. **44**(1), 1–6 (2019)
13. Dai, Z., Sun, J., Sui, P.: Theoretical Study on the Water Surface Transversal Mini-wave due to the Underwater Sound Field. Journal of National University of Defense Technology **26**(1), 95–98 (2004)
14. Lee, M.S., Bourgeois, B.S., Hsieh, S.T., et al.: A Laser Sensing Scheme for Detection of Underwater Acoustic Signals. In: IEEE Conf Southeastcon, pp. 253–257. Knoxville,USA (1988)
15. Tonolini, F., Adib, F.: Networking across Boundaries: Enabling Wireless Communication through the Water-Air Interface. In: Proceedings of the 2018 Conference of the ACM Special Interest Group on Data Communication, pp. 117–131. Budapest, Hungary (2018)
16. Romero, M., Narayanan, R., Lenzing, E.H., et al.: Wireless underwater-to-air communications via water surface modulation and radar detection. In: Radar Sensor Technology XXIV, pp. 114080C.1–114080C.14. (2020)
17. Qu, F., Qian, J., Wang, J., et al.: Cross-Medium Communication Combining Acoustic Wave and Millimeter Wave: Theoretical Channel Model and Experiments. IEEE J. Oceanic Eng. **47**(2), 483–492 (2022)
18. Gai, Y., Tan, Q., Zhang, Y., et al.: A Cross-Medium Uplink Communication Model Merging Acoustic and MillimeterWaves. Journal of Marine Science and Engineering 11(11), (2023)

19. Erzheng, F., Zongru, L.: Water surface capillary wave detection method based on optical devices. Journal of Harbin Engineering University **45**(2), 332–338 (2024)
20. Lurton, X.: An Introduction to Underwater Acoustic: Principles and Applications. Springer Science & Business Media, (2002)
21. Ma, D.: Modern Acoustics Theory Foundations. Science Press, Beijing, China (2004)
22. Zhang, L.: Research on Oprical Heterodyne Detection Technology for Acoustically Induced Water Surface Capillary Waves. Harbin Institute of Technology (2017)

Multi-field Coupling Modeling for Corrosion Damage Prediction of Deepwater Production Pipelines

Rong Lin[1,2], Qiao Zhang[1,2], and Shengnan Wu[1,2](✉)

[1] College of Safety and Ocean Engineering, China University of Petroleum (Beijing), Beijing, China
wushengnan@cup.edu.cn
[2] Key Laboratory of Oil and Gas Safety and Emergency Technology, Ministry of Emergency Management, Beijing, China

Abstract. Subsea production pipelines face increased failure in the complex ocean environments, corrosion is one of the main causes of pipeline failure. This paper focuses on the external corrosion damage mechanism of deepwater production system pipelines subject to different kinds of impact loading in the service process. By introducing electrochemical parametric equations that consider elastic-plastic deformation, stress field and electric field are adequately combined, and the external corrosion damage model of the pipeline is proposed considering the effects of the thermal-force-electrical multi-field coupling. Such model is intended to capture the corrosion electrochemical reaction behavior of a defective pipeline under the influence of various risk factors, including defect size, internal temperature, and axial displacement. The degradation mechanism-driven corrosion damage is summarized by considering the stress at the center of the pipeline defect, the anode current density, the corrosion rate, the amount of corrosion degradation, and the maximum safe operating pressure of the pipeline under different working conditions. The interaction of multi-field coupling in the corrosion damage to the pipeline is demonstrated by a case study. The results indicate that a larger defect size and a greater increase in the temperature inside the pipe will lead to higher stress on the defect, resulting in more serious corrosion damage. Plastic deformation can significantly increase the electrochemical corrosion activity of the pipe to exacerbate the corrosion damage to the pipe, providing support for pipeline risk assessment and integrity management.

Keywords: Deepwater production system · Multi-field coupling · Numerical simulation of pipelines

1 Introduction

Deepwater production systems are important technical equipment for the development of offshore oil and gas resources. In particular, various types of underwater pipelines play an important role in subsea production systems. However, the complexity and harsh

D. Jeng and B. Cai (Eds.): ICCOE 2024, LNCE 423, pp. 386–398, 2025.
https://doi.org/10.1007/978-981-96-0181-3_35

conditions of the marine environment make underwater pipelines highly susceptible to suffer damage [1]. It was studied that corrosion is one of the main causes of failure of subsea pipelines [2–4]. In this case, damage caused by external corrosion of pipelines as well as internal material defects can trigger sudden catastrophic failure of pipelines [5]. Currently, many scholars have conducted research on the external corrosion of pipelines under the coupling of complex factors by using numerical simulations. Considering the effects of submarine environmental loads and pitting corrosion on pipelines, Engin [6] used the finite element method to establish a nonlinear numerical model of the pipeline's external corrosion defects and investigated the effects of the external corrosion defects on the structural performance of the pipelines. A nonlinear finite element method [7] has been proposed to investigate the residual strength of pitting homogeneous pipelines under external pressure by considering the effects of strength, relative depth, corrosion region length and distribution of pitting corrosion. Moreover, based on the numerical study, a mathematical formula for predicting the residual strength of pipelines was proposed. In 2013, Xu et al. [8] firstly combined the force-electricity coupling effect with the corrosion finite element model of buried pipelines. The corrosion damage mechanism of pipelines under multi-field coupling was elucidated. Since then, many scholars [9–11] have also combined the theory of multi-field coupling analysis with numerical simulation and applied it to the analysis of corrosion damage.

At present, the research on pipeline corrosion damage using multi-field coupling theory combined with numerical simulation has been relatively mature. However, most of the studies on external corrosion of submarine pipelines are for pipelines in shallow water areas. Meanwhile, most of the simulation studies on the damage mechanism of external corrosion of pipelines containing defects under multi-field coupling are concerned with the coupling of external force field and corrosion electrochemical field, while ignoring the influence of internal fluid thermal field on corrosion. Moreover, most of them are also about buried pipelines, and the research on deep-water pipelines involves less. This is of limited help in providing decision support and solutions to engineering practice. To address the shortcomings of previous studies, this paper combines the theories of electrode dynamics, heat conduction, and elastic-plastic mechanics in multiple fields. Considering the external environment of the deep-water pipeline and the characteristics of the internal operation, the external corrosion damage analysis model of the pipeline under the action of thermal-force-electrical multi-field coupling is established to realize the analysis of multi-factors influencing mechanism of the corrosion of the pipeline containing defects.

2 Methodology

In this paper, we try to consider the influence of environmental and operational loads during the service period of deep-water pipelines, and use COMSOL Multiphysics multi-field coupling software to establish an analytical model of external corrosion damage under the effect of thermal-force-electrical field multi-field coupling. The thermal field refers to the internal temperature of the pipeline, the mechanical field is the pressure inside and outside the pipe as well as the axial stress caused by seabed activities or uneven terrain, and the electric field refers to electrochemical corrosion. Figure 1 shows the flowchart of the model establishment for analyzing the corrosion mechanism of deep-water pipelines under the action of multi-field coupling.

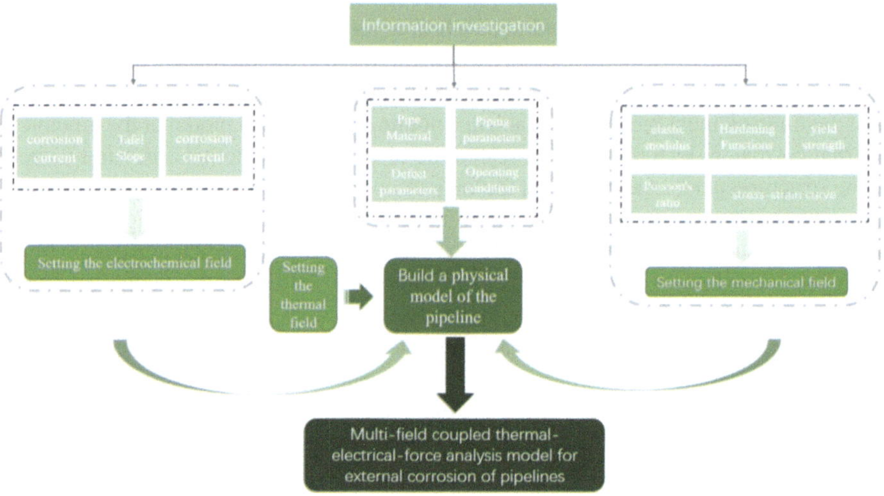

Fig. 1. The concrete steps for establishing a multi-field coupling analysis corrosion model

2.1 Geometric Modeling

The selected research object of this paper is a suspended section of a subsea jumper. The prototype and details of the model is shown in Fig. 2. Among them, (a) shows the various factors affecting the corrosion damage of the pipe section. Stress-strain data and hardening functions for API 5L X65 steel were obtained through literature research [12, 13].

$$\sigma_{yhard}=\sigma_{exp}\left(\varepsilon_{eff}\right)-\sigma_{ys}=\sigma_{exp}\left(\varepsilon_{p}+\frac{\sigma_{e}}{E}\right)-\sigma_{s} \tag{1}$$

where σ_{exp} is the stress-strain curve of API 5L X65 steel, ε_{eff} is total effective strain, σ_{ys} is yield strength of the steel, ε_p is plastic strain, σ_e is effective stress, and E is Young's modulus.

Referring to the relevant literature [14], the outer diameter D of the pipeline is set to 219mm, the wall thickness t is 16mm, and the length is 2 m. The defect shape is an ellipsoidal axial defect. Therefore, the length of the ellipsoidal defect's long axis, namely the defect length (L), is respectively set to 100 mm, 200 mm, 300 mm, and 400 mm for the study of the effect of different defect lengths on pipe corrosion. According to existing studies, the width of the defect has a negligible effect on corrosion [11], so the short axis of the defect is set to a fixed value of 25 mm. According to ASME B31G guideline, the depth of defects should not exceed 80% of the wall thickness, otherwise it is considered as pipe failure. Thus, the depth of the defect (d) is set to 20%, 40%, 60%, and 80% of the wall thickness to investigate the effect of different defect depths on pipe corrosion. Figure 2 (b) shows a schematic of the established geometric model of the pipe section.

After establishing the solid geometry model, it is necessary to model the seawater area with corrosive media in contact with the pipeline. As shown in Fig. 2 (b), the domain is set as an electrolyte in the COMSOL secondary current module with a conductivity of 3.2 S/m and a temperature of 4 °C to simulate deep water condition [15].

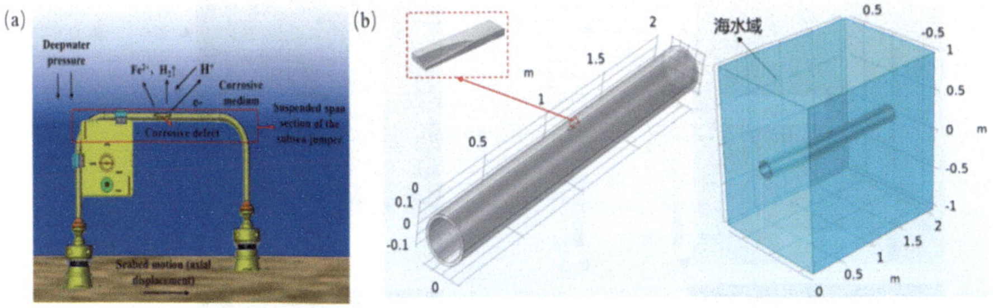

Fig. 2. (a) Prototype of the geometric model (b) Geometric model with the seawater section

2.2 Assumptions and Boundary Condition Definitions

To simplify the analytical model, the simulations in this chapter are carried out based on some assumptions as follows [16, 17]:

(1) The material is assumed to be isotropic.
(2) It is assumed that the fluid temperature inside the pipe is equal to the entire wall of the pipe.

Without considering the axial displacement, the boundary conditions are set to apply temperature and pressure loads inside the pipe. In this case, the volume reference temperature inside the pipe is set to 4 °C, the pressure inside the pipe is 6 MPa, and the seawater pressure outside the pipe is 12 MPa. Moreover, the constraints are set as fixed constraints at both ends of the pipe. In terms of electrochemical corrosion field, the cathodic and anodic equilibrium potentials of the seawater corrosion electrochemical reaction can be derived using equation Nernst's formula [18].

$$E_{O/R} = E^{\circ}_{O/R} - \frac{RT}{zF} lnQ = E^{\circ}_{O/R} - \frac{RT}{zF} ln\frac{a_R}{a_O} \tag{2}$$

where O denotes oxidation and R denotes reduction, participation $E\_(O/R)^{\wedge\circ}$ is the standard electric potential, R is the gas constant, T is the temperature of the reaction, and z is the number of electrons transferred in the reaction; F is the Faraday constant; Q is the activity quotient of the species involved in the reaction; a is the activity of the species involved in the reaction. Calculations show that the cathodic and anodic equilibrium potentials of the electrochemical reaction when the seawater pH is 8 and the concentration of iron ions is 10^{-6}mol/L are respectively −0.44V and −0.605V. Finally, Using the Tafel curve extrapolation method, the final exchange current densities for the cathode and anode are, respectively, found to be 1.464 × 10–4 A/m^2 and 2.512 × 10–2 A/m^2.

3 Results and Discussion

3.1 Effect of Defect Length on Corrosion Damage

Figure 3 shows the effect of initial defect depth of 6.4 mm pipe corrosion parameters by defect length when there is no change in temperature and axial stress in the pipe.

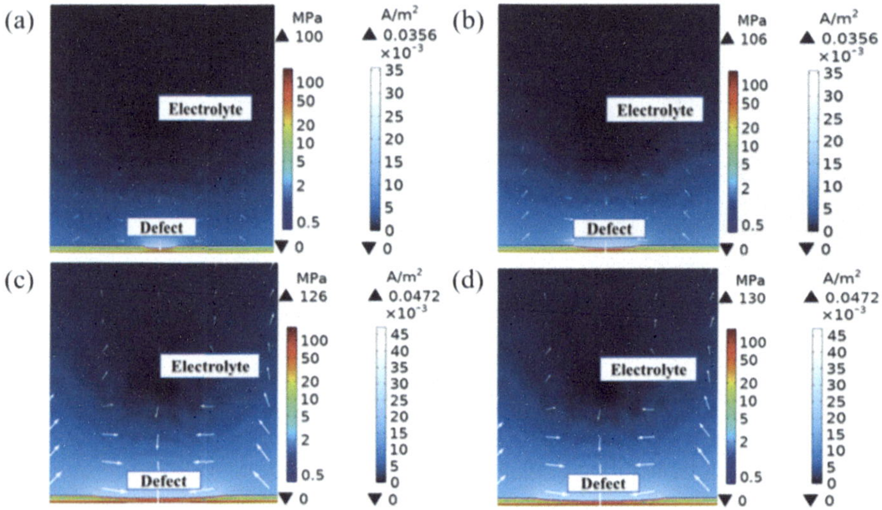

Fig. 3. Distribution of stress and electrolyte current density for different defect lengths, (a) L = 100 mm, (b) L = 200 mm, (c) L = 300 mm, (d) L = 400 mm

As shown in Fig. 3, the white arrow in the figure shows the electrolyte current density vector, which points toward the flow of positive ions in seawater, and its magnitude represents the current intensity [19, 20]. It can be seen from the figure that the electrolyte current density vector flows mainly far away from the defect to the vicinity when the defect length is small. This is because at this point the seawater potential in the vicinity of the defect is low, and the current will flow from the high potential to the low potential due to the potential gradient. As the defects become larger in length, such as 300 mm and 400 mm, more arrows that point outwards from the vicinity of the defect could be seen. It is due to the fact that as the defect increases in size, the electrochemical reaction at the defect becomes more active as it has a larger surface area and reaction potential, which can lead to an increase in the potential at the defect. Therefore, from this point onward, the tendency of the current flowing from the surface of the defect to the surrounding seawater is formed. It can be seen from the figure that the electrolyte current density vector, which is indicated by the white arrow also becomes larger as the defect increases, indicating that the current flowing through the seawater is being enhanced.

In addition, Fig. 4 analyses the variation of relevant electrochemical parameters along the defect length. From Fig. 4 (a), it can be seen that the stress is concentrated near the center of the defect. Moreover, the electrolyte potential rises with increasing defect length. As seen in Fig. 4 (b), the potential near the center of the defect is slightly higher, and the potential at the edge of the defect is slightly lower. This is a result of the stress distribution at the defect location, with higher stress locations having higher electrolyte potentials. Figure 4 (c) and (d) show that as the defect length increases, the anodic and cathodic current densities at the defect exhibit more positive and negative values, respectively. The absolute value is the largest near the center of the defect, which is also basically consistent with the change rule of the defect stress. The changes in anodic and cathodic current densities indicate that the corrosion electrochemistry is more intense and the pipelines are more susceptible to damage.

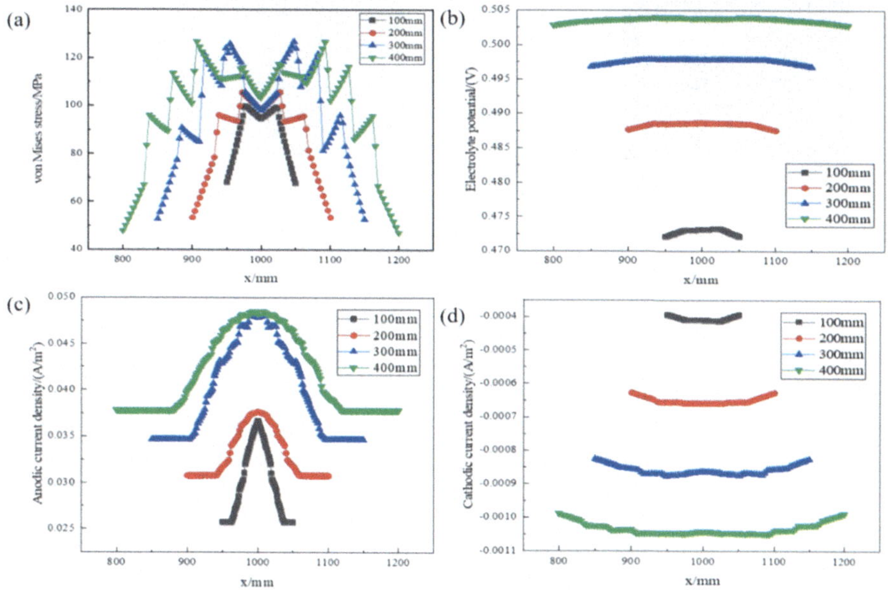

Fig. 4. Distribution of relevant electrochemical parameters (a) stress (b) electrolyte potential (c) anodic current density (d) cathodic current density along the defect length

3.2 Effect of Defect Depth on Corrosion Damage

Figure 5 shows the effect of defect depth on the corrosion parameters of the pipe at an initial defect length of 100 mm when there is no change in the temperature and axial stress in the pipe.

From Fig. 5, it can be seen that as the depth of the defect increases, the stress at the defect and the electrolyte current density also increases. It can be seen from the figure that when the defect depths are 3.2 mm and 6.4 mm, the electrolyte current flows away from the defect towards the vicinity of the defect. As the depth of the defect increases to 9.6 mm, a few outward pointing arrows can be observed near the defect. When the defect depth is 12.8 mm, the electrolyte current flows from the defect to the surrounding seawater, indicating that an increase in the defect size leads to a corresponding increase in the potential at the defect. This, in turn, reverses the direction of the potential gradient, leading to current flow from the defective region to the surrounding seawater. In addition,

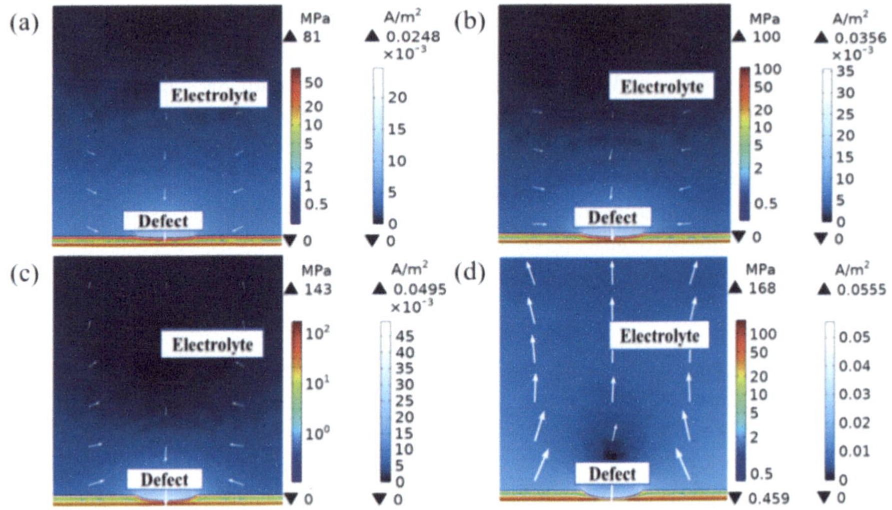

Fig. 5. Distribution of stress and electrolyte current density at different defect depths, (a) d = 3.2 mm, (b) d = 6.4 mm, (c) d = 9.6 mm, (d) d = 12.8 mm

as the defect become deeper, it is observed that the electrolyte current density vector becomes larger. This indicates that the current flowing through the seawater is enhancing.

From Fig. 6 (a), it can be seen that the stress increases as the depth of the defect increases, with the maximum stress being distributed near the center of the defect. Figure 6 (b) shows that the electrolyte potential increases with increasing defect depth

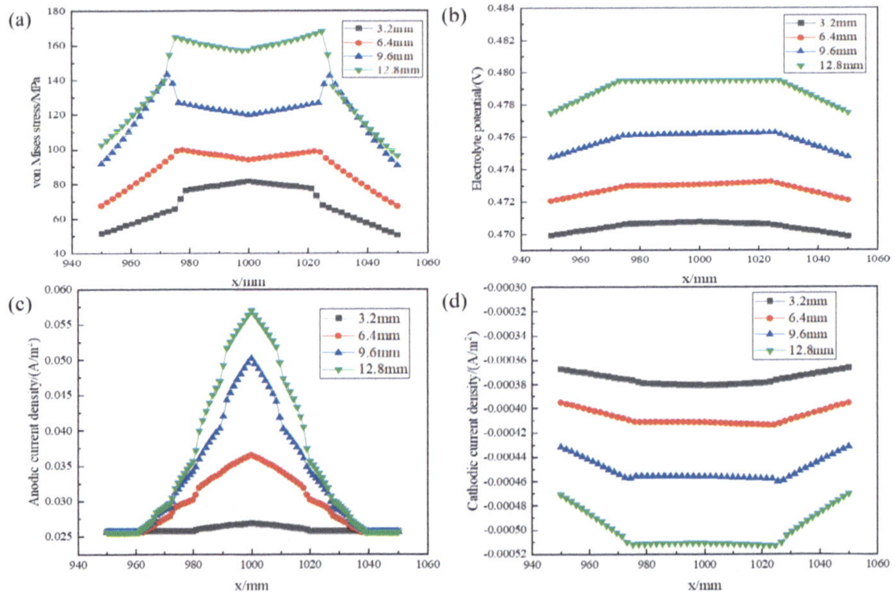

Fig. 6. Distribution of electrochemical parameters (a) stress (b) electrolyte potential (c) anodic current density (d) cathodic current density along the length of the defect

and the potential is higher near the center of the defect as well as lower at the edge of the defect. Figure 6 (c) and (d) show that as the defect become deeper, the anodic and cathodic current densities at the defect respectively show more positive and more negative variations. These values have the largest absolute value near the center of the defect, indicating that deeper defect make the pipeline more susceptible to damage.

3.3 Effect of Internal Temperature and Axial Stress on Corrosion Damage

Figure 7 shows the effects of pipe corrosion caused by thermal stresses generated by a temperature change (ΔT) inside the pipe. The initial defect length is 100 mm, and the depth is 6.4 mm.

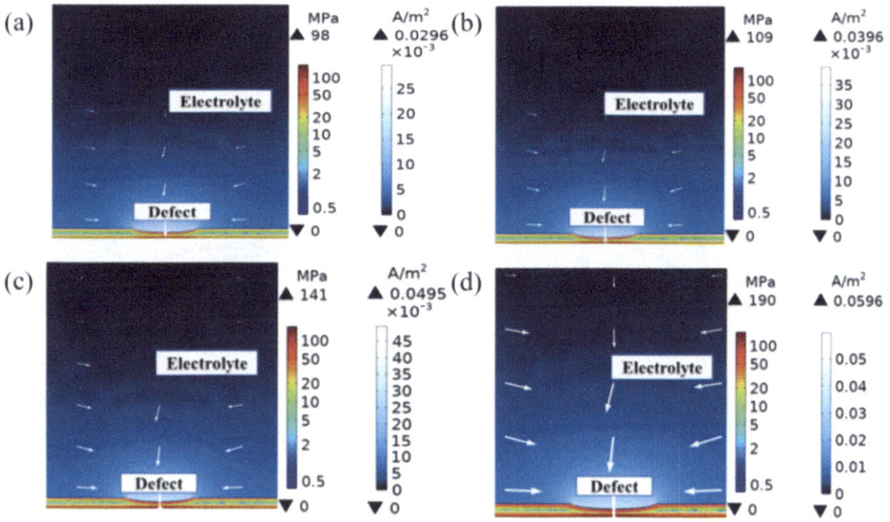

Fig. 7. Distribution of stress and electrolyte current density under different pipe temperatures, (a) ΔT = 20°C, (b) ΔT = 40°C, (c) ΔT = 60°C, (d) ΔT = 80°C

From Fig. 7, it can be seen that the stress at the defect and the electrolyte current density increases as ΔT increases. This is due to the fact that a larger temperature change inside the pipe results in greater thermal stress at the defect. Additionally, as ΔT increases, the white arrows become larger and all point towards the defect. The reason for this phenomenon can be explained in two ways: Firstly, the change in thermal stress increases the concentration of movable electrolyte ions near the defect, leading to an increase in the electrolyte current. Secondly, according to the Nernst equation of the electrode reaction, an increase in temperature results in a higher temperature term in the Nernst equation, leading to a decrease in the potential. This increases the potential difference between the electrolyte near and away from the defect, which in turn increases the electrolyte current.

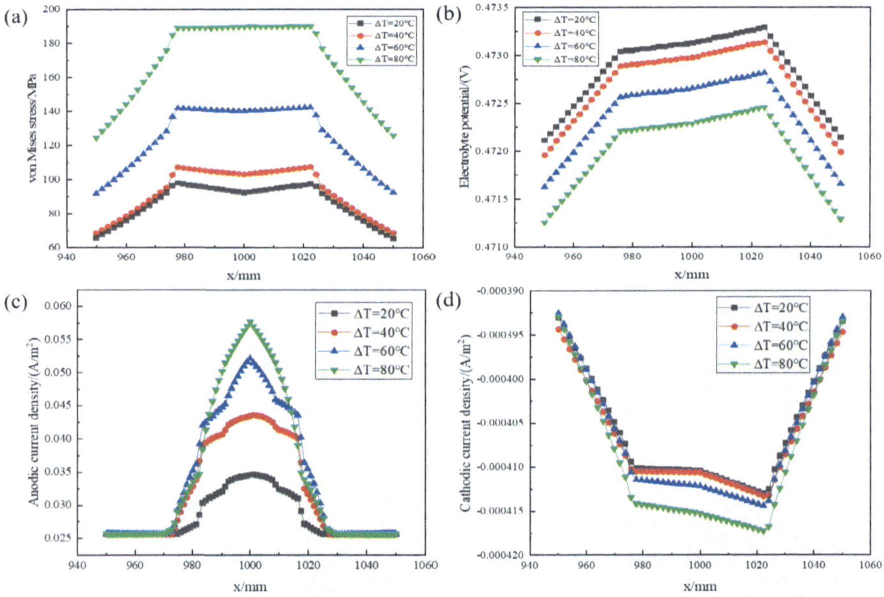

Fig. 8. Distribution of electrochemical parameters (a) stress (b) electrolyte potential (c) anodic current density (d) cathodic current density along defect length

From the Fig. 8, it is evident that as the temperature rise in the pipe increases, the stress, the absolute values of anodic current density and cathodic current density at the defect increase, while the electrolyte potential decreases. The maximum values for all of these parameters are observed near the center of the defect. This suggests that higher temperature rise in the pipe accelerates the corrosion damage process.

3.4 Effect of Axial Stress on Corrosion Damage

Figure 9 illustrates the impact of axial tensile displacements Δx on pipe corrosion, assuming a constant temperature within the pipe and an initial defect length of 100 mm with a depth of 6.4 mm.

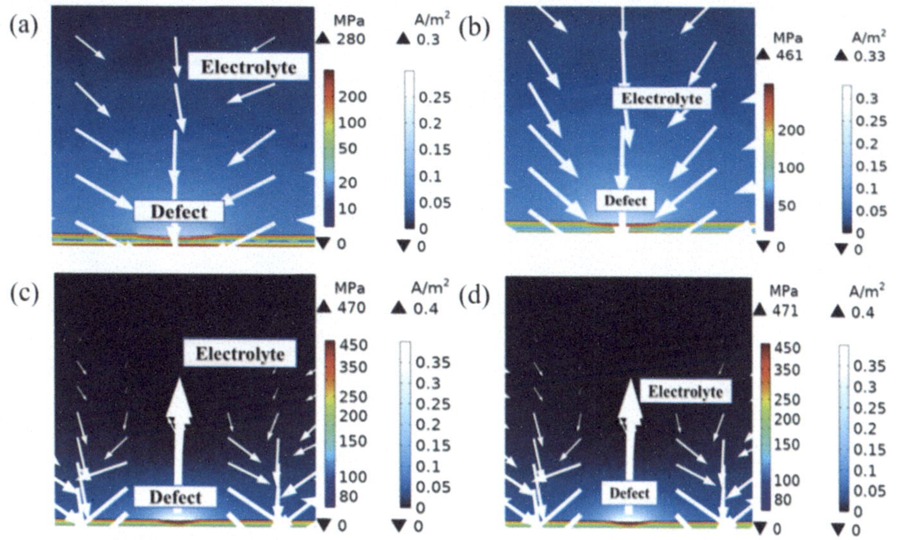

Fig. 9. Distribution of stress and electrolyte current density under different axial displacements, (a) $\Delta x = 2$ mm, (b) $\Delta x = 4$ mm, (c) $\Delta x = 6$ mm, (d) $\Delta x = 8$ mm

From Fig. 9, it is evident that an increase in axial stress results in a significant change in both stress and electrolyte current density at the defect. When the axial tensile displacement is 2mm or 4mm, the maximum stress at the defect is below the yield strength of the pipeline, the pipe undergoes elastic strain. However, when the displacement reaches 6 mm, the maximum stress at the defect exceeds the yield strength of the pipeline, causing plastic deformation. Further increasing the displacement to 8mm does not significantly change the maximum stress or electrolyte current density. The white arrows in the figure show that the electrolyte current is getting stronger as the tensile displacement increases and the high potential at the defect reverses the direction of the current when the tensile displacement increases to 6 mm.

From the Fig. 10, it can be seen that the values of stress, electrolyte potential, anodic current density as well as cathodic current density at the defect increase as the axial tensile displacement increases. Specifically, when the axial displacement increases from 4 mm to 6 mm, the pipeline undergoes a transition from elastic deformation to plastic deformation, leading to significant changes in each parameter. However, when the displacement is further increased to 8mm, the changes in these parameters are no longer significant. This suggests that when the corrosion defect is in localized elastic deformation, the force-electrochemical coupling has less effect on pipeline corrosion. When the tensile strain is further increased resulting in localized plastic deformation, the local corrosion activity increases significantly, aggravating the extent of corrosion damage to the pipeline.

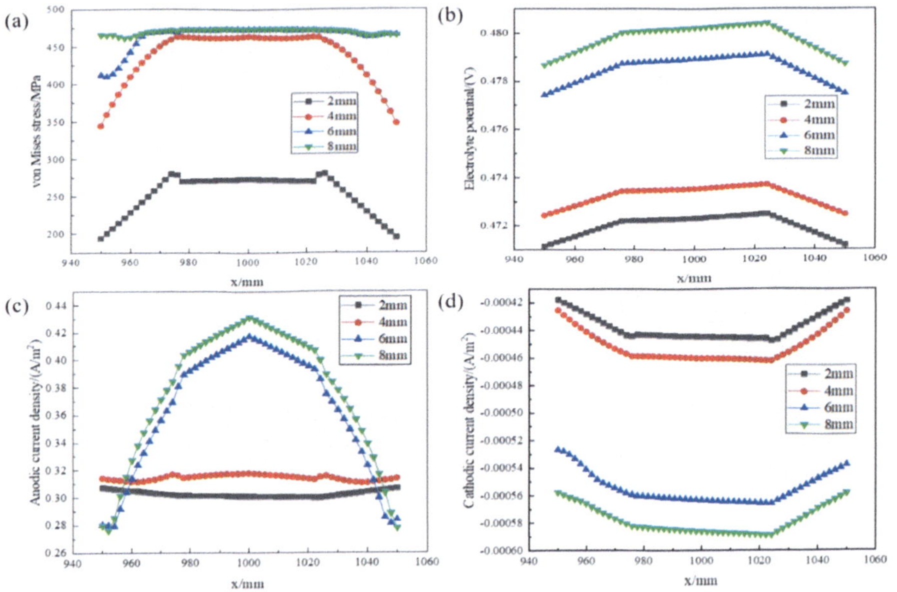

Fig. 10. Distribution of relevant electrochemical parameters (a) stress (b) electrolyte potential (c) anodic current density (d) cathodic current density along defect length for different axial displacements

4 Conclusion

This paper introduces electrochemical equations that take into account the mechanical properties of the pipeline. Combining with the service characteristics of deep-water pipelines, an advanced model of external corrosion damage of pipelines under the action of thermal-force-electrical multi-field coupling was established. Through a systematic and multidisciplinary analysis, this study provides a comprehensive explanation of the mechanism behind external corrosion damage of pipelines under the influence of multi-field coupling. Furthermore, it offers predictions for future corrosion damage of pipelines. The main conclusions are as follows:

(1) As the length and depth of the pipeline defect increase, the stress at the defect also increases. The electrochemical parameters, such as electrolyte current density, cathodic/anodic current density, and electrolyte potential, show similar trends with stress. All of these parameters increase with the rise of stress. This suggests that larger defect sizes intensify the corrosion electrochemical reaction, which in turn worsens the extent of corrosion damage outside the pipeline.
(2) Large temperature changes within the pipeline increase the stresses at the defect by amplifying the thermal stresses at the defect. This increase is also accompanied by a rise in the electrolyte current density and the cathodic/anodic current density. This indicates that the sudden increase in temperature inside the pipe will accelerate the corrosion damage on the outside of the pipeline.

(3) The effect of axial displacement caused by topographic changes on the damage outside the pipeline is investigated. It is found that as the axial displacement increases, the pipeline defect undergoes a transition from elastic deformation to plastic deformation. When the stress of the pipeline reaches the yield strength, the electrochemical activity of corrosion is significantly enhanced.

Acknowledgements. This research is supported by National Key Research and Development Project (2022YFC2806101). The authors will be grateful for the reviewers' and editor's helpful comments.

References

1. Iannuzzi, M., et al.: Materials and corrosion trends in offshore and subsea oil and gas production. NPJ Materials Degradation **1**(1) (2017)
2. Zhu, X.-K..: A comparative study of burst failure models for assessing remaining strength of corroded pipelines. J. Pipeline Sci. Eng. **1**(1), 36–50 (2021)
3. Zhou, W., et al.: Modeling and simulating nonhomogeneous non-Gaussian corrosion fields on buried pipelines and its use in predicting burst capacities of corroded pipelines. Eng. Struct., 245 (2021)
4. Abyani, M., et al.: Predicting failure pressure of the corroded offshore pipelines using an efficient finite element based algorithm and machine learning techniques. Ocean Eng., 254 (2022)
5. Zhang, Z., et al.: Interaction between internal and external defects on pipelines and its effect on failure pressure. Thin-Walled Struct., 159 (2021)
6. Gücüyen, E.: Numerical analysis of deteriorated sub-sea pipelines under environmental loads. Chin. J. Mech. Eng. **28**, 1163–1170 (2015)
7. Feng, L., et al.: Residual ultimate strength investigation of offshore pipeline with pitting corrosion. Appl. Ocean Res. **117**, 102869 (2021)
8. Xu, L.Y., Cheng, Y.F.: Development of a finite element model for simulation and prediction of mechanoelectrochemical effect of pipeline corrosion. Corrosion Sci. **73**, 150–160 (2013)
9. Huang, Y., et al.: Investigation by numerical modeling of the mechano-electrochemical interaction at a dent-corrosion defect on pipelines. Ocean Eng. 256 (2022)
10. Qinglin, X., et al.: Mechanochemical coupling model of corrosion electric field. Harbin Gongye Daxue Xuebao/J. Harbin Institute Technol. **53**(3) (2021)
11. Qin, G., Cheng, Y. F.: Modeling of mechano-electrochemical interaction at a corrosion defect on a suspended gas pipeline and the failure pressure prediction. Thin-Walled Struct. **160**, 107404 (2021)
12. Lo, M., et al.: ANN-and FEA-based assessment equation for a corroded pipeline with a single corrosion defect. J. Marine Sci. Eng. **10**(4), 476 (2022)
13. Xu, L., Cheng, Y.: Corrosion of X100 pipeline steel under plastic strain in a neutral pH bicarbonate solution. Corrosion Sci. **64**, 145–152 (2012)
14. SunGuo-min, Z., Hu, Y., Xu, L., Chun-hong, H.: Study on design analysis of deepwater rigid jumper. Ocean Eng. Equip. Technol. **7**(06), 356–362 (2020)
15. Z. W. ZHANG Guo-qing , QIAN Si-cheng ,YU Xuan, F.: Electrochemical Properties of Al-Zn-In alloy sacrificial anode at 1200 m deepwater. Environ. South China Sea Surface Technol. **51**(05), 177–185 (2022)
16. Soares, E., et al.: Structural integrity analysis of pipelines with interacting corrosion defects by multiphysics modeling. Eng. Failure Anal. **97**, 91–102 (2019)

17. Xu, L.: Assessment of corrosion defects on high-strength steel pipelines. University of Calgary (2013)
18. Kang, F., et al.: Electrochemical synthesis of sulfate graphite intercalation compounds with different electrolyte concentrations. J. Phys. Chem. Solids **57**(6–8), 883–888 (1996)
19. Zhang, D., et al.: Corrosion behavior of X65 steel at different depths of pitting defects under local flow conditions. Exper. Thermal Fluid Sci. **124**, 110333 (2021)
20. Mouloudi, M., et al.: Study of corrosion pit propagation in aluminum by numerical simulation using the finite element method. Int. J. Electrochem. Sci. **16**(11), 21119 (2021)

Key Technologies of Deep-Sea Equipment and Ship Engineering

Physiological Evaluation of Color Schemes on Display Screens of Manned Submersibles Cockpit

Yangyang Li[1], Weizhe Xu[2], and Lu Shi[3,4(✉)]

[1] Shanghai Jiao Tong University Underwater Engineering Institute Co., Ltd., Shanghai, China
[2] State Key Laboratory of Deep-Sea Manned Vehicles (China ShipScientific Research Center), Wuxi, Jiangsu, China
[3] Institute of Underwater Technology, Shanghai Jiao Tong University, Shanghai, China
shilu@sjtu.edu.cn
[4] Shanghai Jiao Tong University-Chiba University International Cooperative Research Center, Shanghai, China

Abstract. The research aims to evaluate the influence of color schemes on the in-cabin display screens of manned submersibles concerning divers' physiological states. By examining the physiological responses of pilots to specific color environments, the study aspires to generate scientific evidence for the design of submersible in-cabin displays, with intentions to optimize the pilot's workspace, minimize physiological stress, and boost the safety and efficiency of diving operations. The study leverages multi-parameter physiological monitoring techniques, including the continuous measurement of pilots' blood pressure, heart rate, heart rate variability, electroencephalogram (EEG), and cerebral blood flow via near-infrared spectroscopy (NIRS). Experiments will be carried out under simulated conditions, exposing pilots to screens with diverse color combinations for set durations. Data collected during visual stimuli will be used to ascertain the impact of color schemes on pilots' physiological indicators. The expected results will demonstrate the specific influences of various color combinations on pilots' physiological parameters. For instance, certain colors may trigger increased blood pressure or heart rate, whilst others might foster enhanced heart rate variability, suggestive of a relaxed state. Alterations in EEG and cerebral blood flow will disclose the effects of different colors on brain activity and hemodynamics. This data will offer quantitative proof of the relationship between color schemes and divers' physiological responses. The study postulates that targeted colors on displays can effectively lessen pilots' mental load and psychological stress. The findings are expected to provide scientific guidance for the ergonomic design of in-cabin environments in submersibles, thus safeguarding the health and safety of pilots.

Keywords: Manned submersible · Mental workload · Ergonomics

D. Jeng and B. Cai (Eds.): ICCOE 2024, LNCE 423, pp. 401–413, 2025.
https://doi.org/10.1007/978-981-96-0181-3_36

1 Introduction

Manned submersibles serve as pioneers in exploring the mysteries of the deep sea, and the design of their in-cabin displays is of paramount importance. This design not only affects the operational efficiency of the submersible but also directly impacts the physiological health and psychological state of divers in the high-pressure deep-sea environment [1–3]. In the enclosed space of the deep sea, divers are exposed to various information displays for extended periods, where the color scheme, information layout, and display technology significantly influence their visual comfort and psychological well-being. Significant advancements have been made in the research of in-cabin displays for manned submersibles worldwide. For instance, the U.S. "Alvin" and French "Nautile" submersibles have invested heavily in human-centric display design to enhance diver efficiency and comfort [4–6]. These designs incorporate color psychology, ergonomics, and the clarity and accuracy of information display, providing a more user-friendly and comfortable work environment for divers.

Although starting later, domestic research on in-cabin displays for manned submersibles has made notable progress. The Chinese submersible has begun to consider the psychological impact of cockpit layout on divers, marking a shift towards prioritizing the physiological and psychological health of divers in deep-sea environments [7–9].The mature research on cockpit displays in aviation and aerospace provides valuable insights for the design of in-cabin displays in manned submersibles [10–12]. For example, appropriate color schemes can significantly reduce visual fatigue and enhance the accuracy of information recognition. Contrasting colors highlight critical information, while soft colors alleviate visual stress from prolonged operations. These findings underscore the importance of color psychology in improving work efficiency and safeguarding operator health. Despite the current focus on display technology upgrades, human-computer interaction optimization, and the application of color psychology, existing research predominantly remains theoretical, with limited studies on physiological assessments and long-term impacts in practical applications. Deep-sea environments pose unique challenges that necessitate the consideration of divers' visual and psychological needs under extreme conditions. Conducting physiological assessments of color schemes for in-cabin displays of manned submersibles not only enhances operational safety and efficiency but also ensures the physical and mental health of divers, offering significant practical value.

The research enhances our comprehension of the effects of colors on human psychology and physiology, especially in extreme conditions. It contributes to advancements in ergonomics, color psychology, and related disciplines, offering scientific and human-centric technical support for deep-sea exploration. As technology for deep-sea exploration advances, the design of in-cabin displays for manned submersibles will encounter new challenges and opportunities. Future studies should explore in greater depth the impacts of color schemes, display technology, and human-computer interaction to ensure a safer, more comfortable, and efficient working environment for divers. Through interdisciplinary research and collaboration, it is expected that the design of future in-cabin displays for manned submersibles will become more human-centric, addressing the real needs of divers more effectively and making a significant contribution to the progress of deep-sea exploration.

2 Methods

2.1 Volunteers

The study included eight male volunteers with an average age of 25.25 years (±1.39), ranging between 23 and 28 years. They had an average height of 1.74 m (±0.04 m), with a range of 1.69 to 1.79 m. The average weight was 67.5 kg (±5.18 kg), with a spectrum from 63 to 75 kg, leading to an average Body Mass Index (BMI) of 22.33 (±1.04), which varied from 21.01 to 24.06. All participants were verified to be in good health via physical examinations and had no history of musculoskeletal disorders of the upper limbs or neck, nor any recent injuries. They volunteered for the study and continued with their usual daily activities and work routines, with the exception of attending experimental sessions as scheduled. Participants were advised to avoid engaging in intense physical activities 72 h before the start of the experiment.

2.2 Experimental Design

The interface design for this study employed standard background colors: white, black, and gray, with numerical stimuli presented in various foreground colors including white, black, red, green, and blue. This design was intended to investigate the effects of background and foreground color combinations on participant response times and accuracy, offering a thorough analysis of color impact in visual processing tasks. Detailed impacts of these specific color combinations are presented in Table 1.

Utilizing a three-dimensional model of the full ocean deep manned submersible's cabin, as illustrated in Fig. 1, a simulated interface layout was meticulously designed for the submersible's cabin. This interface was displayed on a Philips BDM4037UW 40-in. curved monitor, with dimensions 907 × 529 × 72 mm, and strategically divided into seven distinct zones as depicted in Fig. 2. Participants underwent a practice session prior to the commencement of actual testing to acclimate themselves to the experimental environment.

During the formal experiment, subjects were seated 70 cm from the monitor, with their eye level aligned with the display. The primary task involved observing single-digit numbers, ranging from 0 to 9, randomly presented in each of the seven zones. Participants were required to identify whether the number was odd or even and to respond by pressing the "F" and "J" keys on a keyboard. Each digit appeared twice in every zone, resulting in 20 trials per zone and a total of 140 trials across all zones. The sequence of number presentation and zones was randomized. Each experimental phase displayed visual stimuli at an interval of 1200 ms, with the target stimulus shown for 350 ms. The presentation sequence, as outlined in Fig. 3, was orchestrated by E-prime software, which also recorded behavioral metrics such as accuracy (ACC) and reaction time (RT) in real-time during the visual stimulus presentation.

Blood pressure of subjects was measured at rest and under various conditions using the Omron HEM-6111 wrist-type electronic sphygmomanometer. Mean arterial pressure was derived from the recorded systolic and diastolic pressures. The NIRSport portable near-infrared spectroscopy system was employed to monitor real-time changes in cerebral blood flow in the forehead region, analyzing the dynamic changes in oxyhemoglobin across different foreground and background color combinations.

Electrocardiogram (ECG) signals from limb lead II were recorded using the MP150 system from Biopac, USA. Heart rate was extracted from these ECG signals, which were then input into a computer equipped with a heart rate variability (HRV) acquisition card, capturing 256 cardiac cycles as control values. The power spectrum, a frequency-domain method for HRV analysis, was calculated. The HRV power spectrum derived from time series analysis included total power (TP; 0.04 Hz–0.40 Hz), low frequency (LF; 0.04 Hz–0.15 Hz), and high frequency (HF; 0.15 Hz–0.40 Hz). LF reflects both vagal and sympathetic activity, predominantly indicating sympathetic modulation, whereas HF primarily reflects vagal activity. In this study, LF/HF was used as an index of sympathetic nervous system activity, and HF/TP as an index of parasympathetic nervous system activity. Electroencephalogram (EEG) activity was recorded using the MP150 system from Biopac, USA, with a time constant of 0.3 s and a high-cut frequency of 30 Hz. Ag-AgCl disc electrodes were positioned at the Cz and Pz points according to the international 10–20 system for EEG recording. Fast Fourier Transform (FFT) was applied to data sets of 4096 samples, yielding power spectra for each frequency band.

Table 1. Specifications for Foreground and Background Color Combinations.

Background Color	Foreground Color	R	G	B	Color sample
White	Red	255	0	0	
	Green	0	255	0	
	Blue	0	0	255	
	Black	0	0	0	
Gray	Red	255	0	0	
	Green	0	255	0	
	Blue	0	0	255	
	Black	0	0	0	
	White	255	255	255	
Black	Red	255	0	0	
	Green	0	255	0	
	Blue	0	0	255	
	White	255	255	255	

2.3 Data Processing and Analysis

The collected data were processed and analyzed using SPSS 20.0, a comprehensive statistical software. Results are presented as (Mean ± SD). Repeated measures ANOVA was utilized to compare data across different groups. Post-hoc multiple comparisons with Bonferroni correction were performed where significant statistical differences were

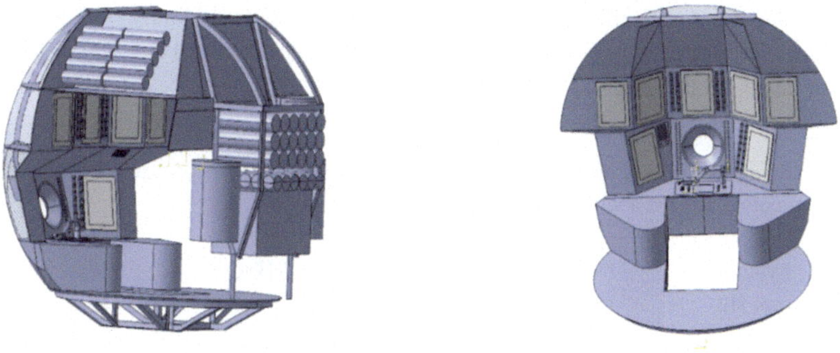

Fig. 1. Overall Design and Display Control of the Full Ocean Manned Submersible Cockpit.

Fig. 2. Control Interface Layout.

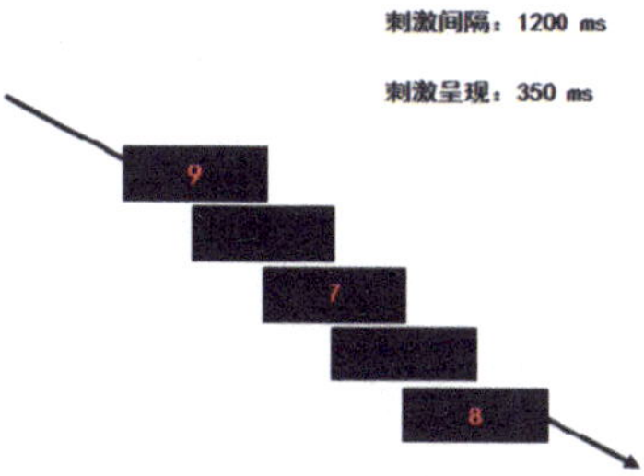

Fig. 3. Visual Stimulus Schematic Diagram.

detected. Paired t-tests were used for comparing mean values between two specific groups. Statistical significance was set at $p < 0.05$.

3 Results

3.1 Blood Pressure, Heart Rate, and Cerebral Blood Flow

Under a white background, there were no significant differences in blood pressure, heart rate, or hemoglobin concentration among participants across various foreground colors (all $P > 0.05$). This suggests that the influence of different foreground colors on subjects under a white background was insignificant. When the background was gray, the oxygenated hemoglobin concentration in the white group was significantly higher than that in the red group, demonstrating a statistically significant difference ($P < 0.05$). This indicates that cognitive load on participants significantly increased when the foreground color was white. Significant changes in blood pressure and heart rate were not observed across different foreground color combinations. With a black background, apart from a significant rise in diastolic blood pressure in the blue group compared to the red group, no other notable differences were observed. However, the oxygenated hemoglobin concentration demonstrated an increasing trend with the foreground colors green, red, white, and blue, pointing to a gradual increase in cognitive load. Comparisons of oxygenated hemoglobin under visual stimuli in the red/white, red/gray, and green/black groups revealed that the red/gray group was significantly higher than the green/black group, showing statistically significant differences (all $P < 0.05$). Detailed data can be found in Tables 2, 3, 4 and 5.

Table 2. Changes in blood pressure, heart rate, and hemoglobin concentration under different foreground colors against a white background.

Group	Sys mm Hg	Dia mm Hg	HR beats/min	HbO 10^{-3}mmol/L	Total HbO 10^{-3}mmol/L
Red	108.50 ± 9.84	67.38 ± 6.14	67.75 ± 11.87	4.08 ± 0.57	2.28 ± 0.71
Blue	110.13 ± 8.94	69.88 ± 4.55	68.75 ± 12.83	4.18 ± 0.38	2.39 ± 0.52
Green	108.50 ± 7.23	65.50 ± 8.49	68.75 ± 10.14	4.12 ± 0.50	2.35 ± 0.76
Black	109.63 ± 9.41	67.13 ± 8.39	68.50 ± 10.09	4.29 ± 0.51	2.66 ± 0.39

Table 3. Changes in blood pressure, heart rate, and hemoglobin concentration under different foreground colors against a gray background

Group	Sys mm Hg	Dia mm Hg	HR beats/min	HbO 10^{-3}mmol/L	Total HbO 10^{-3}mmol/L
Red	106.00 ± 7.76	66.25 ± 7.36	66.75 ± 10.12	4.34 ± 0.46	2.48 ± 0.41
Blue	107.50 ± 8.05	68.75 ± 6.09	66.75 ± 12.35	4.41 ± 0.93	2.46 ± 0.90
Green	109.00 ± 13.09	66.3 ± 6.44	65.63 ± 10.57	4.72 ± 0.60	2.18 ± 0.67
Black	107.25 ± 10.08	68.13 ± 5.82	65.13 ± 8.81	5.19 ± 0.65	2.73 ± 0.58
White	109.38 ± 6.28	72.63 ± 8.12	71.25 ± 13.63	5.58 ± 0.21[a]	2.85 ± 0.40

Comparison with the red group, [a]$P < 0.05$.

Table 4. Changes in blood pressure, heart rate, and hemoglobin concentration under different foreground colors against a black background.

Group	Sys mm Hg	Dia mm Hg	HR beats/min	HbO 10^{-3}mmol/L	Total HbO 10^{-3}mmol/L
Green	108.00 ± 8.05	74.50 ± 13.32	67.88 ± 12.28	3.54 ± 0.37	2.02 ± 0.53
Red	107.50 ± 7.27	65.50 ± 6.82	67.63 ± 10.41	3.67 ± 0.42	2.28 ± 0.42
White	111.25 ± 9.50	71.63 ± 7.71	66.13 ± 13.08	3.98 ± 0.40	2.20 ± 0.33
Blue	109.63 ± 8.78	73.13 ± 5.84[a]	68.13 ± 11.64	4.12 ± 0.51	2.20 ± 0.65

Comparison with the red group, [a]$P < 0.05$.

Table 5. Comparison of blood pressure, heart rate, and hemoglobin concentration among the optimal color combinations under three different background colors.

Group	Sys mm Hg	Dia mm Hg	HR beats/min	HbO 10^{-3}mmol/L	Total HbO 10^{-3}mmol/L
Red/White	108.50 ± 9.84	67.38 ± 6.14	67.75 ± 11.87	4.08 ± 0.57	2.28 ± 0.71
Red/Gray	106.00 ± 7.76	66.25 ± 7.36	66.75 ± 10.12	4.34 ± 0.46	2.48 ± 0.41
Green/Black	108.00 ± 8.05	74.50 ± 13.32	67.88 ± 12.28	3.54 ± 0.37[a]	2.02 ± 0.53

Comparison with the Red/Gray group, [a]$P < 0.05$.

The changes in oxygenated hemoglobin concentration during visual stimulation under different foreground/background color combinations are shown in Fig. 4. This suggests that the oxygenated hemoglobin concentration in the subjects increased significantly during visual stimulation, and more oxygenated hemoglobin was required under the black/gray and white/gray combinations, indicating an increase in cognitive load.

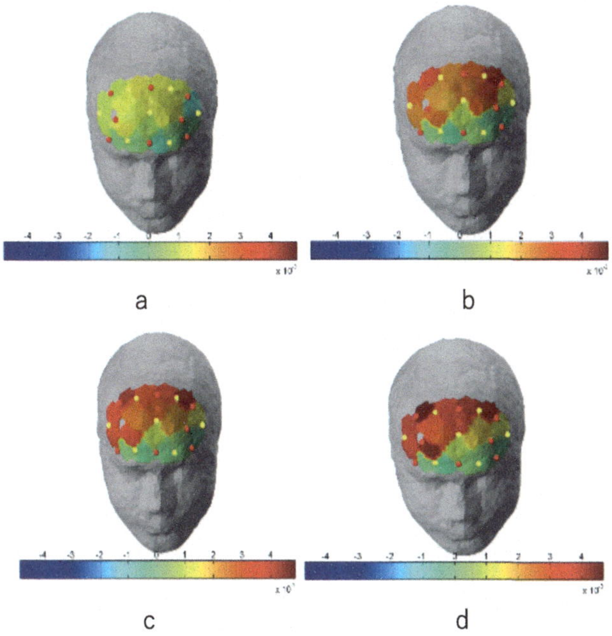

Fig. 4. Visual Stimulus Schematic Diagram. (a. Schematic representation of oxygenated hemoglobin concentration in the prefrontal area at rest; b, c and d respectively showed the oxygenated hemoglobin concentration during visual stimulation with red/gray, black/gray, and white/gray color combinations.)

3.2 Frequency Domain Indices of Heart Rate Variability

With a white background, aside from a significantly lower HF in the black group compared to the blue group, there were no notable differences in frequency domain indices of heart rate variability under various foreground color combinations (all $P > 0.05$). However, there was a trend of increasing LF/HF ratios with the foreground colors red, blue, green, and black, suggesting lower sympathetic nervous system activation in subjects during visual stimulation with red. When the background was gray, no significant differences were observed in the frequency domain indices of heart rate variability among the different foreground color combinations (all $P > 0.05$). Nonetheless, the LF/HF ratio increased, and the HF/TP ratio decreased with the foreground colors red, blue, green, black, and white, indicating higher sympathetic nervous system activation and lower parasympathetic nervous system activation in subjects during visual stimulation with white. With a black background, the LF, HF, and TP in the blue group were significantly higher than those in other groups ($P < 0.05$ or $P < 0.01$). Furthermore, the LF/HF ratio increased, and the HF/TP ratio decreased with foreground colors green, red, white, and blue, suggesting higher sympathetic nervous system activation in subjects during visual stimulation with blue. Comparisons of frequency domain indices of heart rate variability under visual stimulation with red/white, red/gray, and green/black color combinations indicated that the red/white combination resulted in lower sympathetic nervous system activation, reflecting a more relaxed physiological state during this visual stimulation. Detailed data can be found in Tables 6, 7, 8 and 9.

Table 6. Changes in HRV under different foreground colors against a white background.

Group	LF(ms^2)	HF(ms^2)	TP(ms^2)	LF/HF	HF/TP
Red	2237.84 ± 839.54	1284.24 ± 442.39	3522.08 ± 1269.07	1.73 ± 0.22	0.37 ± 0.03
Blue	2716.55 ± 703.12	1553.67 ± 498.07	4270.22 ± 1144.59	1.79 ± 0.27	0.36 ± 0.04
Green	2030.89 ± 532.13	1104.83 ± 339.35	3135.71 ± 818.27	1.91 ± 0.57	0.35 ± 0.06
Black	2078.08 ± 597.24	1049.98 ± 216.42[a]	3128.06 ± 740.39	1.99 ± 0.48	0.34 ± 0.05

Comparison with the blue group, [a]$P < 0.05$.

Table 7. Changes in HRV under different foreground colors against a gray background.

Group	LF(ms^2)	HF(ms^2)	TP(ms^2)	LF/HF	HF/TP
Red	2265.82 ± 743.70	1342.31 ± 410.65	3608.13 ± 1132.01	1.69 ± 0.27	0.38 ± 0.04
Blue	1974.13 ± 720.33	1122.54 ± 446.36	3096.67 ± 1155.67	1.78 ± 0.23	0.36 ± 0.23
Green	2488.24 ± 412.64	1418.75 ± 333.67	3906.98 ± 693.20	1.81 ± 0.41	0.36 ± 0.04
Black	2127.15 ± 612.87	1021.03 ± 279.79	2148.18 ± 829.29	2.11 ± 0.48	0.33 ± 0.05
White	2616.40 ± 449.34	1181.10 ± 366.34	3797.50 ± 749.05	2.31 ± 0.47	0.31 ± 0.04

Table 8. Changes in HRV under different foreground colors against a black background.

Group	LF(ms^2)	HF(ms^2)	TP(ms^2)	LF/HF	HF/TP
Green	1241.39 ± 402.34^b	920.85 ± 332.86	2162.23 ± 729.31^a	1.37 ± 0.15	0.42 ± 0.03
Red	1714.95 ± 501.65	1188.55 ± 395.35	2903.50 ± 887.83	1.47 ± 0.16	0.41 ± 0.03
White	1590.78 ± 386.44^b	1033.29 ± 285.66^a	2624.07 ± 666.97^b	1.56 ± 0.10	0.39 ± 0.02
Blue	2417.31 ± 678.14	1440.06 ± 405.89	3857.37 ± 1034.62	1.70 ± 0.32	0.38 ± 0.04

Table 9. Comparison of HRV among the optimal color combinations under three different background colors.

Group	LF(ms^2)	HF(ms^2)	TP(ms^2)	LF/HF	HF/TP
Red/White	2237.84 ± 839.54	1284.24 ± 442.39	3522.08 ± 1269.07	1.73 ± 0.22	0.37 ± 0.03
Red/Gray	2265.82 ± 743.70	1342.31 ± 410.65	3608.13 ± 1132.01	1.69 ± 0.27	0.38 ± 0.04
Green/Black	1241.39 ± 402.34^a	920.85 ± 332.86	2162.23 ± 729.31	1.37 ± 0.15^b	0.42 ± 0.03^{ab}

Comparison with the red/gray group, $^aP < 0.05$; comparison with the red/white group, $^bP < 0.05$.

3.3 EEG Band Power Ratios

Under a white background, the α/β ratio at Cz and Pz sites was significantly higher in the green group compared to those in the black and red groups, with notable statistical differences ($P < 0.05$ or $P < 0.01$). There were no significant changes observed in the θ/β ratio and (α + θ)/β ratio among groups. This suggests that when the foreground color is green, the power of the α wave decreases, while the β wave power increases, indicating an enhanced state of active thinking and heightened mental tension in subjects. With a gray background, the θ/β ratio, α/β ratio, and (α + θ)/β ratio at Cz and Pz sites were significantly lower than those in the other three groups ($P < 0.05$ or $P < 0.01$). This implies that when the foreground colors are black and white, there is an increase in the degree of thinking, and mental tension escalates. Under a black background, the θ/β ratio and (α + θ)/β ratio at the Pz site in the blue group were significantly lower than those in the red and white groups, with considerable statistical differences ($P < 0.05$ or $P < 0.01$). This suggests that when the foreground color is blue, the level of thinking intensifies, and mental tension increases. Comparisons of EEG band power ratios under visual stimulation with red/white, red/gray, and green/black color combinations revealed that the α/β ratio and (α + θ)/β ratio at Cz and Pz sites in the red/white group were significantly lowered than those in the green/black group, with substantial statistical differences ($P < 0.05$ or $P < 0.01$).This suggests that during visual stimulation with the green/black combination, subjects are relatively more relaxed. Detailed data can be found in Tables 10, 11, 12 and 13.

Table 10. Specifications for Foreground and Background Color Combinations.

Group	Cz			Pz		
	θ/β	α/β	(α + θ)/β	θ/β	α/β	(α + θ)/β
Red	20.66 ± 1.83	25.91 ± 1.51	46.57 ± 2.77	21.19 ± 2.12	25.86 ± 1.72[a]	47.05 ± 3.24
Blue	21.22 ± 1.74	26.53 ± 1.73	47.75 ± 2.76	21.80 ± 1.90	26.80 ± 1.37[c]	48.60 ± 2.75
Green	20.96 ± 1.87	27.41 ± 1.25	48.37 ± 2.05	21.33 ± 1.47	27.61 ± 1.92	48.94 ± 2.80
Black	20.56 ± 0.84	25.60 ± 1.56[b]	46.16 ± 1.77	21.15 ± 1.35	25.75 ± 1.71[a]	46.90 ± 2.59

Comparison with the green group, [a]$P < 0.05$, [b]$P < 0.01$; comparison with the black group, [c]$P < 0.01$.

Table 11. Specifications for Foreground and Background Color Combinations.

Group	Cz			Pz		
	θ/β	α/β	(α + θ)/β	θ/β	α/β	(α + θ)/β
Red	21.64 ± 1.67[bd]	26.72 ± 1.44[c]	48.36 ± 0.96[ad]	22.09 ± 2.17[bd]	27.55 ± 2.09[c]	49.65 ± 1.95[ad]
Blue	21.27 ± 3.75[ad]	26.72 ± 2.09[c]	47.99 ± 3.88[ad]	21.86 ± 3.44[ad]	26.80 ± 1.87[c]	48.66 ± 3.06[ad]
Green	21.13 ± 4.79[bd]	26.56 ± 4.98[c]	47.69 ± 9.21[bc]	21.85 ± 5.02[bc]	25.81 ± 4.29	47.66 ± 8.45[ac]
Black	15.14 ± 3.49	22.76 ± 5.57	37.90 ± 8.42	15.20 ± 3.50	23.42 ± 5.92	38.61 ± 8.58
White	13.72 ± 2.44	21.18 ± 4.67	34.91 ± 6.82	14.25 ± 2.40	21.50 ± 4.62	35.74 ± 6.70

Comparison with the black group, [a]$P < 0.05$, [b]$P < 0.01$; comparison with the white group, [c]$P < 0.05$, [d]$P < 0.01$.

Table 12. Specifications for Foreground and Background Color Combinations.

Group	Cz			Pz		
	θ/β	α/β	(α + θ)/β	θ/β	α/β	(α + θ)/β
Green	22.82 ± 2.22	29.18 ± 1.52	52.00 ± 2.51	22.87 ± 2.14	29.56 ± 1.26	52.43 ± 2.70
Red	22.75 ± 1.71	28.81 ± .51	51.56 ± 2.54	23.55 ± 1.78[b]	29.44 ± 1.56	52.99 ± 2.27[a]
White	22.77 ± 1.58	28.58 ± 1.54	51.35 ± 2.42	23.50 ± 1.65[b]	29.15 ± 1.66	52.65 ± 2.31[a]
Blue	20.10 ± 2.13	24.95 ± 3.47	45.05 ± 5.21	20.60 ± 2.07	25.86 ± 3.34	46.45 ± 4.78

Comparison with the blue group, [a]$P < 0.05$, [b]$P < 0.01$.

Table 13. Specifications for Foreground and Background Color Combinations.

Group	Cz			Pz		
	θ/β	α/β	(α + θ)/β	θ/β	α/β	(α + θ)/β
Red/White	20.66 ± 1.83	25.91 ± 1.51a	46.57 ± 2.77b	21.19 ± 2.12	25.86 ± 1.72 a	47.05 ± 3.24 a
Red/Gray	21.64 ± 1.67	26.72 ± 1.44	48.36 ± 0.96b	22.09 ± 2.17	27.55 ± 2.09	49.65 ± 1.95
Green/Black	22.82 ± 2.22	29.18 ± 1.52	52.00 ± 2.51	22.87 ± 2.14	29.56 ± 1.26	52.43 ± 2.70

Comparison with both the green and black groups, aP < 0.05, bP < 0.01.

4 Discussion

The design of modern transportation vehicles prioritizes not only the aesthetic appeal of the cockpit display and control interface color scheme but also its ergonomic significance. As a visual element, color profoundly influences human psychological and physiological responses. Through rational color combination usage in cockpit display and control interface design, one can maximize operational efficiency, diminish visual fatigue, and enhance the readability and recognition speed of information, thereby bolstering overall safety and comfort.

(1) Color Combination and Visual Recognition: The color combination in display and control interface design is of paramount importance. Diverse color combinations can influence the driver's speed and accuracy in recognizing information. For instance, high-contrast color combinations can accentuate key information, while softer colors can alleviate visual fatigue. Studies pin blue and green as generally soothing, while red can command attention and induce alertness. Hence, the design of display and control interfaces should entail the selection of suitable color combinations, contingent on the information's importance and urgency.

(2) Color and Psychological Responses: Color influences not just visual recognition but is also closely tied to psychological responses. Certain colors can elicit positive emotional reactions like warmth, safety, and comfort, while others can spawn tension or anxiety. In the cockpit environment, the psychological effects of color can be leveraged to either heighten the driver's alertness or induce a relaxing mood to accommodate varying driving conditions and task requirements.

(3) Ergonomic Evaluation Methods: The assessment of color combinations' ergonomic effects in cockpit display and control interfaces necessitates a series of scientific evaluation methods. Such methods may comprise laboratory tests, field trials, simulated driving studies, and physiological and psychological response measurements. These evaluations can help quantify the impact of color combinations on the driver's operational performance, workload, and satisfaction.

(4) Personalization and Adaptive Design: Every driver has unique color preferences and reactions. Therefore, the design of color combinations in display and control interfaces should account for personalization and adaptability. Intelligent systems can dynamically adjust color combinations based on the driver's physiological and psychological states, thus offering a more personalized driving experience.

(5) Technological Development and Future Trends: The advancement of display and human-computer interaction technologies is reshaping the design of color combinations in cockpit display and control interfaces. For instance, the advent of augmented reality (AR) and virtual reality (VR) technologies opens new doors for color combination possibilities. Future display and control interface design will lean more towards interactivity, intelligence, and personalization to keep pace with evolving driving environments and driver needs.

In summation, ergonomic evaluation of color combinations in cockpit display and control interfaces is a multi-dimensional, interdisciplinary research field. It calls for a comprehensive approach that takes into account visual recognition, psychological responses, actual driving tasks, and environmental conditions. Through relentless research and practice, we can pave the way for a safer, more comfortable, and efficient driving environment for drivers.

Funds. The National Key Research and Development Program of China "General Design, Construction and Sea trial of Full Ocean Deep Manned Submersible" (2016YFC0300600); the 13th Five Year Plan for the International Marine Resources Survey and Development of the Oceanic Association (DY135-2017-RS-001); the Technical Standards Project of Shanghai Science and Technology Commission's "Science and Technology Innovation Action Plan" (18DZ2203300, 21DZ2208600).

References

1. Ye, C., Xu, W., Liu, S.: Application of ergonomics to layout design of manned submersible. Xibei Gongye Daxue Xuebao/J. Northwestern Polytech. Univ. **39**(2), 233–240 (2021)
2. Yu-Gang, R., Han-Bin, Z., Xian-Peng, S.: The research of droppable ballast device of manned submersible based on computer-aided ergonomics design. In: 2022 5th World Conference on Mechanical Engineering and Intelligent Manufacturing (WCMEIM), pp. 872–877. IEEE (2022)
3. Wang, Y., Qiao, Y., Guan, X., et al.: Dynamic prediction method for cognitive failure risk of submersible operating environment based on Igra and human error prediction driver. Available at SSRN 4802006
4. Chowdhury, T., Sathianarayanan, D., Palaniappan, M., et al.: Jettisoning systems of deep-sea manned submersibles: a review. OCEANS 2022-Chennai, pp. 1–13 (2022)
5. Wang, W., Zhang, S., Ye, C., et al.: [Retracted] a new method for optimizing the cabin layout of manned submersibles. Complexity **2020**(1), 6626602 (2020)
6. Vamråk Solheim, A., Lesage, M., Asbjørnslett, B.E., et al.: Deep sea mining: towards conceptual design for underwater transportation. In: International Conference on Offshore Mechanics and Arctic Engineering, p. 84386: V06BT06A050. American Society of Mechanical Engineers (2020)
7. Pan, B., Cui, W.: Multidisciplinary Design Optimization and Its Application in Deep Manned Submersible Design. Springer, Singapore (2020). https://doi.org/10.1007/978-981-15-6455-0
8. Chen, D., Zhu, M., Qiao, Y., et al.: An ergonomic design method of manned cabin driven by human operation performance. Adv. Des. Res. **1**(1), 12–20 (2023)
9. Zhou, Y., Chen, D., Xiao, J., et al.: An evaluation method for HMI of deep-sea manned submersible based on human reliability. Sci. Rep. **13**(1), 14507 (2023)
10. Bernard, F., Zare, M., Sagot, J.C., et al.: Using digital and physical simulation to focus on human factors and ergonomics in aviation maintainability. Hum. Fact. **62**(1), 37–54 (2020)

11. Cao, K., Zhang, Y., Jiang, Y., et al.: A civil aircraft cockpit layout evaluation method based on layout design principles. Aerospace **9**(5), 251 (2022)
12. Beiranvand, H., Ayoobi, M., Mortezapour, A., et al.: Applying ergonomics for beyond the earth: integrating participatory ergonomic design for sending a human to space. Work **76**(1), 159–169 (2023)

Research on Oil-Filled Connectors for Deep-Sea Applications

Shenshen Yang[1,2], Yunsheng Shen[1,2], Baolei Lu[1,2], Wentao Yu[1,2], Yao Wang[1,2], Fei Cheng[1,2], and Xuan Wang[1,2](✉)

[1] China Ship Scientific Research Center, Wuxi 214082, China
xuanwangcssrc@126.com
[2] State Key Laboratory of Deep-Sea Manned Vehicles, Wuxi 214082, China

Abstract. In order to improve the reliability and reduce the failure rate of deep-sea underwater equipment, this paper analyses the forms and mechanisms of failures and the destructive hazards arising from ordinary rubber watertight connectors in response to the problems that have arisen on large-depth submersibles in the last two decades. The technical advantages and usage scenarios of Oil-filled connectors are analyzed, which serve as a reference for the development and application of electrical connectors for deep-sea underwater equipment.

Keywords: Watertight connectors · Oil-filled · Failures

With the deepening of human exploration of the ocean, more and more underwater equipment is being applied. Many electrical devices can operate underwater by two main modes. One is that they can withstand pressure by themselves, such as well-insulated mechanical hydraulic components, some electronic chips and so on [1–4]. The other is that most electronic equipment needs a pressure-resistant hull to withstand pressure, providing the normal-pressure internal environment, such as computers, navigation equipment and so on. The watertight connectors are important devices with energy and signal transmission function under ambient pressure, solving the operation problems for a lot of underwater devices. According to the work modes, watertight connectors are divided into dry plugging and wet plugging. According to the materials, watertight connectors are divided into metal connectors, rubber and other non-metal connectors. According to the function modes, watertight connectors are divided into signal, power supply and fiber-optic connectors. According to the operating environment, watertight connectors are divided into pressure-resistant and oil-filled connectors [5, 6]. The Seacon, Impulse and Subconn [7–9] series are currently offered by well-acknowledged international manufacturers. The rubber watertight connectors are used largely on Jiaolong, Shen Hai Yong Shi, and Fen Dou Zhe manned submersibles. Through the experience of accumulated 1000 dives, their failure rate is obviously high, largely belonging to mechanism-based failure (Figs. 1, 2 and Table 1).

D. Jeng and B. Cai (Eds.): ICCOE 2024, LNCE 423, pp. 414–421, 2025.
https://doi.org/10.1007/978-981-96-0181-3_37

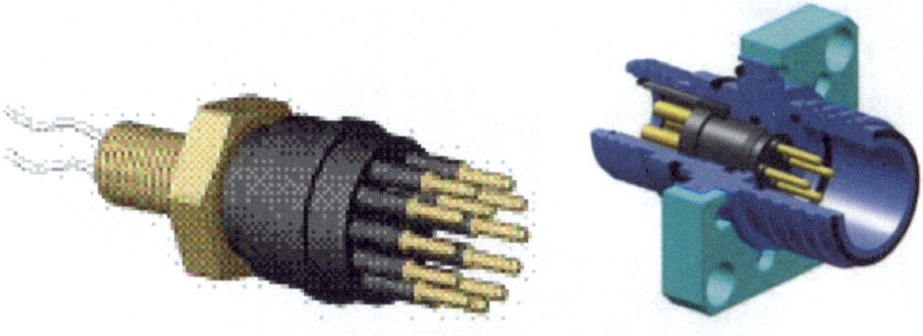

Fig. 1. Seacon Full Ocean Deep Connector

Fig. 2. Three Deep-sea Manned Submersibles

Table 1. The Quantity of Watertight Connectors on Three Deep-sea Manned Submersibles

	Rubber Electrical Watertight Connectors	Metal Electrical Watertight Connectors	Fiber-optic Watertight Connectors
Jiao Long	83	66	3
Shen Hai Yong Shi	150	55	7
Fen Dou Zhe	129	43	5

1 Problems Occurred from Rubber Watertight Connectors

1.1 Deterioration of Rubber Cables with Cracks

Rubber cables from watertight connectors are prone to cause deterioration with cracks due to being exposed to high pressure, irradiation and high salinity for a long time. And the cracks are easily turn to be penetrating ones under high pressure. The sea water can unimpededly immerse into the cables, flowing along the longitudinal direction and penetrating directly into the sockets and pins. Electrochemical corrosion emerges subsequently. The most typical failure phenomenon is that copper green arises on the surface of pins from watertight connectors. Besides, the cracks on the rubber cable are difficult to locate, requiring very careful inspection [10] (Fig. 3).

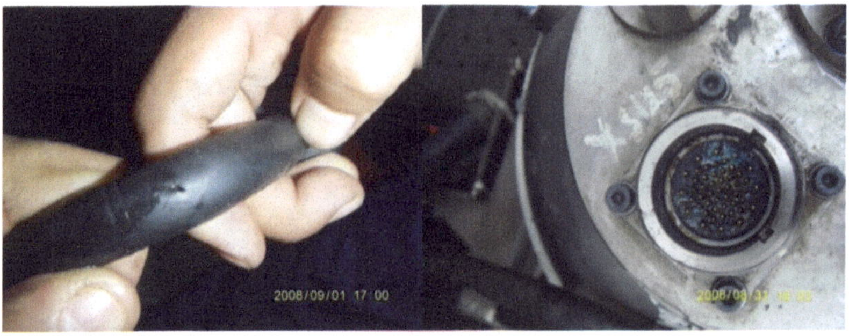

Fig. 3. Cracks Due to Deterioration of Rubber Cables

1.2 Cracking on the Vulcanized Parts of Rubber Cables

Watertight connectors are vulcanized to achieve the bonding and sealing of rubbers and metals. The vulcanization processes are divided into hot and cold vulcanization. As for on-site maintenance, cold vulcanization is generally carried out. When large quantities of watertight connectors are produced in the factory, hot vulcanization is generally carried out. Due to unsuitable environmental conditions or vulcanization process, vulcanized parts of rubber cables are prone to bubbles or gaps due to debonding. Once in seawater, watertight connectors are completely damaged with no possibility of maintenance (Fig. 4).

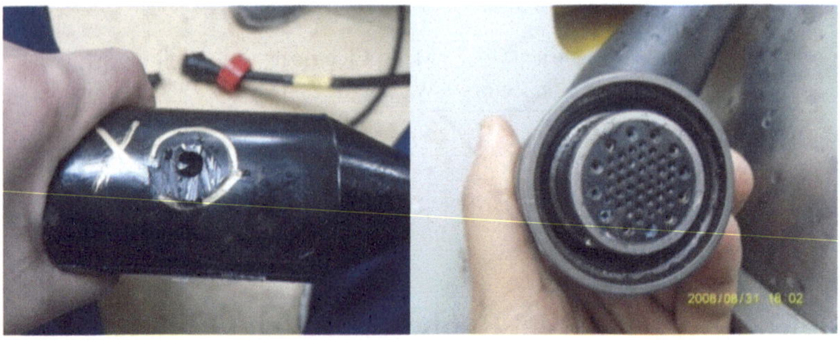

Fig. 4. Cracking on the Vulcanized Parts, Causing Water Leakage

1.3 Seal Failure Due to Reversely Pressurized

Being reversely pressurized means that watertight sockets are installed in the oil-filled junction box or equipment. Not only from the front side but also from the back side are pressurized under deep-sea environment. The constructional form of watertight connectors dictate that they cannot be subjected to large reverse pressure. There is a very small cavity in the middle of the insulator of the plug and socket. This cavity is a process cavity for the installation of the connector in order to allow the sealing surface to be fitted in place. Due to the presence of the cavity, the insulator of the plug socket slips under high pressure at both ends, causing the two insulators to squeeze. On one hand, this can lead to the watertight connector plug not being tightened after re-plugging or replacing the cable with a new one, and the sealing surface not being able to fit tightly enough to affect the seal. On the other hand, the extrusion can lead to the insulator surface falling off, or shattering for severe case, to, which seriously affects the pressure-resistant performance (Figs. 5 and 6).

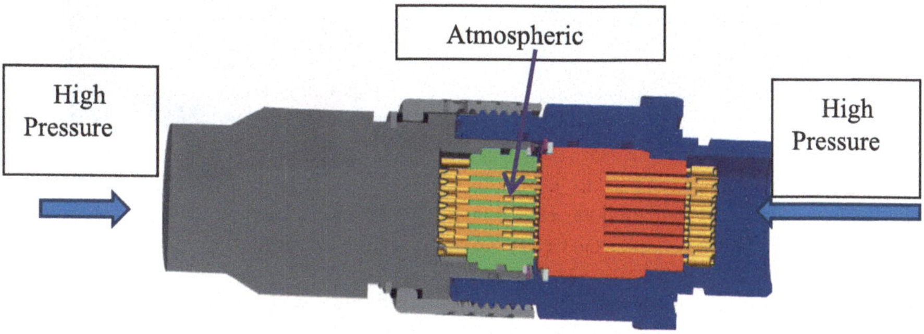

Fig. 5. Schematic Diagram of Reversely Pressurized

Fig. 6. Damage on Insulator Due to Squeezed under Pressure

1.4 Short-Circuiting Occurred Between Core and Shield Wire After Pressurized

The rubber watertight cables are connected to the metal part of the joint in the form of a vulcanized package. When the watertight cables are subjected to high external pressure under water, the rubber part of the cables and the vulcanized package will be extruded and deformed. The internal cable joints will shrink due to the extrusion. And there is a risk of short-circuiting arising, which frequently leads to more serious faults for high-voltage power supply cables (Fig. 7).

1.5 Watertight Connectors Collapsed

Rubber connectors are prone to fatigue damage due to repeatedly pressurized, which can lead to collapsed damage directly to the entire insulation components and rubber.

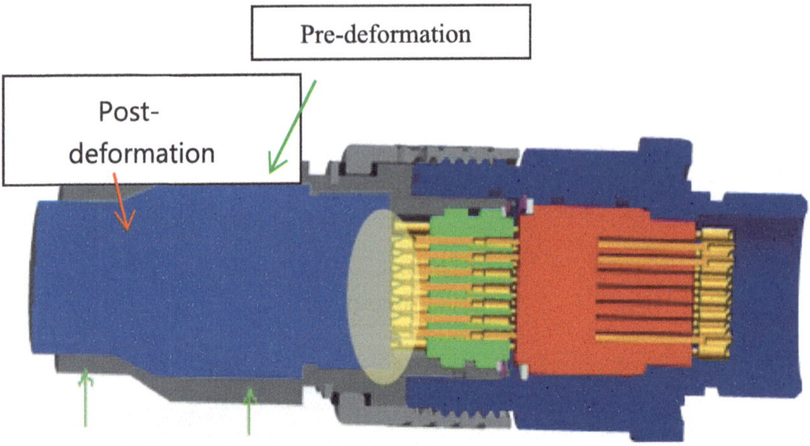

Fig. 7. Schematic Diagram of Post-deformation of Rubber Cables

This kind of damage is the most serious and horrible one. AS for manned submersibles, it is catastrophic. This is the reason that exclusively using rubber watertight cables of BH and other series are prohibited in the design of manned submersibles (Fig. 8).

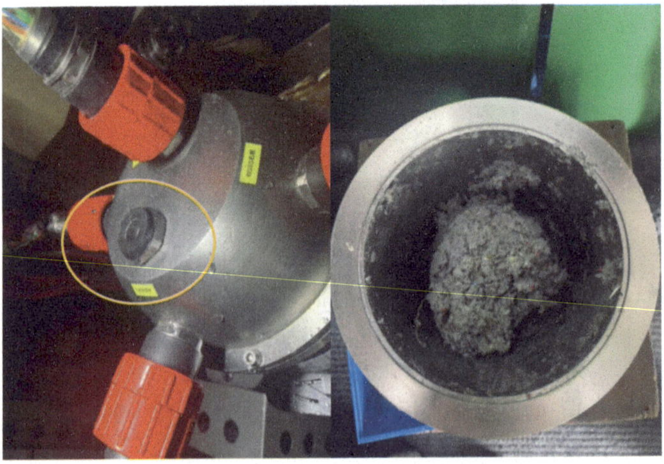

Fig. 8. Collapsed Watertight Connectors

2 Technical Advantages of Oil-Filled Connectors

Oil-filled connectors are the special type of products, such as MSSL_OF, MHDO_OF, where OF is the oil-filled designation. The rubber watertight connectors are used largely on Jiaolong, Shen Hai Yong Shi, and Fen Dou Zhe manned submersibles. Through the related experience their failure rate is obviously high, largely belonging to mechanism-based failure. The oil-filled connectors have an irreplaceable advantage over conventional rubber vulcanized connectors in that they can be used in deep-sea pressure environment.

2.1 Interior Filled with Insulating Oil Without Atmospheric Cavity

No matter installed on pressure-resistant hull or oil-filled devices, pressurized or reversely pressurized, once Oil-filled connectors are assembled, they are filled with flowing insulating oil with high pressure. So that insulators are not squeezed from both sides to the middle (Fig. 9).

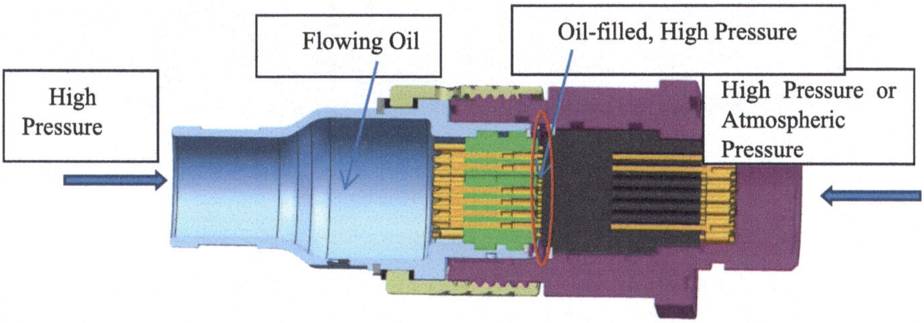

Fig. 9. Schematic Diagram of Oil-filled Connectors

2.2 Wires are Not Stretched

The oil-filled cable uses oil-filled, oil-resistant plastic hoses instead of rubber cables. When high pressure is encountered, the deformation caused by the pressure leads to the flowing of the internal insulating oil. However, the flowing of the insulating oil does not lead to stress on the wires, which effectively solves the problem of stress damage to the wires.

2.3 Oil-Filled Cables Do Not Cause Destructive Water Ingress

As the internal and external pressure of the oil-filled cables are balanced, a small crack on the oil-filled hose will not cause a large amount of seawater to enter. And even if water enters the oil-filled hose, only a small amount of incoming water will sink at the bottom of the hose due to its higher density, which will not make rubber cables penetrate water neither cause electrical insulation problems.

2.4 Oil-Filled Cables Can Be Customized

Due to the transmission of various signals, underwater equipment sometimes needs coaxial cables, shielded cables, twisted pair and other special cables. Cable manufacturers seldomly customize rubber cables, otherwise at the expense of high cost. But oil-filled cables only require oil-filled hoses. And the internal cables can be customized with various wires, making it easy to repair when problems arise (Fig. 10).

Fig. 10. Rubber Cables VS Oil-filled Cables

2.5 Increased Voltage Resistance

The insulation strength of air is generally 3 kV/mm. The insulation strength of oil is generally about ten times that of air, that is 24 kV/mm. Since the bottleneck of the insulation of the connector is air, if it is filled with oil, the insulation strength will be much higher. So the voltage resistance of the oil-filled connector is much higher than that of the general connector. For example, the voltage resistance of MSSL4 is 1000 V, and the voltage resistance of MSSL4_OF is up to 3000 V.

3 Technical Disadvantages of Oil-Filled Connectors

3.1 Complex Production Process

The external pressure will all be concentrated on the insulator of Oil-filled connectors, requiring insulators with excellent pressure resistance and sealing performance.

The production process is complex and difficult. In recent years, there have arisen several methods such as ceramic sintering, glass sintering, PEEK injection and epoxy casting. Based on experience in design and application, we believe that PEEK injection and epoxy casting are more suitable for high pressure environment, due to their impact resistance and process maturity. However, due to the complexity of the process, the corresponding production costs are high.

3.2 Regular Oil Replenishment

Oil-filled connectors will reduce a small amount of oil if they are plugged in and out. So regular oil replenishment is required for frequent plugging. Since some kinds of hydraulic oils are corrosive to the skin, silicone oil and white mineral oil are chosen under the condition of taking no account of the cost. Try to operate with gloves when plug in/out and wipe off the leaking oil after installation to keep the site clean.

4 Summary

Fig. 11. Oil filled connectors have a low failure rate in use

In contrast to rubber watertight connectors, Oil-filled connectors can solve the problem of a range of destructive damages at large depths, especially at depths in excess of 7000 m. The Oil-filled connectors have been tested and proven to be very effective in solving the problem, with an order of magnitude reduction in failure rates. However, at shallow depths, the benefits are not as obvious and may increase the cost and maintenance effort (Fig. 11).

References

1. Daqi, Z.: Research status and prospect of deep-sea underwater vehicle. J. Anhui Normal Univ. (Nat. Sci.) **41**(3), 205–216 (2018)
2. Xu, W.: Overview of present status and development trend of full ocean depth submersibles. Shipbuilding China **57**(2), 206–221 (2016)
3. Zhen, H.: Development and application of manned deep diving technology. China Eng. **21**(6), 87–94 (2019)
4. Jun, C.: Current situation and development of deep-sea submersible equipment. Shipbuilding China **61**(1), 204–218 (2020)
5. Jun, C.: Research on the application of watertight connector component in manned submersible. Mar. Electr. Electron. Eng. **42**(04), 1–6 (2022)
6. Shaobin, C.: The engineering application of general series connectors in deep – ocean. Electromech. Components **33**(4), 32–35 (2013)
7. Birns Homepage. https://birns.com/product-category/marine. Accessed 1 Aug 2023
8. Impulse Homepage. http://www.teledynemarine.com/impulse. Accessed 21 Nov 2022
9. Subconn Homepage. https://www.macartney.com/what-we-offer/systems-and-products/connectors/subconn. Accessed 12 Oct 2023
10. Haibin, Q.: Brief introduction of the application of deep sea watertight connector. Wire Cable **06**, 1–4 (2021)

Numerical Simulation of Flow Around a Submarine Under the Rudder Deflection Condition Based on Different Turbulence Models

Jianlong Huo, Xiaohui Zhang(✉), Lei Zou, and Qimao Xu

Yantai Research Institute of Harbin Engineering University, Yantai 264000, Shandong, China
xh_zhang@hrbeu.edu.cn

Abstract. A numerical study of flow around the submarine model SUBOFF under the rudder deflection condition is performed, considering the complex flow characteristics under the submarine maneuvering condition. Hydrodynamic forces and moments under rudder deflection conditions are calculated based on three different turbulence models, including the Unsteady Reynolds-Averaged Navier-Stokes (URANS) model, the Delayed Detached Eddy Simulation (DDES) model, and DDES with Improved wall-modeling capability (IDDES) model. The results are then compared with simulated values available in the literature. The hydrodynamic forces and moments obtained using the IDDES model are the most accurate. The hydrodynamic characteristics of the submarine are analyzed based on the flow field details such as vortex structures, velocity fields, streamline distribution, turbulence levels and pressure distribution obtained from the three turbulence models. The results show that the IDDES model can capture finer and more realistic flow field information, and can better predict the flow mechanisms under rudder deflection conditions.

Keywords: Rudder deflection · Turbulence modelling · SUBOFF · Vortex structure · Flow field

1 Introduction

Maneuverability is one of the essential hydrodynamic characteristics of a submarine, which is related to its operational safety and mission execution capability. The maneuvering characteristics of submarines are affected by hydrodynamic forces. Therefore, it is necessary to study the hydrodynamic characteristics under the rudder deflection condition. Existing studies have focused on simulating the resistance coefficient and lift coefficient at different rudder angles using steady or unsteady simulations [1–3]. In the maneuvering condition, the flow field around the submarine exhibits an extremely high complexity, with abundant vortex structures formed on the main body and appendages

D. Jeng and B. Cai (Eds.): ICCOE 2024, LNCE 423, pp. 422–432, 2025.
https://doi.org/10.1007/978-981-96-0181-3_38

[4]. Particularly during strenuous maneuver, large-scale flow separation around the rudder may occur, leading to rudder stall and adversely affecting the submarine's maneuverability. Therefore, it is appropriate to use unsteady simulation of hydrodynamic characteristics such as velocity vector field, pressure field, rudder surface pressure coefficient at different rudder angles [5–7].

The existing studies seldom consider the effects of different turbulence models on the numerical simulation of flow around the submarine under the rudder deflection condition. In this paper, the Defense Advanced Research Projects Agency (DARPA) SUBOFF model developed by the David Taylor Research Centre (DTRC) is taken as the object of study, and the numerical simulations of the large rudder angle tests are carried out based on the Unsteady Reynolds-Averaged Navier-Stokes (URANS) model, the Delayed Detached Eddy Simulation (DDES) model, and DDES with Improved wall-modeling capability (IDDES) model. Based on the STAR-CCM+ platform, the validity of the numerical results is verified by comparing the calculated hydrodynamic forces and moments with the simulated values from the existing literature. Then, the hydrodynamic characteristics of the submarine are analyzed based on the flow field details such as vortex structures, velocity fields, streamline distribution, turbulence levels and pressure distribution obtained under different turbulence models.

2 Numerical Method

2.1 Turbulence Modeling

Numerical simulations are performed using STAR-CCM +. The velocity-pressure coupling is done using the SIMPLE algorithm, the control equations are discretized using the finite volume method, the time discretization is done using the second order scheme and the spatial discretization is done using the second order convective format. In addition, the interpolation scheme uses a first-order linear scheme. In the numerical simulation, the fluid medium is water with density $\rho = 997.561$ kg/m^3 and dynamic viscosity $\mu = 8.89 \times 10^{-4}$ Pa·s.

In URANS, the turbulence is modelled using the two-equation model of SST $k-\omega$. The SST $k-\omega$ model uses $k-\omega$ and $k-\varepsilon$ model for the near-wall region and far-field, respectively. The turbulence kinetic energy k and the turbulence specific dissipation rate ω are calculated by the following equation:

$$\frac{\partial(\rho k)}{\partial t} + \frac{\partial(\rho U_i k)}{\partial x_i} = \tilde{P}_k - \beta^* \rho k \omega + \frac{\partial}{\partial x_i}\left[(\mu + \sigma_k \mu_t)\frac{\partial k}{\partial x_i}\right] \tag{1}$$

$$\begin{aligned}\frac{\partial(\rho \omega)}{\partial t} + \frac{\partial(\rho U_i \omega)}{\partial x_i} &= \alpha \rho S^2 - \beta \rho \omega^2 + \frac{\partial}{\partial x_i}\left[(\mu + \sigma_\omega \mu_t)\frac{\partial \omega}{\partial x_i}\right] \\ &\quad + 2(1 - F_1)\rho \sigma_{w2}\frac{1}{\omega}\frac{\partial k}{\partial x_i}\frac{\partial \omega}{\partial x_i}\end{aligned} \tag{2}$$

Here, $\tilde{P}_k$ is the turbulence generation term due to viscous forces; F_1 is the mixing function; more parameters could be found in the published literature [8].

The Detached Eddy Simulation is one of the common hybrid RANS/LES methods. That is, the RANS method is used in the near-wall region to reduce the number of grids and computational effort, and the LES simulation is used in the region away from the wall to ensure that the large-scale separated flow can be captured. The governing equations of the SST-DES turbulence model are:

$$\frac{\partial(\rho k)}{\partial t}+\frac{\partial(\rho U_i k)}{\partial x_i}=\frac{\partial}{\partial x_i}\left[(\mu+\sigma_k\mu_t)\frac{\partial k}{\partial x_i}\right]+\tilde{P}_k-\beta^*\rho\omega k F_{\mathrm{DES}} \tag{3}$$

$$\frac{\partial}{\partial t}(\rho\omega)+\frac{\partial}{\partial x_i}(\rho\omega u_i)=\alpha\rho S^2-\beta\rho\omega^2+\frac{\partial}{\partial x_i}\left[(u+\sigma_\omega u_t)\frac{\partial\omega}{\partial x_i}\right]$$
$$+2(1-F_1)\rho\sigma_{\omega 2}\frac{1}{\omega}\frac{\partial k}{\partial x_i}\frac{\partial\omega}{\partial x_i} \tag{4}$$

Here, F_{DES} is the conversion variable.

For the SST-DDES model, the turbulence length scale l_{DDES} is defined as follows:

$$l_{DDES}=l_{\mathrm{RANS}}-f_d\max(0;l_{\mathrm{RANS}}-l_{LES}) \tag{5}$$

Here, l_{RANS} is the RANS length scale, l_{LES} is the LES length scale, and f_d is the blending function.

For the SST-IDDES model, the turbulence length scale l_{IDDES} is defined as follows:

$$l_{\mathrm{IDDES}}=\tilde{f}_d(1+f_e)l_{\mathrm{RANS}}+\left(1-\tilde{f}_d\right)l_{\mathrm{LES}} \tag{6}$$

Here, $\tilde{f}_d$ is the mixing function; f_e is the gain function.

2.2 Study Object

In this study, the submarine model DARPA SUBOFF is chosen as the subject of the study, which is a conceptual nuclear submarine geometry designed by DTRC with an axisymmetric streamlined hull geometry. In addition, it is attached with a sail and four identical fins, where the horizontal fin is called the sternplane and the vertical fin is called the rudder. The geometrical characteristics of the SUBOFF model [9] are shown in Table 1. Study conditions involve numerical simulation of rudder angle tests with large rudder angles, and the rudder deflections are shown in the schematic diagram in Fig. 1.

2.3 Computational Domain and Grid Generation

A computational domain as shown in Fig. 2 is used with dimensions: $-2.8L_{pp}<x<4.7L_{pp}$; $-1.9L_{pp}<y<1.9L_{pp}$; $-1.9L_{pp}<z<1.9L_{pp}$. Boundary conditions on the outlet side is set to zero-gradient velocity and atmospheric pressure, those on the inlet side are set to fixed inflow velocity and zero-gradient pressure, and those on the other perimeter sides are set to velocity inlet with zero velocity, and the solid surface of the submarine model is set as a no-slip wall. The wall function is used for solving the near-wall flow.

The computational domain is discretized using a trimmer mesh and the near surface of the submarine is discretized using a 12-layer prismatic layer mesh. In order to accurately capture the local flow field, the submarine's hull and appendages are encrypted. The non-dimensional distances from the wall y^+ are concentrated in the range of 30 to 100. Mesh generation is shown in Fig. 3.

Table 1. Geometric features of the model and working conditions.

Term	unit	value
Overall length	m	4.356
Hull diameter	m	0.508
Length between perpendiculars	m	4.261
Fins	—	NACA0020
Rudder area	m^2	0.0248
Rudder shaft longitudinal distance from rudder trailing edge	m	0.137
Rudder tip chord length	m	0.1524
Distance of a transverse section from stern aft end	m	0.6
Distance of the longitudinal section from mid-longitudinal profile	m	0.2
Inflow velocity	m/s	3.3436

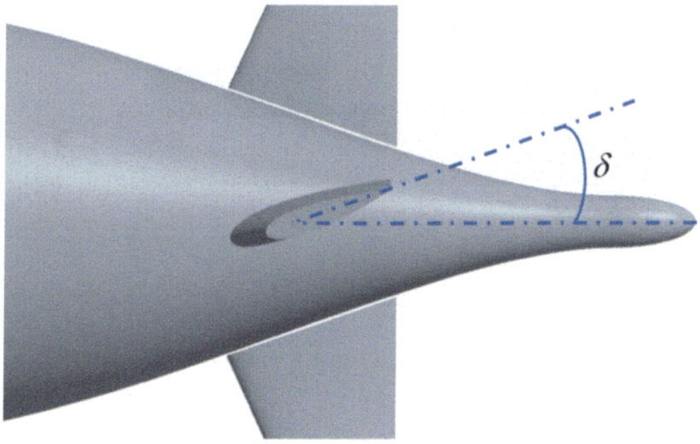

Fig. 1. Schematic diagram of the rudder deflection.

3 Convergence Analysis

In order to critically assess the numerical uncertainty caused by the grids, the convergence study is carried out by using the method provided by ITTC. The case with a rudder angle of 10° is selected for the analysis, and three sets of grids are generated according to the refinement ratio $\sqrt{2}$. The results of the convergence study are shown in Table 2. Referring to reference [7], the total physical simulation time is 10 s, and the time step is 0.002 s. From the perspective of maneuverability, the longitudinal force X and transverse force Y and yawing moment N acting on the submarine are taken into account. The reference point for the forces and moments is 2.106 m from the very end of the submarine's stern [7]. For ease of description, the forces, moments and pressure coefficients are expressed in dimensionless form in the following equations:

$$X' = \frac{X}{0.5\rho U_0^2 L_{PP}^2} \tag{7}$$

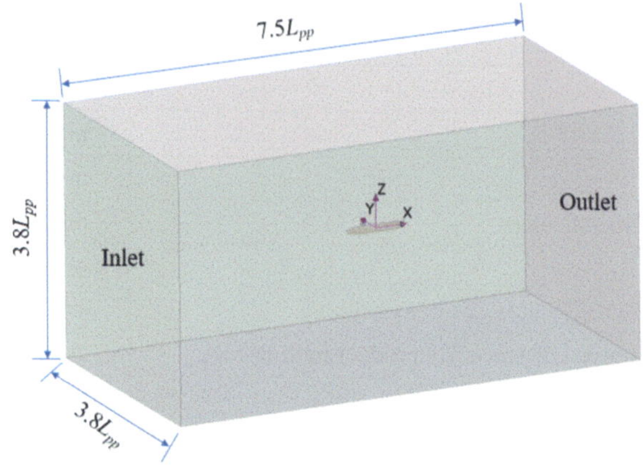

Fig. 2. Schematic diagram of the computational domain.

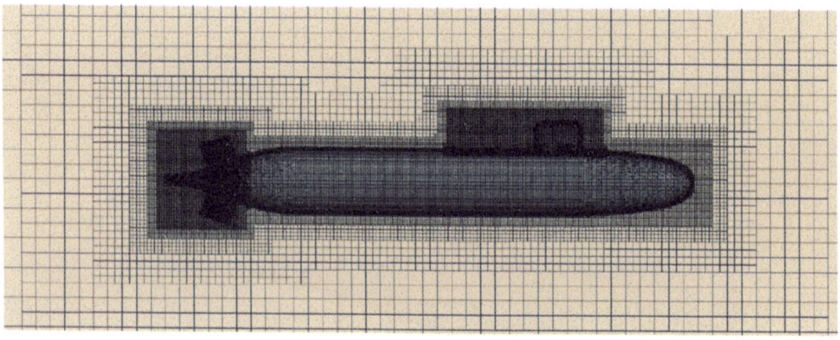

Fig. 3. Mesh Generation Diagram.

$$Y' = \frac{Y}{0.5\rho U_0^2 L_{PP}^2} \tag{8}$$

$$N' = \frac{N}{0.5\rho U_0^2 L_{PP}^3} \tag{9}$$

$$C_p = \frac{p}{0.5\rho U_0^2} \tag{10}$$

It should be noted that the physical model used for the grid independence analysis is the SST $k - \omega$ model with an uncertainty of no more than 1.96%, and the medium grid meets the requirements for use. The number of grids corresponding to coarse, medium and fine grids are 2.52 million, 4.52 million and 9.72 million, respectively.

Table 3 compares the numerical results obtained from the simulation of the three turbulence models with the available simulation data [7]. The IDDES model calculates the lowest overall deviation of the simulated values.

Table 2. Grid independence.

	$\varepsilon_{21}\%S_1$	$\varepsilon_{32}\%S_1$	R_G	P_G	C_G	$U_G(\%S_1)$
X'	−0.007	−0.017	0.429	2.455	1.333	0.904
Y'	0.014	−0.037	−0.391	—	—	1.854
N'	−0.012	−0.044	0.278	3.696	2.600	1.960

Table 3. Accuracy Comparisons.

	X' (10^{-3})	Err1 (%)	Y' (10^{-3})	Err2 (%)	N' (10^{-3})	Err3 (%)
Literature [7]	1.272	—	0.904	—	0.363	—
URANS	1.235	2.91	0.793	12.28	0.407	12.12
DDES	1.245	2.12	0.854	5.53	0.385	6.06
IDDES	1.268	0.31	0.885	2.10	0.371	2.20

4 Results and Discussion

4.1 Vortex Structures

The shedding vortex structure around the stern of the submarine at a rudder angle of 20° is given in Fig. 4. The instantaneous vortex structure is represented by the iso-surface of Q-criterion, which is colored by the non-dimensional axial velocity U_x/U_0. A distinct horseshoe vortex is observed at the junction of the hull and fins, and develops into a U-type structure downstream. Unlike the URANS simulation, the DDES and IDDES simulations show an enhancement of the deflected rudder tip vortices. Significant vortex shedding and a complex vortex structure characterized by a series of secondary vortices of different scales are also detected. It is clear that the URANS simulation does not capture the subtle vortex structures well. Compared with the DDES simulation, more obvious vortex breaking and dissipation phenomena can be observed in the IDDES simulation.

4.2 Velocity Fields

In order to further investigate the flow mechanism, the flow into the rudder and the flow through the upper rudder are selected for the study. The velocity field on the selected section is obtained from the last 1 s calculation. The mean and RMS values of U_x/U_0 on the transverse section for a rudder angle of 20° are given in Figs. 5 and 6. There is almost no difference between the mean values of U_x/U_0 calculated by URANS, DDES and IDDES models. The difference is that there is a significant difference in the RMS values of U_x/U_0. The results indicate that the velocity fluctuations are larger in the IDDES simulation, which may be related to the more pronounced vortex breaking and dissipation phenomena that can be observed in the IDDES simulation.

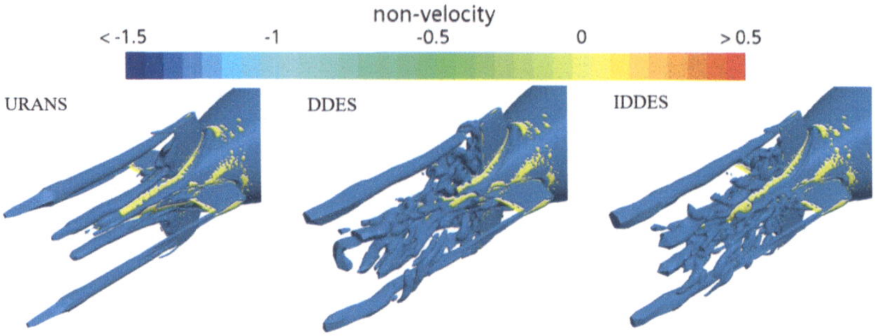

Fig. 4. Instantaneous iso-surface of Q = 100.

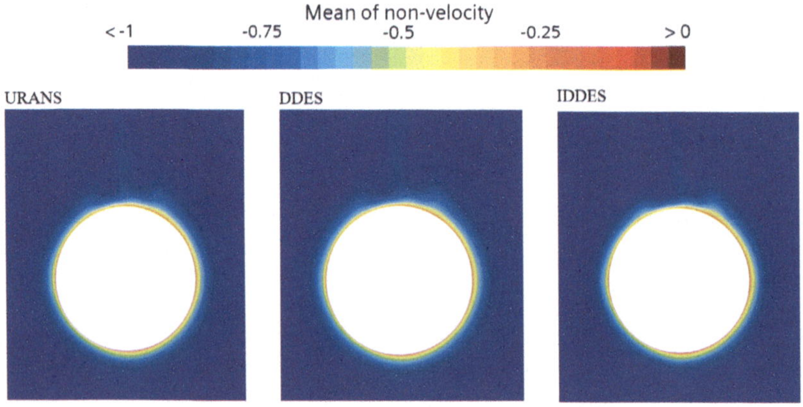

Fig. 5. Mean of U_x/U_0 on the transverse section.

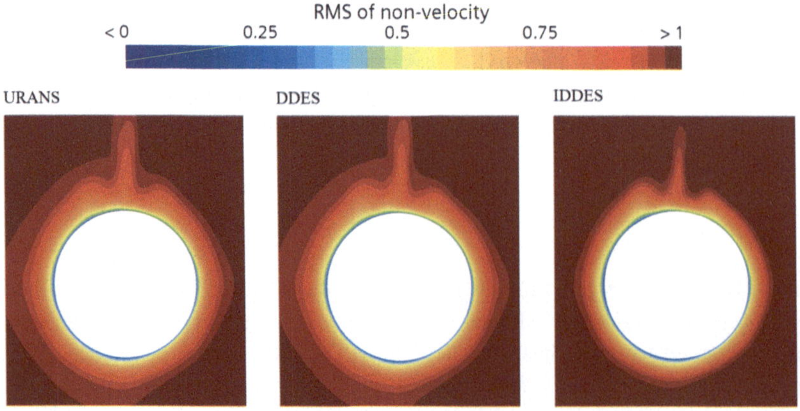

Fig. 6. RMS of U_x/U_0 on the transverse section.

1.0 indicates the position of the leading edge. The pressure distribution in the aft section of the rudder is relatively flat. A large positive pressure region occurs on the windward side of the leading edge, and a large negative pressure region with large wall shear stress occurs on the leeward side of the leading edge. The pressure coefficients and wall shear stress on the rudder section obtained in the IDDES simulation are significantly different compared to the URANS and DDES simulations.

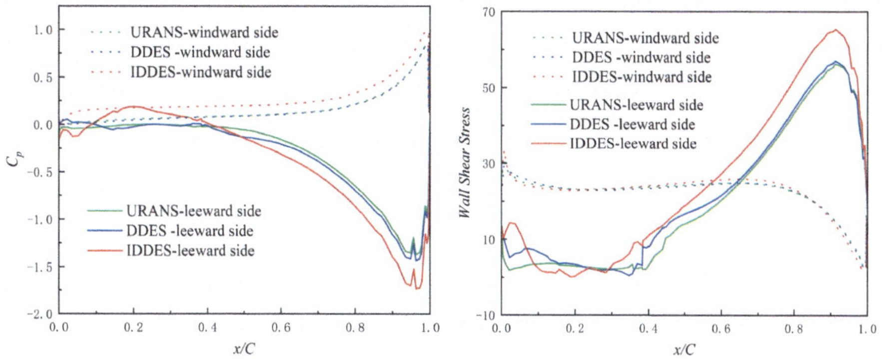

Fig. 11. Rudder section pressure coefficient and wall shear stress (z = 0.20 m).

5 Concluding Remarks

The hydrodynamic characteristics under the rudder deflection condition are investigated based on three different turbulence models, including the Unsteady Reynolds-Averaged Navier-Stokes model, the Delayed Detached Eddy Simulation model, and DDES with Improved wall-modeling capability model. By analyzing the hydrodynamic characteristics and related flow field information under the rudder deflection condition, it can be seen that:

(1) The IDDES simulation can well predict the force and moment on the rudder under the deflection condition; more obvious vortex breaking and dissipation phenomena can be observed; the streamline around the rudder show stronger inhomogeneity; the TKE obtained is smaller; a wider range of low-pressure region and a wider range of pressure fluctuation at the leading edge can be observed.

(2) There is almost no difference in the mean non-dimensional axial velocities obtained for the URANS, DDES and IDDES simulations, but in the case of the IDDES simulation the velocities fluctuate more. The pressure coefficients and wall shear stress on the rudder cross-section simulated by URANS and DDES simulations do not differ much from those simulated by IDDES simulation, which is significantly different from those simulated by IDDES simulation.

Acknowledgments. The work is sponsored by Shandong Province Natural Science Foundation ZR2022QE035 and ZR2022QE106; Heilongjiang Province Natural Science Foundation LH2023E076 and LH2023E075; National Science Foundation of China 52301365.

References

1. Zhao, D., Guo, C., Su, Y.: Hydrodynamic performance of ichthyoid rudder at different rudder angle settings. J. Coastal Res. **32**(5), 1184–1195 (2016)
2. He, L., Kinnas S.A.: Numerical simulation of unsteady propeller/rudder interaction. Int. J. Naval Archit. Ocean Eng. **9**(6), 677–692 (2017)
3. Han, K., Cheng, X., Liu, Z.: Six-DOF CFD simulations of underwater vehicle operating underwater turning maneuvers. J. Mar. Sci. Eng. **9**(12), 1451 (2021)
4. Renilson, M.: Submarine Hydrodynamics. Springer, New York (2015). https://doi.org/10.1007/978-3-319-16184-6
5. Tendou, Y.,Okada, Y., Kajihama, T.: Study on the rudder characteristics of ultimate rudder by numerical calculation. In: Proceedings of the VII International Conference on Computational Methods in Marine Engineering, pp. 932–937. CIMNE (2017)
6. Tong, X.D., Chen, H.Y., Dong, X.Q.: Experimental and numerical investigations on the unsteady thrust of a propeller in presence of an upstream rudder. Ocean Eng. **237**, 109644 (2021)
7. Guo, H., Li, G., Du, L.: Investigation on the flow around a submarine under the rudder deflection condition by using URANS and DDES methods. Appl. Ocean Res. **131**, 103448 (2023)
8. Menter, F.R., Kuntz, M., Langtry, R.:Ten years of industrial experience with the SST turbulence model. Turbulence Heat Mass Transf. **4**(1), 625–632 (2003)
9. Groves, N.C., Huang, T.T., Chang, M.S.:Geometric characteristics of DARPA SUBOFF models (DTRC Model Nos. 5470 and 5471). David Taylor Research Center (1989)

DOB-Based Robust Bounded Control for USV via a L1-Based Triggering Guidance

Zhu Sun[1,2], Guoqing Zhang[1,2](✉), Jiqiang Li[1,2], and Chenfeng Huang[1]

[1] Navigation College, Dalian Maritime University, Dalian 116026, Liaoning, China
zgq_dlmu@163.com

[2] State Key Laboratory of Maritime Technology and Safety, Dalian, China

Abstract. A crucial and challenging issue in existing path-following control of the underactuated surface vessel (USV) is high designed complexity and heavy communication load. In this note, a robust control scheme is proposed to handle the above mentioned issues. The control scheme is divided into two modules: guidance and control. In the former, a L1-based triggering guidance is introduced to solve the overshoot problem at waypoints, and the transmission frequency of the desired heading angle is reduced. In the control module, we present the robust bounded compensating technique to approximate nonlinearities of vessel model. Besides, the dynamic event-triggered control (DETC) technique is used to ease the wear of actuators. Finally, the practicability and superiority of the control system are verified through numerical experiment.

Keywords: USV · Path-following · L1-based guidance · Event-triggered Control

1 Introduction

Along with the advent of the era of intelligent navigation, researches on control of the USV has received more and more attention. In terms of application, the technologies of the USV have been widely used in various marine industries due to its functional particularity, the path-following mission [1], the obstacle avoidance mission [2], the aquaculture farms monitoring mission [3], etc.

With further researches on the USV carried out, the considerable design problems of the USV have gradually emerged. See for example [4–7] and the results therein for the problems of guidance accuracy, uncertain nonlinearity, random disturbances, unknown actuator gain, communication load, etc. To address the above problems, we introduce a robust control scheme with guidance module and control module.

As a critical execution unit in closed-loop system, the guidance module is responsible for generating desired signals to lead the real ship (RS) converge to the reference trajectory. The virtual L1-VS guidance in [8] is an acceleration-based guidance principle, which gets rid of the limitation of time compared with the LVS guidance [9]. Besides, a limited boundary-based virtual guidance [10] is developed to reduce transmission frequency of the signals from guidance module to control module while maintaining guidance accuracy.

D. Jeng and B. Cai (Eds.): ICCOE 2024, LNCE 423, pp. 433–441, 2025.
https://doi.org/10.1007/978-981-96-0181-3_39

When it comes to the control module, one considerable issue is to handle the problem of high designed complexity and heavy communication load. In [11], a robust bounded compensating technique is introduced to approximate the model nonlinearities with concise designed form over the classical radial basis function neural networks (RBF NNs) technique [12]. Besides, heavy communication load is always an significant problem that needs to be solved for USV control. In [13], the DTEC technique is used on the USV to reduce the signals transmission frequency from controller to actuator, which is responsive to the requirement in marine practice.

Motivated by the above considerations, a concise robust path-following control scheme is proposed aiming to handle the narrow channel crossing mission of the USV. The two primary innovations of this note are summarized as follows:

1) The L1-based triggering guidance proposed in this paper eliminates the time-based errors with vertical acceleration command. Besides, the desired heading signals with threshold rule could decrease the usage of signals transmission channel.
2) A disturbance observer based control approach with the robust bounded compensating and DETC technique is introduced to handle the problem of designed complexity and actuator wear, which meets the requirement of marine practice.

2 Problem Formulation and Preliminaries

2.1 Dynamic Model of USV

Based on the theoretical model in [14], the three degrees of freedom (3-DOFs) model could be expressed as:

$$\begin{cases} \dot{x} = u\cos(\psi) - v\sin(\psi) \\ \dot{y} = u\sin(\psi) + v\cos(\psi) \\ \dot{\psi} = r \\ m_u\dot{u} = f_u(\nu) + \tau_u + d_{wu} \\ m_v\dot{v} = f_v(\nu) + d_{wv} \\ m_r\dot{r} = f_r(\nu) + \tau_r + d_{wr} \end{cases} \tag{1}$$

with

$$\begin{cases} f_u(\nu) = m_v vr - d_u u - d_{u2}|u|u - d_{u3}u^3 \\ f_v(\nu) = -m_u ur - d_v v - d_{v2}|v|v - d_{v3}v^3 \\ f_r(\nu) = (m_u - m_v)uv - d_r r - d_{r2}|r|r - d_{r3}r^3 \end{cases} \tag{2}$$

where $\eta = [x, y, \psi]^{\mathrm{T}}$ donates the surge, sway displacement and the yaw angle of the USV, and $\nu = [u, v, r]^{\mathrm{T}}$ are the velocities of the surge, sway and yaw movement. τ_u and τ_r are the control inputs. d_{wi}, $(i = u, v, r)$ indicate the disturbances of marine winds and waves. $f_i(\nu)$, $(i = u, v, r)$ are the nonlinearities of ship model, as shown in (2). The unknown parameters $m_u, m_v, m_r, d_u, d_{u1}, d_{u2}, d_v, d_{v1}, d_{v2}, d_r, d_{r1}, d_{r2}$ donate the hydrodynamic damping and nonlinear damping terms.

Lemma 1. *For a continuously differentiable function $f(\cdot)$, there exists a robust bounded compensating function $f^*(|\cdot|)$, satisfying $|f(\cdot)| \le f^*(|\cdot|)$.*

Assumption 1. *All the signals* η, ν *in (1) and* η_d, ν_d *in (3) are bounded, and their derivations are all available and bounded.*

Assumption 2. *The marine disturbances* $d_{wi}, (i = u, v, r)$ *and their derivations* $\dot{d}_{wi}, (i = u, v, r)$ *are bounded, that is, there exist the unknown constants* $\overline{d}_{wi}(i = u, v, r)$ *and* $C_{i\max}(i = u, v, r)$*, satisfying* $|d_{wi}| \leq \overline{d}_{wi}, (i = u, v, r)$ *and* $|\dot{d}_{wi}| \leq C_{i\max}, (i = u, v, r)$.

Assumption 3. *The sway velocity v is passive-bounded stable under the action of the hydrodynamic damping force.*

2.2 L1-Based Triggering Guidance

As shown in Fig. 1, reference paths are usually planned in the form of designed waypoints, such as $WP_1, WP_2, ..., WP_i$ with $WP_i(x_i, y_i), i = 1, 2, ..., n$. The desired signals are generated by the virtual ship (VS), which could be expressed as (3), where the variables express the same meaning in (1).

$$\begin{cases} \dot{x}_d = u_d \cos(\psi_d) \\ \dot{y}_d = u_d \sin(\psi_d) \\ \dot{\psi}_d = r_d \end{cases} \tag{3}$$

The L1-based triggering guidance divides the reference path into the straight-line path and the curve path. In the straight-line path, ψ_d is calculated in the same way as [9], and $r_d = 0$. In the curve path, the turning radius is generated through the lateral acceleration $a_\perp$, which could be calculated with (4).

$$a_\perp = 2\frac{u_d^2}{l_i} \sin(\phi) \tag{4}$$

where l_i donates the distance between P_i^{in} and P_i^{out}, and the angle of u_d and the connection between P_i^{in} and P_i^{out} is expressed as ϕ.

From Fig. 1, we can construct the guidance law (5) to ensure the trajectory of the RS converge to the reference path.

$$\psi_r = \frac{1}{2}\left[1 - \text{sign}(x_e)\right]\text{sign}(y_e)\pi + \arctan(y_e/x_e) \tag{5}$$

with $x_e = x_d - x, y_e = y_d - y, z_e = \sqrt{x_e^2 + y_e^2}$.

In (5), the desired signal ψ_r updates in real time, which occupies numerous communication resources. To degrade the transmission frequency of signals, we introduce triggering guidance signals, which could be expressed as (6).

$$\psi_r = \psi_d, \ t \in \left[t_k, t_{k+1}\right) \tag{6}$$

with

$$t_{k+1} = \inf\left\{t > t_k \middle| |z_e| \leq \delta_{xy} + d_{\min}\right\} \tag{7}$$

where δ_{xy} is an arbitrarily small positive constant, and $d_{\min}$ is the restricted radius.

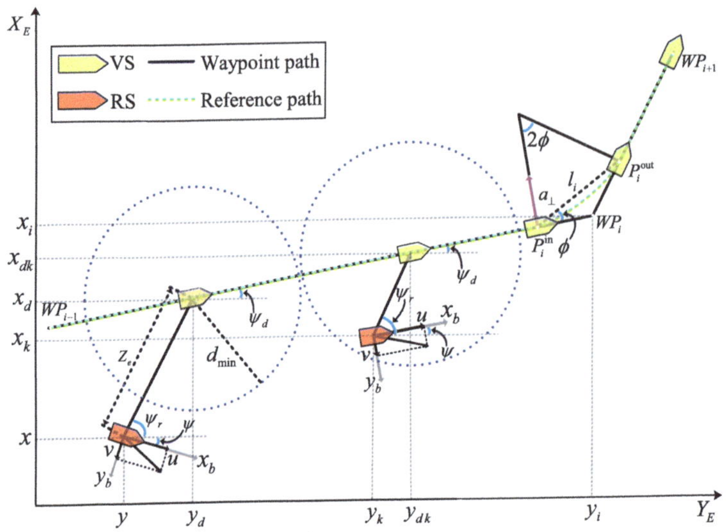

Fig. 1. The framework of L1-based triggering control.

3 Control Design and Stability Analysis

3.1 Control Design

Step 1: In order to stabilise the tracking errors z_e and ψ_e, where $\psi_e = \psi_r - \psi$, we introduce the virtual control laws α_i, $(i = u, r)$ based on the backstepping method.

$$\begin{aligned} \alpha_u &= \cos(\psi_e)^{-1}\left[k_{z_e}\left(z_e - \delta_{xy}\right) + \dot{x}_d\cos(\psi_r) + \dot{y}_d\sin(\psi_r) - v\sin(\psi_e)\right] \\ \alpha_r &= k_{\psi_e}\psi_e + \dot{\psi}_r \end{aligned} \tag{8}$$

where k_{z_e} and k_{ψ_e} are positive designed parameters.

Then, we introduce the dynamic surface control (DSC) technique to eliminate the problem of "calculation explosion" in the derivation of α_i, $(i = u, r)$.

$$\epsilon_i\dot{\beta}_i + \beta_i = \alpha_i,\ \beta_i(0) = \alpha_i(0) \tag{9}$$

where β_i are the first-order filters, and ϵ_i are the time constants, with the filter errors $y_i = \beta_i - \alpha_i$.

Step 2: In this part, the derivation of the dynamic errors could be expressed as (10) along (1).

$$i_e = \dot{\beta}_i - \dot{i} = \dot{\beta}_i - \frac{f_i(\nu)}{m_i} - \frac{\tau_i}{m_i} - \frac{d_{wi}}{m_i} \tag{10}$$

In (10), the signals of control inputs are transmitted in real time, which causes the problem of the communication load. In order to handle this problem, we introduce the DETC technique as (11).

$$\tau_{ik}(t_k) = \tau_i(t),\ \forall t \in \left[t_k, t_{k+1}\right] \tag{11}$$

where the form of t_{k+1} is shown as (12).

$$t_{k+1} = \inf\{t > t_k || \tau_{ik}(t_k) - \tau_i(k)| \geq c_i |\tau_i(k)| + \omega_i\} \tag{12}$$

with the triggered parameters c_i and ω_i satisfying (13).

$$c_i = \max\{\text{sign}(\mu_e)\tanh(\mu_e), c_0\},\ \dot{\omega}_i = -\omega_i^2 \tag{13}$$

where $\mu_e = \left[z_e - \delta_{xy}, \psi_e\right]^{\mathrm{T}}$, and c_0 is a positive small constant.

Then, the control inputs could be rewritten as (14).

$$\tau_i = \frac{\tau_{ik}}{1 + \kappa_1 c_i} - \frac{\kappa_2 \omega_i}{1 + \kappa_1 c_i} \tag{14}$$

with the bounded constant $\kappa_1 = \text{sign}(\kappa_2)$, satisfying $\kappa_1 \in [-1, 1]$.

To observe the external disturbances and allow the controller adaptable to the alterations of disturbances, we design the disturbance observer (DOB) as (15).

$$\begin{cases} \hat{d}_{wi} = \varpi_i + k_{oi} i \\ \dot{\varpi}_i = -k_{oi}\left[\dfrac{\tau_i}{m_i} + \dfrac{f_i^*(\nu)}{m_i} + \varpi_i + k_{oi} i\right] \end{cases} \tag{15}$$

where $\hat{d}_{wi}$ donate the estimate of external disturbances, with the estimated errors $\tilde{d}_{wi} = \hat{d}_{wi} - d_{wi}$, and ϖ_i donate the auxiliary state variables, and k_{oi} are the positive designed parameters.

Then, we design the triggered control laws τ_{ik} as (16).

$$\tau_{ik} = (1 + \kappa_1 c_i)\left(k_{i_e} i_e + \dot{\beta}_i + f_i^*(|\nu|) + \hat{d}_{wi}\right) \tag{16}$$

where k_{i_e} are the positive designed parameters, and $f_i^*(|\nu|)$ are robust bounded compensating function of $f_i(\nu)$ based on Lemma 1.

3.2 Stability Analysis

Based on the above controller design, the main purpose could be stated as Theorem 1.

Theorem 1. For closed-loop system considering Assumption 1 and Assumption 2, we assumed all the above-mentioned signals are defined in a compact set Ω, that is, $\Omega = \left[z_e^2(0) + \psi_e^2(0) + u_e^2(0) + r_e^2(0) + y_u^2(0) + y_r^2(0) + \tilde{d}_{wu}^2(0) + \tilde{d}_{wr}^2(0) \leq 2\Delta\right]$, with arbitrary positive constant Δ, which ensure all signals satisfy semiglobally global uniform ultimate boundedness (SGUUB) by tuning the designed parameters.

Based on the Sect. 3.1, we develop the Lyapunov function as the following form:

$$V = \frac{1}{2}\sum_{\mu = z,\psi} \mu_e^2 + \frac{1}{2}\sum_{i=u,r} m_i i_e^2 + \frac{1}{2}\sum_{i=u,r} y_i^2 + \frac{1}{2}\sum_{i=u,r} \tilde{d}_{wi}^2 \tag{17}$$

Combining with (1), (8), (10), (14), (15) and Lemma 1, the derivation of (17) could be expressed as (18).

$$\begin{aligned}\dot{V} &= \sum_{\mu=z,\psi} \mu_e\dot{\mu}_e + \sum_{i=u,r} m_i i_e \dot{i}_e + \sum_{i=u,r} y_i\dot{y}_i + \sum_{i=u,r} \tilde{d}_{wi}\left(\dot{\hat{d}}_{wi} - \dot{d}_{wi}\right) \\ &\le -\sum_{\mu=z,\psi} k_{\mu_e}\mu_e^2 + \sum_{\substack{\mu=z,\psi \\ i=u,r}} J(\mu_e)\mu_e i_e + \sum_{i=u,r} i_e\left[m_i\dot{\beta}_i + f_i^*(|\nu|) - \frac{\tau_{ki}}{1+\kappa_1 c_i}\right. \\ &\left. + \frac{\kappa_2\omega_i}{1+\kappa_1 c_i} + \tilde{d}_{wi} - \hat{d}_{wi}\right] + \sum_{i=u,r} y_i\dot{y}_i - \sum_{i=u,r} k_{oi}\tilde{d}_{wi}^2 - \sum_{i=u,r} \tilde{d}_{wi}\dot{d}_{wi}\end{aligned} \tag{18}$$

Taking advantage of the Young's inequality, we have:

$$J(\mu_e)\mu_e i_e \le \frac{1}{2}J^2(\mu_e)\mu_e^2 + \frac{1}{2}i_e^2 \le \frac{1}{2}\mu_e^2 + \frac{1}{2}i_e^2 \tag{19}$$

$$(m_i - 1)i_e\dot{\beta}_i \le \frac{m_i+1}{\epsilon_i}i_e y_i \le \frac{m_i+1}{4}i_e^2 + \frac{m_i+1}{\epsilon_i}y_i^2 \tag{20}$$

$$\tilde{d}_{wi}\dot{d}_{wi} \le \frac{1}{2}\tilde{d}_{wi}^2 + \frac{1}{2}\left|\dot{d}_{wi}^2\right| \le \frac{1}{2}\tilde{d}_{wi}^2 + \frac{1}{2}C_{i\max}^2 \tag{21}$$

$$y_i\dot{y}_i = -\frac{1}{\epsilon_i}y_i^2 + y_i B_i \le -\left(\frac{1}{\epsilon_i} - \frac{B_i^{*2}}{4a_i}\right)y_i^2 + a_i \tag{22}$$

where B_i are derivations of α_i, with $|B_i| \le B_i^*$, and a_i are positive constants.

Substituting (16) and (19–22) into (18), then, (23) is obtained.

$$\begin{aligned}\dot{V} \le &-\sum_{\mu=z,\psi}\left(k_{\mu_e} - \frac{1}{2}\right)\mu_e^2 - \sum_{i=u,r}\left(k_{i_e} - \frac{3}{2}\right)i_e^2 - \sum_{i=u,r}\left(\frac{1}{\epsilon_i} - \frac{B_i^{*2}}{4a_i}\right)y_i^2 \\ &- \sum_{i=u,r}(k_{oi} - 1)\tilde{d}_{wi}^2 + \sum_{i=u,r}\left[a_i + \frac{1}{2}C_{i\max}^2 + \frac{1}{2}\left(\frac{\kappa_2\omega_i}{1+\kappa_1 c_i}\right)^2\right]\end{aligned} \tag{23}$$

Furthermore, (23) could be rewritten as (24).

$$\dot{V} \le -2\rho V + \varrho \tag{24}$$

where $\rho = \min\left\{\left(k_{\mu_e} - 1/2\right), \left(k_{i_e} - 3/2\right), \left(1/\epsilon_i - B_i^{*2}/4a_i\right), (k_{oi} - 1)\right\}$ and $\varrho = \sum_{i=u,r}\left[a_i + C_{i\max}^2/2 + \kappa_2\omega_i/4(1+\kappa_1 c_i)^2\right]$.

We have $V(t) \le \varrho/2\rho + \left[V(0) - \varrho/2\rho\right]\exp(-2\rho t)$ through integrating over (24). Subsequently, all signals in the closed-loop system converge to a compact set $\rho/2\rho$, and Theorem 1 is proved.

4 Numerical Experiment

To demonstrate the effectiveness and superiority of the proposed algorithm. An illustrative experiment for path-following control of the USV is conducted under the simulated marine environment. We plan the waypoints as WP_1(450 m, 30 m), WP_2(450 m, 100 m), WP_3(410 m, 400 m), WP_4(410 m, 700 m), WP_5(230 m, 1045 m), WP_6(230 m, 1200 m). The initial position and yaw angle of the USV is (448 m, 10 m) and 30 deg, and the initial values of the other parameters are considered as 0.

The experiment results are shown as follows, where (a) are global images, and (b) are localised enlarged images. Figure 2 shows the result of path-following control of the USV with well-performance. Figure 3 donates the guidance signals with the triggering desired yaw angle. Figure 4 indicates the control inputs under the DETC technique. Figure 5 shows the observing performance of DOB.

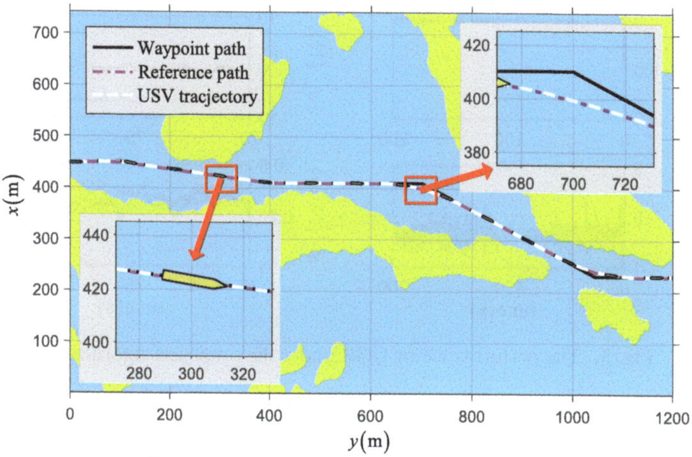

Fig. 2. The path-following trajectory of USV under the proposed algorithm.

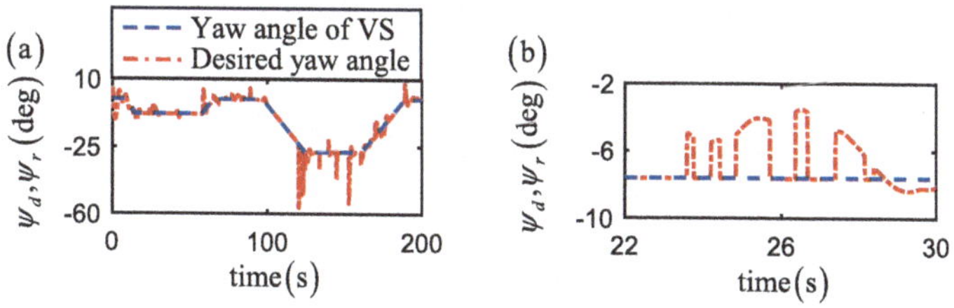

Fig. 3. The guidance signals under the proposed algorithm.

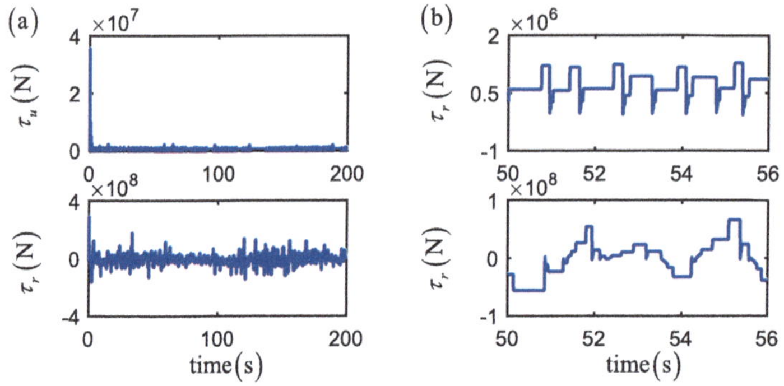

Fig. 4. The DETC control inputs under the proposed algorithm.

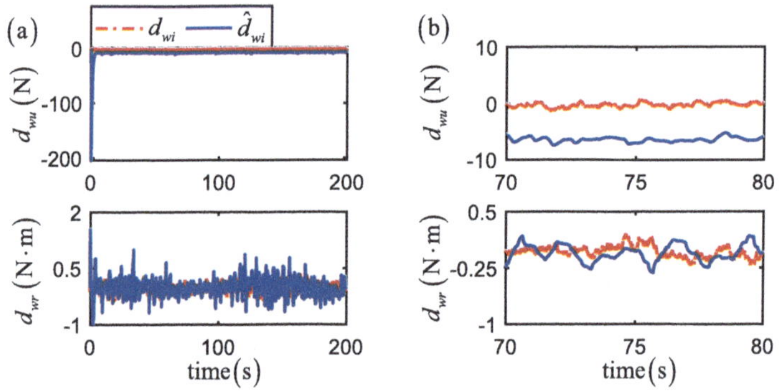

Fig. 5. The performance of DOB under the proposed algorithm.

5 Conclusion

This paper proposes a DOB-based robust control scheme for the path-following control of the USV. The L1-based triggering guidance could eliminate the accumulation error with lateral acceleration order. For control design, the robust bounded compensating technique could reduce the designed complexity compared with the existing methods. Besides, the DETC technique is used to lower the signals transmission frequency with the step-shaped orders. At last, the availability and superiority of the proposed control algorithm is verified through the simulated experiment.

Acknowledgement. This note is partially supported by the National Excellent Youth Science Fund of China (52322111), the National Natural Science Foundation of China (52171291), the Applied Fundamental Research Program of Liaoning Province (2023JH2/101600039), the Youth Talent Support Program of Liaoning Province (XLYC2203129), the Dalian Science and Technology Program for Distinguished Young Scholars (2022RJ07), the Fundamental Research Funds for the Central Universities (3132023502).

References

1. He, X., Geng, Z.: Point stabilization and trajectory tracking of underactuated surface vessels: a geometric control approach. J. Franklin Inst. **358**(14), 7119–7141 (2021)
2. Zhang, G., Han, J., Zhang, X.: APF-based intelligent navigation approach for USV in presence of mixed potential directions: guidance and control design. Ocean Eng. **260**, 111972 (2022)
3. Do, K.D., Pan, J.: Underactuated ships follow smooth paths with Integral actions and without velocity measurements for feedback: theory and experiments. IEEE Trans. Control Syst. Technol. **14**(2), 308–322 (2006)
4. Zhang, G., Gao, S., Li, J., Zhang, W.: Adaptive neural fault-tolerant control for course tracking of unmanned surface vehicle with event-triggered input. Proc. Inst. Mech. Eng. Part I: J. Syst. Control Eng. **235**(9), 1594–1604 (2021)
5. Wang, P., Yong, K., Chen, M., Zhang, X.: Output-feedback flexible performance-based safe formation for underactuated unmanned surface vehicles. IEEE Trans. Intell. Veh. **8**(3), 2437–2447 (2023)
6. Franzè, G., Lucia, W., Venturino, A.: Venturino: A distributed model predictive control strategy for constrained multi-vehicle systems moving in unknown environments. IEEE Trans. Intell. Veh. **6**(2), 343–352 (2021)
7. Zhai, J., Wang, H., Cui, H., Qin, Y.: Fixed-time fuzzy tracking control for high-order nonstrict feedback nonlinear systems with unknown virtual control gains and actuator faults. Inf. Sci. **662**, 120266 (2024)
8. Zhang, G., Bian, W., Zhang, X., Li, J.: Robust adaptive saturation-tolerance control for USVs formation via a novel L1-VS guidance. Ocean Eng. **298**, 117304 (2024)
9. Zhang, G., Zhang, X.: Concise robust adaptive path-following control of underactuated ships using DSC and MLP. IEEE J. Ocean. Eng. **39**(4), 685–694 (2014)
10. Li, J., Zhang, G., Cabecinhas, D., Pascoal, A., Zhang, W.: Prescribed performance path following control of USVs via an output-based threshold rule. IEEE Trans. Veh. Technol. https://doi.org/10.1109/TVT.2023.3338518
11. Li, J., Zhang, G., Zhang, W., Shan, Q., Zhang, W.: Cooperative path following control of USV-UAVs considering low design complexity and command transmission requirements. IEEE Trans. Intell. Veh. **9**(1), 715–724 (2024)
12. Zheng, Z., Zou, Y.: Adaptive integral LOS path following for an unmanned airship with uncertainties based on robust RBFNN backstepping. ISA Trans. **65**, 210–219 (2016)
13. Zhang, G., Liu, S., Huang, J., Zhang, W.: Dynamic event-triggered path following control of underactuated surface vehicle with the experiment verification. IEEE Trans. Veh. Technol. **71**(10), 10415–10425 (2022)
14. Fossen, T.I.: Handbook of Marine Craft Hydrodynamics and Motion Control. Wiley, New York (2011)

Control Parameter Optimization of the Automatic Voltage Regulator for Marine Synchronous Generator Based on Genetic Algorithm

Xuan-Kien Dang(✉), Tien-Diem Nguyen, Viet-Dung Do, and Ngoc-Truc Nguyen

Artificial Intelligent in Transportation Research Group, Ho Chi Minh City University of Transport, Ho Chi Minh City 700000, Vietnam
kien.dang@ut.edu.vn

Abstract. The Proportional-Integral-Derivative (PID) control technique, with advantages such as simple design and fast response, is widely applied in the maritime industry. However, determining the parameters K_P, K_I, and K_D using traditional methods often fails to achieve optimal results due to the influence of noise and errors in measurement. This paper proposes a Genetic Algorithm (GA) to optimize the PID controller parameters, named GA-PID controller, for the Automatic Voltage Regulator (AVR) of marine synchronous generators. The effectiveness of the proposed controller is demonstrated through simulations and comparisons with algorithms such as PID, and Fuzzy-PID (F-PID) under various generator operating modes. Furthermore, the AVR stability is assessed by using a Bode diagram and Root locus to verify the quality of the proposed solution.

Keywords: Automatic Voltage Regulator · Fuzzy Controller · Proportional-Integral-Derivative · Genetic Algorithm · Synchronous Generator

1 Introduction

Ship Power System (SPS) supplies electrical for propulsion, maneuvering, cargo handling, and heating. The power system consists of a combination of generators that convert diesel fuel [1] into electricity, which is uploaded to the ship's power grid and transmitted to the equipment used on the boat. Ship power systems tend to be very complex and perform many functions, with different tasks requiring flexibility, redundancy, and safety (expressed in Fig. 1 [2]). Due to operation at sea, ship power systems must adhere to international standards regarding voltage stability during load changes, backup power requirements, and safety regulations for onboard electrical systems.

Maintaining a stable and continuous power supply for devices on board ships in a maritime environment is a critical task [3]. Electrical incidents can disrupt operations, endanger lives and the environment, and compromise security. The SPS can be

D. Jeng and B. Cai (Eds.): ICCOE 2024, LNCE 423, pp. 442–453, 2025.
https://doi.org/10.1007/978-981-96-0181-3_40

regarded as a kind of nervous system as it encompasses not only power but also automated regimes enabling control over systems formed from multiple factors beyond the perception capabilities of ship operators [4, 5]. The stability of the SPS depends on the provision of electricity from the synchronous generators aboard a ship [4]. Therefore, measures are needed to control the voltage at the terminals of the synchronous generators. The AVR is a critical electrical system component that aims to improve system stability, power quality, and economic efficiency [6, 7].

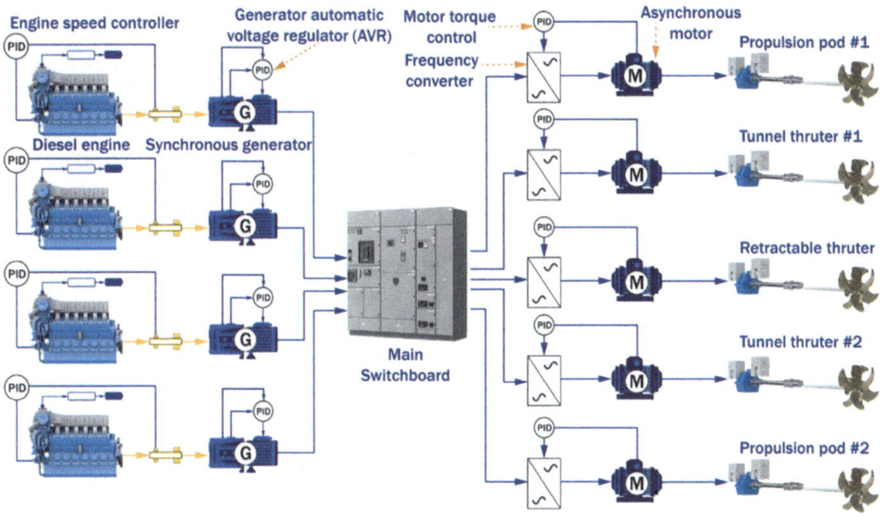

Fig. 1. The framework for the marine energy system.

In the field of ship electrical system stability, recent studies have been carried out to enhance the quality of AVR systems for synchronous generators. These studies have introduced fuzzy adaptive algorithms to improve the AVR system's efficiency and performance [5] and output voltage control for an SPS [6]. The use of neural network predictive control has been implemented, especially in marine synchronous generators, to overcome sudden voltage drops caused by instantaneous fluctuations in the power grid load [8]. Moreover, the simulation of a ship generator excitation system using the power system stabilizer (PSS) plays an important role in improving the performance of the ship power supply based on parameters such as speed, voltage, and generator current [9]. In addition, studies are also underway to develop a model for testing high-load systems on ships operating under sudden load reduction conditions. In this way, the operational parameters are collected to accurately assess the effectiveness of the proposed solutions [10]. Furthermore, other studies have utilized conventional controllers such as PID and LQR (Linear Quadratic Regulator) for generator operation. However, there haven't been many studies focusing on modern theories to enhance the quality of AVR control parameters when dealing with load variations, especially concerning voltage regulation.

The study aims to verify how the proposed genetic optimization strategy impacts the operation of a marine synchronous generator. At the same time, we designed the

GA-PID controller for the AVR to minimize the impact of grid load variations and structural changes. The proposed solution has several notable advantages, including the application of artificial intelligence to a specific field of expertise. The controller is designed to ensure better control quality and optimize the performance of the AVR system compared to traditional solutions. Moreover, the effectiveness of the controller is validated by stability criteria. Here are the main contributions of this paper as follows: *(i)* The authors designed a PID controller and an F-PID controller for the AVR system to minimize errors and increase the performance of the AVR system. However, we confirm that these solutions do not achieve high control performance under conditions of different loads and changes in system parameters; *(ii)* The proposed GA-PID controller is designed based on the error between the response signal and the set-point signal that satisfies the selected fitness function to a minimum value. Besides, we rechecked the controller stability using the Bode and Root locus methods to validate the quality of the proposed solution; and *(iii)* The authors evaluate the effectiveness of the proposed controller with comparisons through actual operations of marine synchronous generators.

2 Problem Formulation of the AVR System

The AVR system maintains a constant voltage at the output of the synchronous generator at various load levels by regulating the excitation voltage [11]. The basic components of an AVR are the amplifier unit, the exciter unit, the generator unit, and the sensor unit [12–14].

Amplifier Unit: This unit is represented by a gain K_A and time constant T_A, in which the typical value is $10 = K_A = 40$ and $0.02\ s = T_A = 0.1$ s. In this paper, K_A is set to 10, and T_A is set to 0.1 s. The transfer function of the amplifier unit is defined as

$$G_A(s) = \frac{v_{ref}(s)}{v_e(s)} = \frac{K_A}{1 + s.T_A} \tag{1}$$

Exciter Unit: The transfer function of the exciter is represented by a gain K_E and a single - time constant T_E, as in Eq. 2. In particular, the typical values are $1 = K_E = 200$, and $0.4\text{ s} = T_E = 1$ s. In simulation, K_E is set to 1, and T_E is set to 0.4 s.

$$G_E(s) = \frac{v_f(s)}{v_r(s)} = \frac{K_E}{1 + s.T_E} \tag{2}$$

Generator Unit: The transfer function relating the generator terminal voltage to its field voltage is expressed by a gain K_G and time constant T_G. The constants depend on the load, from full load to non-load as $0.1 = K_G = 1$ and $1\text{ s} = T_G = 2$ s. K_G is set to 1, and T_G is set to 1 s. The generator unit is expressed by

$$G_G(s) = \frac{v_t(s)}{v_f(s)} = \frac{K_G}{1 + s.T_G} \tag{3}$$

Sensor Unit: The unit is computed by a gain K_R and time constant T_R, therein K_R is usually kept 1, $0.001\text{ s} = T_R = 0.06\text{ s}$, and T_R is 0.05. The transfer function is

$$G_S(s) = \frac{v_s(s)}{v_t(s)} = \frac{K_R}{1 + s.T_R} \tag{4}$$

The open loop transfer function of the AVR system is determined as follows:

$$G(s)H(s) = \frac{K_A K_E K_G K_R}{(1 + sT_A)(1 + sT_E)(1 + sT_G)(1 + sT_R)} \tag{5}$$

Remark 1. In case of the SPS grid continuously switches the load, the exciter dynamically adjusts the excitation voltage to maintain the required grid voltage level. However, if there is a sudden drop in the SPS with a large oscillation amplitude, precise optimization of AVR control parameters is necessary.

3 Optimization of the Automatic Voltage Regulator System

3.1 Fuzzy – Proportional Integral Derivative Controller

The PID controllers are still preferred for high-reliability control systems in the maritime industry. In this study, we build a PID controller as a foundation to develop a fuzzy system that dynamically adjusts control parameters K_P, K_I, and K_D, named a fuzzy proportional-integral-differential (F-PID) [15]. The PID controller is expressed as

$$u_{PID}(t) = K_P\left(e(t) + \frac{1}{T_i}\int_0^t e(t)dt + T_d\frac{de(t)}{dt}\right) \tag{6}$$

where K_P is the amplification coefficient with T_i being the integration time constant, and T_d indicates the differential time constant. The PID controller parameters are determined using the Ziegler-Nichols method (ZN-PID).

To adjust the PID parameters, the authors propose the fuzzy Takagi-Sugeno algorithm that includes a double-input of error (v_e) and error derivative (dv_e/dt) with three outputs (K_{fP}, K_{fI}, K_{fD}) [14]. The fuzzy sets for five input variables are NB (Negative big), NS (Negative small), Z (Zero), PS (Positive small), and PB (Positive big); and for seven output variables are VS (Very small), S (Small), MS (Medium small), M (Medium), MB (Medium big), B (Big), and VB (Very big). The B^i is a binary variable that produces the rule consequence and B^i is expressed [16] as

$$\text{Rule } R_i : \text{If } v_{e1} \text{ is } A_1^i \ldots \text{ and } v_{en} \text{ is } A_n^i \text{ then } u \text{ is } B^i \tag{7}$$

where $A_1^i, A_2^i, \ldots A_n^i$ and B^i denote the fuzzy sets. By applying the center average fuzzified, the fuzzy output is defined as follows:

$$u_f = \frac{\sum_{i=1}^h B^i[\prod_{j=1}^n \mu_{A_j^i}(v_{ej})]}{\sum_{i=1}^h[\prod_{j=1}^n \mu_{A_j^i}(v_{ej})]} \tag{8}$$

for $\mu_{A_j^i}(v_{ej})$ indicates the membership functions, and h is the number of If-Then rules [17]. The F-PID controller parameters are defined as

$$K_P = (K_{Pmax} - K_{Pmin}).K_{fP} + K_{Pmin} \tag{9}$$

$$K_I = (K_{Imax} - K_{Imin}).K_{fI} + K_{Imin} \tag{10}$$

$$K_D = (K_{Dmax} - K_{Dmin}).K_{fD} + K_{Dmin} \tag{11}$$

The limitation values $[K_{Pmax}, K_{Pmin}]$, $[K_{Imax}, K_{Imin}]$, and $[K_{Dmax}, K_{Dmin}]$ are gathered through experimental testing. The PID parameters are obtained after fuzzy logic tuning as $K_P = 0.48$, $K_I = 0.20$, and $K_D = 0.13$.

3.2 Optimal Proportional Integral Derivative Controller

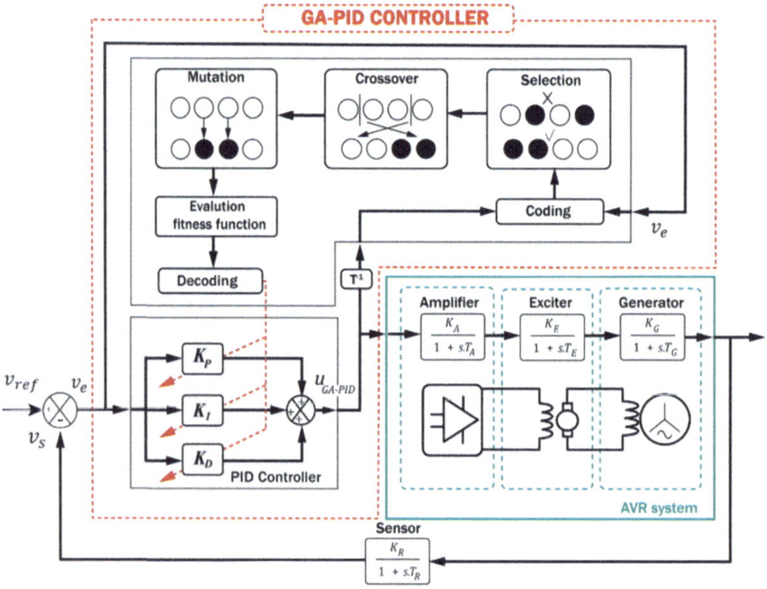

Fig. 2. The structure of the GA-PID controller.

The GA is an evolutionary search algorithm that models genetic principles found in nature [18, 19]. Using a method that combines multiple parameters and combinations, this optimization technique simulates natural selection and the survival of the best gene. The GA algorithm optimally adjusts the AVR control parameters, described in detail in Fig. 2. To filter out optimal values (K_{P-opt}, K_{I-opt}, K_{D-opt}) for the parameters of the PID controller, the results K_P, K_I, and K_D obtained from the Ziegler-Nichols method serve as an initial basis to constrain the searching space.

$$0.2 \le K_{P-opt} \le 1;\ 0.2 \le K_{I-opt} \le 1;\ 0.05 \le K_{D-opt} \le 0.15 \tag{12}$$

The initial population is set by coding the PID parameters K_P, K_I, and K_D into binary strings known as chromosomes. The length of strings depends on the range of the PID parameters. The required bits string is calculated as [20, 21]

$$2^{m_j-1} < \left(b_j - a_j\right) \times 10^4 \le 2^{m_j} - 1 \tag{13}$$

where m_j is a number of bits, b_j and a_j are the max and the min value of each parameter of the PID controller. According to (12), the required number of bits is calculated for 3 parameters K_P, K_I, and K_D as 13, 13, and 10 bits, respectively. Therefore, the total chromosome number length for the GA algorithm is 36 bits. The chromosomes with characteristics of initial parameters are randomly selected within the population and then crossbred with each other to create a new population with characteristics of mutated chromosomes. This process always evaluates the value of the fitness function.

Algorithm 1: Optimizing the PID parameters by using the GA.

Input: K_P, K_I, K_D are chosen by the Ziegler-Nichols method.
Output: K_{P-opt}, K_{I-opt}, K_{D-opt} optimization.

1. **Initialize** Max Generations ← 50, population size ←25, crossover rate ←0.8.
2. **Initialize** A closed-loop PID controller to determine the error.
3. **Coding** K_P, K_I, and K_D input values.
3. **while** Generation ≤ Max Generations **do**
4. Selecting a pair of individuals to prepare for the next generation.
5. Combining selected pairs of individuals to create offspring according to the one-point crossover method.
6. Mutation offspring.
7. Comparing the latest profits to the termination condition.
8. **if** profits are met **then**
9. Decoding the optimal PID parameters.
10. Saving 3 values are K_{P-opt}, K_{I-opt}, and K_{D-opt}.
11. **else**
12. Generation ← Generation +1, return to Step 4.
13. **end if**
14. **end while**
15. Applying the optimal PID parameters for the AVR control system
16. **end**

GA optimizes the PID controller's parameters using a fitness function [22, 23]. The fitness functions are often represented by the integral of time multiplied by absolute error (ITAE), the integral of the absolute magnitude of error (IAE), the integral of squared error (ISE), and the integral of time by squared error (ITSE). The primary objective of the fitness function, mentioned in [24–27], is to reduce the error. It can be inferred that the fitness function's lowest value indicates the best capability to minimize the error [24]. The GA parameters are as follows: evolution process of 50 generations, population size of 25, and crossover rate of 0.8. The search process of GA is illustrated by Algorithm 1 with GA-PID parameters: $K_P = 0.3312$; $K_I = 0.2367$; $K_D = 0.0939$.

3.3 Stability Evaluation of the AVR Controller

To evaluate the stability of the proposed GA-PID, the authors build the Bode and the Root locus diagrams for the AVR control system. The open loop transfer function is defined as

$$G(s)H(s) = \frac{K_A K_E K_G\left(s.K_P + K_I + K_D s^2\right)K_R}{s(1 + sT_A)(1 + sT_E)(1 + sT_G)(1 + sT_R)} \quad (14)$$

$$G(s)H(s) = \frac{a_1 K_D s^2 + a_1 K_P s + a_1 K_I}{as^5 + bs^4 + cs^3 + ds^2 + s} \tag{15}$$

The open loop transfer function of the AVR with GA-PID controller is given as

$$G(s)H(s) = \frac{0.939s^2 + 3.312s + 2.367}{0.002s^5 + 0.067s^4 + 0.615s^3 + 1.55s^2 + s} \tag{16}$$

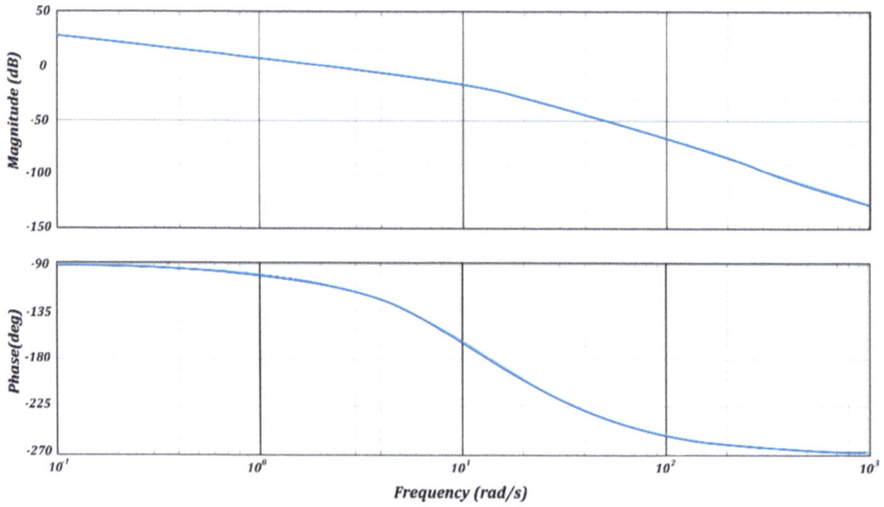

Fig. 3. Bode diagrams of the AVR system with GA-PID controller.

Bode Diagram Stability Analysis: We plot the Bode diagrams of the AVR system in Fig. 3. We obtain as $\omega_C = 2.5$ (rad/s), $\omega_{-\pi} = 1$ (rad/s), $\varnothing(\omega_c) = -109^0$, $L(\omega_{-\pi}) = -21.6$ (dB), $\varnothing M = 180^0 + \varnothing(\omega_c) = 180^0 - 109^0 = 71^0$, and $GM = 21.6$ (dB). Because of $GM > 0$, $\varnothing M > 0$, and the Bode stability, the AVR control system is stable.

Root-Locus Diagram Stability Analysis: The Root locus of the AVR system with GA-PID controller is shown in Fig. 4. From the result, we conclude that the system is stable, with non-closed loop poles on the right half-S plane or the *jw* axis.

4 Simulation Results

4.1 Case Study 1

In independent operation mode, the generator output voltage is maintained at a value of 1 p.u. The simulation results show that the GA-PID controller exhibits the best performance with very small values for overshoot (O_S), rise time (T_r), settling time (T_S), and peak voltage (peak). The output response is shown in Fig. 5, which has corresponding characteristics listed in Table 1.

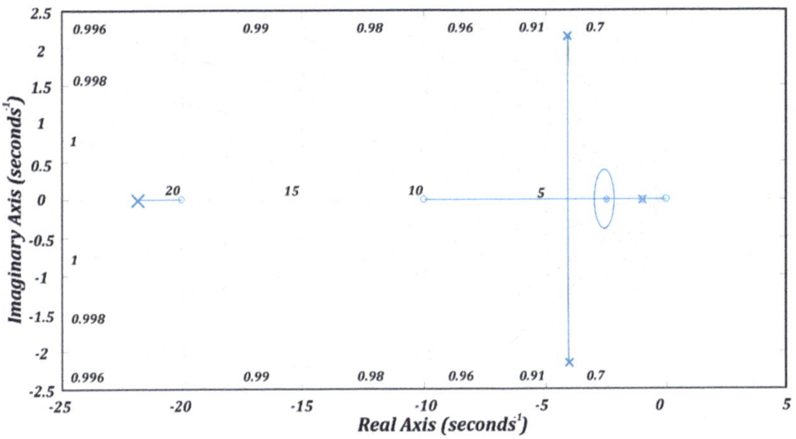

Fig. 4. Root locus of the AVR system with GA-PID controller.

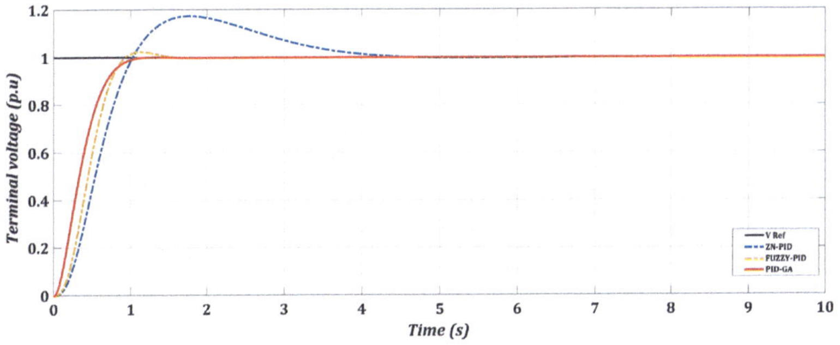

Fig. 5. Response of controllers for the AVR system in case 1.

Table 1. Comparison the responses of controllers.

Algorithm	Os (%)	Tr (s)	Ts (s)	Peak (p.u)
ZN-PID	17.38	0.87	6.8	1.171
F-PID	2.48	0.75	1.63	1.024
GA–PID	**0.20**	**0.65**	**1.09**	**0.199**

4.2 Case Study 2

Load Disturbance: Simulating the designed controllers with a load change of magnitude $\Delta P_L = 0.3$ p.u at time $t = 5$ s for the AVR system. The GA-PID controller is found to exhibit less voltage fluctuation and a shorter time to return to a stable state compared to ZN-PID and F-PID controllers, as illustrated in Fig. 6 and Table 2.

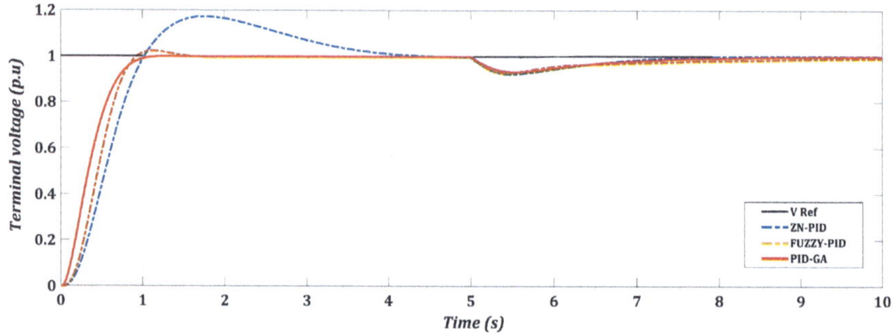

Fig. 6. Comparison response of controllers with load 0.3 p.u.

Table 2. Comparing controllers under load disturbance.

Algorithm	IAE	ITAE	ISE	ITSE
ZN-PID	0.9949	1.490	0.5047	0.2325
F-PID	0.6697	1.247	0.3995	0.1207
GA–PID	**0.5297**	**0.7743**	**0.3009**	**0.0822**

Overload Working Mode: The output voltage of the generator will decrease to 0.8 p.u starting from 10 s to 20 s. The simulation results verified that the GA-PID controller yields the best results, optimizing the voltage error as depicted in Fig. 7.

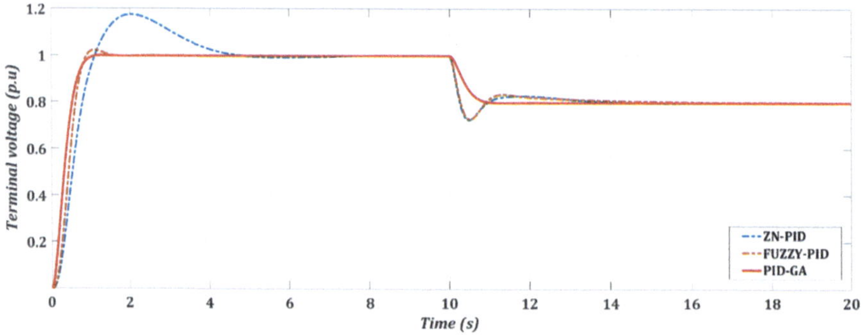

Fig. 7. Comparison response of controllers for AVR system.

4.3 Performance Evaluation

The performance of the controllers is evaluated based on the system's response under system uncertainties. Therefore, the time constants of the AVR system will be adjusted within its limits to investigate the response of the controllers. The deviations in the GA-PID controller are found least for change in time constants of the amplifier (T_A), the exciter (T_E), the generator (T_G), and the sensor (T_R) as compared to the ZN-PID controller and F-PID controller is presented in Table 3. The summary results indicate that the proposed GA-PID controller is the most effective and accurate.

Table 3. Comparison the deviations of controllers.

Values	Algorithm	IAE	ITAE	ISE	ITSE
Variations in Amplifier time constant T_A					
0.06	ZN-PID	0.8666	0.8679	0.4734	0.1856
	F-PID	0.5085	0.1671	0.3738	0.0835
	GA-PID	**0.4226**	**0.1321**	**0.2734**	**0.0500**
0.08	ZN – PID	0.8809	0.8804	0.4862	0.1933
	F-PID	0.5131	0.1680	0.3842	0.0868
	GA-PID	**0.4226**	**0.1237**	**0.2850**	**0.0526**
Variations Exciter time constant T_E					
0.5	ZN-PID	0.9698	0.9655	0.5431	0.2440
	F-PID	0.5853	0.2426	0.4241	0.1059
	GA-PID	**0.4957**	**0.1931**	**0.3283**	**0.0694**
0.7	ZN-PID	1.1260	1.2860	0.6316	0.3506
	F-PID	0.7074	0.4216	0.4802	0.1425
	GA-PID	**0.6523**	**0.4134**	**0.3917**	**0.1086**
Variations Generator time constant T_G					
1.4	ZN-PID	1.1080	1.2790	0.5889	0.30555
	F-PID	0.6672	0.4265	0.4516	0.1225
	GA-PID	**0.6071**	**0.3948**	**0.3624**	**0.0908**
1.6	ZN-PID	1.1720	1.5150	0.6324	0.3661
	F-PID	0.7338	0.5559	0.4783	0.1405
	GA-PID	**0.6935**	**0.5448**	**0.3985**	**0.1150**
Variations Sensor time constant T_R					
0.01	ZN-PID	0.8681	0.8718	0.4719	0.1864
	F–PID	0.55195	0.2291	0.3720	0.0839
	GA-PID	**0.4226**	**0.1321**	**0.2707**	**0.0504**
0.03	ZN-PID	0.8818	0.8824	0.4854	0.1937
	F–PID	0.5147	0.1688	0.3831	0.0869
	GA–PID	**0.4226**	**0.1237**	**0.2835**	**0.0527**

5 Conclusion

This paper has developed a solution to optimize the parameters of the AVR system controller for an SPS, applying the GA optimization algorithm. By designing the control solution and testing it on MATLAB Simulink software, feasible results on system response quality indicators, with the amplitude of voltage fluctuations observed during load changes are maintained at a minimum level. Furthermore, the stability of the AVR system using the GA-PID controller is evaluated according to the Bode and Root locus stability criteria, bringing efficiency to the application of optimization algorithms in the SPS study. However, the proposed solution has only been tested on a single standalone generator. Therefore, further studies should be examined with generators operating simultaneously under different load conditions.

References

1. Lan, H., et al.: Modeling and stability analysis of hybrid PV/Diesel/ESS in ship power system. Inventions **1**(1), 1–16 (2016)
2. Dimopoulos, G.G., Georgopoulou, C.A., Stefanatos, I.C., Zymaris, A.S., Kakalis, N.M.P.: A general-purpose process modelling framework for marine energy systems. Energy Convers. Manag. **86**, 325–339 (2014)
3. Do, V.D., Dang, X.K., Ly, S., Nguyen, N.T.: Positioning the semi-submersible platform using model predictive control with thruster constraints. In: Conference Proceedings - 2023 IEEE Asia Meeting on Environment and Electrical Engineering, EEE-AM 2023, pp. 01–05. Institute of Electrical and Electronics Engineers Inc. (2023)
4. Grzeczka, G., Piłat, T., Polak, A.: The parameters of excitation current of ship synchronous generator as the diagnostic symptoms of the propelling IC engine. J. Mar. Eng. Technol. **16**(4), 344–348 (2017)
5. Dang, X.K., Do, V.D., Do, V.T., Ho, L.A.H.: Enhancing the control performance of automatic voltage regulator for marine synchronous generator by using interactive adaptive fuzzy algorithm. In: Vo, N.S., Hoang, V.P., Vien, Q.T. (eds.) INISCOM 2021. LNICS, SITE, vol. 379, pp. 379–392. Springer, Cham (2021). https://doi.org/10.1007/978-3-030-77424-0_31
6. Park, S.H., Lee, S.K., Lee, S.W., Yu, J.S., Lee, S.S., Won, C.Y.: Output voltage control of a synchronous generator for ships using compound type digital AVR. In: INTELEC, International Telecommunications Energy Conference, pp. 1–6 (2009)
7. Can, Ö., Andiç, C., Ekinci, S., Izci, D.: Enhancing transient response performance of automatic voltage regulator system by using a novel control design strategy. Electr. Eng. **105**(4), 1993–2005 (2023)
8. Orosun, M.M., Orosun, R.O., Adamu, S.S.: Modeling and simulation of automatic generation control system for synchronous generator with model predictive controller. Zimbabwe J. Sci. Technol. **11**, 142–157 (2016)
9. Ibrahim, M.M., Jiya, J., Harrison, I.O.: Modelling and simulation of automatic voltage regulator system. Int. J. Comput. Appl. **178**(1), 24–28 (2017)
10. Salem, A.: Modeling Automatic Voltage Regulator (AVR) Using Artificial Neural Network. International Conference on Innovative Trends in Computer Engineering (ITCE'2019), Aswan, pp. 570–575 (2019)
11. Salih, A.M., Humod, A.T., Hasan, F.A.: Optimum design for PID-ANN controller for automatic voltage regulator of synchronous generator. In: 4th Scientific International Conference Najaf (SICN), pp. 29–30 (2019)

12. Bal, G., Kaplan, O., Yalcin, S.S.: Artificial neural network based automatic voltage regulator for a stand-alone synchronous generator. In: 8th International Conference on Renewable Energy Research and Applications (ICRERA), pp. 1032–1037 (2019)
13. Yavarian, K., Hashemi, F., Mohammadian, A.: Design of intelligent PID controller for AVR system using an adaptive neuro fuzzy inference system. Int. J. Electr. Comput. Eng. (IJECE) **4**(5), 703–718 (2014)
14. Almoshity, S., Gouda, K.A., Younus, S., Mohamed, F.A., Elbreki, A.M.: Comparison study between PID and Fuzzy logic controller to improve performance of AVR system for synchronous generators. In: 13th International Renewable Energy Congress (IREC), pp. 1–4 (2022)
15. Al Gizi, A.J.H., Mustafa, M.W., Al-geelani, N.A., Alsaedi, M.A.: Sugeno fuzzy PID tuning, by genetic-neutral for AVR in electrical power generation. Appl. Soft Comput. J. **28**, 226–236 (2015)
16. Majeed, W.S., Nasser, A.I., Hameed, K.R.: Improve the performance of automatic voltage regulator for power system using self-tuning fuzzy-PID controller. Indones. J. Electr. Eng. Comput. Sci. **29**(3), 1247–1257 (2023)
17. Shi, J.Z.: A fractional order general type-2 fuzzy PID controller design algorithm. IEEE Access **8**, 52151–52172 (2020)
18. Tripti, G., Sambariya, D.K.: Optimal design of fuzzy logic controller for automatic voltage regulator. In: 2017 International Conference on Information, Communication, Instrumentation and Control (ICICIC), Indore, India, pp. 1–6 (2017)
19. Dang, X.K., Do, V.D., Nguyen, X.P.: Robust adaptive fuzzy control using genetic algorithm for dynamic positioning system. IEEE Access **8**, 222077–222092 (2020)
20. Saad, M.S., Jamaluddin, H., Darus, I.Z.M.: Implementation of PID controller tuning using differential evolution and genetic algorithms. Int. J. Innov. Comput. Inf. Control **8**(11), 7761–7779 (2012)
21. Saad, M.S., Jamaluddin, H., Darus, I.Z.M.: PID controller tuning using evolutionary algorithms. WSEAS Trans. Syst. Control **7**(4), 139–149 (2012)
22. Do, V.D., Dang, X.K., Tran, T.D., Pham, T.D.A.: Jacking and energy consumption control over network for jack-up rig: simulation and experiment. Polish Maritime Res. **29**(3), 89–98 (2022)
23. Mousakazemi, S.M.H.: Comparison of the error-integral performance indexes in a GA-tuned PID controlling system of a PWR-type nuclear reactor point-kinetics model. Prog. Nucl. Energy **132**, 103604 (2021)
24. Zahir, A.A.M., Alhady, S.S.N., Othman, W.A.F.W., Wahab, A.A.A., Ahmad, M.F.: Objective functions modification of GA optimized PID controller for brushed DC motor. Int. J. Electr. Comput. Eng. **10**(3), 2426–2433 (2020)
25. Wang, P., Kwok, D.P.: Optimal design of PID process controllers based on genetic algorithms. Control. Eng. Pract. **2**(4), 641–648 (1994)
26. Niu, X.J.: The optimization for PID controller parameters based on genetic algorithm. Appl. Mech. Mater. **513–517**, 4102–4105 (2014)
27. Salman, G.A., Jafar, A.S., Ismael, A.I.: Application of artificial intelligence techniques for LFC and AVR systems using PID controller. Int. J. Power Electron. Drive Syst. **10**(3), 1694–1704 (2019)

Research on Sea Trials and Applications of Deep-Sea Autonomous and Remotely Operated Vehicle

Ren Yugang[1,2](✉), Zhang Xiaoyu[1], Gao Xiang[1,2], and Shi Xianpeng[2]

[1] College of Marine Science and Engineering, Shandong University of Science and Technology, Qingdao 266590, Shandong, China
ytrenyugang@163.com
[2] National Deep Sea Center, Qingdao 266237, Shandong, China

Abstract. Deep submersibles are the "magic weapon" for entering, exploring, developing, and governing the ocean. Their technological level and development directly affect our deep-sea rights, especially in the fields of structure, sensing, dynamics, etc. The advanced level of submersibles greatly represents the development level of deep-sea technology, and has always been the "bull's-eye" pursued by countries around the world for competition. The article focuses on the current development status of all sea deep duplex hybrid deep submersibles, and introduces the development status of ARVs at home and abroad; Based on the project, a detailed description was given of the sea trials of the 3000 m and 8000 m deep duplex hybrid submersible. Finally, based on the experience of sea trials and combined with the operational and maintenance requirements of the national deep-sea base for deep diving technology and equipment, the ARV operation mode was discussed, and the prospects for future commercial applications were analyzed.

Keywords: Deep-sea exploration · Hybrid deep submersibles · Sea trials · Operation mode · Scientific applications

1 Introduction

The deep sea (usually defined as water depth greater than 1000 m) contains a large amount of oil, water, minerals, biological and other strategic resources, which has become a new frontier for the sustainable development of mankind, as well as a new high point for the expansion of strategic corridors, the acquisition of strategic resources and the preparation of strategic advantages for all global players [1–3]. The deep sea continues to engage in robust gaming and competition in areas such as resource exploration and extraction, environmental governance and regulatory negotiations.

Cognition of the mysteries of the deep sea and the development of deep-sea resources cannot be separated from deep-sea technology and engineering. Deep submersibles, as a collection of modern high-end robotic platforms, have become one of the most important technical equipment we rely on to explore the ocean, cognitive the ocean and strategic use of the ocean [4]. Broadly speaking, the deep submersible referred to in this paper is mainly

D. Jeng and B. Cai (Eds.): ICCOE 2024, LNCE 423, pp. 454–467, 2025.
https://doi.org/10.1007/978-981-96-0181-3_41

a robotic vehicle that can carry multiple types of electronic equipment, sensing devices, actuation tools and even personnel to dive into a pre-determined area in the deep sea (>1000 m class) to achieve refined observation, surveillance and detection, and acquire data and samples with high quality [5]. They are usually classified as manned (HOV) and unmanned (UUV). With the rapid development of computer, sensing and material technologies, unmanned deep submersibles have been rapidly developed and evolved into two categories: cable-free autonomous (AUV) and cable-operated (ROV) [6–10]. Autonomous and Remote operated Underwater Vehicles (ARV) in the deep sea integrate the latest ROV and AUV technologies, enabling information exchange and remote control operations through fine optical fibers, as well as cutting off optical fibers for large-area underwater exploration or performing hybrid tasks. The autonomous/remote hybrid mode and integrated detection operation technology represent the development direction of third-generation deep-sea underwater robots that can perform more complex missions.As a new type of composite underwater robot, ARV can achieve visualization of deep-sea target areas through robotic arms and high-definition cameras. It has important scientific research and application value in deep-sea resource investigation and is also a current research hotspot for international underwater robots [11–13].

Internationally, the United States, Japan, France and other countries have carried out research work on hybrid deep submersibles [14]. In particular, the Woods Hole Research Institute (WHOI) of the United States has carried out a great deal of forward-looking work, and completed the development of the world's first Nereus (Poseidon) full-depth hybrid deep submersible in 2008 [15, 16], as shown in Fig. 1, and completed the 10,000-m sea trial in 2009 [17]. Due to the innovative use of ceramic materials in the design concept of pressure-resistant chamber, the relevant mechanism is not mature. It was lost in the deep sea after causing a chain of ceramic balls to implode during a dive in 2014 [18].

a. AUV model.

b. ROV model.

Fig. 1. Poseidon's dual-work mode.

The US company Oceaneering developed the hybrid deep submersible Taser (The Autonomous System for Extended Reach), as shown in Fig. 2. The submersible can achieve three types of movement: autonomous diving, short-range cable-controlled operation and long-range cable-controlled. In 2011, WHOI developed the polar ARV deep submersible Nereid UI [19] for polar exploration needs, as shown in Fig. 3, which can achieve information transmission through fibre-optic microfibre cables and is equipped with a variety of biological and chemical sensors [20].

Fig. 2. Taser composite unmanned submersible.

Fig. 3. Nereid UI composite unmanned underwater vehicle.

The Japan Marine Science and Technology Centre (JAMSTEC) has completed the development of the hybrid deep submersible MR-X1 (Marine Robot Experimental 1), as shown in Fig. 4, with three modes of operation: autonomous, remotely operated, and wireless (radio/acoustic), and with a maximum operating depth of 4,200 m [21]. IFREMER (French Research Institute of Oceanography) has completed the development of Ariane, as shown in Fig. 5, with two modes of operation: AUV and ROV and a maximum working depth of 2011 m. When working in AUV mode, the data collected can be transmitted to the surface through hydroacoustic communication; when working in ROV mode, the data can be transmitted in real time through microfibre optics, which can achieve fine monitoring and sampling under the complex terrain conditions in the deep sea [22–25].

In recent years, deep-sea autonomous/remote-controlled underwater robots have gradually attracted the attention and investment of domestic scientific research institutions, and related technologies and products have been rapidly developed [26–29]. They have been widely used in complex sea areas such as polar ice surveys and abyssal surveys. Since the 13th Five-Year Plan, with the development of many heavy deep diving equipment, the technology gap between us and the West is getting smaller and smaller. The Shenyang Institute of Automation, Chinese Academy of Sciences earlier put forward the ARV submersible design concept, and carried out for the use of the Arctic ARV development (Arctic), as shown in Fig. 6, to meet the needs of polar ice exploration.

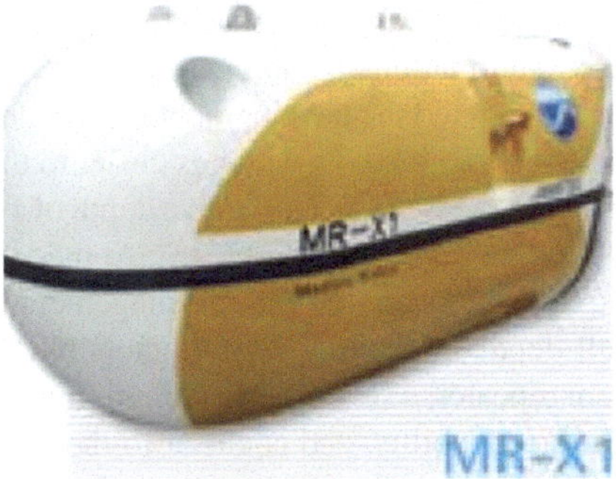

Fig. 4. MR-X1 composite unmanned underwater vehicle.

Fig. 5. Ariane composite unmanned underwater vehicle.

From 2008 to 2010, the Arctic successive completion of the Arctic 84.6° north latitude dive, the submersible is equipped with a variety of types of sensing load, with the Arctic ice underwater multi-physical parameters of the water body detection. In 2014, a new type of "Arctic" ARV realised the long-time sequence polar sea area dive survey [30–35].

Fig. 6. Arctic ARV (Arctic Research Scene 2008).

Under the support of the National Key Research and Development Programme, Shanghai Jiao Tong University and Shenyang Institute of Automation, Chinese Academy of Sciences have respectively taken the lead in the development of the 11,000-m-class hybrid submersible ARV project [36]. On the basis of the development of "Haidou" ARV, Shenyang Institute of Automation, in conjunction with Sanya Institute of Chinese Academy of Sciences, has completed the development of "Haidou No. 1", an all-ocean-depth operational deep submersible (Fig. 7), which is closer to the concept of "AUV". It completed a 10,000-m sea test in May 2020, with a dive depth of 10,907 m, creating a new historical record for an ARV-type submersible [37, 38]. At the same time, Shanghai Jiao Tong University and the National Deep Sea Base Management Centre have completed the development of the full-depth operational deep submersible "Siyuan" (Fig. 8), which is equipped with a 6,000-m repeater, a seven-function hydraulic manipulator, etc.. It has strong near-bottom operation capability, and is closer to the "ROV" in terms of design concept. In 2021, due to the deterioration of the international environment, it was not allowed to go to the Mariana Trench to complete the 10,000-m class sea test, but it has completed the 8000-m class sea test [37–39]. And all the performances and indexes are mature and stable, which is unanimously recognised by the on-site expert group of the Ministry of Science and Technology [40, 41].

Fig. 7. ARV Hai dou No. 1.

Fig. 8. ARV Si yuan.

Under the support of the project "Development of All-Resource Unmanned Vehicle (ARV) System" of the Key R&D Programme of the Ministry of Science and Technology (MOST), Shanghai Jiaotong University (SJTU), as the lead unit, has carried out the development of an 11,000-m-class duplex hybrid Autonomous & Remotely

Operated Vehicle (ARV), which combines the features of AUV and ROV. The vehicle (ARV) can achieve remote control, data transmission and other exploration tasks through high-strength fibre-optic micro-cable, and can also be detached from the cable and dive autonomously to achieve a wide range of exploration tasks. The National Deep Sea Center, as a sea trial unit, has organised and carried out 3000 m and 8000 m sea trials. Based on the application experience of the two sea trials, and combining with the understanding of the deep submersible sea trials carried out in recent years, this paper initially explores the operation mode of the future application of the ARV, and analyses in-depth the possibility of the synergistic operation of the multi-submersible with a view to better servicing the national strategy and scientific exploration, and to play a greater role in the effectiveness of the ARV.

2 System Composition and Operation Mode of Deep-Sea ARV

2.1 Technical Indicators

The Siyuan deep-sea ARV is a composite underwater robot with a maximum operating depth of 11000 m. It fully considers the needs of deep-sea scientific research and has the ability to conduct deep sea biochemical surveys throughout the entire sea, as well as sampling of deep-sea organisms, seawater, and geology. The system includes the ARV body, repeater, water surface monitoring power station, and water surface auxiliary support system; Equipped with three operation modes: autonomous detection, fiber optic micro cable remote control operation, and photoelectric umbilical cable remote control; Capable of underwater sound and inertial positioning, hovering positioning control, and trajectory tracking control systems; Capable of underwater search, observation, data transmission, and recording, providing interfaces for homework tools. Its technological advantages include:

1. Fiber optic micro cables are laid using a 6000 m relay ROV to overcome the impact of shallow flow on fiber optic safety; 2. Installing SBL and sound communication equipment on deep-water relay platforms to improve the stability of positioning systems; 3. Provide two recycling methods, overall recycling and split recycling, to improve recycling safety; 4. Adopting a new type of 100 kg SWL fiber optic micro cable to improve the success rate of submarine operations; 5. Fiber optic micro cable deployment can be equipped with high-precision terrain detection systems, hydraulic robotic arms, and provides a direct deployment mode for armored umbilical cables, with powerful operational functions (Table 1).

2.2 Operation Mode

The deep-sea ARV system generally consists of surface support system, cable control system, and ARV submersible body, equipped with three working modes: autonomous detection, fiber optic micro cable remote control operation, and photoelectric umbilical cable remote control operation. The surface support system provides support and support for the normal operation, release, and maintenance of deep-sea ARV through the mother ship and operation team, mainly including the integrated display and control system, deck release and recovery system, and maintenance support system.

Table 1. Main Technical Indicators of ARV Body at Full Sea Depth.

Name	Technical indicators
Maximum working depth	11000 m
Weight in air	Less than 3.5 tons
Scale	Approximately 4250 mm long × 2240 mm wide × 2180 mm high
Battery capacity	≥16 kwh
Continuous underwater operation time	≥8 h (investigating the chassis, equipped with power, with a speed of 1.5 knots)
Payload	45 kg
Maximum speed	Forward: 3 sections (investigating chassis) Down: 1 section
Working mode	Autonomous detection; Remote control operation of fiber optic micro cables; Remote control operation of photoelectric umbilical cable;
Sampling operation chassis hydraulic drive capability	15 L/min@210bar, 12 L/min@210 b/ar
Mooring thrust	Forward: 100 kg; Downward: 60 kg; Lateral: 40 kg
Automatic control accuracy (relative)	Orientation: 1.5°; Fixed depth/height: 15 cm trajectory; Tracking deviation: 50 cm
Acoustic measurement frequency	Multi beam depth measurement at 100 kHz/450 kHz; Depth measurement side scan 225 kHz/450 kHz
Acoustic detection with resolution along the trajectory direction	Synthetic beam opening angle for transmitting and receiving along the trajectory direction of a depth measurement side scan array
Positioning distance for ARV	<500 m
Acoustic detection vertical trajectory direction resolution	2.5 cm @450 kHz; 8 cm @100 kHz
ARV end fiber optic micro cable storage capacity	≥8 km
7-function robotic arm	Maximum operating range: 1400 mm; Maximum lifting force at full extension: 50 kgf
Optical detection resolution	Camera: 1920 (H) × 1080 (V) Camera: 12 million pixels

The integrated display and control system mainly displays and processes real-time status information, navigation information, attitude information, and video images of deep-sea ARV through multi screen displays, and remotely controls and issues operation instructions to underwater equipment through deck remote control equipment. The deck release and recovery system mainly relies on the A-frame, winch, photoelectric composite cable, and other auxiliary equipment of the mother ship to release and recover ARV. The maintenance and support system mainly performs daily maintenance, repair, and testing on commonly used equipment such as ARV, underwater investigation equipment, and optoelectronic composite cables.

The cable control system is mainly connected to the mother ship and underwater submersible through underwater acoustic communication relays. The upper end of the communication relay of the ARV is connected to the water surface integrated display and control system through an optoelectronic composite cable, as shown in Fig. 9. In ROV mode, the lower end of the communication relay is connected to the deep-sea ARV through ultra-fine optical fibers, completing the rapid transmission of surface and underwater information. In AUV mode, the communication relay cuts off the optical fiber and uses underwater acoustic communication devices to exchange information with the ARV.

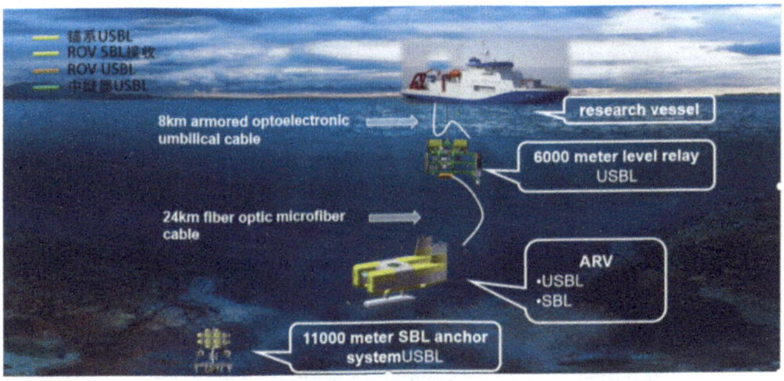

Fig. 9. ARV system operation mode

2.3 Sea Trials of Deep Submersibles

The project "Development of All-Resource Vehicle (ARV) System" is a national key research and development project, with a total funding of more than 90 million RMB. SJTU is the undertaking unit of the project, and the National Deep Sea Center is the undertaking unit of the sea trial, which has organised and carried out the sea trial of ARV in the South China Sea at the middle and shallow water and 10,000-m level.

3,000-m-class Sea Trials in the South China Sea

From February to March 2021, the Deep Sea Centre organised a sea trial voyage in the shallow and medium waters of the South China Sea for the all-ocean depth ARV system. It completed 10 dives in the 1000-m-class and 3000-m-class waters of the South China Sea, as shown in Fig. 10, mainly carrying out functional tests such as joint near-bottom pulling distance between ARV-mounted operational chassis and repeater, overall recovery, split recovery, and large-depth diving, as well as a series of sea trials such as mooring deployment and ultra-short baseline calibration, joint near-bottom pulling distance and overall recovery test of ARV-carrying probe chassis and repeater.

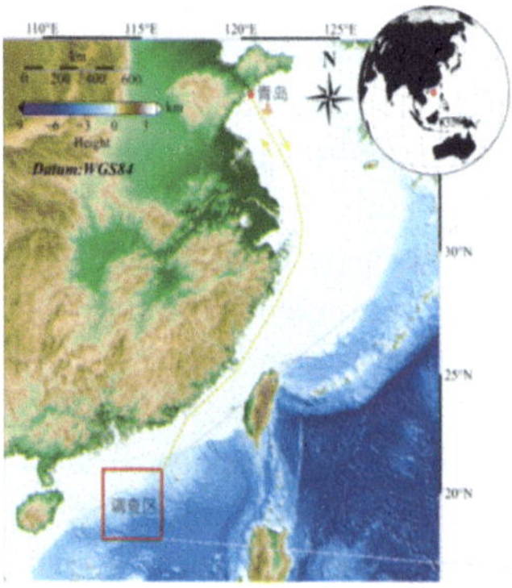

Fig. 10. 3,000-m-class sea trials in the South China Sea.

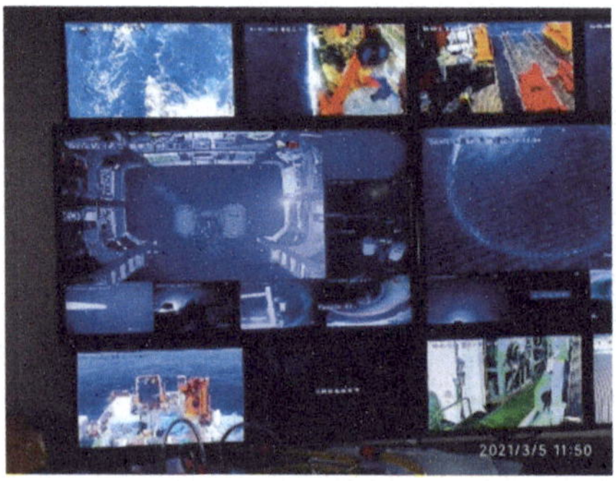

Fig. 11. 3000-m-class dive in the South China Sea.

The sea trial site successfully completed 10 dives in 16 days, especially the first time to achieve the large depth difference, long distance, dual body manoeuvre docking of the ARV body and the dual motion body of the repeater under the 3,000 m deep sea; the first time to validate the 130 m dual-boat acoustic imaging capture and the 120 m visual capture. It completed the ARV carrying fibre optic 6km long-distance voyage, acoustic communication test, underseat video filming, RNA isobaric sampler, manipulator operation and other tests, as shown in Fig. 11. The maximum dive depth of ARV system is 3738.9 m. The on-site witnessing expert group believes that the overall technical performance of ARV system is stable, which lays the foundation for the 10,000-m acceptance sea trial.

West Pacific 8000-m-class Sea Trials

As no permission was obtained from the Republic of Micronesia to enter the Mariana-Yap Trench, an 8000-m level sea trial area was explored in the Philippine Basin under the organisation of the Deep Ocean Centre (DOC). And the 8000-m level acceptance sea trial was completed in the Philippine Basin in the western Pacific Ocean in June-July 2021 by organising a joint effort of 16 units with 60 team members relying on the vessel "Deep Ocean One". The acceptance sea trial was completed, including the autonomous detection of full-depth ARV, remote control of fibre-optic microfibre cable and remote control of optoelectronic umbilical cable. And the ARV-mounted operation chassis carrying seven-function manipulator and four-function manipulator completed the bottom-sitting operation of the sediment pressure preservation device, seawater pressure preservation device, and RNA in-situ fixture. It also acquired samples and data of pressurized seawater, RNA curing liquid, rocks, high-definition visual images, and high-precision topography.

The full-depth ARV system has completed 10 dives in the West Pacific Sea Test Area. The tests fully proved that the system's operation mode is reasonable, its work is normal and its overall performance is stable. The cumulative seafloor operation time of the ARV body is 32 h and 17 min, and the continuous operation time of the single seafloor operation on the operating chassis is 8 h and 16 min. The autonomous seafloor patrolling, multibeam and bathymetry side-scanning detection, and near-bottom observation were accomplished by the ARV body on the probe chassis. The ARV body mounted on an operating chassis carrying a seven-function manipulator completed the sampling of the sediment pressure retaining sampler, the seawater pressure retaining sampler and the RNA in-situ fixture. The sea trial also carried out a large depth free dive of the ARV body and recovery into the repeater test, and the depth difference between the ARV body and the repeater reached 5,560 m. The four-function manipulator and underwater cutting tool tests were carried out under the remote-control operation mode of the photoelectric umbilical cable, as shown in Fig. 12. The four-function manipulator worked normally and the underwater cutting tool functioned normally.

3 Conclusion

3.1 Analysis of Advantages and Disadvantages of ARV Technology

ARV has advantages such as full sea depth, comprehensiveness, and multifunctional mode selection. The maximum homework depth is 11000 m, effectively compensating for the homework depth of Jiaolong, Hailong, and Qianlong. ARV operation technology combines the technical advantages of both ROV and AUV. In practical deep-sea applications, there are many redundant functions to choose from. It can be used as a traditional AUV, with AUV wide underwater detection, patrol, and search functions, as well as a traditional ROV. It uses fiber optic micro cables for manual remote control, which brings more possibilities for the precious operation time window and ship time in the deep sea. The ARV and Sanlong system equipment can be constructed to form a multi platform collaborative detection and operation system that integrates large-scale AUV system detection, medium area ARV system fine detection and fixed-point bottom exploration with manual intervention, small-scale ROV bottom sitting long-term operation, and fine range Jiaolong HOV high-value operation.

(1) Compared with the ROV system, its weakness lies in its weaker operational and load capacity compared to the operational ROV (such as the Hailong 3), but it can be equipped with a 7-function hydraulic manipulator (50 kgf) and other operational

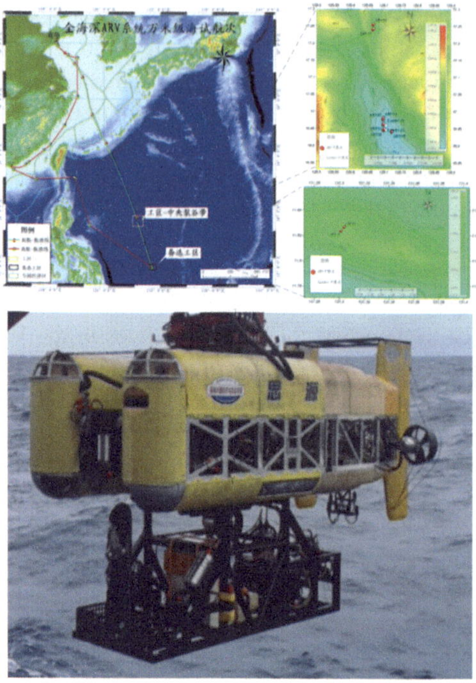

Fig. 12. Fibre optic microfibre cable remote control mode of operation vs. optical umbilical cable mode of operation.

tools, which have strong operational efficiency and can complete most scientific tasks. Advantage: Due to the influence of armored cables, the underwater working area of traditional ROVs is greatly limited. The operating range is only within a radius of 100–200 m, while the underwater activity range of this ARV system can reach a radius of 6000–10000 m. It also has the ability to conduct real-time and precise measurements of the seabed within 2–3 km around the mother ship, as well as long-term sampling operations.

(2) Compared to AUV systems, weaknesses include weaker detection range and degree of autonomy compared to cable less autonomous AUVs (such as Qianlong). Advantages: AUV has a large underwater working area, but it cannot achieve precise fixed-point operations, and it is difficult for mother ship personnel to intervene in real-time. The ARV system can achieve real-time human intervention capabilities and accurately detect sensitive areas.

(3) Compared with the HOV system, its weaknesses include weaker operational functions, payload capacity, and sampling efficiency compared to the Jiaolong. Advantages: It can achieve full sea depth operations and compensate for the depth of 7000 m with the Jiaolong; The safety risk of unmanned platforms is relatively low, and there are no major personal accidents; The actual operation time under deep water at a depth of 7000 m is longer.

3.2 Conclusion and Prospect

(1) Design more intelligent, networked, and autonomous. From the current development trend, the functions of deep-sea ARVs are increasingly tending towards intelligence, networking, and autonomy. Combined with the latest artificial intelligence image recognition technology, deep-sea ARVs have more intelligent and autonomous operation capabilities. Through networked underwater acoustic communication and fiber optic micro cables, they will be able to autonomously navigate in complex deep-sea environments, with the ability to safely survive, accurately detect and operate in complex environments.

(2) Unmanned collaborative networking operations. With the breakthrough of deep-sea communication and networking technology, the trend of multi submersible collaborative operation is becoming increasingly apparent. In the future, with the maturity of various communication methods such as underwater acoustic communication, optical communication, satellite communication, and the successful development of unmanned recovery and release systems that can be adapted to water surfaces and underwater bodies, or unmanned support mother ships, deep-sea ARVs and unmanned ships have the hope of achieving unmanned remote and efficient coordinated operation.

Acknowledgment. The research work presented in this paper which is jointly supported and sponsored by the National Key Research and Development Programme (2022YFC2805200); the National Natural Science Foundation of China (52175018); Major Innovation Project of Shandong Province (2019JZZY010802); Natural Science Foundation of Shandong Province (ZR2020QD041).

References

1. Yoshida, H., Aoki, T., Murashima, T., et al.: A deep sea AUV "MR-X1" for scientificwork. In: Oceans 2003. Celebrating the Past. Teaming Toward the Future (IEEE Cat. No. 03CH3749 2), vol. 2, pp. 1174–1179. IEEE (2003)
2. Johansson, B., Siesjö, J., Furuholmen, M.: Seaeye saber-tooth a hybrid AUV/ROV offshore system. In: Oceans 2010 MTS/IEEE Seattle, pp. 1–3. IEEE (2010)
3. Li, Y.P., Li, S., Zhang, A.Q.: Recent research and development of ARV in SIA. In: Proceedings of the 6th International Symposium on Underwater Technology (2009)
4. Feng, X., Li, Y.: Thirty years of marine robotics. Sci. Bull. **58**(S2), 2–7 (2013)
5. Wu, B., Li, S., Zeng, J., et al.: ARV navigation and control system at Arctic research. In: OCEANS 2009, pp. 1–6. IEEE (2009)
6. Branch, A., Flexas, M.M., Claus, B., et al.: Front delineation and tracking with multiple under water vehicles. J. Field Robot. **36**(3), 568–586 (2019)
7. Sahoo, A., Dwivedy, S.K., Robi, P.S.: Advancements in the field of autonomous under water vehicle. Ocean Eng. **181**, 145–160 (2019)
8. Cui, W.C., Wang, F., Pan, B.B., et al.: Issues to be solved in the design, manufacture and maintenance of full ocean depth manned cabin. Adv. Eng. Res. **11**, 1–29 (2016)

9. Jiang, Z., Hu, Y., Cui, W., et al.: A numerical study on the load-bearing armored cable performance of an 11,000 m ARV for the first phase sea trial. In: International Conference on Offshore Mechanics and Arctic Engineering, vol. 49972, p. V006T05A001. American Society of Mechanical Engineers (2016)
10. Li, Y., Li, S., Zhang, A.: Research status of autonomous/remotely controlled underwater robot. Eng. Res. Eng. Interdisc. Perspect. **8**(2), 217–222 (2016)
11. Bowen, A.D., Yoerger, D.R., Taylor, C., et al.: The Nereus hybrid underwater robotic vehicle for global ocean science operations to 11,000 m depth. In: OCEANS 2008, pp. 1–10. IEEE (2008)
12. Wu, G., Qin, Q.: Analysis of current situation and construction outlook of China's marine scientific research equipment. Prospect. Sci. Technol. **1**(02), 166–182 (2022)
13. Xu, Q., Hu, Z., Ye, C., Wang, S., Liu, S., Cao, J.: Current status and outlook of manned deep submergence technology and application. Foresight Sci. Technol. **1**(02), 36–48 (2022)
14. Li, S., Zhao, H., Feng, X.: Progress of China's deep-sea robotics research and Development proposals. Prospect. Sci. Technol. **1**(02), 49–59 (2022)
15. Li, Y.-P.: Research and application of autonomous/remotely controlled underwater robot. Mod. Phys. Knowl. **33**(1), 19–23 (2021)
16. Shen, X., Wang, Y., Yang, S., et al.: Current status and outlook of underwater glider technology development. J. Underw. Unmanned Syst. **26**(2), 89–106 (2018)
17. Chinese Academy of Sciences Strategic Pilot Project Team for the Research and Development of Frontier Science and Technology Issues of Sea-Doo Abyss. Opening the door to the abyss - progress of the strategic pilot project on research and tackling of frontier scientific and technological problems of sea-doo abyss. Proc. Chin. Acad. Sci. **31**(9), 1105–1111 (2016)
18. China's self-developed "Haidou I" full-depth submersible successfully conducts 10,000-metre sea test. Mar. Eng. **42**(6), 108 (2020)
19. Ren, Y., Liu, B., Ding, Z., Li, Y., Yang, L., Hu, X.: Current status and trend of manned submersible development. J. Mar. Technol. **37**(02), 114–122 (2018)
20. Shi, X., Ren, Y., Tang, J.: Working tools study for JiaoLong manned submersible. Mar. Technol. Soc. **53**(2), 56–64 (2019)
21. Abimbola, F.A., Yang, J., Liu, S., et al.: Survivability study of deepwater drilling riser in freestanding mode. Ocean Eng. **222**(1), 108–114 (2021)
22. Sahoo, A., Dwivedy, S.K., Robi, P.S.: Advancements in the field of autonomous underwater vehicle. Ocean Eng. **181**, 145–160 (2019)
23. Ferreira, C.Z., Cardoso, R., Meza, M., et al.: Controlling tracking trajectory of a robotic vehicle for inspection of under-water structures. Ocean Eng. **149**(2), 373–382 (2018)
24. Sun, D., Zheng, C., Zhang, J., et al.: Development and prospect of hydroacoustic positioning and navigation technology. Proc. Chin. Acad. Sci. **34**(3), 331–338 (2019)
25. Wang, L., Liu, T., Yang, S., et al.: Key technology of deep-sea submersible ARV. Firepower Command Control **35**(11), 6–8 (2010)
26. Bowen, A.D., Yoerger, D.R., Taylor, C., et al.: The nereushybrid underwater robotic vehicle. Underw. Technol. **28**(3), 79–89 (2009)
27. Tang, Y., Wang, J., Lu, Y., et al.: Parametric design method and experimental research on the full-depth autonomous remote-controlled underwater robot 'Haidou.' Robotics **41**(6), 697–705 (2019)
28. Wu, L., Jing, Z., Chen, X., et al.: Current status and future prospects of marine science development in China. Geol. Front. **29**(5), 1–12 (2022)
29. Xu, Q., Ye, C., Wang, S., et al.: Review and prospect of the development of Jiaolong manned submersible in ocean exploration. Chin. J. Nonferrous Metals **31**(10), 8–14 (2021)
30. Kohnen, W.: MTS manned underwater vehicles 2017–2018 global industry overview. Mar. Technol. Soc. J. **52**(3), 125–151 (2018)

31. Bogue, R.: Underwater robots: a review of technologies and applications. Ind. Robot. **42**(3), 186–191 (2015)
32. Yang, J., Cui, Z., Dada, O.A., et al.: Distribution and enrichment of trace metals in surface marine sediments collected by the manned submersible Jiaolong in the Yap Trench, Northwest Pacific Ocean. Mar. Pollut. Bull. **135**(7), 1035–1041 (2018)
33. Perez, J., Kitazato, H., Sumida, P., et al.: Benthopelagic megafauna assemblages of the Rio Grande Rise. Deep Sea Res. Part I **134**(4), 1–11 (2018)
34. Wang, P., Yu, Y., Zhuang, G., et al.: Research and development of all-ocean depth ARV casting system and deep-sea test. Ocean Eng. **41**(1), 82–92 (2023)
35. Cui, X., Kang, J., Hu, Y., et al.: Analysis of cable tension control for passive lift and sink compensation system of all-sea depth ARV. J. Shanghai Ocean Univ. **26**(2), 7 (2017). https://doi.org/10.12024/jsou.20160401725
36. Pan, G., Cao, J., Wu, M., et al.: Progress of underwater gravity measurement technology. Surv. Mapp. Bull. **2**, 1–5 (2019)
37. Ye, C., Cao, J., Liu, S.: Innovation and development of foreign manned deep diving technology. Ship Sci. Technol. **45**(1), 1–7 (2023)
38. Ren, Y., Liu, Y., Ding, Z.: Design of deep-sea non-contact magnetically driven sediment penetration strength measurement device based on Jiaolong manned submersible. China Mech. Eng. **34**(6), 739–745 (2023)
39. Ren, Y., Liu, Y., Ding, Z., Liu, B., Zhang, J.: Design of full ocean depth self-floating samplers and analysis of factors affecting core penetration depth. J. Ocean Univ. China **19**(5), 1094–1102 (2020)
40. Xu, Y., Liu, X., Li, S., Ren, Y.: Core-drilling kinematic modelling and analysis of Jiaolong submersible manipulator. J. Zhejiang Univ.-Sci. A (App. A). Univ.-Sci. A (Appl. Phys. Eng.) **24**(11), 937–948 (2023)
41. Ding, Z., Ren, Y., Zhang, Y., Yang, L., Li, D.: R&D and outlook of deep-sea exploration technology. Ocean Dev. Manag. **36**(04), 73–79 (2019)

Assessing and Analyzing Maritime Pilots' Application of Portable Pilot Units and Offering Management and Guidance at Vietnamese Seaports

Van-Thuc Le[1], Thi-Duyen-Anh Pham[2,3](✉), and Xuan-Kien Dang[2]

[1] Vungtau Maritime Administration, 790000 Ba Ria - Vung Tau, Vietnam
[2] Articial Intelligent in Transportation Research Group, Ho Chi Minh City University of Transport, 700000 Ho Chi Minh, Vietnam
anh.pham@ut.edu.vn
[3] Institute of Languages and Social Sciences, Ho Chi Minh City University of Transport, 700000 Ho Chi Minh, Vietnam

Abstract. Maritime safety has been and remains a prerequisite for ensuring the sustainable development of seagoing safety. At Vietnam's largest and most important seaports, the quality of pilotage services is a top concern for regulatory bodies, ship owners, and port enterprises. To ensure the quality of pilotage services, in addition to human resources, the tools and equipment that support pilots during ship navigation play a crucial role in safety. This paper collects statistics and analyzes the current status of the use of Portable Pilot Units (PPUs) by maritime pilots and proposes solutions for managing and guidance on the use of PPUs at Vietnamese seaports.

Keywords: Portable Pilot Units · Maritime Accident investigation · Maritime Pilot · Ship Maneuvering · Navigational Equipment

1 Introduction

The activity of managing ships at ports, terminals, and berths and guaranteeing their safe passage through congested waterways that pose a number of risks [1], including shallow water and proximity to other navigational hazards, is known as maritime pilotage [2]. Many countries require maritime pilot support for port entry and leave due to Shipmasters' lack of local knowledge of the terrain and the requirement for coordination with mooring services like tugs and linesmen [3]. This remains especially for seaports that are both domestic and international, and maritime piloting is also the practice of maneuvering ships safely in congested waterways. Actually, pilot safety technology has been the subject of numerous scholarly publications [2, 4–6]. There is a knowledge gap about how these technologies affect the pilots' mental tasks because these studies have primarily concentrated on the technology's practical applications. Among them, a Portable Pilot Unit (PPU) is understood to include two components: piloting software

D. Jeng and B. Cai (Eds.): ICCOE 2024, LNCE 423, pp. 468–478, 2025.
https://doi.org/10.1007/978-981-96-0181-3_42

(such as Seaiq, installed on tablet computers running iOS, Android, or Windows platforms) and hardware devices that transmit ship maneuvering data (GPS position, COG, HDG, SOG, ROT) to the tablet computer installed with the piloting software. According to the definition by the International Maritime Pilots' Association (IMPA) [7], the PPU is described as an electronic device that maritime pilots bring on board and use as a tool (for analyzing data on ship position, heading, speed, rate of turn, etc., through sensors to display intuitive ship maneuvering parameters, providing a multi-dimensional perspective of the actual situation) combining both technology and human elements to assist pilots in making decisions during the ship pilotage process.

Since 1980, the PPUs have been studied and produced in the United States, through collaboration between engineers from the U.S. Coast Guard's Research and Development Center, the Applied Physics Laboratory at Johns Hopkins University, and Capt. Joseph F. Bradley from the Bay & River Delaware Pilots' Association. However, it was not until the advent of the Global Positioning System (GPS) in 1991 that the first commercial PPUs became available. In 1994, the Delaware Pilots' Association became the first organization in the U.S. to equip its pilots with PPUs [8]. Since then, PPUs have become more prevalent, with software design and development primarily originating from the U.S., Canada, and Europe with support and professional consultancy from experienced maritime pilots worldwide to ensure that the piloting software is suitable for the specific characteristics of different maritime regions globally and the needs and requirements of pilotage organizations. The International Maritime Pilots Association (IMPA) has updated its guidelines on the design and use of maritime Portable Pilot Units (PPUs) in response to the growing need for them [7]. These modifications aim: Firstly, guarantee that pilots or pilot associations are aware, at the time of purchase, that the PPU satisfies the fundamental requirements of the pilots employed in that district. Secondly, ensure that manufacturers and suppliers of such devices are aware of standards typically used for shipping equipment and pertinent to the needs of pilots. Additionally, ensure that the basic functionality of such devices is easy to use for pilots and simple to explain to others.

It is unrealistic to expect pilots to train every vessel under their control, which poses the question of how to operate new equipment safely [9, 10]. Moreover, pilots operating in more confined waters frequently rely on professional local knowledge, hydrological data, and geotechnical data [11], whereas deep-sea pilots must employ an onboard charting system. There is significantly less use of shipboard systems than may be thought, aside from course reference, radar/ARPA, and AIS when visibility is low, thanks to visual signals, real-time tidal information, and directions. Although this has always been the case, technological devices are gradually taking the role of the assistance pilot's notepad. Since their introduction in 1980, PPUs have seen several stages of development, acceptability, and usage globally. Thus, the issues are (a) the ship's pilot's awareness of the equipment when using it; (b) the navigator's most essential PPU types and features; (c) training and practice; (d) anomalous data collection during usage; (e) error and error assessment method of PPU, etc.). On the one hand, to help marine pilots and seaports in their management work, numerous related studies have been conducted recently [8, 12–19]. These studies aim to shed light on the challenges mentioned above as the vessel maneuvers and maintain safety. On the other hand, artificial intelligence has been suggested as a means of guaranteeing marine safety [20–22], and PPU navigation can be facilitated by utilizing contemporary handheld devices in accordance with COLREG 72 [23].

In addition to human resources, the primary concerns of management agencies, ship owners, and port enterprises in Vietnam are the quality of ship piloting services provided. An essential part of safety work is the use of tools and equipment to assist the pilot in maneuvering the ship. This paper compiles data, examines how maritime pilots are currently using navigational support equipment (PPU) within its tools, and suggests ways to monitor and guide PPU use at Vietnamese seaports. More than 100 marine pilots from four different maritime pilot associations in Vietnam's main southern seaports participated in the poll we conducted to gather their perspectives, as did several high-ranking officials from various maritime state management organizations. Evaluate and assess the state of maritime pilots' use of PPUs and their provision of management and guidance in Vietnamese seaports based on their expertise in the area.

The paper is organized as follows: Sect. 2 analyzes the statistics and evaluation of the use of ship navigation support tools by Vietnamese maritime pilots. Next, surveys intended to advise the administration and usage of PPUs at Vietnamese seaports and to obtain feedback on the shortcomings of PPU devices are clarified in Sect. 3. Finally, Sect. 4 provides conclusions and further direction for this study.

2 Statistics and Evaluation of the Use of Ship Navigation Support Tools by Vietnamese Maritime Pilots

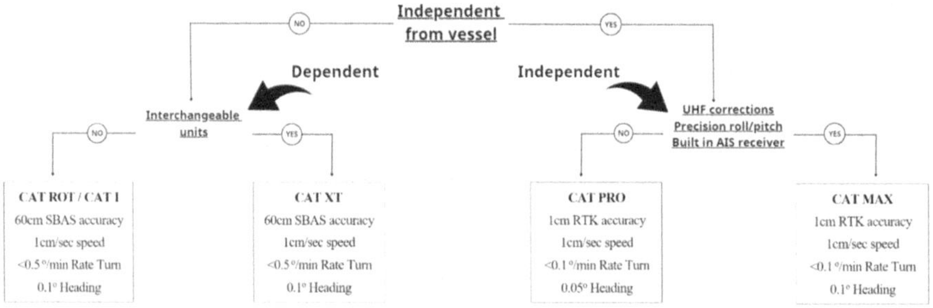

Fig. 1. The diagram illustrates the basic PPUs and the high-end PPUs

In maritime history, PPUs were researched and developed starting in the United States, through collaboration between engineers from the U.S. Coast Guard's Research and Development Center, and the Applied Physics Laboratory at Johns Hopkins University. However, PPUs have become more prevalent, with software design and development primarily originating from the U.S. (Prosertek), Canada (NavSim), France (Cadden Bananas), Germany (Trenz Pilot), Denmark (Trelleborg CAT PRO – Fig. 1), New Zealand (Navicom Dynamics GyroPilot), Poland (Wartsila Pilot Pro), and Norway (AD

Navigation). During pilotage, marine pilots must make certain crucial decisions. Their choice may result in an incident or accident or it could guarantee safe passage. Various ports have implemented various instruments or methods to guarantee the safety of navigation. Among them, PPUs are essential instruments for contemporary marine piloting, particularly while performing close-quarters maneuvers. This is so that they can improve safety and situational awareness by providing useful real-time and predicted data. To fully realize the advantages of PPUs, however, issues with GPS signal stability, chart alignment, data latency, and out-of-date survey data must be resolved. According to the definition by the International Maritime Pilots' Association (IMPA) [7], a PPU is described as an electronic device that maritime pilots bring on board and use as a tool (for analyzing data on ship position, heading, speed, rate of turn, etc., through sensors to display intuitive ship maneuvering parameters, providing a multi-dimensional perspective of the actual situation) combining both technology and human elements to assist pilots in making decisions during the ship pilotage process.

During pilotage, marine pilots must make certain crucial decisions. Their choice may result in an incident or accident or it can guarantee safe passage. Various ports have implemented various instruments or methods to guarantee the safety of navigation. PPUs are essential instruments for contemporary marine piloting, particularly while performing close-quarters maneuvers. This is so that they can improve safety and situational awareness by providing useful real-time and predicted data. To fully realize the advantages of PPUs, however, issues with GPS signal stability, chart alignment, data latency, and out-of-date survey data must be resolved. Challenges with GPS signal stability, chart alignment, data latency, and old survey data must be thoroughly addressed, according to a study of PPU equipment on the situation awareness of marine pilots' viewpoints in Europe and Africa [2] based on:

i. Data needed to make choices that guarantee safe piloting;
ii. Information sources that are consulted while making decisions;
iii. The format of the information that is provided;
iv. The dependability of the data acquired for making decisions;
v. Assessing the information's dependability and failure rate to make informed decisions.

Vietnam may benefit from its communication with PPU-enabled devices without worrying about associated concerns because of the advancements in GPS signal technology that have been made available globally. Although PPUs are not yet mandatory requirements by the International Maritime Organization (IMO) or Vietnam national laws, and most pilots' associations around the world also do not have specific recommendations or guidelines, the reality at major ports globally and particularly at Vietnam's largest and most important ports (especially in the Southeast and Northeast regions) shows that the pilotage process often faces pressure when guiding large vessels. The piloting becomes more challenging and risk-prone when vessels are transiting at night, in poor weather conditions or restricted visibility, or operating in areas with intersecting channels and constrained waters with high traffic density. To mitigate risk factors and promptly utilize all necessary tools to assess collision hazards, in addition to traditional navigational equipment like radar, ECDIS, AIS, etc., the supplementary support of PPUs provides a necessary solution. Pilots' use and reference of PPU-provided information also aligns with Rule 5 and Clause (a) of Rule 7 of COLREG 72 [23], requiring vessels to use all available means appropriate to the prevailing circumstances and conditions to make a full appraisal of the situation and risk of collision; wherein "all available means" can be understood to include PPUs. Moreover, through discussions with several maritime pilots' associations in southern Vietnam ports and experienced pilots, it was found that although PPUs are not mandated, recognizing their benefits, Vietnamese pilots have self-organized groups to share pilotage support technologies over the past decade. Initially using basic software like iNavX and Navionics leveraging tablet GPS chips, most pilots now use Seaiq software with docking functions or PPUs from www.pilotstech.com (China) due to attractive pricing. However, most pilots' use of PPUs relies entirely on self-learning, self-observation, and experience accumulation (some pilots avoid PPUs due to technological aversion or language barriers).

In recent years, some major Vietnamese pilotage organizations like Tan Cang Pilots, Region 1 Pilots, Vung Tau Pilots, Region 5 Pilots, etc. have invested in equipping their pilots with Seaiq software and Pilotstech hardware PPUs within their financial capacities. Nonetheless, after procurement, no pilotage organization or association has compiled data to study and evaluate how frequently their pilots use PPUs and their pros and cons. Furthermore, statistics show that currently, some companies only produce software like Seaiq and Qastor, while others offer integrated software and hardware solutions like Trelleborg (represented by Active Solution in Vietnam), Wartsila, NavSim, and AD Navigation. Some firms focus solely on hardware manufacturing, such as Navicom Dynamics, Trenz, and Pilotstech (the Chinese Pilotstech being widely used by Vietnamese pilotage organizations due to its low cost). Although commercially available for a relatively long time, the accuracy of PPU devices is directly proportional to their cost. The highest accuracy belongs to the high-end, independent PPU generation equipped with precise, independent sensors allowing position accuracy down to centimeters, typically used for highly precise maritime operations like STS, offshore, locks/docking, and mega container/LNG vessel turning in constrained areas (no Vietnamese pilotage unit currently uses this high-end PPU generation).

Mid-range PPUs have accurate sensors but are semi-independent, partially relying on data from the ship's equipment (e.g. Heading). Vietnam has also not widely adopted this mid-range PPU generation yet. Finally, the basic PPUs currently being widely used in Vietnam are fully dependent on the vessel's data, connecting through the AIS Pilot Plug due to their low cost suiting the financial capacities of pilotage organizations. Since these PPUs receive signals from the vessel's AIS Pilot Plug, signal quality entirely depends on the vessel's sensors; if the AIS plug malfunctions, pilots will lose data access, and if the vessel's sensors are inaccurate (delayed, unstable signals or use of low-cost ROT sensors causing predicted positions to deviate from reality, etc.), pilots will consult inaccurate PPU data - especially lacking compatible real-time data sources from modern onboard navigational equipment for cross-verification when piloting older generation vessels. Therefore, when using basic PPUs, caution is advised (particularly piloting large vessels at night in constrained, high-traffic waters), and research should be conducted to evaluate the safety and provide guidance for pilots to assess PPU data deviations from actual field conditions, as well as the pros and cons of the specific PPU models they are equipped with. Recommendations on pilots' responsibilities to harmoniously apply traditional piloting methods and techniques with reasonable use of vessel and PPU navigational data should also be provided. Simultaneously, solutions must coordinate with PPU suppliers to organize internal training programs to develop PPU operational skills.

3 Surveys to Gather Opinions on the Limitations of PPU Devices and Guide the Management and Use of PPUs at Vietnamese Seaports

3.1 Survey Questionnaire

A study was conducted to gather opinions from maritime pilots and state maritime regulatory officials in southern Vietnamese ports regarding their perceptions when PPUs. The study involved discussions, evaluations, and the development of a Survey Questionnaire consisting of 12 questions related to the type and most important features of the PPUs commonly used, training and practical usage, reliability, any abnormal data recordings during use, errors, and methods for assessing PPU errors, among other things. The survey received 133 responses from 126 maritime pilots from four major pilotage organizations and 10 experienced state maritime regulatory officials in the region. Three pilots did not provide their opinions. The summarized feedback results are shown in Table 1.

Table 1. Summary of the results of the expert opinion survey on PPUs

Type of PPU	Type of PPU		Pilot's assessment on the reliability of PPU in use		Ratio of pilots consider PPU as a ship navigation supporting tool	Ratio of PPU supplier conducting training for pilot	Ratio of pilots used to request the software provider to add more function
	Being used	Expectation	Errors understanding	Ratio of pilots checking PPU's errors			
High-end PPU	0	133/133					
Mid-range PPU	0						
Basic PPU	133/133		35/133	21/123	133/133	0	05/123

3.2 The Assessment of the Expert Opinion Survey

During face-to-face interviews, most pilots use the Seaiq Pilot software, while the remaining use the Transas Pilot Pro software (with similar features to Seaiq Pilot); From Table 1, the survey results showed that 100% of pilots use the AIS-based PPU from Pilotstech (KSN55 version); the frequency of PPU usage by pilots is around 80–95%, and most pilots consider PPU use necessary, especially in night pilotage situations with limited visibility or high traffic density. However, pilots regarded the three most important required PPU functions as meeting point (calculating meeting points on the channel), prediction (predicting future ship positions), and record (storing and extracting maneuvering data for review, training, accident/incident investigation). Significantly, 100% of pilots stated that high-end PPUs have very small errors, high reliability, and diverse features, but they have not had opportunities to use them due to the high costs, forcing pilotage organizations and over 90% of pilots to use basic PPUs instead. Last but not least, most pilots recognize and fully understand that basic PPUs receive signals from the vessel's AIS Pilot Plug and the drawback that AIS Pilot Plugs can cause signal delays (lag, smoothing), errors, or signal losses (position, HDG, ROT, offset) compared to reality.

3.3 A Challenge for the Management of Maritime Pilotage Services in Vietnam

In the world, the Fraser River Pilots in British Columbia have collaborated with manufacturers to prioritize PPUs with DGPS positioning capabilities and real-time tidal height display on electronic charts due to the region's challenging, unpredictable tidal regimes [24]. The Columbia River Pilots in Oregon prioritize PPUs capable of automatically calculating and displaying "meeting points" between transiting vessels and targets on the channel, given the area's intersecting waterways. The Halifax Harbor Pilots in Nova Scotia prioritize PPUs designed with 3D positioning systems for overhead bridge clearances, coupled with real-time tidal height data [25]. The Port of Rotterdam Pilots continuously guide VLCCs and LNG carriers along the River Maas, so they prioritize PPUs with highly accurate GPS and ROT sensors integrated with high-resolution electronic charts. In reality, some "port authorities" have also conducted research and initiated steps to establish PPUs as standard tools to enhance maritime safety and security within their areas of responsibility. For instance, the Panama Canal Authority will require all Neo Panamax vessels transiting the canal to be equipped with PPUs capable of Real-Time Kinematic (RTK) positioning starting October 2023 [26]. The Port of Brisbane, Australia, mandates PPU usage for its pilotage force, while the Australian Maritime Pilots Association has adopted a Code of Good Practice for PPUs, providing specific guidance to pilotage organizations on the selection, deployment, and training processes for PPU implementation.

From international sources and survey status in Vietnam, a key issue concerning the management of maritime pilotage services in Vietnam is that pilots and pilotage organizations have recognized the necessity and invested in equipping and using PPUs for over the past decade. However, there are currently no institutional management regulations, safety guidelines, or recommendations from state regulatory bodies or pilots' associations related to the management and recommended use of PPUs. Additionally, there have been no recorded maritime accident cases where the potential cause could be attributed to pilots' misuse of inaccurate PPUs during pilotage operations. Moreover, as maritime pilotage is a profession with distinct local characteristics, the selection of support tools like PPUs also needs to consider compatibility with the specific maritime conditions of the pilotage area, which should be a priority criterion for choosing PPU manufacturers.

It is suggested that the Vietnam Maritime Pilots' Association should take the lead, in coordination with Vietnamese pilotage organizations to comprehensively compile, evaluate the current situation, and conduct a scientific study to establish a "Framework Standard for PPU Feature Selection and Usage in Vietnam". This standard would serve as criteria for pilotage organizations to negotiate and select suitable PPU suppliers aligning with the specific conditions of Vietnam's port areas, avoiding the current situation where organizations lack clear guidance, are passive in negotiations with suppliers, prioritize cost over quality, and are forced to choose basic PPUs with considerable errors. They have also not invested adequate time and resources to understand, compare product lines, assess manufacturer services, and clearly define their actual needs as a basis for negotiating prioritized features with suppliers. Consequently, they may not have selected the most competitive supplier or the most suitable PPU quality for their requirements. Importantly, in contracts, manufacturers are not significantly bound in terms of user training and data update responsibilities, forcing pilots to self-study and accumulate knowledge and experience during PPU operation. On the other hand, to achieve the goals of promoting high-end PPU usage while reducing product costs, mass procurement of high-end PPUs for all pilots is not feasible. Therefore, pilotage organizations need to develop PPU logistics procedures, training processes, and internal usage management protocols to equip pilots assigned to difficult maritime and geographical areas with high-end PPUs on a rotational basis for suitable voyages. From a state maritime regulatory perspective, Maritime Port Authorities should study and establish management and monitoring mechanisms for pilots' PPU usage within their jurisdictions by incorporating specific safety guidance and recommendations into their "Port Regulations." Moreover, to share investment costs and support data sources for developing smart ports in selected areas, pilotage organizations should take the lead in collaborating with port operators, tug service providers, and others to jointly research, pool orders, and select manufacturers capable of providing integrated Smart Port solutions. It would connect smart port infrastructure, with the PPU component for pilots being part of the smart port development package (illustrated in Fig. 2 below). Such solutions would support management, operations, utilization of PPUs, assisting in developing, sharing maneuver plans, and turning solutions for mega container vessels and LNG carriers in constrained waters like Vietnam's CaiMep area.

Through this approach, ship maneuvering data (SafePilot data), tug maneuvering data (SafeTug updating tug movement, thrust/pull forces), and data from meteorological/hydrological monitoring sensor systems (wind speed/direction, currents), laser docking sensors, etc. would be collected and shared in real-time on the SmartPort cloud infrastructure. This aligns with the future vision of sustainable digital transformation at Vietnamese seaports through the critical solution of integrating smart port infrastructure to visually and transparently share operational data in real-time with Maritime

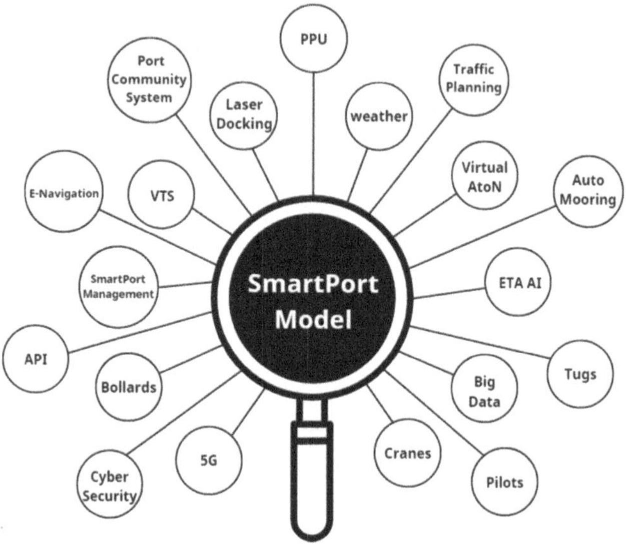

Fig. 2. The illustration of the Smart Port solution and components

Port Authorities, VTS Centers, pilots, tug masters supporting vessel movements, vessels' self-piloting masters, port operators, and relevant parties. This shared data access would enable safety management, monitoring, vessel and port operations, as well as data extraction for training and investigation purposes.

4 Conclusion

It is important to note that PPUs (Portable Pilot Units) should not be relied upon as the only source of information and decision-making during safe pilotage operations. Pilots must have the necessary skills to analyze information from various sources such as human senses (sight and hearing), onboard navigational equipment like radar and ECDIS, and also the ability to effectively operate PPUs. This paper provides a comprehensive perspective on the advantages and limitations of PPU devices, and their current usage in Vietnam. Thus, it is recommended that Maritime Port Authorities incorporate guidelines and safety advisories related to pilots' use of PPUs in the appropriate sections of "Port Regulations" within their respective jurisdictions. Furthermore, it is suggested that the Vietnam Maritime Pilots' Association lead the way in coordination with Vietnamese pilotage organizations for comprehensively compiling and evaluating the current situation. It is also recommended that a scientific study should be conducted to develop a "Standard for PPU Selection and Usage Practices in Vietnam". This standard would serve as criteria for pilotage organizations to select suitable PPU suppliers that align with the specific conditions of Vietnam's port areas. This result will help pilots to effectively and appropriately utilize these devices during pilotage operations.

References

1. Goksu, S., Arslan, O.: Evaluation of dynamic risk factors for ship operations. Aust. J. Marit. Ocean Affairs 1–27 (2024)
2. Bernard, K.: Impact of portable piloting units on the situation awareness of maritime pilots perspectives of Danish and West African pilots. World Maritime University Dissertations (2023)
3. Sharma, A., Nazir, S., Ernstsen, J.: Situation awareness information requirements for maritime navigation: a goal directed task analysis. Saf. Sci. **120**, 745–752 (2019)
4. Alexander, L., Casey, M.J.: Use of portable piloting units by maritime pilots. Can. Hydrogr. Conf. Nat. Surv. Conf. **2008**, 1–9 (2008)
5. Ostendorp, M.C., Lenk, J.C., Lüdtke, A.: Smart glasses to support maritime pilots in harbor maneuvers. Procedia Manufact. **3**, 2840–2847 (2015)
6. Yıldırım, U., Başar, E., Uğurlu, Ö.: Assessment of collisions and grounding accidents with human factors analysis and classification system (HFACS) and statistical methods. Saf. Sci. **119**, 412–425 (2019)
7. IMPA. Guidelines on the design and use of pilot portable unit (2021)
8. APA. Portable Pilot Unit: A Best Practices Summary (2016)
9. Choong, P.A.D., Ahmad, F.A.F.: The development of a conceptual maritime training and assessment framework for Malaysian maritime pilots using the knowledge, skills, and abilities approach. Aust. J. Marit. Ocean Aff. 1–22 (2023)
10. Hwang, H.S., Hwang, T.M., Youn, I.H.: Efficacy evaluation of adaptive collision avoidance systems for autonomous maritime surface ships based on target ships' maneuvering behaviors. J. Int. Marit. Saf. Environ. Affairs Shipping **8**(1–2), 2298250–2298258 (2024)
11. Herath, M.D.D.B.: Augmented reality technology in maritime operations for enhanced visualization in compromised visibility. University of Tasmania, Thesis (2023)
12. Port of Brisbane. Port Procedures and Information for shipping (2023)
13. Canal De Panama. FAQ-Advisory-to-Shipping-No.-A-32–2022-New-Positioning-SystemRev1 (2022)
14. AMPI. PPU Code of Good Practice (2020)
15. EMPA. How to Select the Correct PPU for Your Port (2019)
16. Chiarello, F., Trivelli, L., Bonaccorsi, A., Fantoni, G.: Extracting and mapping industry 4.0 technologies using wikipedia. Comput. Ind. **100**, 244–257 (2018)
17. Hamburg News. HHLA using machine learning to raise productivity in Port of Hamburg (2020)
18. Le, V.T., Dong, X.V., Pham, T.D.A.: Research and propose solutions to form and develop smart ports at Vungtau Seaport. Stais National Conference, pp. 315–323 (2023)
19. Boullauazan, Y., Sys, C., Vanelslander, T.: Developing and demonstrating a maturity model for smart ports. Marit. Policy Manag. **50**(4), 447–465 (2023)
20. Pham, T.A., Dang, X.K., Koboević, Ž, Do, V.D., Pham, T.D.A.: Maritime data mining for marine safety based on deep learning: southern Vietnam case study. NAŠE MORE **71**(1), 21–29 (2024)
21. Pham, T.A., Dang, X.K., Nguyen, S.V.: Optimising maritime big data by K-means Clustering with Mapreduce Model. Industrial Networks and Intelligent Systems. INISCOM 2022: Industrial Networks and Intelligent Systems, pp. 136–151 (2022)
22. Le, V.T., Dang, X.K., Nguyen, D.H., Pham, T.D.A.: A novel maritime risk assessment model of waterway transportation based on Takagi-Sugeno fuzzy logic: Vietnam case study. In: IOP Conference Series: Earth and Environmental Science (ICCOE 2020), pp. 1–8 (2020)
23. SOLAS Chapter V, Regulation 15 & Regulation 17; and Rule 5, Rule 7 COLREG 72 (1972)

24. Wu, Y., et al.: Tidal propagation in the lower Fraser river, British Columbia, Canada. Estuar. Coast. Shelf Sci. **264**(11), 107695 (2021)
25. Shan, S., Sheng, J.: Examination of circulation, flushing time and dispersion in Halifax Harbour of Nova Scotia. Water Qual. Res. J. **47**(3–4), 353–374 (2012)
26. https://www.marineregulations.news/panama-canal-operational-regulations-and-compliance-notices/

Marine Environmental Response and Coastal Protection under Climate Change

Numerical Study of the Morphodynamic Beach Profile with Varying Wave Conditions and Beach Nourished Schemes

Yachao Zhang, Yingxin Wei, Xiaojie Zhang(✉), Jianhua Liu, and Xianghuang Li

Ministry of Natural Resources, East China Sea Survey Center, Beijing, China
zhxjanet@163.com

Abstract. This paper establishes a numerical model of one-dimensional beach profile evolution at a laboratory scale with XBeach, whose ultimate objective is to study the beach profile changes after applying beach nourishment. The prediction data of the model and related physical models are compared and calibrated. The schemes under different wave conditions, sediment volumes, grain sizes, and replenishment methods are numerically simulated on this basis. Some interesting conclusions can be drawn from the schemes. The beach profile does not change regularly with the linear variation of the sediment volume. So, the larger nourished sediment volume of the beach is not the better. Moreover, the results show that the sediment grain size is inversely proportional to the erosion intensity. These provide essential support for beach restoration design, rationality, and stability, as well as for judging the feasibility of beach restoration projects.

Keywords: beach nourishment · numerical model · XBeach · sediment volume · grain size · the morphodynamic of the beach · beach profile

1 Introduction

Approximately 70% of sandy shorelines worldwide are in a state of erosion [1], with average erosion rates greater than 1 m/a, and in some areas, it is as high as 4–5 m/a [2]; thus, the extent of erosion is still increasing. At a globally scale, the erosion of beaches is an important problem. In China, natural embayed beaches are frequently subjected to the effects of storms and typhoon waves, resulting in significant coastal erosion, wave overtopping, beach system instability, inundation, and catastrophic outcomes (Bernabeu et al. 2003; Shimozono et al. 2020; Aránguiz et al. 2020).

Beach nourishment can be considered as an environmentally and economically acceptable method to protect the shoreline from erosion (Dean, 2002). Beach restoration, a technique used to restore an eroding or lost beach or to create a new sandy shoreline, involves the placement of sand fill with or without supporting structures along the shoreline to widen the beach. In addition, nourishments can enhance or replace hard coastal protection structures and subsequently contribute to the renaturation of engineered coastlines (Capobianco and Stive 2000). Many coastal countries around the world are therefore carrying out beach nourishments on a regular basis as a suitable means of erosion mitigation and coastal protection (e.g. Cooke et al. 2012; Hamm et al. 2002).

D. Jeng and B. Cai (Eds.): ICCOE 2024, LNCE 423, pp. 481–491, 2025.
https://doi.org/10.1007/978-981-96-0181-3_43

There are many studies of beach restoration around the world. Karl F. et al. reveal how nourishment projects used for shore protection can restore the space available for dunes to form, increase the likelihood of sediment transfers inland and increase the diversity of topography and vegetation in Italy (Karl F. et al. 2009). M. Paula Bunicontro et al. propose mainly restoration, prevention, and adaptation strategies, such as the use of sand fences and new techniques for beach cleaning, all of which are aimed at promoting sedimentary balance within the dune-beach system in Argentina (M. Paula Bunicontro et al. 2021). SHU Fangfang et al. analyzed the evolution of its profile configuration and plane morphology, and its storm response characteristics in Tianquan Bay, Xiamen, China (SHU Fangfang et al. 2018). Studies must take into account numerous variables that contribute to the processes and morpho-dynamic behavior of nourishment beaches to comprehensively understand these phenomena. The purpose of this paper is to elucidate a beach profile sequence with varying nourishment conditions and provide a useful understanding and reference, especially for beach nourishment. The results are of general interest for the design of beach nourishment projects.

2 Model Setup and Validation

2.1 Model Description

XBeach is a representative model for short-period evolution simulation of beach restoration projects recommended by the Technical Guide for Beach Conservation and Restoration. It was developed jointly by TUDelft (Delft University of Technology) and Deltares Institute (Delft Delta Research Centre). It can be used to model coastal dynamics such as tidal waves, tsunami waves, storm surges and the corresponding evolution of coastlines, seabed evolution and sediment transport. The model solves coupled equations of wave action equation, roller energy equation, shallow water equations, sediment transport, morphological updating. The descriptions of model formulations and manual can be referred to the documents (XBeach 2010; Lescinski, J., 2009; Nederhoff, 2015). Simulating hydrodynamic processes such as wave propagation and transformation with minor errors has fully validated the model. (Vries, 2011). It includes a morphodynamic component that reproduces the known forms of morphological storm response under various geometric and hydrodynamic conditions, suggesting that among the available modeling options, XBeach is a valuable tool for morphodynamic investigations of beaches. (Karunarathna et al., 2018; Dissanayake et al., 2014; Harter and Figlus, 2017).

2.2 Model Validation

The model of this paper was verified by physical model test, and the physical model test data were collected from published journals -- Roelvink and Stive test data. Roelvink and Stive tested the evolution of bar profile under wave action in an experimental tank. The initial terrain of the experiment was a uniform slope with a gradient of 1:40, and the waves were irregular waves generated according to the JONSWAP spectrum. The test time was 12 h. The significant wave height is 0.174 m, the spectral peak period is 2 s, and the median particle size is 0.1 mm. In the XBeach model, the number of horizontal grids is 230, and the step size is 0.2 m. The krone number is 0.97, and the terrain acceleration factor is 6.

At the beginning of the beach face, erosion occurred quickly, and a submerged bar quickly formed near the coastline to protect the beach from further erosion. After a 12-h simulation, the comparison between the terrain calculation results and the measured results is shown in Fig. 2. The erosion area coincides; the sand bars formed by numerical simulation are not as prominent as those measured, but the trend is consistent, indicating a suitable fitting.

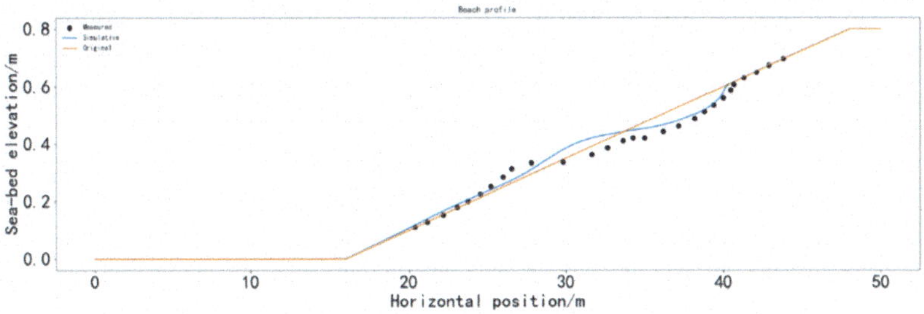

Fig. 1. Model validation

At the same time, the error relative to the measured value of bed elevation is calculated, and the BSS is 0.88, R2 is 0.83, and SCI is 0.26. Brier Skill Score (BSS) is the empirical coefficient. This parameter relates the variance of the difference between the measured data and the model results to the variance of the data. A BSS equal to 1 indicates that the simulation performance of the model is good, and a BSS equal to 0 indicates that the fitting result is poor, and the model cannot be used for calculation. The coefficient of determination R2 describes the correlation between the measured and the model results, and evaluates the goodness of fit. The closer R2 is to 1, the greater the fit between the model results and the measured data. The Scatter Index (SCI) normalized the errors to avoid abnormal results in data with small averages and large variability. The smaller the SCI value, the better the model performance.

$$\mathrm{BSS} = 1 - \frac{\mathrm{Var}(c - m)}{\mathrm{Var}(m)} = 0.88$$

$$R^2 = \frac{\mathrm{Cov}(m, c)}{\sigma_m \sigma_c} = 0.83$$

$$\mathrm{SCI} = \frac{\sqrt{\frac{1}{N}\Sigma_{i=1}^{N}(c_i - m_i)^2}}{\frac{1}{n}\Sigma_{i=1}^{N} m_i} = 0.26$$

where m is the measured value; c is the calculated value of the model; σ_m is the standard deviation of the measured value; σ_c is the standard deviation of the calculated value of the model.

Although the model does not accurately simulate the shape of the offshore bar, the position of erosion and accumulation is consistent, and the statistical error analysis results are reasonable. The comparison between the experimental and numerical beach profiles indicated good agreement. Meanwhile this paper focuses on the relative stability of different sediment replenishment schemes, and the error analysis shows that the model is highly reliable and can be used to simulate the spatio-temporal evolution of overall sediment redistribution.

2.3 Model Setup

Although near-shore sediment transport and landform changes are three-dimensional, lateral (perpendicular to the shoreline) and coastal (parallel to the shoreline) profile changes are traditionally considered separately in beach conservation design. At the same time, the results of the two-dimensional and one-dimensional models are consistent, indicating that coastal sediment transport can be ignored in the short term.

In order to study the restoration stability of different design schemes, the XBeach model is used to simulate the response of the beach under different conditions, including different sediment volumes, sediment grain sizes, and replenishment methods on the stability of the beach profile. The evolution of the 12 h, which is enough time for the beach to reach an equilibrium profile, is simulated respectively. The specific simulation scheme is shown in the table below. The profiles of beaches nourished can be seen in Fig. 2.

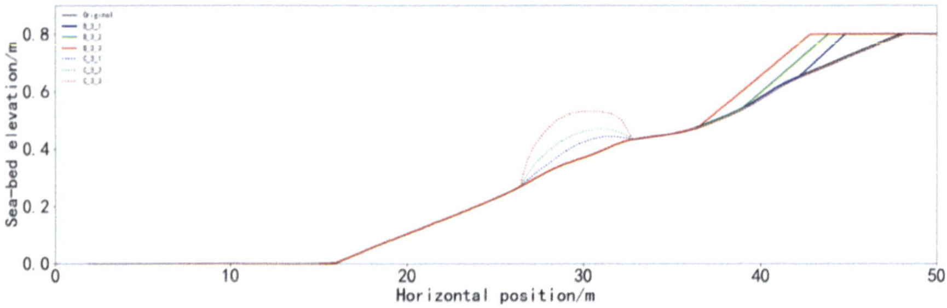

Fig. 2. Beach restoration

3 Beach Profile Changes

3.1 Beach Profile Changes with Wave Conditions

Waves significantly impact the topographic changes of beaches, subsequently causing coastal erosion and management crises. According to the calculation results of the beach profile, a wave height of 0.2 m with a period of 3 s, a wave height of 0.15 m with a period of 2.5 s, and a wave height of 0.1 m with a period of 2 s are used as wave conditions, respectively. The beach beam and bar are divided and the simulation calculation is 12h.

Figure 3 and Fig. 4, respectively, show the evolution results of beach profiles under different wave conditions of beam and bar replenishment. It can be seen that for the same

Table 1. Numerical study schemes.

	SCHEMES	wave height (m)	Period (s)	Median diameter (mm)	Time (h)	Sediment volume (m^3/m)
ORIGINAL	A	0.25	2	0.2	0.5	0
BEACH RESTORATION ON BERM	B_1_1	0.2	3	0.2	12	0.2
	B_1_2	0.15	2.5	0.2	12	0.2
	B_1_3	0.1	2	0.2	12	0.2
	B_2_1	0.15	2.5	0.1	12	0.2
	B_2_2	0.15	2.5	0.2	12	0.2
	B_2_3	0.15	2.5	0.4	12	0.2
	B_2_4	0.15	2.5	0.6	12	0.2
	B_3_1	0.15	2.5	0.2	12	0.2
	B_3_2	0.15	2.5	0.2	12	0.4
	B_3_3	0.15	2.5	0.2	12	0.8
BEACH RESTORATION ON BAR	C_1_1	0.2	3	0.2	12	0.2
	C_1_2	0.15	2.5	0.2	12	0.2
	C_1_3	0.1	2	0.2	12	0.2
	C_2_1	0.15	2.5	0.1	12	0.2
	C_2_2	0.15	2.5	0.2	12	0.2
	C_2_3	0.15	2.5	0.4	12	0.2
	C_2_4	0.15	2.5	0.6	12	0.2
	C_3_1	0.15	2.5	0.2	12	0.2
	C_3_2	0.15	2.5	0.2	12	0.4
	C_3_3	0.15	2.5	0.2	12	0.8

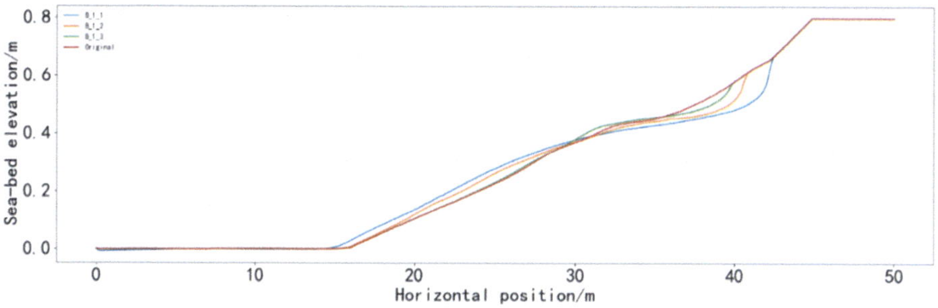

Fig. 3. Beach profile changes under different wave conditions (replenishment on beach beam)

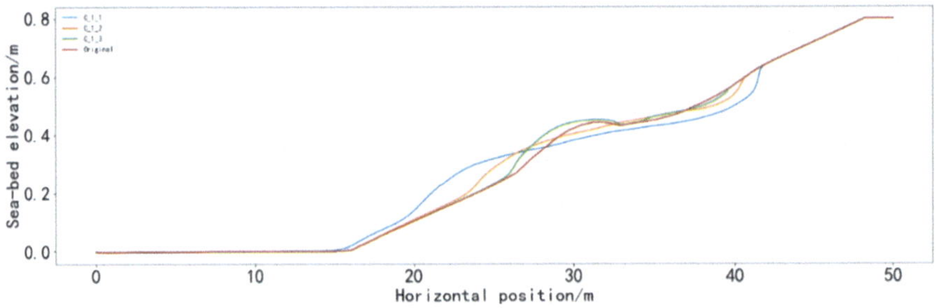

Fig. 4. Beach profile changes under different wave conditions (replenishment on beach bar)

design profile, whether beam replenishment or bar replenishment, the larger the wave height, the more pronounced the change of beach surface morphology, which means that more sediment is transported farther offshore by waves and indicates that beach profile is inversely correlated with wave height. At the same time, in the case of small waves, the shape of the replenishment profile is good, so the stability of the restoration scheme should be studied in combination with the actual wave situation.

3.2 Beach Profile Changes with Sediment Volume

In order to analyze the influence of different amount of sand replenishment on the restoration stability, the single width of the sand discharging capacity of the beam replenishment and the bar replenishment were set as 0.2 m^3/m, 0.4 m^3/m and 0.8 m^3/m, respectively, with other factors being the same. The results (Fig. 5 and Fig. 6) show that the higher the amount of sediment in the profile, the better the erosion protection effect on the beach profile, and the profile shape is easier to maintain. However, the change of the beach profile does not change regularly with the linear change of the amount of sand thrown, especially the way of the sand dam. When the single width of the sand thrown is 0.2 ~ 0.4 m^3/m, the shape change is not obvious. Therefore, from the economic point of view, in the perspective of profile stability, the larger the sediment volume is not the better.

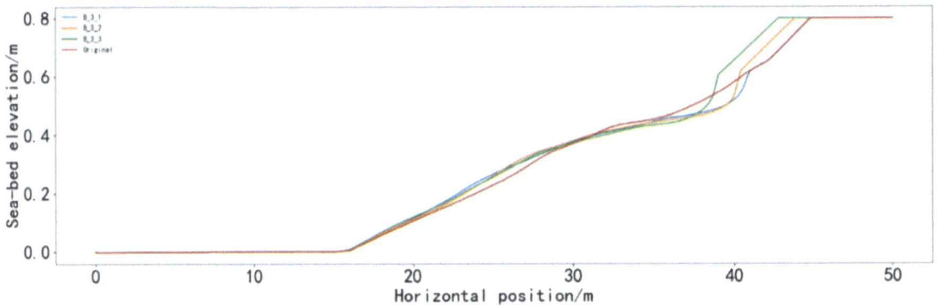

Fig. 5. Beach profile changes under different sediment recharge conditions (replenishment on beach beam)

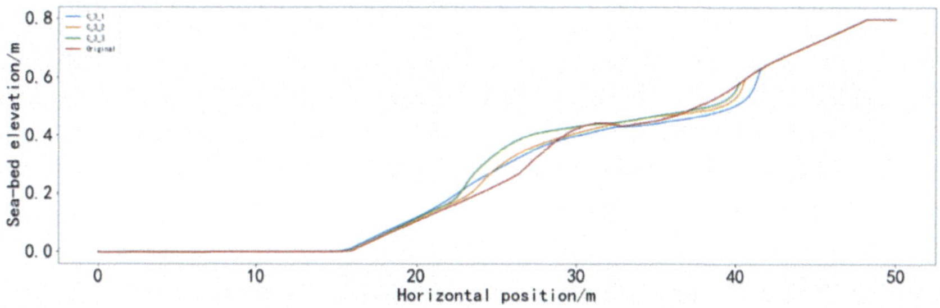

Fig. 6. Beach profile changes under different sediment recharge conditions (replenishment on beach bar)

3.3 Beach Profile Changes with Sediment Grain Size

This study simulated different median particle sizes (0.1 mm, 0.2 mm, 0.4 mm, 0.6 mm) of the design section. In addition to the different particle sizes of the design section, other factors such as terrain, wave, and sand fill amount were the same in each group to determine the influence of the median particle size of the sand fill on the stability of the beach.

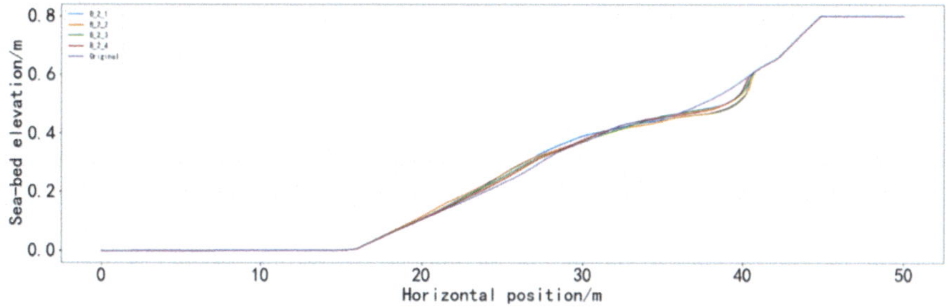

Fig. 7. Beach profile under different particle size conditions (replenishment on beach beam)

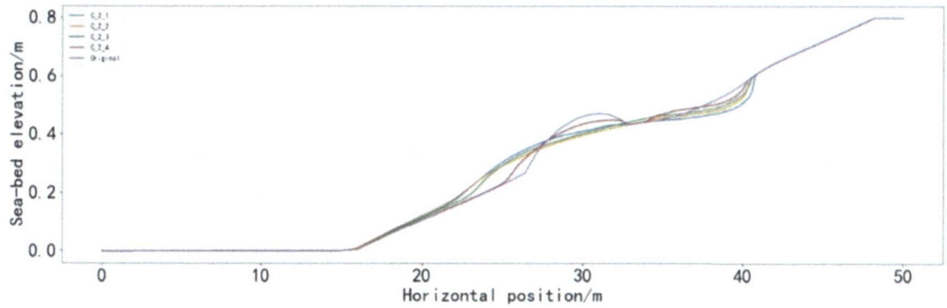

Fig. 8. Beach profile under different particle size conditions (replenishment on beach bar)

Figure 7 and Fig. 8 show that the trends of profile shape change of the four particle sizes are the same. However, the profile of the particle size of 0.1 mm has more erosion and the beach beam has receded; the profile changes of the particle sizes of 0.2 mm and 0.4 mm are almost the same; and the particle size of 0.6 mm has the best retention ability for the width and height of the beach beam. Under this dynamic condition, it can be qualitatively considered that the sediment particle size is inversely proportional to the erosion intensity, and the design section with a particle size greater than 0.6 mm is more favorable to the stability of the beach.

3.4 Beach Profile Changes with Sand Replenishment Method

At present, there are four kinds of artificial sand replenishment sites used in the project: (1) Dune replenishment, which places the sand above the average spring tide, does not increase the width of the dry beach and resists the storm surge by strengthening the dune; (2) The beam sand is mainly filled above the average water level, which can directly increase the width of the dry beach; (3) The profile sand replenishment makes the short-term effect obvious, with the sediment directly blowing and filling the profile; (4) Shore (bar) to the sea side of the sand replenishment, with the supplied sediment dumped near the near-shore underwater crushing area, can reduce wave energy and shore sand supply. And the fourth is a more active and environmentally friendly way of artificial beach cultivation. Comparatively speaking, Beach restoration on bar is a more commonly used beach maintenance scheme because of its low construction difficulty. This method can directly increase the width of beach beam with remarkable effect. Nearshore (sand bar) replenishment has little impact on the environment, which is also one of the key issues in current beach conservation research.

Therefore, in order to analyze the influence of different sand replenishment methods on the restoration stability, the morphological changes of the sand replenishment profiles of the beach beam and the bar were compared under the condition of a certain amount of sand discharge (0.2 m^3/m, 0.4 m^3/m, 0.8 m^3/m in a single width). The results show that, in general, the morphological stability of the underwater profile is better with different replenishment methods. However, it will be different from other conditions. For example, when the amount of sand replenishment is 0.2 m^3/m, sand replenishments of the beach beam and bar are more consistent. Therefore, all factors are not independent in restoring the beach, which needs comprehensive consideration and research (Figs. 9, 10 and 11).

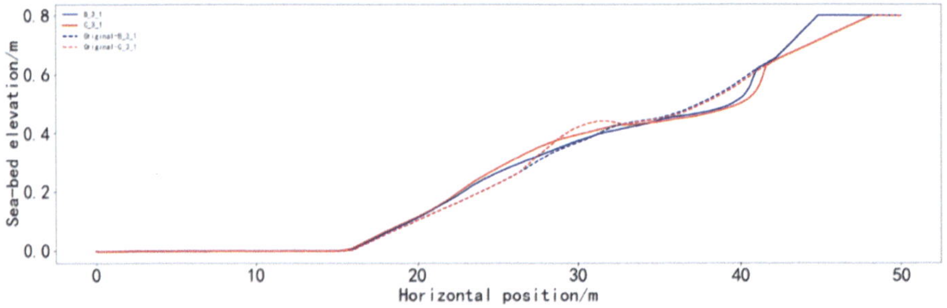

Fig. 9. Beach profile under different replenishment modes (sediment volume 0.2 m^3/m)

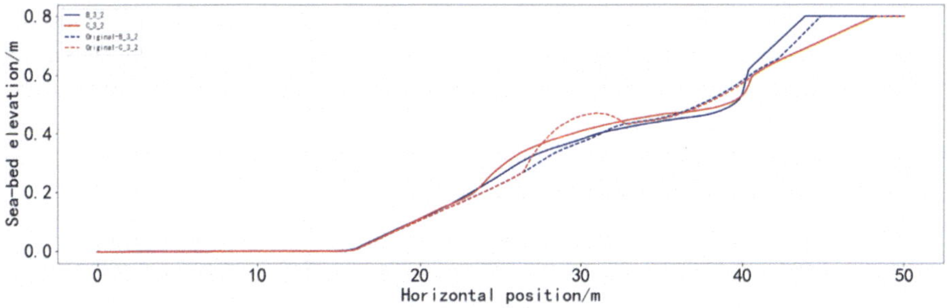

Fig. 10. Beach profile under different replenishment modes (sediment volume 0.4 m^3/m)

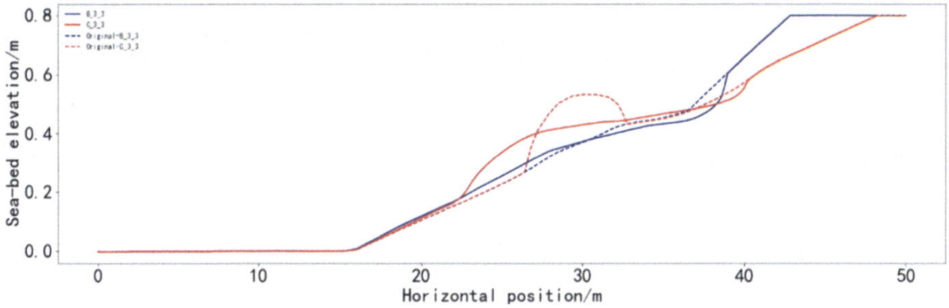

Fig. 11. Beach profile under different replenishment modes (sediment volume 0.8 m^3/m)

4 Conclusions

In this paper, a numerical model of the evolution of XBeach one-dimensional beach profile at laboratory scale was established, and the measured data of the model and related physical models were compared and verified. On this basis, the sand stability scheme under different wave conditions, sand recharge amounts, sediment particle sizes, and sand recharge methods were numerically simulated. The results are as follows:

(1) The more significant the wave height, the more pronounced the change of beach surface morphology. When the wave is small, the shape of the replenishment profile can be maintained well. Therefore, the stability of the restoration scheme should be studied in combination with the actual wave situation during beach restoration.
(2) The erosion protection effect of the beach profile is better, and the profile shape is easier to maintain with the increased sediment recharge. However, the beach profile does not change regularly with the linear change of sand discharge and will appear in a specific range of single-width sand discharge (0.2 ~ 0.4 m^3/m), and the morphological change is not obvious. Therefore, from the economic point of view and the perspective of profile stability, the larger the single-width sand discharge of the beach is not the better.

(3) The changing trends of beach surface under the same wave conditions are almost the same for the beach restorations with different particle sizes of sand filling, and it can be qualitatively considered that the sediment particle size is inversely proportional to the erosion intensity.
(4) All factors are not independent in the process of beach restoration and need to be comprehensively considered and studied.

In engineering practice, intense storms may occur, and the ability of artificial sand replenishment to withstand storms should be evaluated in the short and rapid storm process. Therefore, it is necessary to simulate the changing pattern of bed elevation after the storm in the sand replenishment profile.

At the same time, the numerical simulation in this paper supplements the small-scale physical model test, and then the law of sediment redistribution is discussed. However, due to the complexity and variability of Marine environmental conditions and the complexity of sediment movement patterns, it is still necessary to conduct a large number of field measurements and large-scale physical model tests for the evolution of the beach at actual scale. Combined with numerical simulation, a more in-depth study of the sediment movement patterns after beach maintenance can be carried out to achieve the best effect and improve engineering efficiency. It provides a theoretical basis and technical support for sand recharging projects.

Acknowledgments. This work was supported by the Youth Marine Science Fund of East China Sea Bureau, Ministry of Natural Resources (No. 2023180504).

References

Cai Feng, S., Xianze, L.J., et al.: Coastal erosion in China under the condition of global climate change and measures for its prevention. Prog. Nat. Sci. **19**(4), 415–426 (2009). https://doi.org/10.1016/j.pnsc.2008.05.034

Komar, P.D.: Beach Processes and Sedimentation, 2nd edn. Prentice-Hall, Engle wood Cliffs (1998)

Dongxing, X., Wang Wenhai, W., Guiqiu, et al.: Coastal erosion in China. Acta Geographica Sinica (in Chinese) **48**(5), 468–476 (1993)

Allen, J.R.: Beach erosion as a function of variations in the sediment budget, Sandy Hook, New Jersey, USA. Earth Surf. Proc. Landforms **6**, 139–150 (1981)

Bernabeu, A.M., Medina, R., Vidal, C.: A morphological model of the beach profile integrating wave and tidal influences. Mar. Geol. **197**(1–4), 95–116 (2003)

Shimozono, T., et al.: Coastal impacts of super typhoon Hagibis on Greater Tokyo and Shizuoka areas. Jpn. Coast Eng. J. **62**(2), 129–145 (2020). https://doi.org/10.1080/21664250.2020.1744212

Aránguiz R, et al.: The 2018 Sulawesi tsunami in Palu city as a result of several landslides and coseismic tsunamis. Coast Eng. J. **62**(4), 445–459 (2020). https://doi.org/10.1080/21664250.2020.1780719

Dean, R.G.: Beach nourishment: Theory and practice. In:Advanced Series on Ocean Engineering, Volume 18. Singapore: World Scientific, 399p (2002).

Capobianco, M., Stive, M.J.F.: Soft intervention technology as a tool for integrated coastal zone management. J. Coast. Conserv. **6**, 33–40 (2000)

Cooke, B.C., Jones, A.R., Goodwin, I.D., Bishop, M.J.: Nourishment practices on Australian sandy beaches: a review. J. Environ. Manag. **113**, 319–327 (2012)

Hamm, L., Capobianco, M., Dette, H., et al.: A summary of European experience with shore nourishment. Coast Eng. **47**, 237–264 (2002). https://doi.org/10.1016/S0378-3839(02)00127-8

Nordstrom, K.F., Gamper, U., Fontolan, G., Bezzi, A., Jackson, N.L.: Characteristics of coastal dunetopography and vegetation in environments recently modified using beach fill and vegetation plantings, Veneto, Italy. Environ. Manag. **44**(6), 1121–1135 (2009)

Bunicontro, M.P., Marcomini, S.C., López, R.A.: Assessment of beach erosion susceptibility based on geo-indicators and soft management policies: a Patagonian case study (Argentina.) Bulletin of Engineering Geology and the Environment **80**, 2805–2824 (2021)

Fangfang, S., Feng, C., Hongshuai, Q., et al.: Morphodynamics of an Artificial Cobble Beach in Tianquan Bay, Xiamen, China. J. Ocean Univ. China **18**(04), 868–882 (2019)

XBeach Model Description and Manual, Unesco-IHE Institute for Water Education, Deltares and Delft University of Technology, June 21, 2010 version 6

Lescinski, J.: Modelling storm impacts on beaches, dunes and barrier islands. Coast. Eng. **56**(11–12), 1133–1152 (2009)

Nederhoff, K.. Modeling the Effects of Hard Structures on Dune Erosion and Overwash. Delft University of Technology (2015)

Vries, B.D.: Dune Erosion Near Sea Walls-XBeach Validation. Delft University of Technology (2011)

Karunarathna, H., Brown, J., Chatzirodou, A., Dissanayake, P., Wisse, P.: Multitimescale morphological modelling of a dune-fronted sandy beach. Coast. Eng. **136**, 161–171 (2018)

Dissanayake, P., Brown, J., Karunarathna, H.: Modelling storm-induced beach/ dune evolution: sefton coast, Liverpool Bay. UK. Mar. Geol. **357**, 225–242 (2014)

Harter, C., Figlus, J.: Numerical modeling of the morphodynamic response of a low-lying barrier island beach and foredune system inundated during Hurricane Ike using XBeach and CSHORE. Coast. Eng. **120**, 64–74 (2017)

Effects of Water-Cement Ratios on Degradation of High-Water Material During Weathering

Fangnan Cui[1,3] and Bangwen Lu[1,2](✉)

[1] Zhengzhou Institute of Multipurpose Utilization of Mineral Resources, CAGS, Zhengzhou 450006, China
bluad@connect.ust.hk

[2] Key Laboratory for Polymetallic Ores' Evaluation and Utilization of Ministry of National Resources, Zhengzhou 450006, China

[3] School of Management, Jinan University, Guangzhou 510632, China

Abstract. High-water material (HWM) is a new cementing material with many excellent properties and is apt to get weathering. The objective of this study is to explore the degradation mechanism of HWM during weathering and the effects of w/c ratios on the weathering process of HWM were explored by laboratory tests. HWM samples of w/c = 3:1 and 6:1 were prepared and conducted to a 10-day indoor weathering test, and microstructure, chemical component, and physical and mechanical characteristics were investigated. The results showed that: The degradation of HWM during weathering is mainly due to the loss of water and carbonization of Ettringite. Free and crystal water loses rapidly and the carbonization reaction of Ettringite occurs and Ettringite crystals are easily corroded by CO_2 in air. The degradation of HWM during weathering highly depends on w/c ratio and HWM with a lower w/c ratio has a better weathering resistance, and loses its water and strength easier during weathering. After 10-day weathering, HWM with w/c = 3:1 and 6:1 lost 20% and 70% of strength respectively.

Keywords: High-water material · Water-cement ratio · Weathering · Strength · Microstructure

1 Introduction

High-water material (HWM), also known as high-water-content and quick-setting material, or high-water rapid-setting material, is a new cementing material with many excellent advantages, such as high water content, good pumpability, rapid setting, high early strength, recrystallization recovery strength after the early failure of the stone body, simple material production process and low cost [1–3].

Initially, HWM was mainly used for backfilling, roadway support, shotcrete anchor support, etc. in coal underground mining [4, 5]. Recently, with the development of technology and the improvement of the performance of HWM, the application fields of HWM have expanded from the underground, where it is below the groundwater level and filled with groundwater, to near ground surface or even above the groundwater level. Yan

D. Jeng and B. Cai (Eds.): ICCOE 2024, LNCE 423, pp. 492–503, 2025.
https://doi.org/10.1007/978-981-96-0181-3_44

et al. (2000 and 2004) carried out a feasibility study on the application of HWM in soft soil foundation reinforcement and investigated the physical and mechanical properties of the consolidated soft soil mixed HWM [6, 7]. Hou et al. (2007 and 2012) studied the feasibility of HWM in the maintenance and reinforcement of coral sand islands and reefs and port reinforcement [8, 9].

The physical and mechanical properties of HWM are highly dependent on the existing environment [1]. He et al. (2014) compared physical and mechanical parameters of HWM curing in various environments and revealed the environmental effects on the strength characteristics of HWM [10]. Wang et al. (2014) found that HWM loses strength more quickly in the air environment than in water, and the water environment helped to improve the stability of its inner structure and bearing capacity [11]. Therefore, when HWM is used in environments above the groundwater level, its weathering resistance properties become one of key factors affecting the physical and mechanical properties and its applicability.

Several previous researches were conducted to study the degradation mechanism of HWM during weathering. Sun and Song (1994) and Song et al. (1999) analyzed the chemical composition and microstructure of the weathered layer of HWM stones and found that the weathering of HWM was mainly due to the carbonization reaction of Ettringite and CO2 in the air, which disintegrates Ettringite [1, 12]. Xie et al. (2013) tested the influence of water content on the physical and mechanical properties of HWM, and reported that HWM with higher w/c ratios lost water more easily [13]. However, the above researches only carried out the qualitative analysis, and few ones conducted quantitative studies on the weathering mechanism of HWM. It is necessary to do further researches on the effects of w/c ratios on the weathering resistance of HWM.

In this paper, the degradation mechanism of HWM during weathering and the effects of w/c ratios on the weathering characteristics of HWM were investigated by combining macroscopic and microscopic test methods.

2 Materials and Methods

2.1 Test Materials

Table 1 shows the chemical compositions of HWM. The raw materials of HWM are mainly comprised of four ingredients, namely A, AA, B, and BB. A and B are the main raw materials, and AA and BB are additives. A mainly contains bauxite and B is mainly gypsum. The main functions of the two additives AA and BB are retarder and accelerating agent respectively. HWM is generated by mixing Grouts A and B, where Grout A is made by mixing A and AA with water and Grout B is made by mixing B and BB with water. The proportion of four ingredients used in this study is A: AA: B: BB = 1:0.1:1:0.04 and samples with various w/c ratios were prepared by changing the amount of water. In this study, five groups of samples of HWM with various w/c ratios were prepared, namely w/c = 3:1, 4:1, 5:1, 6:1, and 7:1.

According to Sun and Song (1994), the main component of HWM stone is Ettringite ($3CaO{\bullet}Al_2O_3{\bullet}3CaSO_4{\bullet}32H_2O$). The hydration and hardening of HWM generally follows the below reaction:

$$3CA \cdot SO_4 + 8CaSo_4 \cdot 2H_2O + 6CaO + 80H_2O \rightarrow 3C_3A \cdot 3CaSO_4 \cdot 32H_2O \quad (1)$$

Table 1. Composition of HWM.

Ingredients	Minerals	Chemical composition	Remarks
A	Bauxite	$3CaO{\cdot}3Al_2O_3{\cdot}CaSO_4$, $2CaO{\cdot}SiO_2$ etc	Retarder; suspending agent; dispersant etc
AA	Additives	Na_2CO_3, $BaBiO_3$ etc	
B	Lime, gypsum	$CaSO_4$, $CaSO_4{\cdot}2H_2O$ $CaSO_4{\cdot}0.5H_2O$ etc	Accelerating agent; suspending agent; activator etc
BB	Additives	SiO_2, $CaSO_4$ etc	

During the process of hydration and hardening, a large amount of free water in the mixed slurry participates in and transforms into the bound water of Ettringite, which eventually leads to the formation of HWM stones and the property of high water content.

2.2 Test Procedures

Firstly, samples were made with a standard cast iron mold (ϕ50 × 100 mm), and the preparation procedures included weighing, adding water, mixing, injection molding, and demolding. After that, samples were put into clean water for curing for 28 days to reduce the influence of the curing period. Then samples were taken out of the water and the weathering tests got started.

Fig. 1. Schematic diagram of the soil column (unit: mm).

Two groups of samples with w/c = 3:1 and 6:1 respectively were used to do the indoor natural weathering test, as such w/c ratios are widely used in field [1]. Figure 1 showed the initial photo of samples of the weathering test. Both two groups had 12 standard cylindrical samples (i.e. ϕ50 mm*100 mm). All samples were put in an environment controlling room. The room temperature was kept 26–30 °C and the air humidity was 40%–50%.

The weathering test lasted for 10 days. During the test, the surface morphology and weight were monitored, and compressive strengths of samples were tested on the 4th, 6th, 8th, and 10th days. 3 samples of each group were taken out for each uniaxial compression test and the average value was regarded as the uniaxial compressive strength of the weathered samples.

2.3 Test Equipment

In this study, the mechanical characteristic parameters of weathered HWM samples were obtained by using RMT150B multi-function testing machine. Scanning electron microscopy (SEM) was used to get the micro-morphology of HWM. Chemical components of HWM were determined by XRD analysis. The above tests were conducted in Geotechnical Engineering Laboratory, Sichuan University, China.

3 Interpretation of Experimental Results

3.1 Test Phenomenon During Weathering Process

Figures 1, 6, and 7 show the photos of HWM with w/c = 3:1 and 6:1 before weathering and after 4- and 10-day weathering respectively. The initial appearances of HWM samples with various w/c ratios were almost similar. All the samples were dark khaki-colored. During the weathering process, the surface color of all samples turned lighter with time and gradually became light khaki. Few white powders appeared on the surface, and the amount of powder increased with time. Xie et al. (2014) reported that the white powder was identified as calcium carbonate ($CaCO_3$) and calcium carbonate ($CaSO_4$) by XRD analysis, which indicates that the Ettringite in the surface layer of HWM stones had become gypsum, calcite, and other minerals.

During the weathering process, more and more small shrinkage cracks appeared on the surface. With increase of surface water loss, surface shrinkage cracks became wider and deeper with weathering time. At the same time, the surface structure became sparse and soft, and exfoliation phenomenon was observed in the partial surface of HWM samples.

Comparing Figs. 2 and 3, it could be seen that HWM with w/c = 6:1 lost water much more quickly than w/c = 3:1. The surface of w/c = 3:1 turned white earlier. With the loss of surface water, shrinkage cracks on samples with w/c = 6:1 became denser, wider, and deeper. After 4-day weathering, white powders and small shrinkage cracks appeared on samples with w/c = 6:1, and the surface turned almost white. However, at the same time, fewer white powders and cracks appeared only on the top surface of HWM with w/c = 3:1, and the surface of the other parts had no change. After 10-day weathering, no obvious cracks appeared on the surface of samples with w/c = 3:1, and the surface structure was still very tight and strong; while the surface structure of samples with w/c = 6:1 became loose and fully covered with tiny shrinkage cracks. It shows that the weathering degree of samples with w/c = 6:1 is significantly higher than that of 3:1.

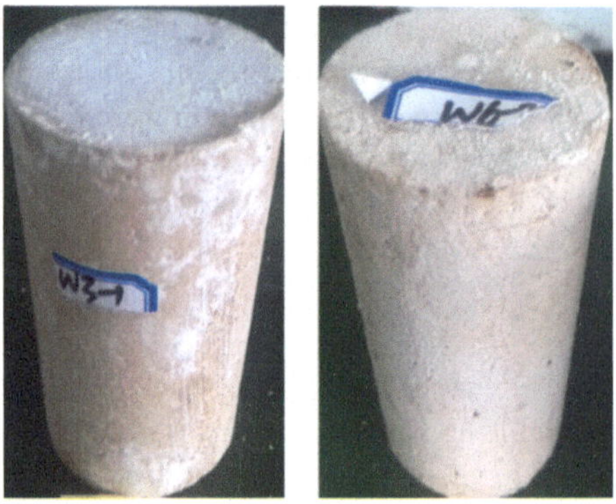

Fig. 2. Photos of HWM with w/c = 3:1 (left) and 6:1 (right) after 4-day weathering.

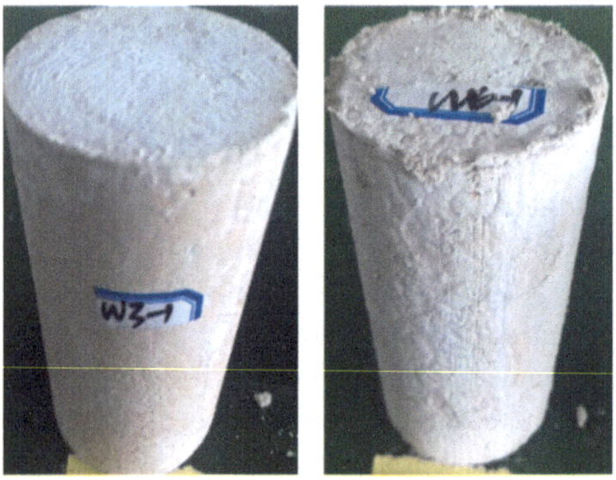

Fig. 3. Photos of HWM with w/c = 3:1 (left) and 6:1 (right) after 10-day weathering.

3.2 Variations of Water Content and Density of HWM During Weathering

During the weathering process, weights of HWM samples were measured in real-time and water contents were then calculated. Figure 8 shows the water content of HWM samples during weathering. Water content of two group of samples decreased rapidly with time. HWM of w/c = 6:1 has a higher water content than that of 6:1. After 10-day weathering, water contents of samples of w/c = 3:1 and 6:1 decreased from the initial 74.0% and 84.6% to 30.7% and 34.4%, respectively.

Figure 9 shows the rate of water loss of samples with w/c = 3:1 and 6:1 with weathering time. It indicated that the initial water loss rates of the two groups were small (i.e. lower than 2%/day). In the first 4 days, the rates of both two groups kept stable. On the 4th day, the water loss rates of samples with w/c = 3:1 and 6:1 increased to 2.2% and 1.8% per day, respectively. After that, the rates of the two groups increased rapidly. The rate of w/c = 6:1 sharply rose to 16.8%/day on the 10th day. The rate of w/c = 3:1 increased from 2.2%/day on the 4th day to 8.9%/day on the 8th day. Afterward the water loss rate slowed down and dropped to 6.9%/day on the 10th day.

Comparing Fig. 8 and Fig. 9, in general, in the natural air-weathering conditions, the water loss rates of samples with w/c = 6:1 were much faster and easier than that of 3:1.

There are two main reasons: Firstly, HWM with w/c = 6:1 have a higher moisture content and more free water inside, and the free water dissipates faster and more easily in natural air-weathering conditions. Secondly, microscopically, HWM with a higher w/c ratio has fewer Ettringite crystals, a looser network structure, and larger and more internal pores inside, and the free water inside the sample is easier to lose.

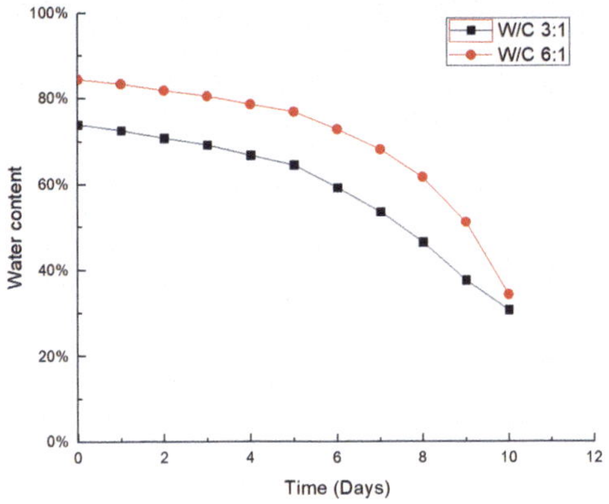

Fig. 4. Water content of HWM with weathering duration.

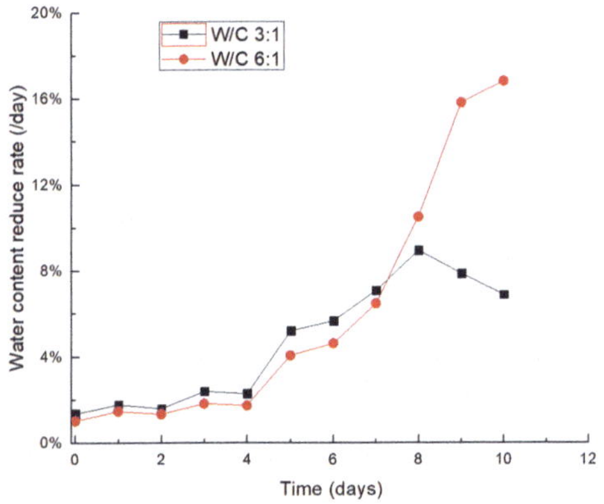

Fig. 5. Water content reducing rate of HWM with weathering duration.

3.3 Measured Variations of Compressive Strength

Uniaxial compressive strengths of HWM samples with various weathering duration were measured. Figure 6 shows the test results. It indicated that strengths of HWM samples of both w/c = 3:1 and 6:1 gradually decreased with weathering duration. Initial uniaxial compressive strengths of samples with w/c = 3:1 and 6:1 were 1.68 MPa and 0.40 MPa respectively. In the first 4 days, strengths of all samples decreased gradually. The strength decreasing rates of samples with w/c = 3:1 and 6:1 were 0.01 and 0.02 MPa per day respectively. On the 4th day, the strengths of samples with w/c = 3:1 and 6:1 decreased to 1.64 MPa and 0.32 MPa respectively. After that, the strength decrease rates of the two groups accelerated slightly. The strength decrease rates of samples with w/c = 3:1 and 6:1 were 0.04 and 0.03 MPa per day respectively. On the 10th day, the strengths of samples with w/c = 3:1 and 6:1 decreased to 1.40 MPa and 0.13 MPa respectively, which means that HWM with w/c = 3:1 could still maintain about 80% of the strength after 10-day weathering, while samples with w/c = 6:1 could still maintain about 30% of the strength under the same weathering conditions. It indicates that a higher w/c ratio helps to promote the weathering resistance of HWM.

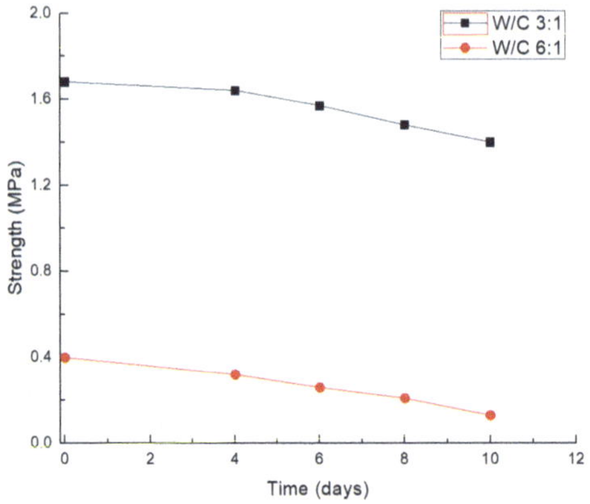

Fig. 6. Measured variations of strength of HWM with weathering duration.

In the uniaxial compression test, the stress-strain curve of the sample can reveal the whole process of the deformation of the sample and its development to failure under the external load. Figure 7 compared stress-strain curves of samples with w/c = 3:1 and 6:1 after 10-day weathering.

According to the classification standard of rock mechanics, the stress-strain curve of HWM can be divided into four stages: namely, the compression stage (Stage I), the elastic stage (Stage II), the broken stage (Stage III), and the post-peak residual stage (Stage IV).

In Stage I, pores and cracks in HWM samples would be compressed and closed under loading. Stage I of samples with w/c = 3:1 was very short, and the strain was quite small. However, that of w/c = 6:1 was longer and the final strain was about 0.01, which was much larger than that of w/c = 3:1. It indicates that the pores and cracks in samples with w/c = 6:1 are much larger than that of 3:1.

In Stage II, the stress-strain curve of samples with w/c = 3:1 was extremely steep and short. Its elastic modulus was about $E = 0.066$ GPa. The strain at the end of Stage II was about 0.04, and the corresponding stress was 1.40 MPa, which was almost equal to the peak stress. As for samples with w/c = 3:1 in Stage II, the stress-strain curve was relatively gentle, and the elastic modulus was about only $E = 0.0015$ GPa. The strain at the end of Stage II was about 0.05, and the corresponding stress was 0.10 MPa.

In Stage III, the partical stress concentration increased significantly. Stage III of samples with w/c = 3:1 was extremely short. The σ-ε curve increased sharply and reached the peak stress (i.e. 1.41 MPa). Then the samples with w/c = 3:1 got failure suddenly. While Stage III of samples with w/c = 6:1 was much longer and the strain increased from ε = 0.05 to 0.75, and its peak stress was σ = 0.35 MPa.

In Stage IV, the stress of samples with w/c = 3:1 dropped sharply from the peak stress of 1.41MPa (ε = 0.02) to 0.76MPa (ε = 0.05). Then samples entered the residual strength stage, and the residual strength was about 30%-40% of the peak stress, and the corresponding strain range was ε = 0.1~0.35. As for samples with w/c = 6:1, the stress decreased slightly with the increase of strain, the residual strength maintained between 0.08–0.1MPa, which was about 60%-75% of the peak stress, and the corresponding strain range was 0.1–0.4.

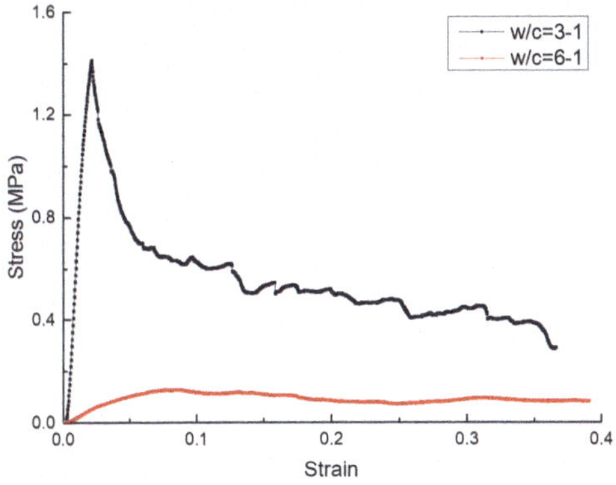

Fig. 7. Stress-strain curves of weathered HWM with w/c 3:1 and 6:1.

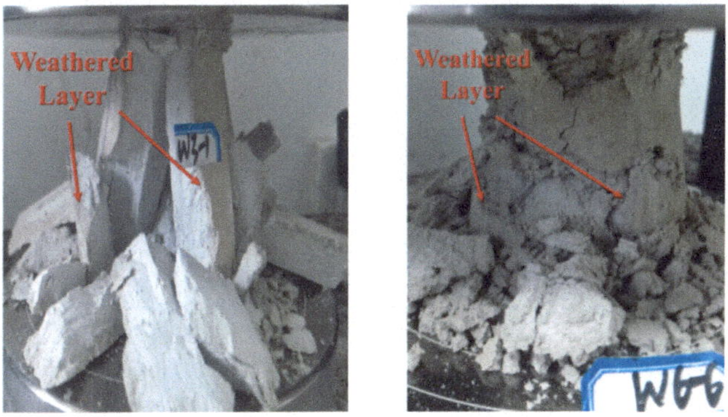

(a) w/c=3:1 (b) w/c=6:1

Fig. 8. Failure modes of HWM with w/c = 3:1 and 6:1 after 10-day weathering.

Figure 8 shows the photos of the compression failure mode of HWM samples with w/c = 3:1 and 6:1 after 10-day weathering. It indicated that the failure mode of samples with w/c = 3:1 was a split failure. The specimen was split into several pieces along the axial direction from the inside, and then the strength dropped rapidly, which was consistent with its stress-strain curve. While the failure mode of samples with w/c = 3:1 was a ductile failure. Under the axial loading, the axial strain of the sample continued to increase, and the sample was compressed vertically and expanded laterally. As the axial compression increased, chalking occurred on the sample, and the loose weathered layer on the outer surface was constantly peeling off. The sample looked like a "compressed biscuit". At this time, although the sample had been destroyed, its residual stress still maintained a high level. This was consistent with the stress-strain curve of the specimen (see Fig. 7).

3.4 Microstructure of HWM

Figure 9 and Fig. 10 are SEM images (×1000 times) of HWM with w/c = 3:1 and 6:1 respectively. As shown in Fig. 4, the hydration product of HWM is dominated by Ettringite crystals, whose microscopic morphology is needle-like, rod-like, or network-like structure, and mainly rod-like. The Ettringite crystals stagger and connect, and finally form a dense network structure, which plays the role of skeleton and support. Fibrous hydrated silica gel and pom-shaped aluminum glue are filled in the network structure, which results in few pores and a high density of HWM. Due to the staggered and connected network structure of Ettringite, there is a significantly larger contact surface and more inner pores for HWM than other cementing materials, which results in the high ability to absorbe water (i.e. normally several times of water than its own mass). It could explain the reasons for the high water-cement ratio and even super high water-cement ratio of HWM from the point of view of microstructure.

(a) w/c=3:1 (b) w/c=6:1

Fig. 9. SEM (×1000) of HWM with w/c = 3:1 and 6:1.

Compared with Fig. 9, the microstructure of HWM with w/c = 6:1 has much fewer Ettringite crystals, a looser network structure, and more internal pores, which helps samples with w/c = 6:1 could absorb more water and own high water content from the microstructural view. Furthermore, the Ettringite crystals in samples with w/c =

6:1 are significantly slender. There are also much less fibrous hydrated silica gel and pom-shaped aluminum glue filled in the network structure. Therefore, the density of the microstructure in Fig. 5 is much smaller, and the microstructure has more pores.

3.5 Chemical Components of Weathering Product of HWM

As shown in Fig. 8, after 10-day weathering, a 5- and 8-cm thick weathered layer appears on the surface of HWM with w/c = 3:1 and 6:1 respectively. The outer surface of HWM samples with w/c = 6:1 was full of shrinkage cracks, and the structure became extremely loose, which resulted in losing almost all the strength.

Chemical components of weathering product of HWM were determined by XRD analysis as shown in Table 2. The main components are calcium carbonate ($CaCO_3$) and calcium carbonate ($CaSO_4$), which indicates that the Ettringite in the HWM has been carbonized into minerals such as gypsum and calcite.

Table 2. Compositions of weathered HWM.

Material Components	Semi-Quantitative
$CaCO_3$	47%
$CaSO_4$	53%

4 Discussions

Sun and Song (1994) reported that the degradation of HWM during weathering is mainly due to two aspects: loss of water and carbonization of Ettringite [14].

1) Loss of water during weathering process

Water inside HWM is mainly divided into free water and Ettringite crystal water. Free water is the main water type and it is also the largest component in HWM. Free water is mainly stored in pores and gaps of Ettringite crystals. During the weathering process, free water would quickly volatilize along internal pores of HWM, causing loss of water rapidly. The loss rate of free water depends on the internal pores of HWM [1].

Comparing Figs. 9 and 10, HWM of higher w/c ratio have looser inner structure and more internal pores. This is consistent with the results in Figs. 4 and 5, namely HWM of higher w/c ratios have a lower density, higher water content, and faster water loss rate under natural weathering. With water lose, the internal structure of HWM becomes looser and its compressive strength also decrease rapidly.

Therefore, HWM of a lower w/c lost fewer water due to fewer inside pores and gaps during the weathering process, hence having a better weathering resistance.

2) Carbonization of Ettringite during weathering process

During weathering, CO_2 in air diffuses through the inner pores and dissolves in free water in the inner pores of HWM. The Ettringite crystals are easily corroded by CO_2 in the pore solution, and carbonization reaction occurs, then calcium carbonate is formed. The carbonization reaction follows Eqs. (2) and (3).

$$3CA \cdot SO_4 + 8CaSO_4 \cdot 2H_2O + 6CaO + 80H_2O \rightarrow 3C_3A \cdot 3CaSO_4 \cdot 32H_2O \quad (2)$$

$$3CA \cdot SO_4 + 8CaSO_4 \cdot 2H_2O + 6CaO + 80H_2O \rightarrow 3C_3A \cdot 3CaSO_4 \cdot 32H_2O \quad (3)$$

The carbonization reaction causes the microscopical and macroscopical degradation of HWM. During weathering, crystal lattices of Ettringite became longer and the macromolecular structure turn into a small one, which results in volume swelling. Besides, HWM stones become loose due to loss of free water. Eventually, HWM loses its strength and even becomes chalking.

According to Xie et al. (2014), the speed of carbonization of HWM highly depends on the inner porosity. HWM with a lower w/c has denser inner structure and fewer internal pores, which means that diffusion of CO_2 inside HWM becomes slower and the carbonization speed becomes slower, hence the anti-weathering ability is also stronger.

Besides, $CaCO_3$ and $CaSO_4$ produced by the carbonization reaction are insoluble minerals, which would block the gel pores and capillary pores inside HWM, thereby delaying or even preventing further carbonization inside the stone body. This is the inherent reason that the weathering resistance of HWM with w/c = 3:1 is far better than that of w/c = 6:1.

Based on the above discussions, it can be concluded that degradation of HWM during weathering highly depends on the w/c ratio, and HWM with a lower w/c ratio has a better weathering resistance.

5 Conclusions

In this study, HWM samples of w/c = 3:1 and 6:1 were prepared and conducted to a 10-day indoor natural weathering test. Microstructure, chemical components, and physical and mechanical characteristics of HWM were investigated. The following conclusions were obtained:

(1) The main component of HWM is Ettringite ($3CaO{\bullet}Al_2O_3{\bullet}3CaSO_4{\bullet}32H_2O$). HWM with a higher w/c ratio has higher water content, lower density, fewer Ettringite crystals, looser and more porous microstructure, and higher uniaxial compressive strength.
(2) The degradation of HWM during weathering is mainly due to the loss of water and carbonization of Ettringite. Free and crystal water loses rapidly and the carbonization reaction of Ettringite by CO_2 in air occurs. The weathering products of HWM are $CaCO_3$ and $CaSO_4$.
(3) The degradation of HWM during weathering highly depends on the w/c ratio. HWM with lower w/c has a denser inner structure and fewer internal pores, which means that diffusion of CO_2 inside HWM becomes slower and the carbonization speed becomes slower.
(4) HWM with a lower w/c ratio has a better weathering resistance, and loses its water and strength easier during weathering. After 10-day weathering, HWM with w/c = 3:1 and 6:1 lost 20% and 70% of strength respectively.

This study revealed the degradation mechanism of HWM during weathering. The findings of this research provide insights for engineers to design potential engineering measures to protect HWM from weathering. It is suggested to conduct further research on the anti-weathering properties of HWM in order to improve the service performance of HWM in practice.

Acknowledgement. This research was substantially supported by the Natural Science Foundation of China (Grants No. 51369001) and the Geological Survey Program of China Geological Survey (Grants No. DD20190269, DD20221782).

References

1. Sun, H., Song, C.: Application of high-water solidifying materials. China University of Mining and Technology Press. Xuzhou, China (1994)
2. Peng, M., Jiang, J., Zhang, X.: Effect of composition on the performance and microstructures of mining high-water solidified materials. Min. Eng. Res. **26**(3), 56–59 (2011)
3. Xie, S., Wang, E., Chen, D., Li, H.: Stability analysis and control technology of gob-side entry retaining with double roadways by filling with high-water material in gently inclined coal seam. Int. J. Coal Sci. Technol. **9**(1), 1–1 (2022)
4. Hu, H., Cui, M.: Characteristics of high water content hardening body and analysis on the mechanical mechanism of fill body. Min. Res. Dev. **21**(5), 23–26 (2001)
5. Zhang, F., Wang, X., Bai, J.: Post-peak mechanical characteristics of the high-water material for backfilling the gob-side entry retaining: from experiment to field application. Arab. J. Geosci. **13**, 386 (2020)
6. Yan, Z., Zhu, Z.: Research on the mechanical characteristics of the consolidated soft soil mixed with the high-water-content and quick-setting materials. J. S. China Univ. Technol. (Natural Science Edition). **28**(06), 79–84 (2000)
7. Yan, Z.: SEM study of microstructure of soft soil mixed with high-water-content and quick-setting materials. Rock Soil Mech. **25**(02), 275–278+282 (2004)
8. Hou, L., Tang, J., Jiang, K., Wang, L.: Basic characteristics of superabsorbent polymers and its application in port engineering. Port Waterw. Eng. **05**, 54–56 (2007)
9. Hou, L.: Applied experiments of superabsorbent polymers on repaired and strengthened calcareous sands island. Concrete **12**, 122–124 (2012)
10. He, T., Liu, C., Ye, D., Wang, Y.: Experimental study on environmental effects on the compressive strength of the high-water-content and quick-setting stone. Adv. Eng. Sci. **46**(Z1), 86–91 (2014)
11. Wang, D., Zeng, J., Liu, C., Xu, Y.: Creep law and rheological model of high-water rapid-setting material in water environment. Adv. Eng. Sci. **46**(3), 1–6 (2014)
12. Song, C., Cheng, X., Wang, Z.: Weathering Mechanism of Ettringite. J. Univ. Sci. Technol. Beijing **05**, 459–461 (1999)
13. Xie, H., Liu, C.: Analysis on influence of moisture content on deformation characteristics of the high-water-content material stone. J. Sichuan Univ. Eng. Sci. Ed. **45**(S1), 1–6 (2013)
14. Xie, H., Liu, C.: Analysis on raw material of high-water and quick-setting material stone. Sci. Technol. Eng. **14**(30), 254–257 (2014)

Seepage Treatment in the Structural Transverse Joint Between Dam Sections 12# and 13# at Jinghong Power Plant

Wenhua Jiang(✉)

Shanghai Jiao Tong University Underwater Engineering Institute Co., Ltd., Shanghai City, China
13795269037@163.com

Abstract. The project addresses the issue of seepage in the transverse joint of the dam at Jinghong Power Plant through a comprehensive 50-day treatment plan. The project is divided into five critical phases: construction preparation, underwater inspection, treatment of the upstream face of the transverse joint, drilling and grouting within the gallery, and final acceptance and equipment organization. The construction site is located upstream of the dam, with the on-site project team and personnel accommodations arranged in nearby houses to ensure convenience and safety for the workers. Overwater operations will be conducted from a 10*10 m construction platform, which is movable and positioned by ropes fixed to the dam, facilitating rapid deployment of personnel and equipment. The underwater work platform is constructed from 10# channel steel with rollers for flexible movement on the dam surface. The detailed construction process for treating the seepage in the transverse joint includes meticulous planning from the preparation phase, through underwater re-inspection and crack marking, opening a "U" shaped groove, drilling grouting holes, sealing the joint surface, conducting air pressure tests, grouting, removing grouting pipes, installing SR waterproof modules, and finally, underwater camera inspection for acceptance, ensuring each step meets the expected technical standards. Drilling and grouting operations within the gallery follow a sequence from low to high and from upstream to downstream to ensure the continuity and integrity of the grouting effect.

Keywords: Seepage · Power Plant · Transverse joint · Underwater

1 Introduction

In the great practice of harmonious coexistence between humans and nature, dams, as outstanding representatives of hydraulic engineering, not only demonstrate human's ability to transform nature but also reflect a profound understanding and respect for natural laws. However, with the passage of time, the transverse joints in dam structures have gradually revealed their vulnerability, with seepage issues acting as hidden dangers that constantly test the durability and safety of dams. Seepage through transverse joints can not only lead to structural damage in dams, affecting their normal functions, but also potentially trigger secondary disasters, profoundly impacting the ecological balance

D. Jeng and B. Cai (Eds.): ICCOE 2024, LNCE 423, pp. 504–515, 2025.
https://doi.org/10.1007/978-981-96-0181-3_45

downstream and the lives of the people [1, 2]. Therefore, the research and practice of handling seepage in dam transverse joints are not only significant topics in the field of engineering technology but also crucial for ensuring water resource security and maintaining social stability.

Globally, the issue of seepage in dam transverse joints has been a research hotspot in the field of hydraulic engineering. Internationally, developed countries and regions such as the United States and Europe have accumulated rich experience in the construction and maintenance of large dams. For example, institutions such as the US Bureau of Reclamation and the European Commission on Large Dams have conducted in-depth research on the prevention and treatment of seepage in dam transverse joints, forming a series of standards and norms [3, 4].

These research outcomes have not only enhanced the safety performance of dams but also provided valuable experience for global dam engineering.In China, with the rapid development of hydraulic engineering construction, the technology for handling seepage in dam transverse joints has also made significant progress, especially in major projects such as the Three Gorges Dam and the Xiaolangdi Water Control Project, where the application of seepage treatment technology has achieved good results [5, 6]. Research institutions such as the Chinese Academy of Hydraulic and Hydroelectric Sciences and the Yangtze River Water Resources Commission, as well as universities such as Tsinghua University and Hohai University, have conducted extensive research in the field of seepage treatment in dam transverse joints, promoting the innovation and development of China's hydraulic engineering technology [7, 8].

This article aims to delve into the causes, impacts, and treatment methods of seepage in dam transverse joints. By analyzing the current research status both domestically and internationally, and by combining the unique characteristics of seepage in different types of dam transverse joints, we propose targeted treatment strategies and technical solutions. Initially, we will explore the primary causes of seepage. These include material aging, which leads to the degradation of the dam's structural components over time, compromising their ability to effectively seal transverse joints. Structural deformation, another significant factor, can occur due to the differential settlement of the dam foundation or the varying loads exerted by the water body. Temperature changes can also induce thermal expansion and contraction, causing cracks and openings in the joints. Additionally, poor construction practices and inadequate maintenance can exacerbate these issues, leading to increased seepage. The impacts of seepage on dam structures and their functions are profound and multifaceted. Seepage can weaken the structural integrity of the dam, reducing its ability to withstand hydrostatic pressures and increasing the risk of failure. It can also lead to internal erosion, known as piping, which progressively enlarges seepage paths and further undermines the dam's stability. Furthermore, seepage can cause the saturation and weakening of the dam's foundation materials, thereby affecting the overall stability of the structure. Through practical case studies, we will analyze these impacts in detail, highlighting the critical need for effective seepage management.

To address these challenges, we propose several targeted treatment strategies and technical solutions. These may include the use of advanced materials for joint sealing, such as expansive waterproof grout and hydrophilic polyurethane, which can adapt to movements within the dam structure and provide long-lasting seals. Structural reinforcement techniques, such as the installation of additional drainage systems and the use of geosynthetic materials, can help manage and redirect seepage flows. Regular monitoring and maintenance regimes, incorporating modern technologies like remote sensing and automated monitoring systems, are essential for early detection and timely intervention. In addition to these technical solutions, we emphasize the importance of an integrated approach that combines engineering practices with environmental and socio-economic considerations. Effective management of seepage in dam transverse joints not only enhances the structural safety and reliability of dams but also ensures the protection of downstream communities and ecosystems. By promoting sustainable dam management practices, we can contribute to the broader goal of sustainable development.

The issue of seepage in dam transverse joints is a complex and multifaceted challenge that requires a comprehensive and multidisciplinary approach. By continuing to advance our understanding through research and applying innovative solutions in practice, we can effectively control seepage and enhance the safety and reliability of dam engineering. This, in turn, provides a solid foundation for the sustainable development of human society, ensuring that our water resources are managed responsibly and efficiently for future generations.

2 Methods

2.1 Overview and Inspection of Transverse Joint Seepage

The 12# dam section at Jinghong Power Plant is an overflow section (dam longitudinal 0–014.000 to 0–034.000) featuring the No. 7 overflow surface opening and the No. 5 diversion bottom opening. The surface opening measures 15.0 m × 21.0 m (width × height), while the bottom opening is 8.0 m × 14.0 m (width × height), with a base elevation of 534.00 m. A gate slot for closure is installed after the inlet section, which is sealed during reservoir impoundment. The overflow weir surface adopts a WES curve, with the crest elevation at 581.000 m, connecting to a stilling basin at the downstream reverse arc end. The 13# dam section, spanning 23.5 m (dam longitudinal 0 + 009.500 to 0–014.000), is located between the powerhouse and overflow sections, serving as the right sediment flushing bottom opening section. The inspection scope covers an elevation range from EL.551 m to EL.581 m, and extends from the transverse joint between the 11# and 12# dam sections to the left wall of the No. 7 overflow surface opening in the lateral direction. In the upstream-downstream direction, the area extends from the upstream dam face to the maintenance gate of the No. 7 overflow surface opening, see Fig. 1. The inspection utilizes remotely operated vehicles (ROVs), two-dimensional imaging sonar, and tracer ink cartridges to conduct underwater checks on the transverse joints and surrounding areas of the 12# and 13# dam sections.

2.2 Performance of Materials for Transverse Joint Seepage Treatment

Commonly used materials for treating seepage in dam transverse joints encompass sealants and grouting materials. Examples of sealants include hardy substances such

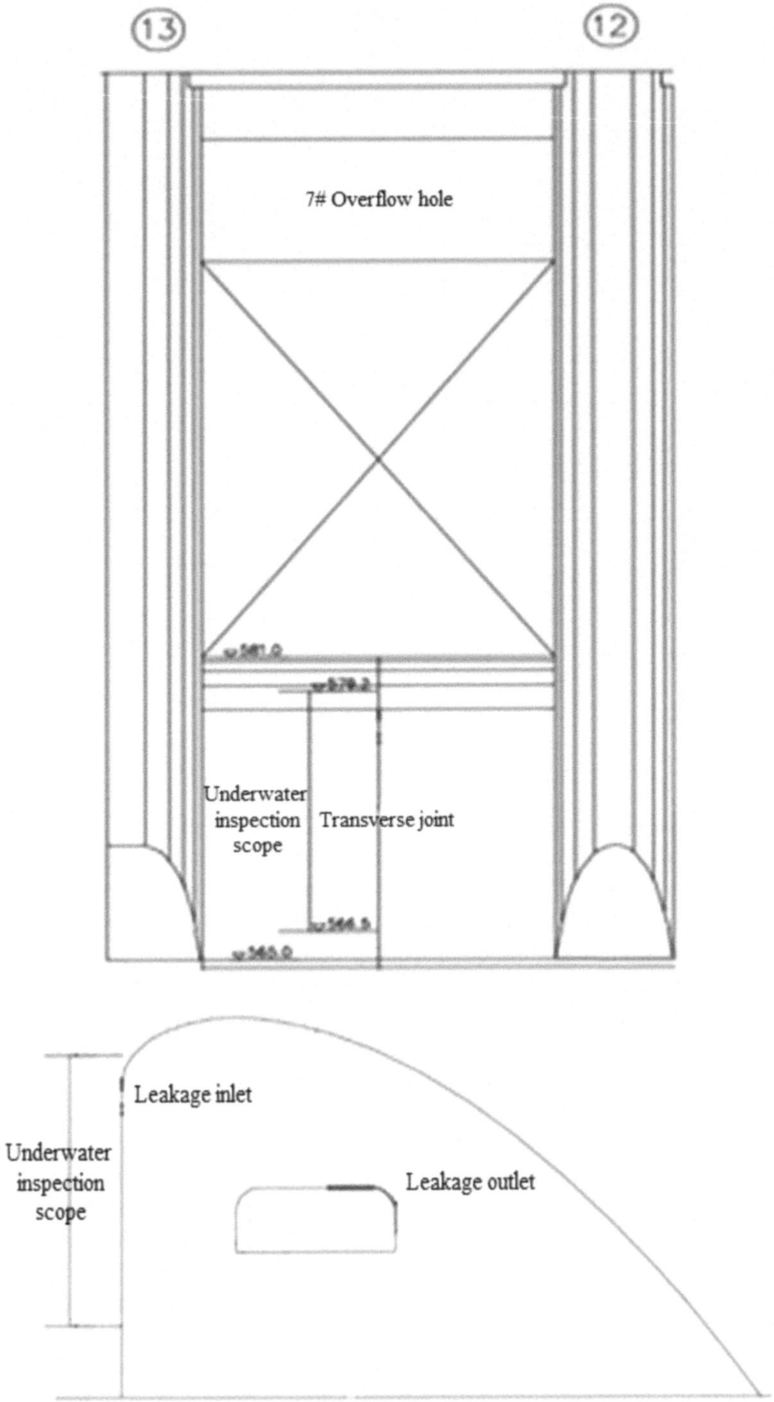

Fig. 1. Schematic of Dam Inspection Scope.

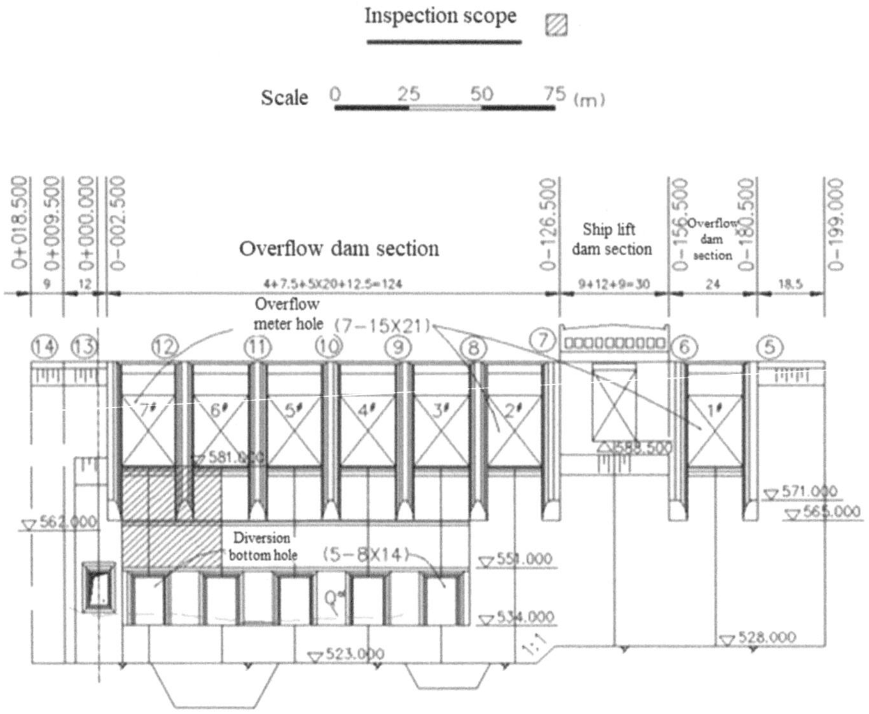

Fig. 1. *(continued)*

as polyurethane or epoxy resins. These materials are highly effective in their ability to fill and seal small cracks, preventing water infiltration which could lead to larger structural issues over time. These sealants have great adhesive properties, ensuring that they form a tight bond with the surrounding material, thus providing a reliable defense against water penetration. On the other hand, grouting materials are another category of substances frequently used in this context. These encompass a variety of substances such as cement grout, chemical grout (for instance, silicate grout), and polymer grout. These materials are typically injected under high pressure into the cracks, where they solidify and form a formidable barrier against further water penetration. The process of injecting these grouting materials not only helps to fill the crack but also compact the surrounding material, thereby increasing the overall structural integrity of the dam. The selection of appropriate treatment materials is not a decision to be taken lightly. Several factors must be taken into consideration such as the severity of the seepage, the material and structural type of the dam, and the prevailing environmental conditions. It's vital to select materials that are compatible with the existing structure and can withstand the specific environmental conditions of the dam's location. All these factors, when carefully considered, will ensure the safety and stability of the dam structure, thereby prolonging its useful life and keeping the associated human settlements and ecosystems safe.The chemical grouting materials and plastic waterproofing materials are detailed in Table 1 and 2.

Table 1. Performance indicators for chemical grouting materials.

Material	Item	Indicator
LW Water-Soluble Polyurethane	Grout Density (g/cm^3)	1.05 ± 0.05
	Grout Viscosity (mPa·s)	150 ~ 400
	Gel Time (s, Grout:Water = 1:10)	≤1.5
	Water Entrainment (times)	≥25
	Swelling Rate in Water (%, 28d)	≥100
	Tensile Strength (MPa)	≥1.8
	Elongation at Break (%)	≥80
HW Water-Soluble Polyurethane	Grout Viscosity (mPa·s)	75 ± 20
	Grout Density (g/cm^3)	1.1 ± 0.05
	Gel Time (s, Grout:Water = 1:10)	≤30
	Bond Strength (wet surface) (MPa)	≥2.0
	Compressive Strength (MPa)	≥20
	Water Swelling Multiplier (%)	2~4

Table 2. Key performance indicators for plastic waterproofing materials.

<table>
<tr><th>NO</th><th colspan="3">Item</th><th>Indicator</th></tr>
<tr><td>1</td><td colspan="3">Density (g/cm^3)</td><td>≥1.15</td></tr>
<tr><td>2</td><td colspan="3">Workability (Penetration) 0.1 mm</td><td>≥80</td></tr>
<tr><td>3</td><td colspan="3">Flowability (Sag) mm</td><td>≤2</td></tr>
<tr><td rowspan="6">4</td><td rowspan="6">Tensile Bond Performance</td><td rowspan="2">Ambient Temperature and Dry Conditions</td><td>Elongation at Break (%)</td><td>≥250</td></tr>
<tr><td>Failure Mode</td><td>Cohesive Failure</td></tr>
<tr><td rowspan="2">ow Temperature and Dry Conditions (−20 °C)</td><td>Elongation at Break (%)</td><td>≥50</td></tr>
<tr><td>Failure Mode</td><td>Cohesive Failure</td></tr>
<tr><td rowspan="2">Freeze-Thaw Cycling (300 Cycles)</td><td>Elongation at Break (%)</td><td>≥250</td></tr>
<tr><td>Failure Mode</td><td>Cohesive Failure</td></tr>
<tr><td>5</td><td colspan="3">Permeability Resistance (MPa)</td><td>≥1.5</td></tr>
<tr><td>6</td><td colspan="3">Flow Stoppage Length (mm)</td><td>≥135</td></tr>
</table>

2.3 Key Construction Machinery and Equipment

In the demanding task of treating seepage in dam transverse joints, diving equipment is crucial, serving as both a bridge to the underwater world and a lifeline for divers. Its safety and reliability directly influence the success of the treatment and divers' safety, especially in deep or high-pressure conditions. The equipment must offer a stable life support system, resist water pressure, and adapt to variable underwater conditions. Precision and operability are essential for delicate tasks like joint inspection and reinforcement. Modern diving gear, incorporating advanced technologies like communication and navigation systems, enhances underwater efficiency and safety, supporting effective seepage treatment in dam transverse joints [9, 10]. The following Table 3 gives a summary of machinery and equipment in this project.

Table 3. Machinery and equipment.

NO	Machinery Name	Model	Quantity
1	Diving Equipment	KMB-28	3
2	Air Compressor	3 m^3/min Electric	1
3	Air Compressor	0.6 m^3/min Electric	1
4	Air Compressor	40MPa/100L	1
5	Diving Cylinders	12L	4
6	Underwater Color Camera	SXD-IV BFK	2
7	Dive Depth Gauge	Scubapro	3
8	Decompression Chamber	Double Chamber, 4 Doors	1
9	Air Storage Tank	6m^3	1
10	Diesel Generator	5KW	1
11	High-Pressure Cleaning Pump	20MPa	2
12	Grouting Pump	Electric Dual Liquid	2
13	Hydraulic Station	HPP18EF	1
14	Hydraulic Chain Saw	HDC13C	2
15	Hydraulic Pick	HH15	2
16	Hydraulic Drill	HRD10	1
17	Hoist	2T	1
18	Surface Work Platform	10 m*10 m	1

3 Results

3.1 Leak Points and Leakage Volume

At the dam Sect. 12#–13#, five leak points were identified within the transverse joint elevation range of EL.576.5 m–578.1 m, specifically at elevations EL.576.5 m, 576.9 m, EL.577.7 m, 577.9 m, and 578.1 m. The leakage volume in the EL.570.0 gallery of the transverse joint at dam Sects. 12#–13# is approximately 5.36L/s, which is relatively small. Given the presence of multiple leak entrances at the dam front, the individual leak inflow is even smaller, making it difficult to observe ink-sucking phenomena at some leak points. It is speculated that there may be other minor leak entrances within the elevation range of EL.576.5 m–578.1 m at dam Sects. 12#–13#. Additionally, the appearance of the transverse joint at elevations EL.566.5 m–579.2 m is poor, with multiple local damages and openings. To ensure comprehensive sealing of subsequent leak entrances, it is recommended to expand the sealing range of the transverse joint to EL.566.5 m–579.2 m.

No significant water seepage was observed in the foundation galleries (including curtain grouting gallery, first foundation drainage gallery, and second foundation drainage gallery) near the transverse joint at dam Sects. 12#–13#. The drainage from the drainage holes in the foundation gallery of the dam sections on both sides of the joint was 0 L/s, 0.06L/s, and 0 L/s (see Table 2.2.2–1), indicating very little foundation drainage. The total drainage volume from the dam body on both sides of the joint was essentially 0 L/s. The WES weir crest flow surface, after grouting treatment, showed no signs of leakage. From the on-site inspection of the flow surface of the reverse arc section at the power plant, although there was water seepage at the joint, the seepage was not significant. It is possible that the dam body upstream surface transverse joint seal, the surrounding seal strips of the transverse joint at EL.570 m gallery, are locally damaged or the concrete near the seal strips was not densely poured, forming a seepage pathway. This gradually fills the internal seal of the transverse joint below the EL.570 m gallery, leading to seepage points appearing in the EL.570 m gallery.

3.2 Upstream Face Treatment of Transverse Joints

The treatment involves cutting slots along the joint surface, drilling grouting holes, embedding grouting pipes, sealing, and conducting chemical grouting. Polyurethane grouting material is injected to fill and seal the leaking joint surface. After grouting, the slots are cleaned again, filled with SR sealing material, and covered with anti-seepage SR cover plate modules, which are securely fixed to establish the entire waterproof system on the joint surface. The treatment top elevation is EL.579.2 m, and the bottom elevation is EL.566.5 m. Since the reservoir reached near its normal water level at the end of 2008, the water level has fluctuated between 590.97 m and 602.14 m, with an annual maximum fluctuation of 10.58 m. The upstream face treatment of the transverse joint is underwater construction, with a maximum treatment head of 35.5 m. Underwater construction uses a floating raft as a support work platform, equipped with corresponding diving equipment, air compressors, and specialized construction equipment. A submerged operating basket serves as a work platform for divers, with a winch on the support platform controlling the basket elevation, allowing divers to complete their tasks within the basket. The main construction process for transverse joint leakage treatment includes: construction preparation → underwater review and crack marking → opening "U" shaped slots → drilling grouting holes → joint surface sealing → air pressure test → grouting → removal of grouting pipes → installation of SR anti-seepage modules → underwater camera inspection and acceptance. The standard cross-section of the transverse joint surface waterproof system is shown in Fig. 2.

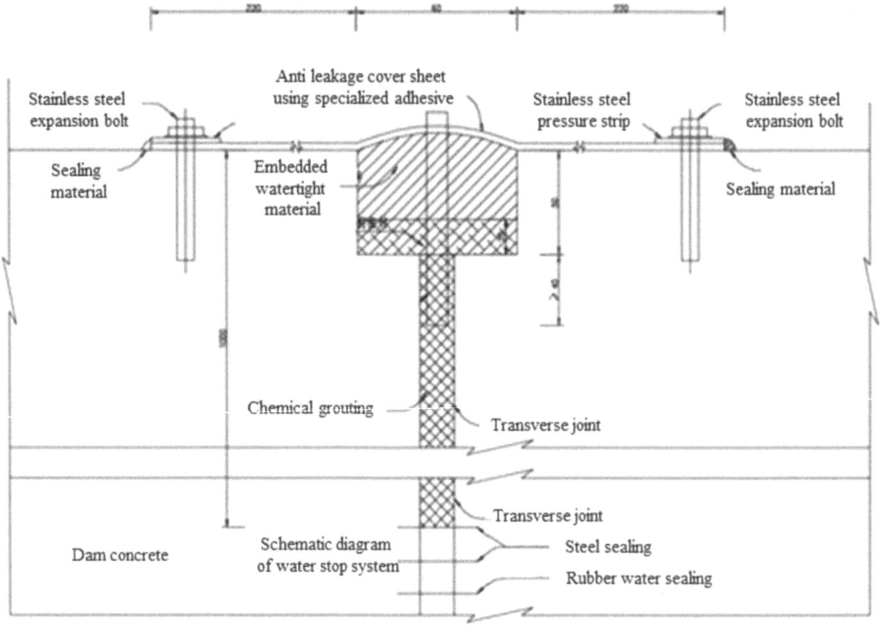

Fig. 2. Standard Cross-Section of the Transverse Joint Surface Waterproof System.

3.3 Drilling and Grouting in the Gallery

The treatment elevation ranges from EL.566.5 m to 579.2 m. Above elevation 570 m, inclined or horizontal holes are drilled in the gallery, while below this elevation, downward inclined holes are drilled. The goal is to distribute the drilling locations uniformly across the transverse joint. Two rows of grouting holes are arranged on either side of the joint, with one side having holes 1.9 m deep (first-layer grouting holes), and the other side adjusting the hole depth based on the joint depth, ensuring the holes penetrate the crack by at least 50 cm. Post-treatment, pressure water tests and core sample inspections are conducted on the dam's transverse joints (5% of the grouting holes are used for this). The permeability rate (Lu) should not exceed 1Lu, and the core samples must meet requirements, showing complete integrity and dense filling of chemical grout within the cracks. After a certain period following grouting, the embedded sealing materials are removed, and the surface is treated with epoxy mortar spray. The main process sequence includes: joint inspection → drilling grouting holes → slotting, drilling drainage and pressure relief holes, pipe embedding, and joint sealing → using borehole TV to determine joint positions, installing grouting pipelines, and sealing holes → connectivity check → chemical grouting → cutting pipes, sealing holes → surface cleaning → quality inspection. Detailed Drilling Parameters in Table 4.

Table 4. Drilling parameters.

NO	Dam Transect	Dam Longitude	Elevation at Orifice (m)	Depth (m)	Azimuth (°)	Inclination (°)
1	D0 + 006.000	0–015.500	571.000	3.110	35	0
2	D0 + 006.000	0–013.300	570.700	1.900	330	0
3	D0 + 006.000	0–015.500	571.750	5.632	35	62
4	D0 + 006.000	0–013.300	571.450	1.900	330	62
5	D0 + 007.250	0–013.300	572.800	4.530	270	80
6	D0 + 007.550	0–014.700	572.800	1.900	90	60
7	D0 + 009.250	0–013.300	572.800	4.530	270	80
8	D0 + 009.550	0–014.700	572.800	1.900	90	60
9	D0 + 011.250	0–013.300	572.800	4.530	270	80
10	D0 + 010.950	0–014.700	572.800	1.900	90	60
11	D0 + 013.250	0–013.300	572.800	4.530	270	80
12	D0 + 012.950	0–014.700	572.800	1.900	90	60
13	D0 + 014.500	0–012.500	571.750	4.600	200	62
14	D0 + 014.500	0–014.700	571.450	1.900	150	62
15	D0 + 014.500	0–013.300	571.000	4.772	201	0
16	D0 + 014.500	0–014.700	570.700	1.900	150	0
17	D0 + 014.500	0–012.500	570.000	6.000	200	−45
18	D0 + 014.500	0–014.700	570.300	1.900	150	−45
19	D0 + 013.250	0–013.300	570.000	4.530	270	−80
20	D0 + 012.950	0–014.700	570.000	1.900	90	−60
21	D0 + 011.250	0–013.300	570.000	4.530	270	−80
22	D0 + 010.950	0–014.700	570.000	1.900	90	−60
23	D0 + 009.250	0–013.300	570.000	4.530	270	−80
24	D0 + 009.550	0–014.700	570.000	1.900	90	−60
25	D0 + 007.750	0–013.300	570.000	4.530	270	−80
26	D0 + 008.050	0–014.700	570.000	1.900	90	−60
27	D0 + 006.800	0–013.300	570.000	5.000	35	−45
28	D0 + 007.100	0–014.700	570.000	1.900	330	−45

4 Discussion

In the critical task of managing seepage in dam transverse joints through underwater diving operations, the foremost consideration is the safety of the divers. A comprehensive pre-dive assessment is conducted to understand the site's conditions, including detailed measurements of water depth, flow rate, temperature, geology, and wind speed. This data is meticulously recorded and analyzed to gain insights into the site's hydrological, meteorological, and geological characteristics. This understanding is crucial for formulating a robust diving plan and establishing emergency safety measures. It is imperative that all personnel involved in the diving operations, such as divers, phone operators, and signal line handlers, are certified professionals. The use of non-certified individuals in diving activities is strictly prohibited, and any diver who is unwell is not assigned to work, thereby ensuring the highest level of operational safety.

The integrity of underwater inspections is vital, and to achieve this, all personnel must thoroughly familiarize themselves with the power station plans before each inspection. This ensures a clear understanding of the inspection scope and allows for the development of specific diving inspection methods and diver routes tailored to the actual site topography. Daily pre-job meetings are held to detail the tasks for the day and to clarify the diver's routes. On-site technicians further elaborate on the diver's route immediately before the dive, ensuring that each diver is fully aware of their responsibilities. Once underwater, surface technicians guide the divers according to a pre-established inspection plan, ensuring that the results meet the contractual requirements.

Considering the significant impact of water quality on underwater inspections, it is recommended to conduct these inspections when visibility is optimal, such as before floods or after a period when the water has cleared post-flood. This approach enhances the effectiveness and accuracy of the inspections. Underwater terrain surveying requires careful attention to reference points and weather conditions to maintain accuracy. Precision instruments that have been calibrated are essential for this task, and continuous checks on reference points and water levels are conducted throughout the surveying process. To minimize the influence of surface waves, surveys are ideally conducted on clear weather days. These stringent measures ensure the accuracy and reliability of underwater terrain surveys, providing essential scientific data support for the effective management of seepage in dam transverse joints.

References

1. Adamo, N., Al-Ansari, N., Sissakian, V., et al.: Dam safety problems related to seepage. J. Earth Sci. Geotech. Eng. **10**(6), 191–239 (2020)
2. Mozafari, M., Milanović, P., Jamei, J.: Water leakage problems at the Tangab Dam Reservoir (SW Iran), case study of the complexities of dams on karst. Bull. Eng. Geol. Env. **80**(10), 7989–8007 (2021)
3. Ball, G.P., Robertson, A.J., Morales, K.M.: Seepage investigation of the Rio Grande from below Leasburg Dam, Leasburg, New Mexico, to above El Paso, Texas, 2018. US Geological Survey (2020)
4. Garcia, D.M.G., Ramos, A.N.L., Santos, M.V.M., et al.: Seepage simulation analysis for isotropic soils of homogeneous embankment dams. In: 2021 IEEE 13th International Conference on Humanoid, Nanotechnology, Information Technology, Communication and Control, Environment, and Management (HNICEM). IEEE, pp. 1–5 (2021)
5. Zhang, C., Chai, J., Cao, J., et al.: Numerical simulation of seepage and stability of tailings dams: a case study in Lixi, China. Water **12**(3), 742 (2020)
6. Yang, J., Hu, J., Wu, Y., et al.: Numerical simulation of seepage and stability of tailing dams: a case study in Ledong, China. Sustainability **14**(19), 12393 (2022)

7. Yang, C., Shen, Z., Xu, L., et al.: A simplified method for leakage estimation of clay core dams with different groundwater Levels. Water **14**(12), 1961 (2022)
8. Miao, K., Bai, Z., Huang, Y., et al.: Research on seepage control of Jurong pumped storage hydroelectric power station. Water **14**(2), 141 (2022)
9. Xiang, Y., Sheng, J.B., Wang, L., et al.: Research progresses on equipment technologies used in safety inspection, repair, and reinforcement for deepwater dams. Sci. China Technol. Sci. **65**(5), 1059–1071 (2022)
10. Adamo, N., Al-Ansari, N., Sissakian, V., et al.: Dams safety: inspections, safety reviews, and legislations. J. Earth Sci. Geotech. Eng. **11**(1), 109–143 (2021)

Author Index

D. Jeng and B. Cai (Eds.): ICCOE 2024, LNCE 423, pp. 517–519, 2025.
https://doi.org/10.1007/978-981-96-0181-3

The manufacturer's authorised representative in the EU is Springer Nature Customer Service Centre GmbH, Europaplatz 3, 69115 Heidelberg, Germany. If you have any concerns regarding our products, please contact ProductSafety@springernature.com

Printed and bound by CPI Group (UK) Ltd, Croydon, CR0 4YY
01/10/2026
02225941-0017